MILITARY DICTIONARY:

COMPRISING

TECHNICAL DEFINITIONS;

INFORMATION

ON RAISING AND KEEPING TROOPS;

ACTUAL SERVICE,

INCLUDING

MAKESHIFTS AND IMPROVED MATÉRIEL;

AND

LAW, GOVERNMENT, REGULATION, AND ADMINISTRATION
RELATING TO LAND FORCES.

BY

COLONEL H. L. SCOTT,

INSPECTOR-GENERAL, U. S. A.

THE LAWBOOK EXCHANGE, LTD.
Clark, New Jersey

ISBN-13: 978-1-58477-579-9 (hardcover)
ISBN-10: 1-58477-579-3 (hardcover)
ISBN-13: 978-1-58477-990-2 (paperback)
ISBN-10: 1-58477-990-X (paperback)

Front cover illustration page 259.
Rear cover illustration page 522.

The quality of this reprint is equivalent to the quality of the original work.

Lawbook Exchange edition 2006, 2009

THE LAWBOOK EXCHANGE, LTD.
33 Terminal Avenue
Clark, New Jersey 07066-1321

Please see our website for a selection of our other publications and fine facsimile reprints of classic works of legal history:
www.lawbookexchange.com

Library of Congress Cataloging-in-Publication Data

Scott, H. L. (Henry Lee), 1814-1886.
  Military dictionary : comprising techincal definitions : information on raising and keeping troops; actual service, including makeshifts and improved materiel; and law, government, regulation,
and administration relating to land forces / by Colonel H. L. Scott.
      p. cm.
  Originally published: New York : D. Van Nostrand, 1863.
  Includes bibliographical references.
  ISBN 1-58477-579-3 (alk. paper)
  1. Military art and science--Dictionaries. 2. English language--Terms and phrases. I. Title.
U24.S42 2005
355'.003--dc22   2005002802

MILITARY DICTIONARY:

COMPRISING

TECHNICAL DEFINITIONS;

INFORMATION

ON RAISING AND KEEPING TROOPS;

ACTUAL SERVICE,

INCLUDING

MAKESHIFTS AND IMPROVED MATÉRIEL;

AND

LAW, GOVERNMENT, REGULATION, AND ADMINISTRATION
RELATING TO LAND FORCES.

BY

COLONEL H. L. SCOTT,

INSPECTOR-GENERAL, U. S. A.

NEW YORK:
D. VAN NOSTRAND, 192 BROADWAY.
LONDON:
TRÜBNER & CO.
1863.

ENTERED, according to Act of Congress, in the year 1861, by
HENRY L. SCOTT,
In the Clerk's Office of the District Court of the United States for the
Southern District of New York.

JOHN F. TROW,
PRINTER, STEREOTYPER, AND ELECTROTYPER,
No. 50 Greene Street, New York.

PREFACE.

A MILITARY dictionary which, with technical definitions, comprises information on actual service; on law, government, regulation, and administration; on raising and keeping troops, and on makeshifts and improved *matériel*, is much needed; and the design of the present work is in some measure to occupy that gap in military literature.

In legal articles, plain decisions from constitutional exponents of law have been accepted as conclusive; but when without such a guide, an endeavor has been made to set forth the true intent and meaning of laws in dispute, by simple, clear, and logical annotations. Much interesting law matter has been abridged from Prendergast's *Law* relating to officers of the army; and in respect to courts-martial, actual service, improved *matériel*, &c., &c., the author is indebted to many standard authorities, sometimes only designated by name in different articles; but, in such cases, referred to fully by the titles of their works in the list of abbreviations which follows this preface.

It is only deemed necessary to add, that the work was not prepared in view of existing disturbances, but was begun some years ago, and that the few additions made since it was put in the hands of the publisher in January last, refer only to improvements in *matériel*.

TITLES OF WORKS

REFERRED TO BY ABBREVIATIONS IN THE TEXT, AND EXPLANATIONS
OF OTHER ABBREVIATIONS USED.

Act.—Act of Congress of the United States. Reference embraces date of act.

Aide Memoire—to the military sciences framed from contributions of officers of different services, and edited by a Committee of the Corps of Royal Engineers in Dublin.

Aide Memoire d'Artillerie—à l'usage des Officiers d'Artillerie. Paris, 1855.

Art.—(Articles of War,) included in an act of Congress for establishing rules and articles for the government of the armies of the United States, approved April 10, 1806. Reference embraces the number of the article.

BARDIN.—Dictionnaire de l'Armée de Terre, ou Recherches Historiques sur l'Art et les Usages Militaires des Anciens et des Modernes. Par le Général Bardin, &c. Ouvrage terminé sous la direction du Général Oudinot de Reggio. 5,337 pp. Paris, 1851.

BAUCHER.—Method of Horsemanship. Philadelphia, 1851.

BENTON.—Ordnance and Gunnery. By Capt. J. G. Benton, U. S. Ordnance.

BLACKSTONE.—Commentaries, with Notes. 4 vols. London, 1844.

BOUVIER.—Law Dictionary adapted to the Constitution and Laws of the United States. By John Bouvier. Philadelphia, 1839.

BRANDE.—Encyclopedia of Science, Literature, and Art.

BUGEAUD.—Aperçus sur quelques Détails de la Guerre. Par le Maréchal Bugeaud.
Ibid.—Instructions Pratiques. Bugeaud.

BURNS.—Naval and Military English-and-French Technical Dictionary. By Lieut.-Colonel Burns, Royal Artillery. London, 1852.

CAVALLI.—Mémoire sur divers Perfectionnements Militaires. Par J. Cavalli, Colonel d'Artillerie, &c., &c. Traduit de l'Italien. Paris, 1856.

COUTURIER.—Dictionnaire Portatif et Raisonné. Par le Général Le Couturier. Paris, 1825.

DE HART.—Courts-martial. By Captain W. C. De Hart, 2d U. S. Artillery.

ABBREVIATIONS, AND TITLES OF AUTHORITIES.

DECKER.—De la Tactique des Trois Armes: Infanterie, Cavalerie, Artillerie. Par C. Decker, Lieut.-Colonel, &c., &c.

DOUGLAS.—Naval Gunnery. By Gen. Sir Howard Douglas.

DUFOUR.—Cours de Tactique. Par le Général Dufour.

DUNLOP.—Digest of Laws of the United States.

Experiments, &c.—By officers of the Ordnance in Small-Arms. 1856, (official.)

FAVÉ—Histoire et Tactique des Trois Armes, et plus Particulièrement de l'Artillerie de Campagne. Par Ild. Favé, Capitaine d'Artillerie.

FONBLANQUE.—The Administration and Organization of the British Army, with especial reference to Supply and Finance. By Edward Barrington de Fonblanque, Asst. Commissary-General. London, 1858.

GALTON.—The Art of Travel. By Francis Galton. London, 1860.

GIBBON.—The Artillerist's Manual. By Capt. John Gibbon, 4th U. S. Artillery.

GORDON.—Digest of Laws of the United States.

GUILLOT.—Législation et Administration Militaire, ou Programme Détaillé des Matières Enseignées à l'Ecole Impériale de l'Etat Major. Par M. Léon Guillot, &c.

HAILLOT.—Statistique Militaire, et Recherches sur l'Organization des Armées Étrangères. Par C. T. Haillot, Chef-d'Escadron d'Artillerie.

HETZEL.—Cross' and Hetzel's Military Laws of the United States.

HOUGH.—Military Law Authorities. By Lieut.-Colonel Hough, Deputy Judge-advocate General, &c.

HYDE.—Elementary Principles of Fortification. By John Hyde, Professor Military College, Addiscombe.

JEBB.—Practical Treatise on Attack and Defence. By Colonel Jebb, Royal Engineers.

JOMINI.—Tableau Analitique.

KINGSBURY.—Artillery and Infantry. By Captain Kingsbury, Ordnance Department.

LE GRAND.—Dictionnaire Militaire Portatif. Par Le Grand.

MACOMB.—Courts-martial. By Major-General Macomb. New York, 1841.

MCCLELLAN.—Military Commission in Europe. Report by Captain McClellan, U. S. Army.

MAHAN.—Field Fortifications. By Professor Mahan, U. S. Military Academy.

MAYO and MOULTON.—Army and Navy Pension Laws. Washington, 1852.

Mémorial—des Officiers d'Infanterie et de Cavalerie. Paris, 1846.

MORDECAI.—Digest of Military Laws. By Major Mordecai, U. S. Army.

NAPOLEON.—Maxims of War.

PETERS.—Digest of Decisions of Federal Courts.

PRENDERGAST.—The Law relating to Officers in the Army. By Harris Prendergast of Lincoln's Inn, Esq., Barrister-at-Law.

ROUVRE.—Aide Memoire de l'Officier d'Etat Major en Campagne. Par M. De Rouvre, Chef-d'Escadron d'Etat Major, Aide-de-camp de son Ex. le Maréchal Magnan.

RUFFIN.—Manuel d'Administration et de Comptabilité à l'usage des Officiers des Compagnies ou Escadron des Corps d'Infanterie et de Cavalerie. Par M. Ruffin.

SCOTT.—Orders and Correspondence of Gen. Winfield Scott, Congressional Documents, &c.

SKINNER.—Youatt on the Horse. By Skinner.

VATTEL.—Law of Nations. Philadelphia, 1817.

WHEATON.—Elements of International Law. Philadelphia, 1846.

YOUATT.—Youatt on the Horse. By Skinner.

MILITARY DICTIONARY.

ABANDONING A POST, OR MISBEHAVIOR BEFORE AN ENEMY. Punishable with death, or otherwise, as a court-martial shall direct; (*Art.* 52.)

ABATIS (*French*)—are rows of felled trees deprived of their smaller branches, the remainder sharpened to a point, and employed for defence. Abatis should be placed so as not to be exposed to the fire of artillery. In redoubts or intrenchments, they are usually fixed

FIG. 1.

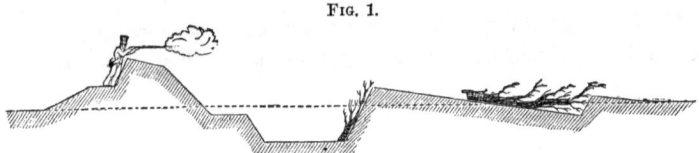

in an upright position against the counterscarp, or at the foot of the glacis, the plane of which is broken so as to conceal the abatis from the view of the enemy, and to guard against obstructing the musketry fire from the parapet in their rear.

FIG. 2.

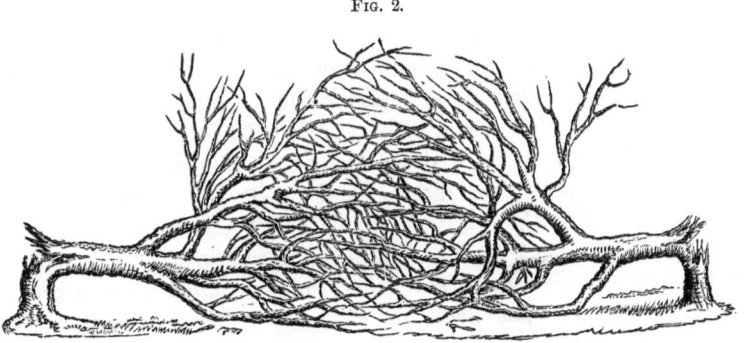

Abatis are also an excellent means of blocking up a road, when trees grow on either side. If branches are properly placed, and intertwined one within another, their disengagement is extremely difficult. An abatis will always be found a very useful and effective auxiliary to the defence of houses or isolated posts, if judiciously placed within range of musketry. When close in front of the windows on the ground floor, or used as a cover to the entrance door, it will be extremely difficult for the enemy to force his way into the building.

ABSENCE, WITH LEAVE. Every colonel or other officer commanding a regiment, troop, or company, and actually quartered with it, may give furloughs to non-commissioned officers or soldiers in such numbers, and for so long a time, as he shall judge to be most consistent with the good of the service; and a captain or other inferior officer, commanding a troop or company, or in any garrison, fort, or barrack of the United States, (his field-officer being absent,) may give furloughs to non-commissioned officers or soldiers for a time not exceeding twenty days in six months, but not more than two persons to be absent at the same time, excepting some extraordinary occasion should require it; (*Art.* 12.)

The law does not specify by whom leaves of absence may be given to commissioned officers, and the omission has been supplied by orders of the President.

ABSENCE, WITHOUT LEAVE, FROM CAMP, PARADE, OR RENDEZVOUS. Punished, by sentence of a court-martial, according to the nature of the offence; (*Articles* 41, 42, 43, and 44.)

ABUSES AND DISORDERS. Every commanding officer shall keep good order, and, to the utmost of his power, redress all abuses and disorders which may be committed by any officer or soldier of his command. If, upon complaint made to him of officers or soldiers beating, or otherwise ill-treating, any person, of disturbing fairs or markets, or of committing any kinds of riots, to the disquieting of the citizens of the United States, &c., the said commander shall refuse or omit to see justice done to the offender or offenders, and reparation made to the party or parties injured, as far as part of the offender's pay shall enable him or them, he shall, upon proof thereof, be cashiered, or otherwise punished, as a general court-martial shall direct; (*Art.* 32.)

ACADEMY. The Military Academy of the United States is located at West Point, N. Y. The students, called cadets, are subject to the rules and articles of war. They are appointed from each congressional district, upon the nomination of the representative of the district in Congress. Each district is allowed but one representative at the Military Academy; but besides the number so appointed, the

President of the United States annually appoints ten cadets from at large. The Academy furnishes about forty graduates a year, who receive commissions of the lowest grade in some one of the different corps of the army, provided vacancies exist. If there be no vacancies, the graduates are attached to different corps as supernumerary officers of the lowest grade, not exceeding one to each company. The Military Academy was founded by act of Congress in 1802. Its present high reputation is mainly due to Colonel Sylvanus Thayer, who did not become Superintendent until 1817.

At the breaking out of the war of 1812, there were about seventy graduates of the Academy holding commissions, and but little knowledge of the military art and of the science of war prevailed. At the breaking out of the Mexican war, the officers of our army were mostly graduates of the Academy. Every branch of the service was filled with men of talent and military information; volunteer corps raised during the war sought and obtained as their commanders graduates of the Military Academy. General officers from political life appointed staff officers from the same class. In all positions which the graduates held during that brilliant war, the honor and glory of the United States were sustained, and the great usefulness of an institution, which annually costs little, if any more than the maintenance of one frigate afloat, was satisfactorily demonstrated to the people of the United States. (*See* SUPERINTENDENT.) Military Academies, modelled upon that at West Point, have also been established within their respective limits by the States of Virginia, Kentucky, North Carolina, South Carolina, and Alabama, and perhaps others.

ACCOUNTS. Officers accountable for public money or property render quarterly accounts to the Treasury Department, if resident in the United States; and every six months, if resident in a foreign country. Additional returns may be required by the Secretary of War, if the public interest requires it; (*Act* Jan. 31, 1823.) Every officer or agent offending against the foregoing provisions may be dismissed by the President of the United States; (*Act* Jan. 31, 1823.) The method of rendering accounts by Administrative Agents of the application of all public money and material passing through their hands, has been prescribed by regulations made pursuant to law. The object of a system of accountability should be, in respect to the army, to obtain plain statements of the operations and results of Military Administration. The system should be neither complex nor cumbrous, but should be adapted to a state of war; and while carefully guarding against losses to the Government, should, at the same time, by prompt settlements, through government agents, present with armies in the field, dispense with accumulations of papers, which manifestly subject administrative officers to great losses, even if they were not frequently obliged to wait years before obtaining a settlement of their accounts.

By the present system of accountability it is prescribed: 1. That all accounts whatever in which the United States are concerned shall be settled and adjusted in the Treasury Department; (*Act* March 3, 1817.) 2. It is made the duty of the second and third auditors of the Treasury, to receive and examine all military accounts; to receive from the second comptroller the accounts which shall have been finally adjusted; to preserve such accounts; to record all warrants drawn by the Secretary of War; and make such reports on the business assigned to them as the Secretary of War may deem necessary, and require for the service of his Department; (*Act* March 3, 1817.) 3. It is the duty of the second comptroller to examine all accounts settled by the second and third auditors, and certify the balances arising thereon to the Secretary of War; to countersign all legal warrants drawn by the Secretary of War; *to report to the Secretary of War the official forms to be issued in the different offices for disbursing the public money, and the manner and form of keeping and stating the accounts of the persons employed therein;* and it shall also be the Comptroller's duty to superintend the preservation of the public accounts subject to his revision; (*Act* March 3, 1817.)

The great obstacles to the simplification and prompt settlement of army accounts interposed by law consist: 1. In the requirement that military accounts shall be adjusted and settled at the Treasury Department, instead of being settled by the War Department, and reported to the Treasury; 2. In making the second and third auditors and second comptroller officers of the Treasury instead of officers of the War Department; 3. In authorizing the second comptroller to establish forms for keeping and stating military accounts, instead of requiring him in those matters to conform to the directions of the Secretary of War; and, 4. In withholding from the War Department the power of appointing agents to accompany armies in the field for the prompt settlement of accounts. With the changes of law here suggested, it would be easy for the War Department, through the various grades in the several administrative staff departments, to establish a simple system of accountability with requisite means of control and supervision, which would operate advantageously to the government, and to individual agents. Under the present system there is, and must be, a remarkable similarity in the duties of all grades of the staff administrative departments. (Consult

Cours d'Administration, par Vauchelle, *Intendant Militaire; Cours d'Etudes sur l'Administration Militaire, par* Odier: *Memorial des Officiers d'Infanterie et de Cavalerie,* 1846.)

ACCOUTREMENTS. Black leather belts, &c., furnished by the ordnance department.

PARTS.	Infantry.	Artillery.	Cavalry.	Rifle.
	$ cts.	$ cts.	$ cts.	$ cts.
Cartridge box	1 10	....	....	95
Cartridge box plate	10	....	10	10
Cartridge box belt	69	....	....	....
Cartridge box belt plate	10	....	....	....
Bayonet scabbard and frog	56	....	....	....
Waist belt, private's	25	....	....	37
Waist belt plate	10	....	....	10
Cap pouch and pick	40	....	40	40
Gun sling	16	....	....	16
Sabre belt	....	1 03	1 35	....
Sabre belt plate	....	60	60	....
Sword belt	....	1 00	....	....
Sword belt plate	....	10	....	....
Sword belt, non-commissioned officer's and musician's	62	....	....	62
Sword belt plate do. do.	10	....	....	10
Waist belt do. do.	37	....	....	37
Waist belt plate do. do.	60	....	....	60
Carbine cartridge box	....	....	87	....
Pistol do.	....	....	75	....
Holsters, with soft leather caps	....	....	2 63	....
Carbine sling	....	....	95	....
Carbine swivels	....	....	88	....
Sabre knot	....	....	30	....
Bullet pouch	....	....	....	53
Flask and pouch belt	....	....	....	40
Powder flask	....	....	....	1 20
Waist belt, sapper's, with frog for sword bayonet, $1.				

Infantry accoutrements for 100 men, including non-commissioned officers' shoulder-belts and plates, weigh 330 lbs.; rifle accoutrements for 100 men, including non-commissioned officers' shoulder-belts and plates, weigh 329 lbs.; 100 carbine slings and swivels, 110 lbs. (*See* Arms.)

Mr. Dingee's directions for reblacking Belts.—Brush them with a hard brush, to clean the surface; if they are very greasy, use a wire scratch-brush. Then, with a soft brush or sponge, apply the following mixture, viz.: one gallon soft water, two pounds extract of logwood, half a pound of broken nutgalls, boiled until the logwood is dissolved. When cold, add half a pint of the pyrolignite of iron—made by dissolving iron filings in pyroligneous acid, as much as the acid will take up. The dye thus made should be well stirred, and then left to settle. When clear, bottle it free from sediment, and keep it well corked for use. Dye the belts in the shade; then apply a little sperm or olive oil, and rub well with a hard brush. Should any bad spots appear, scratch up the surface with the wire brush, and wet two or three times with a simple decoction of gallnuts or sumach, and again apply the dye. Logwood is not essential, and a solution of copperas may be used instead of the acetate of iron.

ADDRESS. An address to a court-martial, by either party, must be in writing. (Consult *Hough's Law Authorities.*)

ADJUTANT, (Latin *adjutor,* aid.) An officer selected by the colonel of the regiment from the subalterns. He communicates the orders of the colonel, and has duties in respect to his regiment assimilated to those of an adjutant-general with an army.

ADJUTANT-GENERAL. The principal organ of the commander of an army in publishing orders. The same organ of the commander of a division, brigade, geographical division, or department, is styled Assistant Adjutant-general. The laws of the United States, however, provide for but one Adjutant-general with the rank of colonel, (made by regulations chief of a bureau of the War Department, and charged with the recruiting service, records, returns, &c.,) one Assistant Adjutant-general with the rank of lieutenant-colonel, and twelve other assistants with the rank of major and captain. (*See* ARMY ORGANIZATION.)

The bureau duties of Adjutants-general and assistants are: publishing orders in writing; making up written instructions, and transmitting them; reception of reports and returns; disposing of them; forming tables, showing the state and position of corps; regulating details of service; corresponding with the administrative departments relative to the wants of troops; corresponding with the corps, detachments, or individual officers serving under the orders of the same commander; and the methodical arrangement and care of the records and papers of his office. The active duties of Adjutants-general consist in establishing camps; visiting guards and outposts; mustering and inspecting troops; inspecting guards and detachments; forming parades and lines of battle; the conduct and control of deserters and prisoners; making *reconnaissances;* and in general discharging such other active duties as may be assigned them.

ADJUTANT-GENERAL OF A STATE. (*See* MILITIA.)

ADJUTANT-GENERAL, DEPUTY, &c. An act making further provision for the army, and for other purposes: Approved July 6, 1812, provides: Sec. 2, That to any army of the United States, other than that in which the adjutant-general, inspector-general, quartermaster-general, and paymaster of the army, shall serve, it shall be lawful for the President to appoint one deputy adjutant-general, one deputy inspector-general, one deputy quartermaster-general, and one deputy paymaster-general, who shall be taken from the line of the army, and who shall, each, in addition to his pay and other emoluments, be entitled to fifty dollars per month, which shall be in full compensation for his extra services. And that there shall be, to each of the foregoing deputies, such number of assistant deputies (not exceeding three to each department) as the public service may require, who shall in like manner be taken from the line, and who shall each be entitled to thirty dollars per month, in addition to his pay and other emoluments, which shall be in full compensation for his extra services, &c.

ADMINISTRATION, ADMINISTRATIVE. These words are derived from *ministrare, administrare,* to *serve.* Administration is a branch of political economy; it is the action of administrative agents in executing laws or regulations conformable to law. The aim of a system of administration is to secure the performance of public duties, either directly, ministerially, or through the intervention of sub-agents. It is exercised over individuals or things, in civil matters, in courts of law, in political bodies, in the army and in the navy, and in general in all *financial* matters of government. Administration consists in establishing the ways and means of public receipts and expenditures; in watching over such employments; in the collection, care, and distribution of material and money; and in rendering and auditing accounts of such employments. Army Administration also embraces in war the means by which an army is supported in foreign countries by a general in campaign, when without regular supplies, without resorting to pillage. The wars of the French revolution brought into use REQUISITIONS, a moderate kind of marauding, weighing more heavily upon countries than upon individuals. Requisitions are, however, an uncertain and unequal means of supply, and only enable an army to live from hand to mouth, and although practicable in offensive wars, are only justifiable in rapid movements, where time does not admit the employment of more certain means of supply. The system is less odious than pillage.

Bonaparte skilfully adopted another method, in harmony with the spirit of wars of invasion, and also more reliable as a means of support. He substituted himself in place of the supreme authorities of the invaded country, and exacted *pecuniary contributions,* paying, or promising to pay, for all provisions and other supplies needed for his army. Some writers think that even this modified system can only succeed in gigantic operations, where an army upon a new soil successively gives repose to that previously occupied. Such a system was, however, well executed by Marshal Suchet in Spain, and a similar system was also matured and published in orders by General Scott while in Mexico. A treaty of peace, however, soon after was made, which put an end to military operations, and the system was therefore only partially executed. But with a sufficient army in a fertile country, the experience of the world has shown that if the inhabitants are protected from injuries, they will very generally sell to the best paymasters. It is therefore the interest of an invading army not to interfere with the ordinary avocations of citizens, and such is the modern usage.

Bonaparte (according to Las Casas) thought that an entire revolution in the habits and education of the soldier, and perhaps also in those of the officers, was essential to the formation of a veritable self-subsisting army. Such an army (he said) cannot exist with present ovens, magazines, administration, wagons, &c., &c. Such an army will exist when, in imitation of the Romans, the soldier shall receive his corn, shall personally carry his mill and cooking utensils, cook his own bread, &c., &c., and when the present frightful paper administration has been dispensed with. He added that he had meditated upon all those changes, but a period of profound peace was necessary to put them in practice. If he had been constrained to keep a large army in peace, he would have employed it upon the public works, and given it an organization, a dress, and a mode of subsistence altogether special. If such a scheme be practicable, no approach to it yet exists.

The French have made some progress in developing a system of administration suited to a large army, but hardly a step in the direction pointed out by Napoleon. The French administrative service is a powerful means of moving armies in unforeseen emergencies. Its foresight provides resources, and the adversary soonest ready has the greatest chance of success. Not a century since, the French government required six months' preparation before an army could move; now, in the language of Gen. Lamarque, "The cannon is loaded, and the blow may be given at the same moment as the manifesto, and, if necessary, the blow may precede it." Ordinary army administration consists in the organization and other means by which various administrative duties are performed, necessary to provide for the wants of troops, and for all the foreseen demands of a state of war, including labor and the supplies for garrisons, sieges, &c. Such duties embrace subsistence magazines, daily rations, forage, dress, encampments, barracks, hospitals, transportation, &c., &c., the administrative duties of engineers, and of the ordnance department, estimates, accountability, payments, recruiting, and in general the receipt and proper application of money. The Secretary of War, under the orders of the President, is the head of military administration in the United States. The object of such administration is to provide, through the resources placed by law at his disposition, for the constant wants, regular or accidental, of all who compose the army. Good administration embraces a foreknowledge of wants, as well as the creation, operation, and watchfulness of the ways and means necessary to satisfy them; the payment of expenses, and the settlement of accounts.

Army administration is divided into several branches determined by law. These different branches constitute the administrative service of an army, the operations of which should be so regulated that the Secretary of War will be always informed of the condition of each, and be able to exercise, subordinate to law, a complete financial control over each. These different branches of administration are: 1. The recruiting service, and the custody of records and returns of personnel; 2. The administrative service of engineers and topographical engineers; 3. The ordnance department; 4. The quartermaster's department; 5. The subsistence department; 6. The pay department; 7. The administrative service of the medical department; and, 8. The settlement of army accounts. Bureaux of the War Department charged with these different matters have been organized by the President and Secretary of War, under the joint authority given these functionaries by the act of Congress of 1813 (*See* REGULATION) to make regulations better defining the powers and duties of certain staff officers. The adjutant-general of the army and the heads of administrative corps have each been assigned a bureau in the War Department, under the direction of the Secretary of War, for the management of the administrative duties with which they have been respectively charged. Administration and Command are distinct. Administration is controlled by the head of an executive department of the government, under the orders of the President, by means of legally appointed administrative agents, with or without rank, while Command, or the discipline, military control, and direction of military service of officers and soldiers, can be legally exercised only by the military hierarchy, at the head of which is the constitutional commander-in-chief of the army, navy, and militia, followed by the commander of the army, and other military grades created by Congress. (*See* ACCOUNTS; ACCOUNTABILITY; ADJUTANT-GENERAL; ALLOWANCES; AMBULANCES; APPROPRIATIONS; ARREARS OF PAY; ARMY OF THE UNITED STATES; ARMY REGULATIONS; AUDITORS; BAGGAGE; BAKING; BARRACKS; BED; BOOKS; BONDS; BOUNTY; BRIDGE; CALLING FORTH MILITIA; CARPENTRY; CASEMATE; CLERKS; CLOTHING; COMMISSARY; COMMISSION;

Comptroller; Congress; Conscription; Contracts; Councils of Administration; Damage; Deceased; Defaulters; Delinquents; Department of War; Depôt; Disbursing Officers; Discharge; Embezzlement; Engineer Corps; Engineers, (Topographical;) Enlistments; Exchange of Prisoners; Execution of Laws; Exempts; Extra Expenses; Extra Allowances; Gratuity; Indian; Insurrection; Laws (Military) and References; Losses; Logistics; Marshals; Measures; Medical Department; Mileage; Militia; Muster; Nitre; Obstruction of Laws; Ordnance Department; Ordnance Sergeants; Organizing; Oven; Passports; Pay; Pay Department; Paymaster-general; Pension; Ponton; Posse Comitatus; President; Purchasing; Quarters; Quartermaster's Department; Quartermaster-general; Raise and References; Ration; Recruiting; Reenlisting; Regulation; Remedy; Returns; Roads; Sale; Sanitary Precautions; Sappers; Saw-mill; Secretary of War; Service; Staff; State Troops; Standards; Storekeepers; Stoppage of Pay; Subsistence Department; Suit; Superintendent; Telegraph; Tent; Tools; Trade; Transfers; Travelling Allowances; Treaty; Uniform; Utensils; Value; Veteran; Veterinary; Volunteers; Wagon; War; Weights; Wills, (Nuncupative); Wounds. (Consult Bardin, *Dictionnaire de l'Armée de Terre*; *Legislation et Administration Militaire*, par M. Leon Guillot; *Military Laws of the United States*; Gen. Scott's *orders in Mexico*; Suchet's *Memoirs*.)

ADMISSIONS. The judge advocate is authorized, when he sees proper, to admit what a prisoner expects to prove by absent witnesses.

ADOBES—are unburnt brick made from earth of a loamy character containing about two-thirds fine sand mixed intimately with one-third or less of clayey dust or sand. Stiff clay will not answer, as the rays of the sun would crack it in pieces. The adobe, under the action of the sun, becomes a compact mass. Upon our Indian frontiers in New Mexico, in Mexico, and in Central America, adobe houses and adobe defences against the Indians are common structures. Four men usually work together in making adobe brick. One mixes the mass in a hole, and loads the barrow, two carry it on a common hand-barrow, and the fourth moulds the brick. The moulder has a double mould, or one which forms two adobes, each eighteen inches long, nine inches wide, and four inches thick. The partition between the two compartments should be of one and a half inch stuff, the other parts of inch board; a cleat on either outer side, extending the length of the mould, permits the mould to be easily handled. It must be well morticed together so as not to wabble. The moulder has no bottom, the adobe being deposited on the surface of the ground, made tolerably level, and without reversing, as in brick making. The mould is raised gradually and slowly away from the moulded masses. Before placing it on the ground to mould another couple, the inner sides of the mould are washed with water, kept at hand; this is all that is required to preserve the mud from sticking and thus breaking the adobe. The mould is emptied a second time on the ground at about three inches from the first couple, and in refilling, the balance of the mud left over from the first moulding is cast in the compartments, and the two men with the barrow of mud throw their load directly upon the mould, and all that is over and above what is necessary to fill it is scraped off by the moulder's hands toward where his next couple is to be. The dumping of the mud from the barrow is facilitated by casting into the barrow a little finely powdered dry manure or dust.

An adobe eighteen inches long, nine inches wide, and four inches thick, is the best average size for moulding and for building. They are sometimes made sixteen inches long and twelve inches wide; in such cases they are all laid as *headers;* but with the eighteen inch adobe they afford the means of binding the wall strongly by alternating headers and stretchers, as in brick-laying. In the hot spring and summer suns two or three days uninterrupted drying is sufficient at the first; the adobes are then carefully turned up on edge, so as to expose the under or still wet face to the southern and western sunshine. They should be left in this position from a week to fifteen days to dry thoroughly, when, if not wanted for immediate use, they may be stacked on edge and covered from the weather. Houses in New Mexico are seldom built over one story high. This enables the builder to place on the roof-covering at once, if necessary. But in all cases, intervals in the work must be allowed, or the house will not only be unsafe, but, if immediately occupied, damp and disagreeable. The inside plastering with mud is most frequently done before the roof is covered in, so as to dry with the wall. If the wall must be left unfinished through the fall rains or the winter, the top of it is covered with a bushy weed called *cachanilla,* and this is covered with earth, to exclude water and protect it till the ensuing year. If door and window frames are at hand, the Mexicans prefer to put them in as they build; but oftener they leave gaps for doors and windows, unfilled with the frames, till the whole is finished. The adobes are laid with mud mortar made from the earth at the base of the wall; the holes thus formed are readily filled again with the rubbish from the house when completed. When the wall is ready to receive the roof-covering, heavy joists are laid, about two feet apart, on the top of the walls, strong enough to bear near a foot of earth all over the roof; the joists, as they rest upon the wall, are supported upon boards, or plates, as they are called, to distribute the weight of the roof, and prevent the joists from crushing into the walls. Across the joists, and over the whole roof, averaging about two inches in diameter, poles are now placed, the largest on the highest side of the roof to begin the slope, and on this is placed a close covering of the *cachanilla*, which is aromatic and keeps out bugs; it is evergreen, and a plant of the most suitable length to fill the interstices in the poles. Small willow brush is often used in the absence of *cachanilla*. The earth-covering of the roof is now put on, extending all round the roof to the parapet above the joists, which is only one-half the width of the wall below; this brings the dirt roof to cover over one-half the width or thickness of the wall, by which leaks in the room below are prevented. An adobe house, if well secured, is warmer in winter, and cooler in summer, than one of wood or brick. The brick is cold and damp, the adobe is dry and a much worse conductor of heat—no furrowing nor lathing is necessary—and the rough inside can be whitewashed or slapped with plaster. The durability of adobe walls is extraordinary. The Pecos Church, not far from Santa Fé, is doubtless one hundred years old; its mud walls (adobe) are as firm to this day as a rock, and they cannot be less than fifty feet high.

ADVANCED. Any portion of an army which is in front of the rest. It is figuratively applied to the promotion of officers and soldiers.

ADVANCED COVERED WAY—is a *terre plein*, on the exterior of the advanced ditch, similar to the first covered way.

ADVANCED DITCH—is an excavation beyond the glacis of the *enceinte*, having its surface on the prolongation of that slope, that an enemy may find no shelter when in the ditch.

ADVANCED GUARD. A detachment of troops which precedes the march of the main body.

ADVANCED LUNETTES—are works resembling bastions or ravelins, having faces and flanks. They are formed upon or beyond the glacis.

ADVANCED WORKS—are such as are constructed beyond the covered way and glacis, but within the range of the musketry of the main works.

ADVANCES of public money may be authorized by the President of the United States to persons in the military or naval service employed on distant stations. Prohibited otherwise; (*Act* Jan. 31, 1823.)

ADVISING TO DESERT. Punishable with death or otherwise, as a court-martial may direct; (*Art.* 23, *Articles of War.*)

AFFAIR. Any slight action or engagement. Affair of outpost; affair of rear-guard, &c.

AFFIDAVITS, being admissions upon oath, are evidence as such against the parties who made them, (*Hough.*) In the trial of cases not capital, the deposition of witnesses not of the staff or line of the army, taken before a Justice of the Peace in presence of the prosecutor and person accused, may be read in evidence; (*Art* 10.)

AIDES-DE-CAMP—are *ex-officio* assistants adjutant-general; (*Act* March 2, 1821.) They are confidential officers selected by general officers to assist them in their military duties. A lieutenant-general appoints not exceeding four in time of war, and two in peace, with the rank of lieutenant-colonel; a major-general two, and a brigadier-general one. Attached to the person of the general, they receive orders only from him. Their functions are difficult and delicate. Often enjoying the full confidence of the general, they are employed in representing him, in writing orders, in carrying them in person if necessary, in communicating them verbally upon battle-fields and other fields of manœuvre. It is important that Aides-de-Camp should know well the positions of troops, routes, posts, quarters of generals, composition of columns, and orders of corps: facility in the use of the pen should be joined with *exactness* of expression; upon fields of battle they watch the movements of the enemy; not only grand manœuvres but special tactics should be familiar to them. It is necessary that their knowledge should be sufficiently comprehensive to understand the object and purpose of all orders, and also to judge in the varying circumstances of a battle-field, whether it is not necessary to modify an order when carried in person, or if there be time to return for new instructions.

AIM. (*See* FIRING; TARGET.)

ALARM-POST—is the place appointed for every regiment or detachment to assemble, in case of a sudden alarm.

ALARMS, FALSE. Any officer who shall occasion false alarms in camp, garrison, or quarters, shall suffer death or other punishment as a court-martial may direct; (*Art.* 49.)

ALIBI. Elsewhere. An Alibi is the best of all defence if a man is innocent; but if it turns out to be untrue, it is conclusive against those who resort to it; (*Hough.*)

ALLOWANCES. The receipts of an officer consist of pay and allowances, sometimes called pay and emoluments. Allowances are regular and occasional; they consist of money for servants, forage, rations, and travelling expenses; and of fuel and quarters, stationery, straw for bedding, transportation of baggage, and forage in kind under certain circumstances. An allowance for servants and forage is only given where the servants and horses allowed are actually kept in service by the officer. Double rations are given to the commander of the army, the commander of an army in the field, a geographical division, department, military post and arsenal; and ten dollars per month is allowed to the actual commander of a company. Armies have always been paid by means of pay and allowances. It is the least expensive mode of supporting an army, and it is at the same time the most just method of graduating the pay according to circumstances. In the United States army, however, the allowances made are not sufficient, and not properly graduated. Several of the allowances given in European armies, are withheld from our own; and of those withheld, some are charges which press very heavily upon officers in campaign, when all their energies are needed for the service of the country. Of the allowances given in European armies, but withheld from the United States army, the following are the most important: Allowance, as equipment money at the beginning of a campaign, marching allowance, indemnity for losses in the field, prize money, and barrack furniture allowance. (*See* INDEMNIFICATION.)

AMBULANCES (*French*)—are flying hospitals so organized that they can follow an army in all its movements, and are intended to succor the wounded as soon as possible. Other sick are also placed in Ambulance, but the Ambulances are emptied as soon as fixed hos-

FIG. 3.

AMBULANCE CART PROPOSED FOR THE U. S. SERVICE.

pitals are at hand. In the French army, an Ambulance of infantry is composed of five wagons containing cases of instruments for amputating and trepanning, bandages for divers fractures, utensils of all kinds, medicines, and 8,900 dressings. The Ambulance of cavalry is composed of three wagons, containing the articles above enumerated, with 4,900 dressings. The Ambulances are distributed as follows: Each division of infantry has one Ambulance of infantry, and each division of cavalry an Ambulance of cavalry. The headquarters of an army corps is allowed two Ambulances; the grand park of artillery one Ambulance of cavalry; the reserve of the army at general headquarters

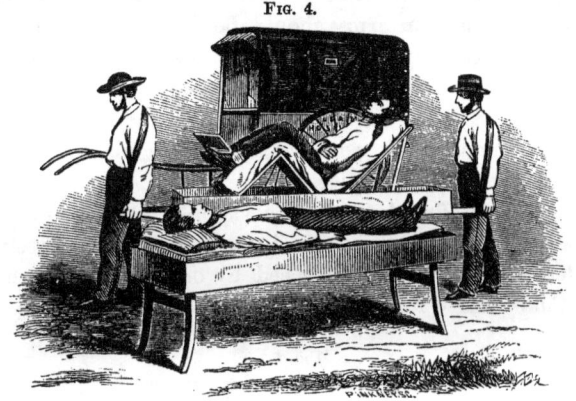

FIG. 4.

AMBULANCE CART PROPOSED FOR THE U. S. SERVICE.

six Ambulances; four of infantry, and two of cavalry. The number of Ambulance carts and wagons recently ordered for the United States service, in case of war, greatly exceeds the foregoing allowance, and would be doubtless required in operations of small detachments, or wherever, from any cause, it is impracticable to establish fixed hospitals, or leave wounded to the care of inhabitants. (*See* SURGERY; WAGON.)

AMBUSCADE. A body of men lying in wait to surprise an enemy.

AMICUS CURIÆ. Counsel, or at least Amici Curiæ, (friends of the court,) are allowed to prisoners in all cases, but no person is permitted to address the court, or interfere in any manner with its proceedings, except the parties themselves. (*Hough's Law Authorities*.)

AMMUNITION—is a term which comprehends gunpowder, and all the various projectiles and pyrotechnical compositions and stores used in the service.

Any commissioned officer convicted at a general court-martial of having sold without a proper order, embezzled, misapplied or, through neglect, suffered provisions, forage, army clothing, ammunition, or other military stores belonging to the United States to be spoiled or damaged, shall at his own expense make good the loss or damage, and shall forfeit his pay and be dismissed from the service; (*Art.* 36.) Any non-commissioned officer or soldier who shall be convicted at a regimental court-martial of having sold, or designedly, or through neglect, wasted ammunition delivered to him, shall be punished at the discretion of such court; (*Art* 37.)

The quantity of ammunition with troops is usually fixed at two hundred rounds for each piece of ordnance. These supplies are transported in caissons, and an army should be followed, in all cases, by a second supply at least equal to the first. The ammunition which cannot be carried in the caissons attached to pieces will be kept in boxes in reserve.

Additional supplies of ordnance stores are placed in convenient depots, according to circumstances.

Ammunition for Small Arms.—This supply consists of one hundred rounds to each man: forty rounds in cartridge box, and sixty in reserve. Percussion caps should exceed by one-half the number of cartridges. Cuts 5 and 6 represent the bullets of new arms.

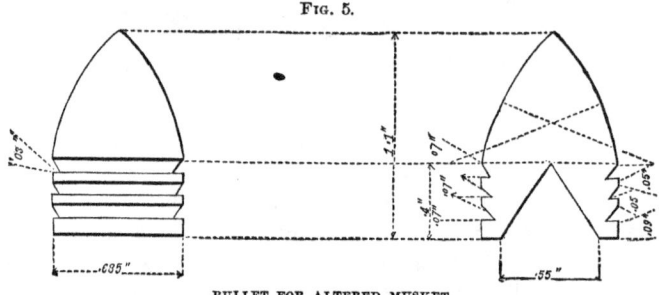

Fig. 5.

BULLET FOR ALTERED MUSKET.
Weight of ball, 730 grains; weight of powder, 70 grains.

To use the new cartridge carrying the powder and elongated ball attached to each other, tear the fold and pour out the powder; then seize the ball end firmly between the thumb and forefinger of the right hand, and strike the cylinder of the cartridge a smart blow across the muzzle of the piece; this breaks the cartridge and exposes the bottom of the ball; a slight pressure of the thumb and forefinger forces the ball into the bore clear of all cartridge paper. In striking the cartridge, the cylinder should be held square across, or at right angles to the muzzle; otherwise, a blow given in an oblique direction would only bend the cartridge without rupturing it.

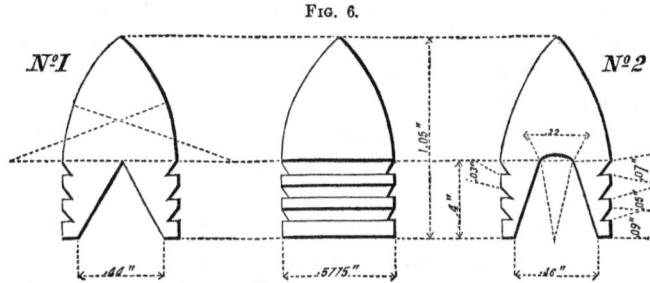

BULLETS FOR NEW RIFLE-MUSKET AND PISTOL-CARBINE.

Weight of No. 1, 500 grains.   Weight of No. 2, 450 grains.
Weight of powder, 60 grains.   Weight of powder, 40 grains.
No. 1, section of musket bullet.   No. 2, section of pistol-carbine bullet.
Both bullets have the same exterior.

Ammunition for a siege train of one hundred pieces, consisting of the following:

Guns	24-pounder about one-third the whole number	32
	18-pounder, one-tenth the whole number	10
	12-pounder, " " "	10
Howitzers.—8-inch siege, one-eighth " "		13
Mortars	10-inch siege, one-seventh " "	14
	8-inch siege, one-fourteenth " "	7
Stone Mortars, one-seventh " "		14
Coehorn Mortars (in addition to the 100 pieces)		6
Wall Pieces, for the attack of one front		40

The 18 and 24-pounders should be furnished with one thousand rounds each, the 12-pounders with twelve hundred rounds, the 8-inch howitzers and mortars with six hundred rounds. In addition to the above, fifty rounds of spherical-case shot should be furnished to each gun. Powder magazines, containing from fifty to one hundred thousand pounds of powder, must be accessible.

Cartridges for siege and garrison service are usually one-fourth the weight of the shot; but the charge varies according to circumstances from one-third the weight of the shot (for a breaching battery) to one-sixth of that weight for firing double shot, or hot shot, and still less for ricochet firing. The charges for mortars and howitzers vary according to the required range. For columbiads and sea-coast howitzers, the cartridge should always occupy the whole length of the chamber; for this purpose, in firing with reduced charges a cartridge *block* is placed in the bag over the powder. For mortars, cartridge bags may be made in the same manner as for guns, but the charge is usually poured loose into the chamber. Charges vary for mortar shells from 11 lbs. to 4 oz. according to the size of the mortar, and whether the intention be to fill the shell, to burst it, or simply to blow out the fuse. For *hot shot*, cartridge bags are made double by putting one bag free from holes within another. (*For full details concerning ammunition, including its preparation, &c.*, consult ORDNANCE MANUAL, 1850; consult also Experiments with small arms by Ordnance Officers, 1856. *See* ARMS; CANISTER; CARTRIDGE; FRICTION TUBES; FUZE; GRAPE SHOT; GUN POWDER; ORDNANCE AND ORDNANCE STORES; RIFLED ORDNANCE; SABOT; SHELLS; SOLID SHOT; SPHERICAL-CASE.)

AMNESTY. An act of oblivion, or forgiveness of past offences.

ANGLE OF DEFENCE—is that formed by the meeting of the flank and line of defence, or the face of the bastion produced.

ANGLE OF THE POLYGON—is that formed by the meeting of two of the sides of the polygon; it is likewise called the polygon angle.

APOLOGY—when made and accepted, debars the officer who accepts from bringing forward the matter as a substantive accusation, (*Hough*.)

APPEAL. Any officer or soldier who may think himself wronged by his colonel or the commanding officer of his regiment, and after due application to him, is refused redress, may appeal to the next higher commander, who is to examine into said complaint, and take proper measures for redressing the wrong complained of, and transmit, as soon as possible, to the Department of War, a true statement of such complaint, with the proceedings had thereon; (*Art* 34.) If any inferior officer or soldier shall think himself wronged by his captain, or other officer, he is to complain thereof to the commanding officer of the regiment, who is required to summon a regimental court-martial for doing justice to the complainant; from which regimental court-martial, either party may, if he thinks himself still aggrieved, appeal to a general court-martial. But if, upon a second hearing, the appeal shall appear vexatious and groundless, the person so appealing shall be punished at the discretion of the said court-martial; (*Art.* 35.) (*See* REMEDY.)

The wrongs here alluded to, have reference chiefly to matters of accounts between the captain, or commander of the company, and the soldier, relating to clothing and other supplies, as well as to pay; and the regimental court, in examining into such transactions, may be considered more as a court of inquiry than a court-martial; or, it may be viewed as an arbitration board, called on to adjust and settle differences arising in the settlements of accounts between the captain and his men. One reason why a power of appeal is declared to be a matter of absolute right to inferior officers, or soldiers, complaining of being wronged by their officers, doubtless is, that a regimental or garrison court-martial has not the power of inflicting any punishment on commissioned officers. It can do no more than express its opinion that the complaint is just, or the contrary, and where it is practicable and proper, relieve the sufferer as to any existing grievance; but, the injury complained of, however flagrant, must still have remained unredressed, as far as punishment is concerned, if an appeal to a general court-martial had not been declared to be a matter of right to the party aggrieved.

APPOINTING POWER, &c. It has been contended by advocates of executive discretion, that army appointments are embraced in the power granted to the President in the 2d section of the Constitution, to nominate, and, by and with the advice and consent of the Senate, appoint " all other officers of the United States, whose appointments are not herein otherwise provided for, and which may be established by law. But the Congress may, by law, vest the appointment of such inferior officers as they think proper in the President alone, in the courts of law, or in the heads of departments." If due regard, however, be paid to the words, "*whose appointments are not herein otherwise provided for,*" the pretension set up in favor of Executive power, will receive no support from the terms of the Constitution. The powers granted to Congress to *raise* and support armies, and to make all *rules* for the *government* and *regulation* of the land and naval forces, are necessarily so comprehensive in character, as to embrace all means which Congress, according to circumstances, may deem proper and necessary in order to raise armies, or to govern them when raised. Rules of appointment to office, rules of promotion—another form of appointment—and all rules whatever in relation to the land and naval forces, save the appointment of the commander-in-chief of those united forces, who is designated by the Constitution, are hence within the competency of Congress.

It is true that this great power vested in Congress has been exercised by them, in most cases, by giving to the President a large discretion in appointments and other matters connected with the army. But the principle itself—that supreme command is vested in Congress—has been often asserted in our military legislation. Contemporaneously with the foundation of the government laws have been passed, giving to general and other officers the right of appointment to certain offices; in other cases, the President has been confined in his selection to classes designated by law; again, rules have been made by Congress for the promotion of officers, and in 1846 an army of volunteers was raised by Congress, the officers of which Congress directed should be appointed, according to the laws of the States in which the troops were raised, excepting the general officers, who were to be appointed by the President and Senate—a clear recognition that the troops thus raised were United States troops, and not militia. (*See* CONGRESS; PROMOTION; VOLUNTEERS.)

APPOINTMENT—is Office, Rank, Employment, Equipment.

APPROACHES—are the first, second, and third parallels, trenches, saps, mines, &c., by which the besiegers approach a fortified place.

APPROPRIATIONS—for the support of armies, are limited by the Constitution to a term not to exceed two years. The President is authorized to transfer appropriations for subsistence, forage, the medical and quartermaster's department, from one branch of military expenditure to any other of the above-mentioned branches; (*Act* May 1, 1820.) (*See* TRANSFERS.)

APRON. A piece of sheet lead used to cover the vent of a cannon.

APPUI, POINT D'. A term applied to any given point upon which a line of troops is formed.

ARDENT SPIRITS. The introduction of ardent spirits into Indian Territory, under any pretence, prohibited; (*Act* July 9, 1832.) The President of the United States may take such measures as he may deem expedient to prevent or restrain the vending or distributing of spirituous liquors among Indians. Goods of traders introducing it forfeited; (*Acts* March 30, 1802, and May 6, 1832.)

ARM. Infantry, artillery, and cavalry, are arms of the service.

ARMISTICE, *Armistitium*, i. e. *sistere ab armis*. A temporary truce, or suspension of hostilities.

ARMORER. The person who makes, cleans, or repairs arms.

ARMORY. A manufactory or place of deposit for arms. (*See* ARSENAL; ORDNANCE DEPARTMENT.)

ARMS, SMALL. Casting away arms and ammunition punishable with death or otherwise according to the sentence of a general court-martial; (*Art.* 52.) Officers, non-commissioned officers, and soldiers should be instructed and practised in the nomenclature of the arms, the manner of dismounting and mounting them, and the precautions and care required for their preservation. Each soldier should have a screw-driver and a wiper, and each squad of ten a wire and a tumbler punch, and a spring vice. No other implements should be used in taking arms apart or in setting them up. In the inspection of arms, officers should attend to the qualities essential to service, rather than a bright polish on the exterior of the arms. The arms should be inspected in the quarters at least once a month, with the barrel and lock separated from the stock.

PRINCIPAL DIMENSIONS, WEIGHTS, ETC., OF SMALL ARMS.

Dimensions.		Rifle muskets.			Rifles.		Pistol carbine.
		1822.	1840.	1855.	1841.	1855.	1855.
		Inches.	Inches.	Inches.	Inches.	Inches.	Inches.
Barrel.	Diameter of bore	.69	.69	.58	.58	.58	.58
	Variation allowed, more	.015	.015	.0025	.0025	.0025	.0025
	Diameter at muzzle	.82	8.85	.78	.90	.90	.82
	Diam'r at breech between flats.	1.25	1.25	1.14	1.15	1.14	1.
	Length without breech screw.	42.	42.	40.	33.	33.	12.
Bayonet.—Length of blade		16.	18.	18.	21.7	21.7	......
Ramrod.—Length		41.96	41.70	39.60	33.	33.	12.
Arm complete.	Length without bayonet	57.64	57.80	55.85	48.8	49.3	17.6
	With bayonet fixed	73.64	75.80	73.85	71.3	71.8	......
	With butt-piece	......	......	......	......	......	28.2
Grooves.	Number	3	3	3	3	3	3
	Twist	6.	6.	6.	6.	6.	4.
	Width	.36	.36	.30	.30	.30	.30
	Depth at muzzle	.005	.005	.005	.005	.005	.005
	Depth at breech	.015	.015	.015	.013	.013	.008
WEIGHTS.		Lbs.	Lbs.	Lbs.	Lbs.	Lbs.	Lbs.
Barrel, without breech screw		4.	4.19	4.28	4.8	4.8	1.4
Lock, with side screws		* 95	.95	.81	.55	.81	.6
Bayonet		.73	.64	.72	3.05	3.05	......
Arm complete.	Without bayonet	9.06	9.51	9.18	9.68	9.93	3.56
	With bayonet	9.82	10.15	9.90	12.72	12.98	......
	With butt-piece	......	......	......	......	......	5.09

HEIGHTS OF HAUSSE, ETC.

Table of approximate heights for rear sights of new arms, measured from the line of metal of the barrel. Pieces fired from the shoulder and rest.

Distance.	New Rifle musket. Weight of ball, 500 grains. Weight of powder, 60 grains.	Rifle musket (altered). Weight of ball, 730 grains. Weight of powder, 70 grains.	Remarks.
Yards.	Inches.	Inches.	
100	.40	.42	The top of the front sight is seen "fine" through the notch of the rear sight.
200	.54	.62	
300	.70	.82	
400	.88	1.08	
500	1.10	1.34	
600	1.35	1.65	
700	1.63	1.96	
800	1.94	2.28	
900	2.28	2.61	
1000	2.63	2.94	

* Maynard primer.

PENETRATIONS.

Table of penetrations in a target made of seasoned white pine plank one inch thick, and placed one and a half inches apart.

Arm.	Weight of ball.	Weight of powder.	Diameter of bullet.	Planks penetrated.	Distance.
	Grains.	Grains.	Inch.	Number.	Yards.
Altered rifle............................	500	60	.5775	9¼	200
Altered musket.....................	730	70	.685	10½	200
New rifle-musket..................	500	60	.5775	11	200
Pistol-carbine.......................	450	40	.5775	5¾	200
Altered rifle............................	500	60	.5775	5⅔	600
Altered musket.....................	730	70	.685	6⅛	600
New rifle-musket..................	500	60	.5775	6¼	600
Pistol-carbine.......................	450	40	.5775	3	500
Altered musket.....................	730	70	.685	3½	1000
Altered rifle...........................	500	60	.5775	3	1000
New rifle-musket..................	500	60	.5775	3¼	1000

At 1,000 yards, a bullet from the new rifle-musket passed completely through the frame of the target, which was made of solid white pine, three inches thick.

The elongated musket bullets do not cease to ricochet on level ground, at the distance of 1,000 yards. A strong wind blowing perpendicularly to the direction of the rifle-musket bullet, will deflect it from its course 12 feet in 1,000 yards, about 3 feet in 500 yards, and about ½ foot in 200 yards. The effect of wind on the pistol-carbine bullets is somewhat greater, for the same distance. When two oblong bullets are fired from the new rifle-musket, or altered rifle, with the ordinary service charge of 60 grains, they separate from each other and from the plane of fire about 4 feet in a distance of 200 yards. If the piece be held firmly against the shoulder, no serious inconvenience will be felt in firing this increased charge; the only precaution necessary to be observed in aiming, is to give the barrel greater elevation than for the single bullet, in the proportion of 6 feet for 200 yards. In cases of emergency, firing with two bullets might be effectively employed against masses of infantry and cavalry, if the distance does not exceed 300 yards. Muzzle-loading small arms can be discharged two or three times in a minute, and breech-loading arms about ten times. Rapidity of loading and discharging fire-arms is however of doubtful advantage in actual service, as soldiers are apt to discharge their pieces without proper aim, and thus waste ammunition.

ARM.] MILITARY DICTIONARY. 31

Nomenclature descriptive of the Rifle Musket.
MODEL OF 1855.

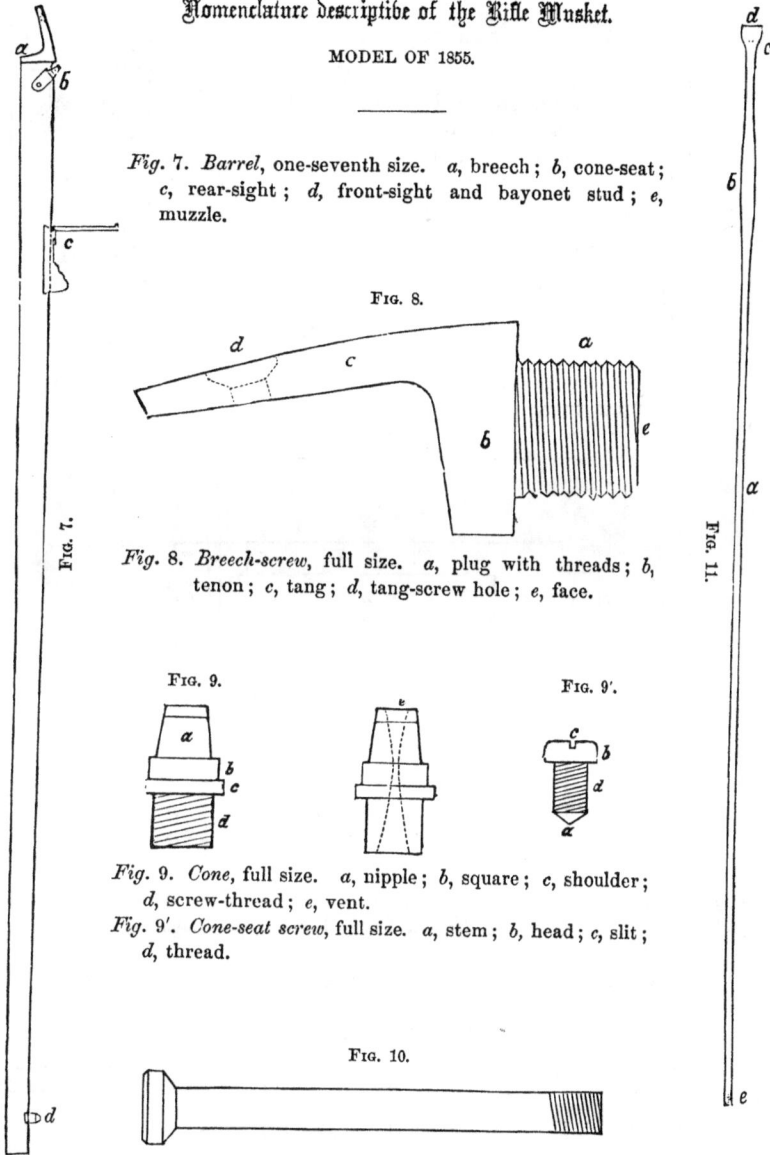

Fig. 7. *Barrel*, one-seventh size. *a*, breech; *b*, cone-seat; *c*, rear-sight; *d*, front-sight and bayonet stud; *e*, muzzle.

Fig. 8. *Breech-screw*, full size. *a*, plug with threads; *b*, tenon; *c*, tang; *d*, tang-screw hole; *e*, face.

Fig. 9. *Cone*, full size. *a*, nipple; *b*, square; *c*, shoulder; *d*, screw-thread; *e*, vent.

Fig. 9'. *Cone-seat screw*, full size. *a*, stem; *b*, head; *c*, slit; *d*, thread.

Fig. 10. *Tang-screw*, full size.

Fig. 11. *Ramrod*, one-seventh size. *a*, stem; *b*, swell; *c*, head; *d*, cup; *e*, screw.

32   MILITARY DICTIONARY.   [ARM.

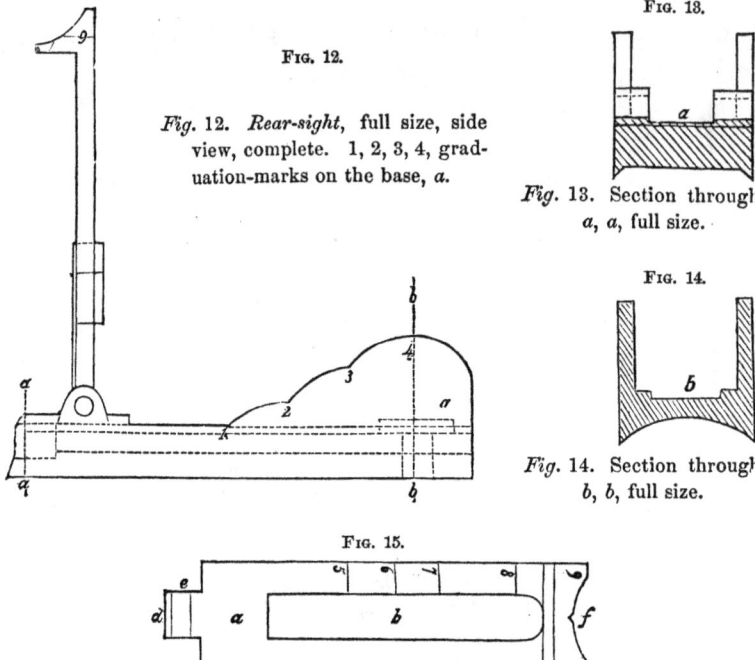

Fig. 12. *Rear-sight*, full size, side view, complete. 1, 2, 3, 4, graduation-marks on the base, *a*.

Fig. 13. Section through *a*, *a*, full size.

Fig. 14. Section through *b*, *b*, full size.

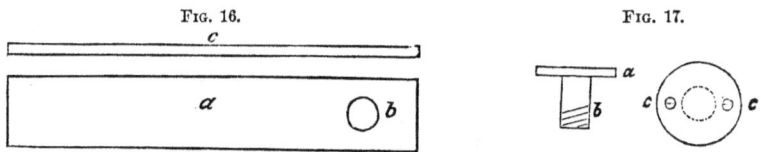

Fig. 15. *Leaf*, full size. *a*, frame; *b*, slot; *d*, tongue; *e*, joint-pin hole; *f*, sight-notch; 5, 6, 7, 8, 9, graduation-marks.

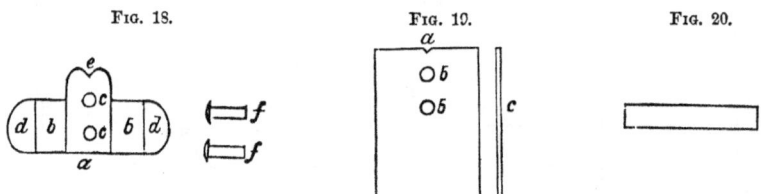

Fig. 16. *Leaf-spring*, full size. *a*, blade; *b*, screw-hole; *c*, thickness.
Fig. 17. *Leaf-spring screw*, full size. *a*, head; *b*, stem; *c*, *c*, holes for screw-driver.

Fig. 18. *Slide*, full size. *a*, back-piece; *b*, *b*, grooves; *c*, *c*, rivet-holes; *d*, *d*, handles; *e*, sight-notch; *f*, *f*, rivets.
Fig. 19. *Slide-spring*, full size. *a*, sight-notch; *b*, *b*, rivet-holes; *c*, thickness.
Fig. 20. *Joint-pin*, full size.

FIG. 21.

FIG. 22.

FIG. 23.

FIG. 24.

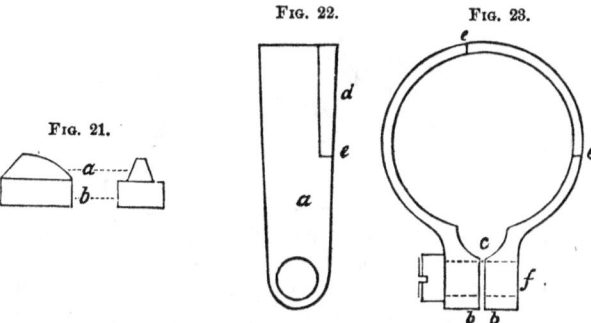

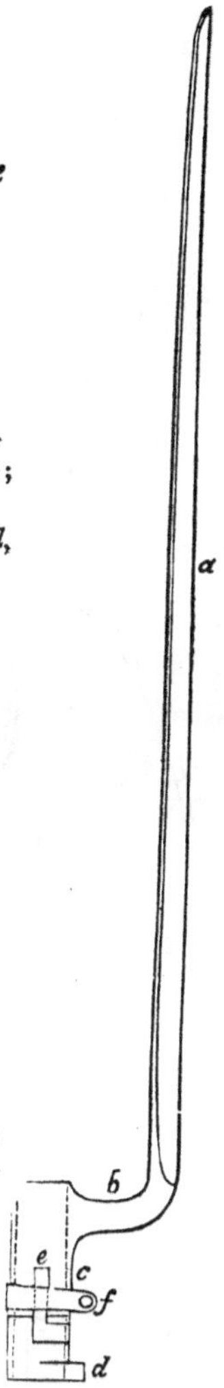

Fig. 21. *Front-sight* and *bayonet-stud*, full size. *a*, sight; *b*, stud.

Figs. 22, 23. *Bayonet-clasp*, full size. *a*, body; *b*, *b*, stud; *c*, bridge; *d*, groove; *e*, *e*, stops; *f*, screw.

Fig. 24. *Bayonet*, quarter size. *a*, blade; *b*, neck; *c*, socket; *d*, bridge; *e*, stud mortise; *f*, clasp.

FIG. 25.

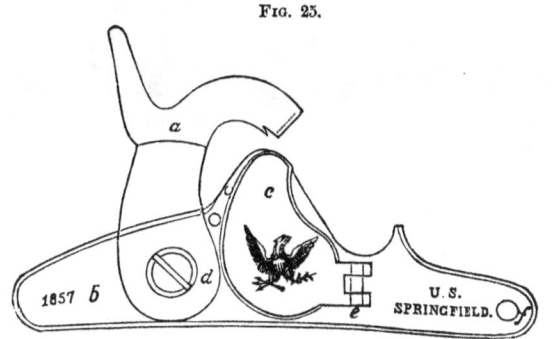

Fig. 25. *Lock*, outside view, half size. *a*, hammer; *b*, lock-plate; *c*, magazine-cover; *d*, tumbler-screw; *e*, joint-pin; *f*, side-screw hole.

FIG. 26.

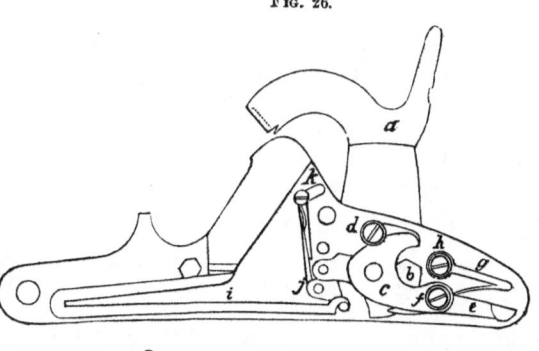

Fig. 26. *Lock*, inside view, half size, showing the parts with the hammer at half cock. *a*, hammer; *b*, tumbler; *c*, bridle; *d*, bridle-screw; *e*, sear, ; *f*, sear-screw; *g*, sear-spring; *h*, sear-spring screw; *i*, mainspring; *j*, swivel; *k*, cover-catch.

FIG. 27.

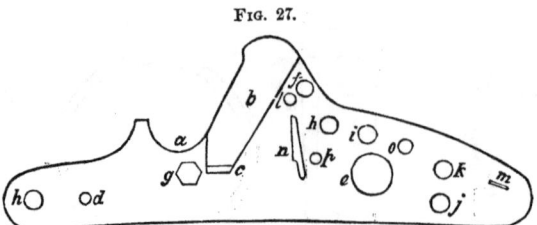

Fig. 27. *Lock-plate*, half size, showing the position of the holes, &c. *a*, cone-seat notch; *b*, bolster; *c*, mainspring notch; *d*, hole for mainspring pivot; *e*, hole for arbor of tumbler; *f*, hole for cover-catch; *g*, hole for cover hinge stud; *h*, *h*, side-screw holes; *i*, hole for bridle-screw; *j*, hole for sear-screw; *k*, hole for sear-spring; *l*, hole for catch-spring screw; *m*, sear-spring stud-mortise; *n*, feed-finger slot; *o*, bridle pivot hole; *p*, feed-finger-spring-screw hole.

FIG. 28.

FIG 29.

FIG. 30.

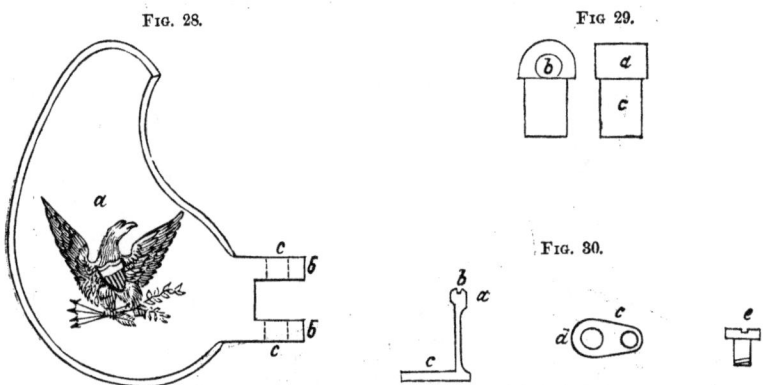

Fig. 28. *Magazine-cover*, full size. *a*, body; *b*, *b*, jaws; *c*, *c*, holes for joint-pin.
Fig. 29. *Cover-hinge stud*, full size, two views. *a*, head; *b*, joint-pin hole; *c*, stem.
Fig. 30. *Cover-catch* and *screw*, full size, two views. *a*, head; *b*, notch; *c*, *c*, foot; *d*, screw-hole; *e*, catch-screw.

FIG. 31.

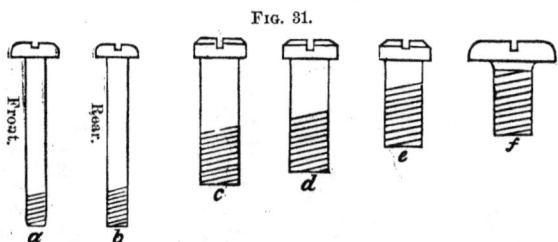

Fig. 31. *Lock-screws*, full size, and *side-screws*, half size. *a*, *b*, side-screws; *c*, sear-screw; *d*, bridle-screw; *e*, sear-spring screw; *f*, tumbler-screw.

NOTE.—*In all the screws*, the parts are the stem, the head, the slit, the thread.

Fig. 32. *Mainspring-swivel,* full size. *a, a,* body; *b,* axis; *c,* tumbler-pin hole; *d,* finger-pivot hole.

Fig. 33. *Feed-finger,* full size, two views. *a, a,* eye-pivot; *b,* crook; *c, c,* finger.

Fig. 34. *Feed-finger spring,* full size. *a,* eye; *b,* long branch; *c,* short branch; *d,* screw.

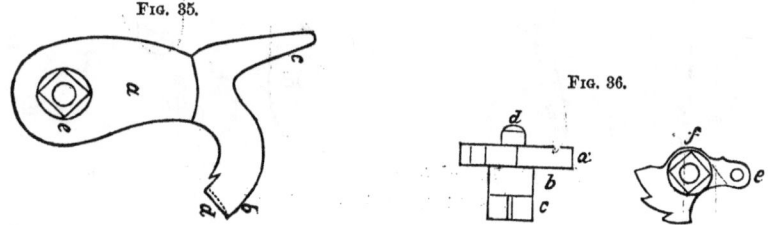

Fig. 35. *Hammer,* half size. *a,* body; *b,* head; *c,* comb; *d,* countersink, slit, and knife-edge; *e,* tumbler-hole.

Fig. 36. *Tumbler,* half size, two views. *a,* body; *b,* arbor; *c,* squares; *d,* pivot; *e,* swivel-arm and pin-hole; *f,* tumbler-screw hole.

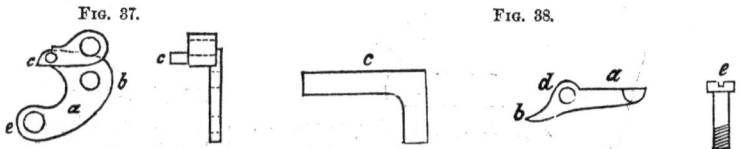

Fig. 37. *Bridle,* half size, two views. *a,* body; *b,* eye for tumbler-pivot; *c,* pivot; *d,* hole for bridle-screw; *e,* hole for sear-screw.

Fig. 38. *Sear,* half size, two views. *a,* body; *b,* nose; *c,* arm; *d,* screw-hole; *e,* screw.

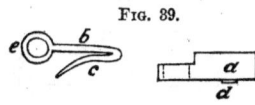

Fig. 39. *Sear-spring,* half size, two views. *a,* blade; *b,* upper branch: *c,* lower branch; *d,* stud; *e,* screw-hole.

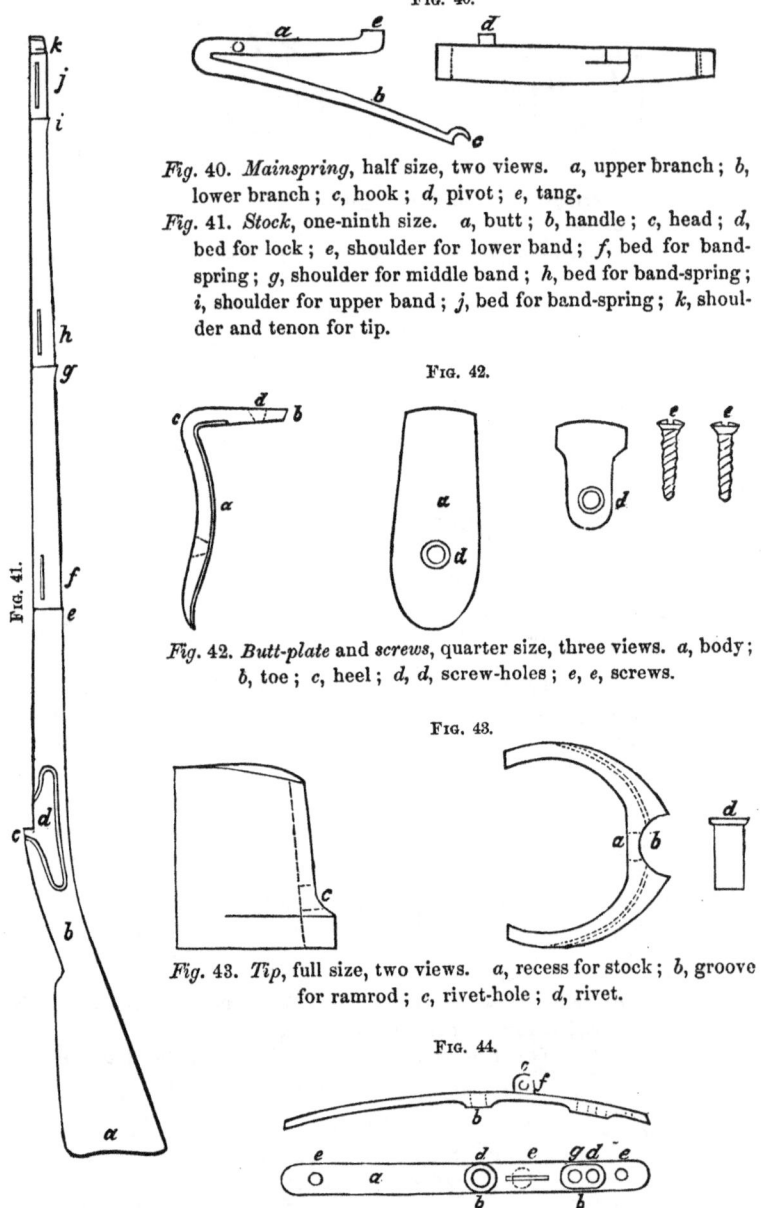

Fig. 40. *Mainspring,* half size, two views. *a,* upper branch; *b,* lower branch; *c,* hook; *d,* pivot; *e,* tang.

Fig. 41. *Stock,* one-ninth size. *a,* butt; *b,* handle; *c,* head; *d,* bed for lock; *e,* shoulder for lower band; *f,* bed for band-spring; *g,* shoulder for middle band; *h,* bed for band-spring; *i,* shoulder for upper band; *j,* bed for band-spring; *k,* shoulder and tenon for tip.

Fig. 42. *Butt-plate* and *screws,* quarter size, three views. *a,* body; *b,* toe; *c,* heel; *d, d,* screw-holes; *e, e,* screws.

Fig. 43. *Tip,* full size, two views. *a,* recess for stock; *b,* groove for ramrod; *c,* rivet-hole; *d,* rivet.

Fig. 44. *Guard-plate,* quarter size. *a,* body; *b, b,* bolsters; *c, c,* trigger-stud and mortise; *d, d,* holes for guard-bow; *e, e,* for wood screws; *f,* for trigger-screw; *g,* for tang-screw.

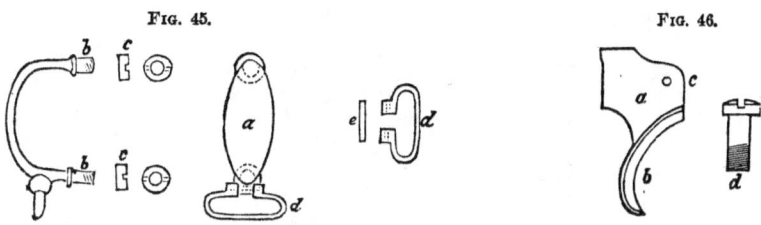

Fig. 45. *Guard-bow*, quarter size, two views. *a*, body; *b, b*, stems; *c, c*, nuts; *d, d*, swivel; *e*, rivet.
Fig. 46. *Trigger*, half size. *a*, blade; *b*, finger-piece; *c*, hole for screw; *d*, screw, full size.
Fig. 48. *Guard-screws*, half size.

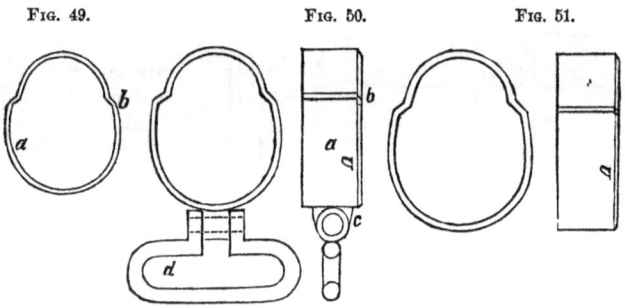

Fig. 49. *Upper band*, half size.
Fig. 50. *Middle band*, half size.
Fig. 51. *Lower band*, half size. *a*, body; *b, b*, creases; *U* denotes the upper edge; *c*, swivel-stud (on middle band only); *d*, swivel.

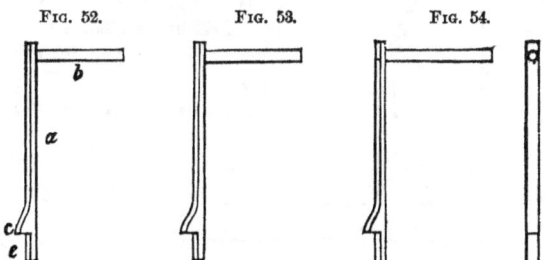

Figs. 52, 53, 54. *Upper, middle,* and *lower band-springs,* half size. *a*, stem; *b*, wire; *c*, shoulder; *e*, tang.

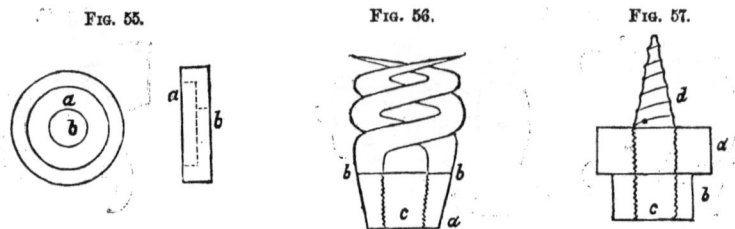

Fig. 55. *Side-screw washer,* full size. *a,* countersink ; *b,* hole for screw.
Fig. 56. *Wiper,* full size. *a,* body ; *b, b,* prongs ; *c,* screw-hole for rod.
Fig. 57. *Ball-screw,* full size. *a,* body ; *b,* tang ; *c,* screw-hole for rod ; *d,* screw to draw the ball.

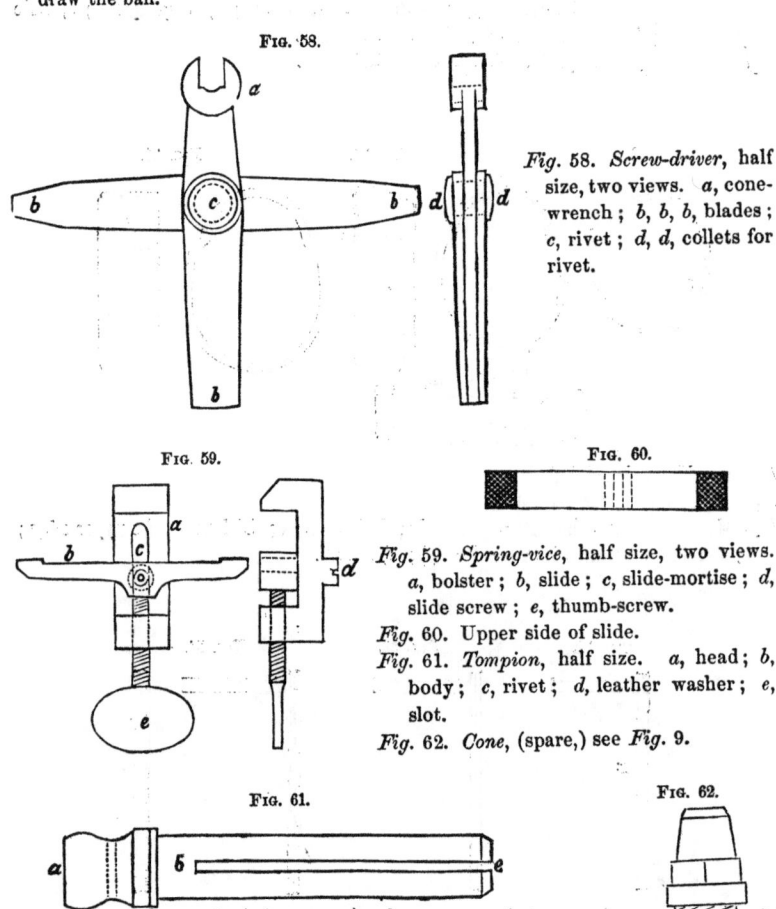

Fig. 58. *Screw-driver,* half size, two views. *a,* cone-wrench ; *b, b, b,* blades ; *c,* rivet ; *d, d,* collets for rivet.

Fig. 59. *Spring-vice,* half size, two views. *a,* bolster ; *b,* slide ; *c,* slide-mortise ; *d,* slide screw ; *e,* thumb-screw.
Fig. 60. Upper side of slide.
Fig. 61. *Tompion,* half size. *a,* head ; *b,* body ; *c,* rivet ; *d,* leather washer ; *e,* slot.
Fig. 62. *Cone,* (spare,) see *Fig.* 9.

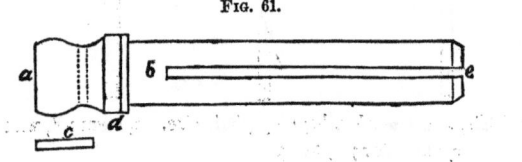
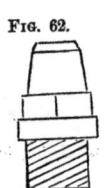

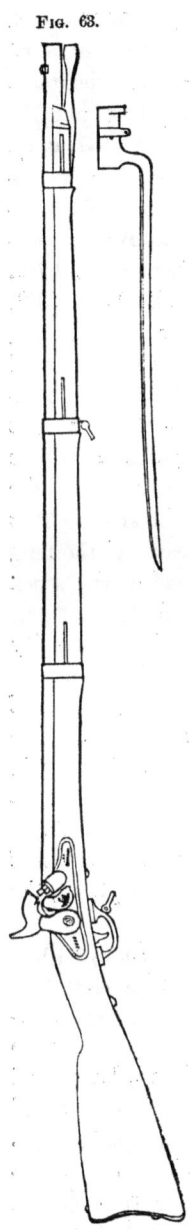

Fig. 63.

Rifle Musket and Appendages, Model 1855.

Wiper.   Ball-screw.   Screw-driver.   Spring-vice.

Tompion.   Spare cone.

Tumbler and Wire Punch.

MATERIALS OF WHICH THE PARTS ARE MADE.

Steel.

Tumbler; Lock-swivel, Feed-finger; Finger-spring; Cover-catch; Sear; Sear-spring; Mainspring; Band-springs; Ramrod; Rear-sight (except the screw); Screw-driver; Wiper; Ball-screw; Cone; Tumbler, and Wire Punch.

Brass.

Tip for Stock; head of Tompion.

Wood.

Stock; Tompion.

Iron.

Socket of the Bayonet, and all other parts not enumerated.

Rules for Dismounting the Rifle Musket, model of 1855.—1st. Unfix the bayonet (24). 2d. Put the tompion (60) into the muzzle of the barrel. 3d. Draw the ramrod (11). 4th. Turn out the tang-screw (10). 5th. Take off the lock (25): to do this, first put the hammer at half-cock, then unscrew partially the side-screws (31, *a*, *b*), and, with a slight tap on the head of each screw with a wooden instrument, loosen the lock from its bed in the stock; then turn out the side-screws, and remove the lock with the left hand. 6th. Remove the side-screws (31, *a*, *b*), taking care not to disturb the washers (55). 7th. Take off the upper band (49). 8th. Take off the middle band (50). 9th. Take off the lower band (51). (*Note.*—The letter U, on bands, is to indicate the upper side in assembling.) 10th. Take out the barrel (7): in doing this, turn the musket horizontally, with the barrel downward, holding the barrel loosely with the left hand below the rear sight (12), the right hand grasping the stock by the handle; and if it does not leave the stock, tap the tompion in the muzzle gently against the ground or floor, which will loosen the breech end from the stock. This is preferable to lifting the barrel out by the muzzle, because if the tang of the breech-screw (8) should bind in the wood, the head of the stock (41 *c*) would be liable to be split by raising the muzzle first.

The foregoing parts of the rifle musket are all that should usually be taken off or dismounted. The soldier should never dismount the *band-springs, guard, side-screw washers, butt-plate, rear-sight, cone, and cone-seat screw*, except when an officer considers it necessary. The breech-screw should be taken out only by an armorer, and *never* in ordinary cleaning. The lock should not be taken apart, nor the bayonet-clasp taken off, except when absolutely necessary in the opinion of an officer. *If proper and regular care be taken of the arm, this will be very seldom necessary.* The musket being thus taken to pieces, the soldier, under ordinary circumstances, will—

To clean the barrel—1st. Stop the hole in the cone (9, *e*) with a peg of soft wood; pour a gill of water (warm, if it can be had) into the muzzle; let it stand a short time, to soften the deposit of the powder; put a plug of soft wood into the muzzle, and shake the water up and down the barrel well; pour this out and repeat the washing until the water runs clear; take out the peg from the cone, and stand the barrel, muzzle downwards, to drain, for a few moments. 2d. Screw the wiper (56, *c*) on to the end of the ramrod (11, *e*) and put a piece of *dry cloth*, or *tow*, round it, sufficient to prevent it from chafing the grooves of the barrel; wipe the barrel quite dry, changing or drying the cloth two or three times. 3d. Put no oil into the vent (9, *e*), as it will clog the passage, and cause the first primer to miss fire; but, with a slightly oiled rag on the wiper, rub the bore of the barrel, and the face of the breech-screw (8, *e*), and immediately insert the tompion (61) into the muzzle. 4th. To clean the exterior of the barrel, lay it flat on a bench, or board, to avoid bending it. The practice of supporting the barrel at each end and rubbing it with a strap or buff-stick, or with the ramrod, or any other instrument, *to burnish it*, is pernicious, and should be strictly forbidden. 5th. After firing, the barrel should always be washed as soon as practicable; when the water comes off clear, wipe the barrel dry, and pass into it a rag moistened with oil. Fine *flour of emery*-cloth is the best article to clean the exterior of the barrel.

To clean the lock.—Wipe every part with a moist rag, and then a dry one; if any part of the interior shows rust, put a drop of oil on the point or end of a piece of soft wood dipped into flour of emery; rub out the rust clean and wipe the surface dry; then rub every part with a slightly oiled rag.

To clean the mountings.—For the mountings, and all iron and steel parts, use fine flour of emery moistened with oil, or flour of emery-cloth. For brass, use rotten-stone moistened with vinegar, or water, and avoid oil or grease. Use a hard brush, or a piece of soft pine, cedar, or crocus-cloth. Remove dirt from the screw-holes by screwing a piece of soft wood into them. Wipe clean with a linen rag, and leave the parts slightly oiled. In cleaning the arms, the aim should be to *preserve the qualities essential to service*, rather than to obtain a bright polish. *Burnishing* the barrel (or other parts) should be strictly avoided, as it tends to crook the barrel, and also to destroy the uniformity of the exterior finish of the arm.

It is not essential for the musket to be dismounted every time that it is cleaned; for, after firing in fine weather, or when dampness could not get between the barrel and the stock, it can be perfectly cleaned as follows: Put a piece of rag or soft leather on the top of the cone, and let the hammer down upon it; pour a gill of water into the muzzle carefully, so that it cannot run down the outside; put a plug of wood into the muzzle, and shake the gun up and down, changing the water repeatedly until it runs clear. Then withdraw the leather, and stand the musket on the muzzle a few moments; then wipe out the barrel (as told in the second rule for cleaning), and also wipe the exterior of the lock and the outside of the barrel around the cone and cone-seat, first with a damp rag, and then with a dry one, and lastly with a rag that has been slightly oiled. In this way, all dirt from firing may be removed without taking out a screw. If, however, the hammer works stiffly, or grates upon the tumbler, the lock must immediately be taken off, and the parts cleaned and touched with oil.

To re-assemble the musket.—The parts of the musket are put together in the inverse order of taking them apart, viz.: 1st. The barrel. Drop the barrel into its place in the stock, and squeeze it down with the hand; give the butt of the stock a gentle tap against the floor to settle the breech end of the barrel against the head of the stock (41, *c*). 2d. Put on the lower band with the letter U upward, being careful not to mar the stock, or barrel, in sliding it into its place; apply the thumb to the band-spring to see that it plays freely. 3d. Put on the middle, and, 4th. The upper band, in the same manner. 5th. The lock. Half-cock the hammer; take the lock in the right hand, with the main spring and sear toward you, holding the stock with the left hand by the swell, with the butt between the knees. Enter the lock fairly into the lock-bed, taking care to keep the arm of the sear clear of the trigger; press the plate well down into the wood, and then turn the musket over, holding the lock and stock together with the left hand. 6th. With the right hand, turn in the side-screws, after having touched their screw-threads with oil. Observe that the point of the rear-screw is *flat*, and should not project beyond the plate, to interfere with the hammer. The front screw has a round point. 7th. Turn in the tang-screw, after having oiled the screw-thread. Be careful to see that each of these screws are turned firmly home, *but not forced*. Observe that the lock plays freely, without friction, and that no limb is bound by the wood. 8th. Return the ramrod. 9th. Refix the bayonet, after having oiled the clasp and socket to prevent chafing. 10th. Replace the tompion. *Oil the stock* well with sperm or linseed oil; let it stand a few hours, and then rub it with a woollen rag until the wood is perfectly dry. Repeat this from time to time, and it will produce a polish which moisture will not affect. Linseed oil is the best for this purpose, and it should be used while the arm is dismounted.

Rules for the more complete dismounting of the rifle-musket, when cleaned by an armorer.—1st. The parts which should be dismounted by an experienced armorer will be given in their regular order following No. 10, viz.: 11th. Unscrew the cone, keeping the wrench well down on the square of the cone, to prevent the corners from being injured. 12th. Take out the cone-seat screw (9′). 13th. Take out the upper, middle, and lower band-springs (52, 53, 54), using a wire punch of proper size. 14th. Take out the guard-screws (48). *Note.*— The guard, butt-plate, and side-screw heads have concave slits, for which the screw-driver is adapted: this lessens the danger of the stock being marred by accident or carelessness in letting the screw-driver slip out, while in the act of turning the screw: great care should be used to prevent such injuries. 15th. Take out the guard, and be careful not to injure the wood at each end of the guard-plate (44). 16th. Take out the side-screw washers (55) with a drift-punch. 17th. Take out the butt-plate screws (42) with the largest blade of the screw-driver, and remove the butt-plate (42). 18th. Remove the rear sight (12), by turning out the leaf-spring screw (17), which will release the sight from the barrel. 19th. Turn out the breech-screw (7), by means of a "breech-screw wrench" suited to the tenon (b) of the breech-screw (8). No other wrench should ever be used for this purpose, and the barrel should be held in clamps fitting neatly the breech (7, a).

In re-assembling the parts, the armorer is to observe the inverse order of taking them apart, viz.: 1st. Breech-screw to be screwed into the barrel after being oiled; 2d. Rear-sight to be affixed; 3d. Butt-plate and screws; 4th. Side-screw washers; 5th. Guard; 6th. Guard-screws; 7th. Lower, middle, and upper-band springs; 8th. Cone-seat screw; 9th. Cone. The remaining parts follow as given for the soldier, commencing with the barrel (see page 42).

Order in which the Lock is taken apart.—1st. Cock the piece, and put the spring-vice (59) on the mainspring; give the thumb-screw a turn sufficient to liberate the spring from the swivel (32) and mainspring notch (27, c). Remove the spring; 2d. The sear-spring screw: Before turning this screw entirely out, strike the elbow of the spring with the screw-driver, so as to disengage the pivot from its mortise: then remove the screw and spring; 3d. The sear-screw and sear; 4th. The bridle-screw and bridle; 5th. The tumbler-screw; 6th. The tumbler. This is driven out with a punch inserted in the screw-hole, which at the same time liberates the hammer. 7th. Detach the mainspring swivel from the tumbler with a drift punch. 8th. Take out the feed-finger and spring. The magazine-cover should never be taken off except when absolutely necessary; 9th. The catch-spring and screw. The lock is re-assembled in the inverse order of taking apart, viz.: 1st. The catch-spring; 2d. The feed-finger and spring; 3d. Mainspring swivel; 4th. Tumbler and hammer; 5th. Tumbler-screw; 6th. Bridle and screw; 7th. Sear and screw; 8th. Sear-spring and screw; 9th. Mainspring.

Before replacing the screws, oil them slightly with good sperm-oil, putting a drop on the point of the screw; also on the arbor and pivot of the tumbler; between the movable branches of the springs, and the lock-plate; on the hook and notches of the tumbler. After the lock is put together, avoid turning the screws in so hard as to make the limbs bind: to insure this, try the motion of each limb before and after its spring is mounted, and see that it moves without friction. When a lock has, from any cause, become gummed with oil and dirt, it may be cleaned by being boiled in soapsuds, or in pearlash or soda water, to loosen the thick oil; but heat should never be applied to any part of it in any other way. As rust and dirt are produced by exploding caps or primers, although no charge be fired, the parts of the barrel and cone exposed should be carefully wiped and oiled after such exercise. Besides the precautions in dismounting, remounting, and cleaning, which have been pointed out in the foregoing pages, habitual care in handling arms is necessary to keep them in good and serviceable condition. In *ordering arms* on parade, let the butt be brought gently to the ground, especially on pavements or hard roads. This will save the mechanism of the lock from shocks, highly injurious to it, from the loosening of screws and splitting the wood-work.

Rifled arms should not have the *ramrod sprung* in the bore with unnecessary force. It batters the head of the rod and wears injuriously the grooves. The soldier should let the rod slide down gently, supported by the thumb and finger; and the inspecting officer can satisfy himself of the condition of the bottom of the bore by gently tapping with the rod. The face of the breech can be polished, after washing, by means of a cork fixed on the wiper or ball-screw; the polished surface can be seen if the muzzle is turned to the light.

In *stacking arms*, care should be taken not to injure the bayonets by forcibly straining the edges against each other. The stack can be as well secured without such force being used. No cutting, marking, or scraping, in any way, the wood or iron should be allowed; and no part of the gun should be touched with a file. Take every possible care to prevent water from getting in between the lock, or barrel, and stock. If any should get there, dismount the gun as soon as possible, clean and oil the parts as directed, and see that they are perfectly dry before re-assembling them.

To place a coil of primers in the magazine.—Let down the hammer; open the magazine, by pulling back the head of the cover-catch with the thumb-nail of the left hand, while the thumb-nail of the right hand is pushed under the cover at the bottom. Remove the covering paper from the coil of primers; separate any parts that may happen to stick together; unwind about one inch; place the coil in the magazine, and the free end of it in the groove, flat-side towards the cone, and one primer beyond the end of the feed-finger; close the magazine. Should an exploded primer fail to ignite the charge, there must be moisture, or some obstruction, in the vent; or the gun may be improperly loaded. After a night in a damp place, a drop of moisture sometimes collects in the vent, and, unless removed, prevents the first primer, or cap, from igniting the charge. If, by accident, a coil of primers becomes softened by dampness, it can be made good again by a short exposure to a dry warm atmosphere. Should the cocking of the hammer fail to feed out properly the primer, open the magazine and notice, while working the hammer, the cause of the difficulty. It can generally be readily corrected.

RIFLE-MUSKET (1842).—This arm differs from the original model in the following particulars: 1st. The bore is grooved. 2d. It has a *rear sight* similar to that for the new musket, and a *front sight* of iron attached to the upper strap of the upper band. To prevent the band from moving sideways, a short stud is attached to the under side of the strap, which fits into a groove in the barrel. 3d. The head of the ramrod is reamed out to fit the pointed end of the ball. 4th. The lock is altered to the Maynard principle, differing from the one described for the new rifle-musket of 1855, by its size, the absence of the swivel, and the facts, that the mainspring is fastened by a screw, and the finger spring by a pin. 5th. To adapt the cone seat to this modified lock, a portion of the breech of the barrel is cut off, and a new breech piece with cone seat attached, is screwed on in its place. *Breech piece:* body, shoulder, screw thread, chamber (conical), tang, tenon, tang screw hole, chamfer, notch for side screw, cone seat, vent, vent screw, vent screw thread, cone thread.

RIFLE-MUSKET (1822).—The bayonet of this arm has no clasp, or ramrod spring; in all other respects the nomenclature is the same as that of the rifle-musket (1842).

PERCUSSION-RIFLE (1841).—The bore of this arm is reamed up and re-rifled; it also has a rear sight similar to the rifle-musket of 1855, and a stud and guide attached for a sword bayonet.

RIFLE (1855).—The exterior size of the barrel is nearly the same as that of the model of 1841. The barrel has a stud and guide for attaching a sword bayonet. The breech and cone seat are finished like the same parts of the new rifle-musket. *Lock:* Identical with that of the new rifle-musket. *Rear sight:* Similar to that of the new rifle-musket. *Mountings:* Similar to those of the new rifle-musket, with the addition of a catch box, smaller than the one on the rifle of 1841. *Ramrod:* Similar to the new rifle-musket. *Sword bayonet: Blade*—

shoulder, back, edge, bevel, point, curvature, groove, *tang* riveting, rivet hole, rivet. *Hilt: Gripe*—ridges, back, beak, slot for stud, slot for guide, hole for finger piece, hole for spring screw, hole for rivet (tang), mortice for tang: *Finger piece*—head, notch. *Finger piece spring*—blade, screw hole, boss: *Guard*—long and short branch, knobs, muzzle socket. *Scabbard:* Black leather, with brass band and tip.

Materials. Steel.—Tumbler, lock swivel, feeding finger, cover catch, sear, all the springs, ramrod, blade of sword bayonet, finger piece, rear sight, except screw, cone, screw driver, ball screw and wiper. *Brass.*—Sword bayonet handle, front sight, and all the mountings. *Wood.*—Stock (black walnut). *Iron.*—All the remaining parts.

PISTOL-CARBINE (1855).—*Barrel:* Muzzle, front sight, breech, breech pin threads, flats, bevels and oval, cone seat, vent, vent screw, bore, grooves, lands. This barrel tapers with a straight line from breech to muzzle. The portion of the flat in rear of the cone seat is parallel to the axis of the bore. *Breech screw:* Plug, with threads (16 to the inch), tenon, shoulders, tang, tang screw hole, bevel sight mortice. *Cone:* Same as for musket. *Rear sight:* Base, ears, joint screw, screw hole, 1st, 2d, and 3d leaves, 4 sight notches, eye joint, screw holes. *Tang screw:* Shoulder. *Lock:* Same as for rifle-musket, 1855, except in size, which is reduced to conform to a magazine capable of holding one-half a strip of primers. *Mountings: Band, swivel, and spring,* correspond to the middle band, swivel, and spring of the new musket. *Guard plate: Butt cup*—screw hole, tang. *Butt strap*—holes for catch spring and hook, tang, strap, and guard plate screws, shoulders for breech screw tang, and butt cup tang, reinforces for hook, and catch spring. *Cup screw*—head, eye. *Swivel ring.* The remaining mountings are similar to the corresponding parts of the new rifle-musket. *Ramrod:* Head (riveted on), cup, foot with a female screw. *Ramrod swivel:* Two side bars, screw, cross bar, riveted into the side bars. *Stock:* Butt, handle, curve, facings, reinforce, chase; *shoulders* for band and tip, *grooves* for barrel and ramrod; *beds* for tang and tenon, lock, washers, guard plate, nuts for guard bow and trigger stud, butt plate, band spring, tip, butt cup and strap, butt piece cap, and catch spring, hook nut; *mortices* for trigger, hook, and catch spring; *holes* for rod, tip rivet, band spring, side screws, tang screw, cup screw, strap screw, butt plate screws, and cap screws. *Butt piece: Plate*—two wood screws; *cap,* hollow, upper and lower tang, screw holes, two wood screws, cavity for pistol handle, hook, stem, nut; *spring catch,* screw, head, blade; *finger piece,* loop for spring, screw thread, rivet and nut.

Materials. Steel.—Cone, tumbler, lock swivel, finger, sear, lock springs, band springs, ramrod, except the head, rear sight except screw, spring catch, screw driver, wiper and ball screw. *Brass.*—Butt plate, butt cup, cup, guard plate and bow, band, and tip. *Wood.*—Stock and butt piece. *Iron.*—Head of ramrod, and remaining parts (Consult ORDNANCE MANUAL; ALLIN'S MANAGEMENT OF RIFLE-MUSKET, &c.; SMALL ARMS, 1856.)

ARMY. In its widest signification, Army is the military force of the state. It is the active and paid portion of the militia. It is an assemblage of agents and instruments proper and necessary to carry on war abroad, or suppress insurrection and repel invasion at home. The MILITARY ART organizes and combines its elements, and gives force and activity to armies.

In the United States, Congress raises, supports, governs, and regulates armies. RAISING is the prescribed means of organizing and collecting; SUPPORTING is the system of administration; GOVERNMENT consists in the creation of a hierarchy, with rules for rewarding and punishing; and REGULATION embraces the precise determination of methodical rights and duties, including the systems of tactics to be practised. Different armies are designated as follows: Standing or Regular Army; Army in the field; Army of Observation; Army of Invasion; Army of Occupation; Besieging Army; Covering Army; Offensive Army; Defensive Army; Army of the East; Army of Mexico; Army of Reserve, &c. The military art divides Armies into different ARMS; upon the theatre of war, it assembles an army in one or in many camps or cantonments; it links the army to a BASE by means of a LINE OF OPERATIONS; during the course of its movements, the army rests upon fortresses or entrenched camps; marches in combined columns, or columns in mass; for battle, it is distributed into Army Corps, Divisions, Brigades, and Battalions, and upon the day of action it is assembled between an advanced and rearguard, and flanking parties. The advance guard clears away the front, and secures all defiles; the rearguard watches over the safety of communications, and the flanking parties secure the flanks. The military art ranges an army according to circumstances; it determines the calibre of the ordnance, and the manner of using it. Laws and lawful orders are the basis of the daily duties of troops. Orders of the day direct movements; breaking up camps; maintain discipline; and provide for, and watch over, the distribution of supplies.

ARMY OF THE UNITED STATES—(ORGANIZATION OF THE).

MILITARY DICTIONARY. [ARMY OF U. S.

	Major-general.	Brigadier-generals.	Adjutant-general.	Assistant Adj. Gen., (Lieut. Col.)	Asst. Adjts. Gen., (Majors bvt.)	Asst. Adjts. Gen., (Captains bvt.)	Judge-advocate.	Inspector-general.	Quartermaster-general.	Assist. Quartermasters-general.	Dep. Quartermasters-general.	Quartermasters.	Assistant Quartermasters.	Com. General of Subsistence.	Asst. Com. Gen. of Subsistence.	Commissaries of Sub., (Majors.)	Commissaries of Sub., (Captains.)	Surgeon-general.	Surgeons.	Assistant Surgeons.	Paymaster-general.	Deputy Paymasters-general.	Paymasters.	Colonels.
General Officers...............	1	3	..	..	..	..	..	..	..	..	..	..	..	..	..	..	..	..	..	..	..	..	..	..
Aids-de-camp to General Officers.	..	..	..	..	..	..	..	..	..	..	..	..	..	..	..	..	..	..	..	..	..	..	..	..
Adjutant-general's Department..	..	..	1	1	4	a 8	..	..	..	..	..	..	..	..	..	..	..	..	..	..	..	..	..	..
Judge-advocate's Department....	..	..	..	..	..	..	a 1	..	..	..	..	..	..	..	..	..	..	..	..	..	..	..	..	..
Inspector-general's Department..	..	..	..	..	..	..	..	2	..	..	..	..	..	..	..	..	..	..	..	..	..	..	..	..
Quartermaster's Department.....	..	..	..	..	..	..	..	..	1	2	2	4	a 28	..	..	..	..	..	..	..	..	..	..	..
Subsistence Department.........	..	..	..	..	..	..	..	..	..	..	..	..	..	1	1	2	a 8	..	..	..	..	..	..	..
Medical Department........ ...	..	..	..	..	..	..	..	..	..	..	..	..	..	..	..	..	..	1	26	80	..	..	..	..
Pay Department.................	..	..	..	..	..	..	..	..	..	..	..	..	..	..	..	..	..	..	..	..	1	2	25	.
Corps of Engineers..............	..	..	..	..	..	..	..	..	..	..	..	..	..	..	..	..	..	..	..	..	..	..	..	(
Corps of Topographical Engineers	..	..	..	..	..	..	..	..	..	..	..	..	..	..	..	..	..	..	..	..	..	..	..	1
Ordnance Department...........	..	..	..	..	..	..	..	..	..	..	..	..	..	..	..	..	..	..	..	..	..	..	..	(
Two Regiments of Dragoons.....	..	..	..	..	..	..	..	..	..	..	..	..	..	..	..	..	..	..	..	..	..	..	..	2
Two Regiments of Cavalry......	..	..	..	..	..	..	..	..	..	..	..	..	..	..	..	..	..	..	..	..	..	..	..	2
Regiment of Mounted Riflemen.	..	..	..	..	..	..	..	..	..	..	..	..	..	..	..	..	..	..	..	..	..	..	..	1
Four Regiments of Artillery.....	..	..	..	..	..	..	..	..	..	..	..	..	..	..	..	..	..	..	..	..	..	..	..	4
Ten Regiments of Infantry......	..	..	..	..	..	..	..	..	..	..	..	..	..	..	..	..	..	..	..	..	..	..	..	10
Non-commissioned Staff unattached to Regiments.........	..	..	..	..	..	..	..	..	..	..	..	..	..	..	..	..	..	..	..	..	..	..	..	..
Grand aggregate.........	1	3	1	1	4	a 8	a 1	2	1	2	2	4	a 28	1	1	2	a 8	1	26	80	1	2	25	22

(a) *One* of the eight Assistant Adjutants-general (captains by brevet), *four* of the twenty-eight Assistant Quartermasters, and *one* of the eight Commissaries of Subsistence (captains), belonging also to regiments, and being included in their strength, are, to avoid counting them *twice*, excluded, as *Staff* officers, from the columns, "total commissioned," and "aggregate," of their respective Departments. The Regimental and Staff commissions, held by these officers, are of *unequal* grades; and hence they are not affected by the provisions of the 7th section of the act of June 18, 1846. The like remark is applicable to the *Judge-advocate* of the army, who is, at the same time, a Captain in the Ordnance Department.

MILITARY DICTIONARY.

Lieutenant Colonels.	Majors.	Captains.	Aids-de-camp.	Adjutants.	Regimental Quartermasters.	First Lieutenants.	Second Lieutenants.	Brevet Second Lieutenants.	Military Storekeepers.	Sergeant Majors.	Quartermaster Sergeants.	Principal or Chief Musicians.	Chief Buglers.	Ordnance Sergeants.	Hospital Stewards.	Sergeants.	Corporals.	Buglers.	Musicians.	Farriers and Blacksmiths.	Artificers.	Privates.	Enlisted men of Ordnance.	Total commissioned.	Total enlisted.	Aggregate.
..	..	..	c 5	..	..	..	..	..	..	..	..	..	..	..	..	..	..	..	..	..	..	..	..	4	......	4
..	..	..	..	..	..	..	..	..	..	..	..	..	..	..	..	..	..	..	..	..	..	..	..	......	......	......
..	..	..	..	..	..	..	..	..	..	..	..	..	..	..	..	..	..	..	..	..	..	..	..	13	......	13
..	..	..	..	..	..	..	..	..	..	..	..	..	..	..	..	..	..	..	..	..	..	..	..	......	......	......
..	..	..	..	..	..	..	..	..	..	..	..	..	..	..	..	..	..	..	..	..	..	..	..	2	......	2
..	..	..	..	..	..	..	..	7	..	..	..	..	..	..	..	..	..	..	..	..	..	..	..	40	......	40
..	..	..	..	..	..	..	..	..	..	..	..	..	..	..	..	..	..	..	..	..	..	..	..	11	......	11
..	..	..	..	..	..	..	..	..	..	..	..	..	..	g 68	..	..	..	..	..	..	..	..	..	107	68	175
..	..	b ..	..	..	..	..	..	e ..	..	..	..	..	..	..	..	..	..	..	..	..	..	..	..	28	......	28
2	4	b 13	..	..	..	12	11	e 3	..	..	..	..	..	..	..	10	10	..	2	..	..	78	..	46	100	146
1	4	b 17	..	..	..	10	3	e 3	..	..	..	..	..	..	..	..	..	..	..	..	..	..	..	39	......	39
1	4	b 17	..	..	..	12	1	e 3	15	..	..	..	..	..	..	..	..	..	..	..	..	......	400	54	400	454
2	4	20	..	2	d 2	20	20	e 4	..	2	2	2	4	..	..	80	80	40	..	20	..	1,000	..	74	1,230	1,304
2	4	20	..	2	d 2	20	20	e 4	..	2	2	2	4	..	..	80	80	40	..	20	..	1,000	..	74	1,230	1,304
1	2	10	..	1	d 1	10	10	e 1	..	1	1	1	2	..	..	40	40	20	..	20	..	640	..	36	765	801
4	8	48	..	d 4	d 4	96	48	e 5	..	4	4	..	..	..	..	192	192	..	96	..	96	h 2,148	..	213	2,732	2,945
10	20	100	..	d 10	d 10	100	100	e 4	..	10	10	20	..	..	..	400	400	..	200	..	..	4,200	..	344	5,240	5,584
..	..	..	..	..	..	..	..	..	..	..	..	..	..	f 73	..	..	..	..	..	..	..	..	..	......	73	73
23	50	b 245	c 5	d 19	d 19	280	213	e 27	22	19	19	25	10	f 73	g 68	802	802	100	298	60	96	i 9,066	400	1,085	11,838	12,923

(b) By the act of March 3, 1853, section 9, a Lieutenant of Engineers, Topographical Engineers, and Ordnance, having served "fourteen years' continuous service as Lieutenant," is entitled to promotion to the rank of Captain; but such promotion is not to increase the whole number of Officers, in either of said corps, beyond the number previously fixed by law.

(c) The *five* Aids-de-camp, being taken from regiments, in the strength of which they are included, are, to avoid counting them *twice*, excluded, as *Staff* officers, from the columns, "total commissioned," and "aggregate."

(d) The Adjutants of Artillery and Infantry (14), and *all* the Regimental Quartermasters (19),

Organization of Regiments and Companies.	Colonel.	Lieutenant-colonel.	Majors.	Adjutant.	Regimental Quartermaster.	Captains.	First Lieutenants.	Second Lieutenants.	Serjeant-major.	Quartermaster Serjeant.	Principal or Chief Musician.	Chief Buglers.	Sergeants.	Corporals.	Buglers.	Musicians.	Farriers and Blacksmiths.	Artificers.	Privates.	Total commissioned.	Minimum strength.	Maximum strength.	Minimum strength.	Maximum strength.
																					Total enlisted.		Aggregate.	
Regiment of Dragoons and Cavalry	1	1	2	1	1	10	10	10	1	1	1	2	40	40	20	..	10	..	500	35	615	855	650	890
Company of Dragoons and Cavalry	..	..	..	..	..	1	1	1	..	..	..	..	4	4	2	..	1	..	50	3	61	85	64	88
Regiment of Mounted Riflemen	1	1	2	1	1	10	10	10	1	1	1	2	40	40	20	..	20	..	640	35	765	865	800	900
Company of Mounted Riflemen	..	..	..	..	..	1	1	1	..	..	..	..	4	4	2	..	2	..	64	3	76	86	79	89
Regiment of Artillery	1	1	2	1	1	12	24	12	1	1	..	..	48	48	..	24	..	24	526*	52	672*	1,034	724	1,086
Company of Light Artillery	..	..	..	..	..	1	2	1	..	..	..	..	4	4	..	2	..	2	64	4	76	86	80	90
Company of Artillery	..	..	..	..	..	1	2	1	..	..	..	..	4	4	..	2	..	2	42	4	54	86	58	90
Regiment of Infantry	1	1	2	1	1	10	10	10	1	1	2	..	40	40	..	20	..	..	420	34	524	844	558	878
Company of Infantry	..	..	..	..	..	1	1	1	..	..	..	..	4	4	..	2	..	..	42	3	52	84	55	87

* The regiment being understood to consist of *one* Light and *eleven* Heavy companies.

being taken from the Subalterns, and accounted for in their several regiments as belonging to Companies, are excluded, as regimental *Staff* officers, from the columns "total commissioned," and "aggregate."

(*e*) Under the 4th section of the act of April 29, 1812, "making further provision for the Corps of Engineers," one Brevet Second Lieutenant is allowed to every "company." The number authorized is, consequently, *one hundred and ninety-nine*. The number, now attached to the Army, is *twenty-seven*.

(*f*) By the act of April 5, 1832, section 2d, "providing for the organization of the Ordnance Department," the number of Ordnance Sergeants cannot exceed "*one* for each military post." The number, actually in service, is *seventy-three*.

(*g*) By the act of August 16, 1856, section 2d, "providing for a necessary increase and better organization of the Medical and Hospital Department of the Army," the number of Hospital Stewards cannot exceed "*one* for each military post." The number, actually in service, is *sixty-eight*.

(*h*) *Two* companies in the 1st and 2d, and *one* in each of the other regiments of artillery, being equipped as Light Artillery, are allowed, in consequence, "*sixty-four*," instead of "*forty-two*" privates per company. See act "to increase the rank and file of the Army," &c., approved June 17, 1850, section 1st.

(*i*) By the act of June 17, 1850, "to increase the rank and file of the Army," &c., section 2d, the President is authorized, whenever the exigencies of the service require it, to increase to *seventy-four*, the number of privates in any company, "serving at the several military posts on the Western frontier, and at remote and distant stations." In the table, the minimum, or fixed, organization is given, viz.: *fifty* privates to a company of Dragoons, *sixty-four* to a company of Light Artillery and Riflemen, and *forty-two* to the Artillery and Infantry. If *all* the companies belonging to "regiments" (198) were serving at distant stations, the "total enlisted" would be 17,502, and the "aggregate" 18,587.

The organization by corps limits the number of officers in the army, but not their rank; the President, by and with the advice and consent of the Senate, being authorized by law to confer rank by brevet for gallant and meritorious services (*see* BREVET). Four Surgeons and four Assistants have been added to the Medical Department, and one Signal Officer created, with the rank of Major, since the preparation of these tables.

The most glaring deficiency in the military legislation of the United States, is the want of a GENERAL LAW, regulating the organization of all troops that Congress may see fit to raise, so that, upon adding to, or diminishing, the public force in any emergency, it will be only necessary to prescribe what number of men are to be added or taken away. This general law should embrace general officers, staff corps, and departments, engineers, and regiments of cavalry, artillery, and infantry; it should establish rules of promotion and appointment; it should regulate the recruiting service; it should provide for the repression of military crimes and disorders; it should not fail to stimulate the appetite for rewards; it should make just rules concerning captures, which would recognize the rights of captors; it should regulate the indemnification for losses; and it should provide for the organization of a suitable board, which would take advantage of all improvements in the military art and suggest, from time to time, such modifications of the general law as might appear just and proper. In respect to Army Organization, there are two acts of Congress of the general character here suggested. One, an act to regulate the medical establishment, approved March 2, 1799; and the second, an act for the better organizing of the troops of the United States, and for other purposes, approved March 3, 1799. Both of these acts were drawn by Alexander Hamilton, as he explained in a letter to the Secretary of War, "as permanent rules to attach to all provisions of law for the increase or diminution of the public force." Subsequent legislation has, however, without providing any other permanent rule regulating the organization in respect to general officers, staff corps, and departments, &c., according to the increase or diminution of force, almost entirely superseded the provisions of the remarkable acts here referred to. (*See* ARTICLES OF WAR.)

ARMY REGULATIONS—a book so called, published in the name of the President of the United States "for the government of all concerned." The Constitution provides that "Congress shall have power to make rules for the government and *regulation* of the Land and Naval forces." The only acts of Congress in force, authorizing the President to make regulations, better defining the powers and duties of officers, are contained in the 5th section of the act of March 3, 1813, and the 9th section of the act approved April 26, 1816. The first of these acts is an act for the better organization of the general staff of the army, and the second relates (with the exception of the last section, concerning forage and private servants) to the same subject. By the 5th section of the act of 1813, it is provided, "That it shall be the duty of the Secretary of the War Department, and he is hereby authorized, to prepare general regulations, better defining and prescribing the respective duties and powers of the several officers in the adjutant-general, inspector-general, quartermaster-general, and commissary of ordnance departments, of the topographical engineers, of the aides of generals, and generally of the general and regimental staff; which regulations, when approved by the President of the United States, shall be respected and obeyed, until altered or revoked by the same authority. And the said general regulations, thus prepared and approved, shall be laid before Congress at their next session."

Remarking here, that the regulations to be prepared and approved refer *only* to the powers and duties of the officers of the several staff departments, enumerated in the act, it follows that no other regulations made by the President can derive any force whatever from this act. The 9th section of the act of 1816 therefore only continued this then existing power of the President in providing " That the several officers of the staff shall respectively receive the pay and emoluments, and retain all the privileges, secured to the staff of the Army, by the act of March 3, 1813, and not incompatible with the provisions of this act: and that the regulations in force before the reduction of the Army be recognized, as far as the same shall be found applicable to the service; subject, however, to such alterations as the Secretary of War may adopt, with the approbation of the President." It would seem, therefore, that whatever may be contained in the President's Army regulations of a legislative character concerning officers of the Army, not belonging to staff departments, must, if valid, be a legitimate deduction from some positive law, or depend for its legality upon the exercise of authority delegated to the constitutional commander-in-chief or other military commander, in the rules made by Congress for the government of the Army. Congress has delegated to the President, authority to prescribe the uniform of the Army; authority to establish the ration; and besides the authority given by law to other military commanders, he also has been authorized to relieve, in special cases, an inefficient military commander from duty with any command; to assign any senior to duty with mixed corps, so that the command may fall by law on such senior in rank; to limit the discretion of commanding officers in special cases, in regard to what is needful for the service; and hence also he has been given authority to carve out special commands from general commands, in particular cases; (62d *Article of War.*) These are all-important functions, but they do not authorize *special cases* to be made general rules, and it is much to be regretted that the lines of separation between regulations and the orders of the commander-in-chief have not been kept distinct. (*See* COMMAND; CONGRESS; OBEDIENCE; ORDERS. Consult opinions of Attorneys-general, particularly the opinion of Mr. Berrien, July 18, 1839.)

ARREARS OF PAY. The troops shall be paid in such manner that the arrears shall, at no time, exceed two months, unless the circumstances of the case shall render it unavoidable; (*Act* March 16, 1802; *Act* March 3, 1813.) This provision of law has been strangely executed by never paying troops oftener than once in two months, and not unfrequently neglecting to pay them for a much longer time.

ARREST IN ORDER TO TRIAL. Before an officer or soldier, or other person subject to military law, can be brought to trial, he must be charged with some crime or offence against the rules and articles of war, and placed in arrest. The articles of war direct that whenever any officer shall be charged with a crime, he shall be arrested and confined in his barracks, quarters, or tent, and deprived of his sword by the commanding officer. And that " non-commissioned officers and soldiers, charged with crimes, shall be confined until tried by a court-martial, or released by proper authority;" (ARTS. 77, 78.) The arrest of an officer is generally executed through a staff-officer; by an adjutant, if ordered by the commanding officer of a regiment; or by an officer of the general staff, if ordered by a superior officer; and sometimes by the officer with whom the arrest originates. On being placed in arrest, an officer resigns his sword. If this form be sometimes omitted, the custom is invariably observed, of an officer in arrest not wearing a sword. By the custom of the army, it is usual, except in capital cases, to allow an officer in arrest the limits of the garrison or even greater limits, at the discretion of the commanding officer, who regulates his conduct by the dictates of propriety and humanity. A non-commissioned officer or soldier is confined in charge of a guard; but, by the custom of the service, the non-commissioned staff and sergeants may be simply arrested. The articles of war declare, " that no officer or soldier, who shall be put in arrest or imprisonment, shall continue in his confinement more than eight days, or until such time as a court-martial can be conveniently assembled; (ART. 79.) The latter part of this clause evidently allows a latitude, which is capable of being abused; but, as in a free country there is no wrong without a remedy, an action might be brought against the offender in a civil court, (*See* INJURIES,) if the mode of redress for all officers and soldiers, who conceive themselves injured by their commanding officer, be not sufficient. (ARTS. 34, 35.)

It is declared by the articles of war, that "no officer commanding a guard, or provost-marshal, shall refuse to receive or keep any prisoner committed to his charge, by any officer belonging to the forces of the United States; provided, the officer committing shall, at the same time, deliver an account in writing, signed by himself, of the crime with which the said prisoner is charged;" and it is also declared, that "no officer commanding a guard, or provost-marshal, shall presume to release any prisoner committed to his charge, without proper authority for so doing, nor shall he suffer any person to escape, on the penalty of being punished for it by the sentence of a court-martial. Every officer or provost-marshal, to whose charge prisoners shall be committed, shall, within twenty-four hours after such commitment, or as soon as he shall be relieved from his guard, make report in writing, to the commanding officer, of their names, their crimes, and the names of the officers who committed them, on the penalty of being punished for disobedience, or neglect, at the discretion of a court-martial; (ARTS. 80, 81, 82.) Thus the liberty of the citizen, under military law, so far as is consistent with the ends of justice, seems to be guarded with precautions little inferior to those which secure personal liberty under the civil laws of the state. The penalty of an officer's breaking his arrest, or leaving his confinement before he is set at liberty by his commanding officer, or by a superior officer, is declared to be cashiering by sentence of a general court-martial; (ART. 77.) A court-martial has no control over the nature of the arrest of a prisoner, except as to his personal freedom in court; the court cannot, even to facilitate his defence, interfere to cause a close arrest to be enlarged. The officer in command is alone responsible for the prisoners under his charge. Individuals placed in arrest, may be released, without being brought before a court-martial; by the authority ordering the arrest, or by superior authority. It is not obligatory on the commander to place an officer in arrest, on application to that effect from an officer under his command. He will exercise a sound discretion on the subject. But in all applications for redress of supposed grievances inflicted by a superior, it will be his duty, in case he shall not deem it proper to order an investigation, to give his reasons in writing, for declining to act; these reasons, if not satisfactory, the complaining party may, should he think fit so to do, forward to the next common superior, together with a copy of his application for redress. An officer has no right to demand a court-martial, either on himself, or on others; the general-in-chief or officer competent to order a court, being the judge of its necessity or propriety. Nor has any officer, who may have been placed in arrest, any right to demand a trial, or to persist in considering himself under arrest, after he shall have been released by proper authority. An officer under arrest will not make a visit of etiquette to the commanding officer, or other superior officer, or call on him, unless sent for; and in case of business, he will make known his object in writing. It is considered indecorous in an officer in arrest to appear at public places.

ARREST BY CIVIL AUTHORITY. By section 21, Act January 11, 1812, no non-commissioned officer, musician, or private, can be arrested on *mesne process*, or taken or charged in execution for any debt contracted before enlistment under twenty dollars, nor for any debt whatever, contracted after enlistment. (*See* MESNE PROCESS.)

ARSENAL. A place of deposit for ordnance and ordnance stores. There are also arsenals of construction and repairs. (*See* ORDNANCE.)

ARTICLES OF WAR. There can be no doubt that the prerogative to command and regulate the whole military force of the kingdom, whether consisting of the feudal tenants, or of the militia, or of paid troops, resided in the Crown of England. Nevertheless the power of the sovereign was restricted by a provision, that he should exercise his military jurisdiction only "according to the laws and usages of the realm." In the reign of Edward VI., however, parliament asserted authority over military matters by passing an act for the government of the army; various offences, as losing, selling, or fraudulently exchanging horses or armor; desertion; detaining the pay of soldiers; and taking rewards for granting them discharges, were put under the jurisdiction of the civil magistrate. It was also provided that the act should be read once a month by every field officer to the soldiers under his command, and once a quarter by the governor or captain of every garrison or fortress. At this period, however, there was no standing army, the feudal system was still in force, every man in the realm was more or less a soldier; military law was accordingly restricted to such persons as were actually serving in the field, the process of civil judicature being obviously inapplicable to their case—but directly the soldier ceased to belong to the force in actual campaign, the civil power stepped in and claimed cognizance of his offences.

Until the Civil War in the reign of Charles I., it is probable that no regular permanent code of rules or articles for enforcing military discipline was in existence; the ruling authority had promulgated its orders for the government and regulation of the army as occasion required. Each war, each expedition, had its own edict, which fell into disuse again upon the disbanding of the army, which inevitably followed the cessation of hostilities. Several instances, indeed, of rules and ordinances for military government by the ancient kings are still extant; one of Richard I., for the government of those going by sea to the Holy Land, is to be found in Rymer's Fœdera. An elaborate code of "statutes, ordonnances, and customs to be observed in the army," made in the 9th year of Richard II., is to be found among the Cottonian MS. in the British Museum—and those of Henry V., Henry VII., and Henry VIII., have not been lost.

The experience of ages and the precedents of former wars, therefore, enabled the authorities to frame a sufficiently comprehensive code in case of need; accordingly, soon after the outbreak of the civil war, the necessities of the case compelled the parliament to enact ordinances or articles of war. The first complete "*Lawes and Ordinances of Warre*" (as he called them) were issued by Essex, the commander-in-chief of the parliamentary army in 1642. These articles are remarkable and interesting, as undoubtedly forming the groundwork of those now in use. Two years after the publication of Essex's ordinances, on the marching of the Scottish army into England, soon after the ratification of the solemn league and covenant, "Articles of War" were issued for its government. These articles, although very dissimilar to those of Essex, considering that both were in force in the same kingdom at the same time, and were applicable to armies fighting on the same side, nevertheless treat mainly of the same offences. The form of judicature established, consisted of two courts of justice, called "Councils of War," the one superior, and the other inferior. The superior court, also called the "Court of War," took cognizance of the more serious offences, and likewise heard appeals from the decision of the lower court, called the "Marshal Court." No trace of the constitution of these courts is now to be found except that "the judges were sworn to do justice." Within a few months of the promulgation of the latter, (August, 1644,) the same parliament that was the author of the petition of right, passed an ordinance, establishing a system of martial law, applicable not only to soldiers, but to all persons alike. By this ordinance, the Earl of Essex, captain-general of the parliamentary forces, together with fifty-six others named therein, (among whom were peers, members of the House of Commons, gentry, and officers of the army,) were constituted "commissioners," and any twelve of them authorized to hear and determine all such causes as "belong to military cognizance," according to the articles mentioned in the ordinance, and to proceed to the trial, condemnation, and execution, of all offenders against the said articles, and to inflict upon them such punishment, either by death or otherwise, *corporally*, as the said commissioners, or the major part of them then present, should judge to appertain to justice, according to the measure of the offence. Under cover of this ordinance, which, after one refusal by the peers, was subsequently renewed, parliament proceeded to issue a variety of orders for the conduct of the war, and the regulation of the army; and many persons were tried by court-martial and executed. After the expiration of this last ordinance, the absolute executive power, in all matters of military law, fell into the hands of Cromwell, who claimed it as his right, in virtue of his office of general-in-chief. "The general," says Whitlocke, "sent his order to several garrisons, to hold courts-martial, for the punishment of soldiers offending against the articles of war; provided that if any be sentenced to lose life or limb, that then they transmit to the judge-advocate the examinations and proceedings of the court-martial, that the *General's* pleasure may be known thereon." On one occasion, deeming it necessary for the sake of discipline, to make an immediate example, Cromwell seized several officers with his own hand, called a court-martial on the field, condemned them to death, and shot one forthwith at the head of his regiment. It will thus be seen, that the administration of martial law was almost invariably in the hands of the most considerable power in the state—it alternated between king and parliament, and between parliament and dictator, as each became uppermost in the realm. On the restoration of Charles II., the army, with the exception of about five thousand men, consisting of General Monk's regiment called "the Coldstream," the first regiment of foot, the royal regiment of Horse Guards, called "the Oxford Blues," and a few other regiments, was disbanded. The force kept on foot was the first permanent military force, or "standing army," known in England; and from it the present army dates its origin.

A statute passed in the reign of Charles II., intituled, "An act for ordering the forces in the several counties of this kingdom," recites that, "within all his majesty's realms and dominions, the sole and supreme power, government, command, and disposition of the militia, and of all forces by sea and land, and of all forts and places of strength is, and by the laws of England ever was, the undoubted right of his majesty, and his royal predecessors, kings and queens of England." With the exception of some slight encroachment on the part of the Crown, and protests on the part of the parliament, matters remained in very much the same state till the revolution, at which period military law assumed a permanent and definite form, as it now exists. The only allusions to the military power of the Crown, in the Bill of

Rights, are, "that the raising and keeping of a standing army in time of peace, *without consent of parliament,* is contrary to law;" and that "subjects, if Protestants, may have arms for their defence, suitable to their condition, and as allowed by law." In the first year, however, of the reign of William and Mary, British regiments, jealous of the supposed preference shown by William for his Dutch troops, mutinied at Ipswich. The king suppressed the mutiny with a strong hand, at the same time communicating the event to parliament. Parliament, anxious to devise means for the convenient application of a code of laws for the regulation and management of the army, and at the same time determined to place a check upon the exercise of the military power of the king, passed, on the 3d April, 1689, for a period of six months only, the first mutiny act, the preamble of which is as follows:

"Whereas, the raising or keeping a standing army within this kingdome, in time of peace, unlesse it be with the consent of Parlyament, is against law; and whereas it is judged necessary, by their majestyes and this present parlyament that, during this time of warr, severall of the forces which are now on foote should be continued and others raised, for the safety of the kingdome, for the common defence of the Protestant religion, and for the reducing of Ireland. And whereas no man can be prejudged of life or limb, or subjected to any kinde of punishment by martiall law, or in any other manner than by the judgment of his peeres, and according to the knowne and established lawes of this realme; yet, nevertheless, it being requisite for retaining such forces as are or shall be raised during this exigence of affaires in their duty, that an exact discipline be observed; and that soldiers who shall mutiny or stirr up sedition, or who shall desert their majestye's service, be brought to more exemplary and speedy punishment than the usual formes of law will allow."

The act provides for the assembling and constitution of courts-martial, for the oath of members, for the punishment of desertion, mutiny, sedition, false musters, &c.; for the regulation of billets; and is ordered to be read at the head of every regiment, troop, or company, at every muster, "that noe soldier may pretend ignorance." No power is, however, reserved to the sovereign to make articles of war. This act was renewed soon after its expiration; and with the exception of about three years only, viz., from 10th April, 1698, to 20th February, 1701, has been annually re-enacted (with many alterations and amendments) ever since. The first statutory recognition of articles of war, occurs in the 1st Anne, statute 2, c. 20, in a clause, which saves to her majesty the right of making articles of war, for the regulation of her forces "beyond the seas in time of war." It is not until the 3d Geo. 1, c. 2, that we find the sovereign distinctly empowered by the mutiny act to make articles of war for the government of the troops at home. A clause in that act, after reciting that no effectual provision has been made for the government of his majesty's land forces, empowers the king to make and constitute, under his sign manual, articles for the better government of his majesty's forces, "as well within the kingdoms of Great Britain and Ireland as beyond the seas." This privilege has been annually re-enacted, and annually exercised by the Crown to the present day.

Under the Constitution of the United States, Congress only can make rules of government and regulation for the land forces, and those rules, commonly called *Articles of War*, were originally borrowed jointly from the English mutiny act annually passed by parliament, and their articles of war established by the king. The existing articles for the government of the army of the United States, enacted April 10, 1806, are substantially the same as those originally borrowed July 30, 1775, and enlarged by the old Congress from the same sources, Sept. 20, 1776. The act consists of but three sections. The first declares: The following shall be the rules and articles by which the *armies* of the United States shall be governed;" and gives one hundred and one articles, all noticed in these pages. Each article is confined, in express terms, to the persons composing the army. The *second* SECTION contains the only exception in the cases as follows: "In time of war, all persons, *not* citizens of, or owing allegiance to, the United States of America, who shall be found lurking, as *spies*, in or about the fortifications or encampments of the armies of the United States, or any of them, shall suffer death, according to the *law and usage of nations*, by sentence of a general court-martial." The third section merely repeals the previous act for governing the army.

The Articles of War, therefore, are, and under the Constitution of the United States can be, nothing more than a code for the government and regulation of the army. Or, in other words, within the United States, these articles are " a system of rule superadded to the common law, for regulating the citizen in his character of a soldier," and applicable to no other citizens. Beyond the United States another code is essential; for, although armies take with them the Rules and Articles of War, and the custom of war in like cases—in a foreign country, the soldier must be tried by some tribunal for offences which at home would be punishable by the ordinary courts of law. It is impossible to subject him to any foreign dominion, and hence, in the absence of rules made by Congress for the government of the army under such circumstances, the will of the commander of the troops, *ex necessitate rei*, takes the place of law, and the declaration of his will is called MARTIAL LAW. (*See* MARTIAL LAW.)

The most casual reader of our Articles of War will be struck by the fact, that whereas the mutiny act of Great Britain is annually subjected to the supervision of parliament, and altered or modified according to circumstances, yet the Rules and Articles of War, passed in 1806, have remained upon our statute book from that day to the present without any general revision. Another fact equally important is, that while the king of Great Britain not only *commands*, but *governs* the British army, and therefore modifies the government of the army at his pleasure, the President of the United States is simply the commander of our army, under such rules for raising, supporting, governing, and regulating it, as Congress may appoint. The necessity of attention to the military establishment on the part of Congress is therefore manifest, and it is most earnestly to be hoped that, in their wisdom they will, at some early day, fulfil their constitutional obligations of raising, governing, and regulating armies: 1. By establishing a system of recruiting which will bring into the ranks, soldiers who will make good officers; 2. By providing that all commissioned officers shall be appointed from enlisted soldiers, or from military academies, and making rules precisely regulating the manner in which such appointments shall be made; 3. In making rules for a system of promotion partly by seniority, and partly by merit; 4. In passing other remunerative laws, such as prize money, field allowances, indemnification for losses, &c.; 5. In accurately defining the powers, rights, and duties of all officers and soldiers; 6. In providing remedies for wrongs, including appeals to federal civil courts, to determine the true exposition of military laws in dispute; and 7. In revising the penal code, and better adapting it to a system of government which will provide rewards for good conduct, and not simply punishments for bad. *See* ABANDONING A POST; ABSENCE WITHOUT LEAVE; ABSENCE WITH LEAVE; ABUSES AND DISORDERS; ALARMS; AMMUNITION; APPEAL; ARMS, (CASTING AWAY;) ARREST; BREACH OF ARREST; BREVET; BRIBE AT MUSTER; BOOTY; CASTING AWAY; CERTIFICATES OF MUSTER; CERTIFICATES, (FALSE;) CHALLENGES, (DIFFERENT KINDS;) CHAPLAIN; COMMAND; CONDUCT UNBECOMING AN OFFICER AND A GENTLEMAN; CONFINEMENT; CONNIVING; CONTEMPT; CORPORAL; CORRESPONDENCE, (WITH AN ENEMY;) COURTS-MARTIAL, AND REFERENCES UNDER THAT HEAD; COURTS OF INQUIRY; COWARDICE; CRIMES; CUSTOM OF WAR; DEATH;

Deceased; Department; Deposition of Witnesses; Detachment; Desertion; Discharge; Dismission; Disobedience; Disorders; Disrespect; Drunkenness; Duels; Embezzlement; Engineers; Enlistments; Enticing; Exactions; False; Frauds; Frays; Furloughs; General Officers; Grievances; Harboring an Enemy; Hiring of Duty; Injuring private Property; Judge-advocate; Jurisdiction; Leave; Line; Lying out of Camp or Quarter; Menacing; Militia; Misbehavior; Mitigation; Money; Monthly Returns; Musters; Mutiny; Oath; Obedience; Offences not Specified; Officers; Orders; Pardon; Parole: Pillage; Post; President; Prisoner; Proceedings; Promulgation; Provost-marshal; Quarrels; Rank; Redressing Wrongs; Re-enlisting; Refusal to receive Prisoners; Releasing Prisoners; Relieving an Enemy; Reproachful Speeches; Retainers; Returns; Safeguard; Secretary of War; Selling; Sentence; Sentinel; Spies; Staff; State Troops; Stores; Stripes; Standing Army; Subscribing; Suspension; Sutlers; Trials; Upbraiding; Violence; Waste or Spoil; Watchword; Witness; Worship; Wrongs; and references under the heading of Law, all military laws being rules for the government and regulation of the army, although they may also include other matters. (Consult Pipon's Manual of Military Law.)

ARTIFICER. Military workman; two allowed to each company of artillery.

ARTILLERY. The word is more ancient than the use of powder, and was applied to machines of war, and all projectiles that the masters of artillery had under their direction. In foreign armies the word Artillery is still indifferently applied to an arm of the service, the material used, and branch of science. By Artillery in the U. S. army is usually, but not always, meant an arm of the service, designed to use mountain, field, and heavy ordnance, and the knowledge requisite for such use. There are four regiments of Artillery in our army, in each of which the law authorizes two companies to be equipped as harnessed batteries; (*See* Army, for their organization.) The remaining companies are, from supposed necessities of service, usually employed as infantry, but their name, and liability at any time to become artillerists, must cause officers not to neglect such knowledge of their arm as may be derived from books, and the establishment of the school of practice at Fort Monroe cannot fail to have the happiest effects in making skilful artillerists. The instructions for field artillery, and heavy and mountain artillery, are contained in books published by the War Department, one called

"Instruction for Field Artillery, Horse and Foot," and another "Heavy Artillery" being "a complete system of instruction for Siege, Garrison, Sea coast and Mountain Artillery," and a third "Evolutions of Field Artillery," by Major Robert Anderson.

Composition of a field battery on the war establishment.—Four 12-pounders or four six-pounder guns, and two 24-pounders or 12-pounder howitzers. Six pieces mounted to each battery. Carriages including caissons, spare gun-carriages, forges, and battery wagons, accompany each battery, together with implements and equipments specified in the ordnance manual. Draught horses, six to each battery wagon, and 12-pounder gun-carriage, four to other carriages, and one twelfth spare. Harness corresponding to the number of horses to the carriage.

Tactics.—A battery going into line with other troops, is usually formed in column of sections, and deployed into line as the enemy is approached. Under ordinary circumstances the best formation is the column doubled on the centre section, as the deploy is then toward both wings at the same time, and more promptly performed. Unless in extreme cases, the cannoneers should never be mounted on the boxes when the battery is within range of the enemy, as the explosion of a caisson might destroy nearly every cannoneer belonging to a piece. When several batteries are united, they are formed by sections in one or several parallel columns, or in double columns on the centre, or still better, in two columns joined, and presenting a front of four pieces with the same intervals as in line. Sometimes they are formed in close column with a front of four or six pieces, and the batteries being spaced a distance apart equal to the interval between two pieces. When deployed, the distance between the batteries is double this. When horse-artillery and mounted batteries are placed together, the former are placed on the wings, and the distances and intervals of the whole conform to those of horse-artillery; as in manœuvring no regard is paid to inversions, it frequently happens that the batteries change their relative positions, and it is necessary that each space should be large enough to contain a horse-artillery battery. A close column of several batteries is deployed in the same manner as a column of cavalry; the leading battery moving off at an increased gait, and the others, obliquing to the right or left, gain their intervals and form in line or battery to the front as usual. The changes of front to fire to the right and left are made on the wings in the same manner as with a single battery; but it is better to make these changes on the centre battery. But four of these changes are practicable, viz., two to fire to the right by throwing the left wing to the front or rear, and two to fire to the left by throwing the right wing to the front or rear. In the other four changes of front, the pivot pieces would be masked by the rest of the carriages, and could not commence their fire soon enough. On this account the pivot carriages, in these changes, should be on the side towards which the fire is to be delivered. In defensive battles, the contour of the ground is of the first importance, and if properly taken advantage of, may be made to double the force and importance of artillery.

Artillery, held in reserve, arriving in mass or deployed upon the field of battle, occupies positions determined by circumstances and localities. Heights and commanding positions should be secured, and those positions, also, from which an oblique fire may be obtained upon the enemy. In a *defensive* position, those points are sought from whence the enemy may be discovered at the greatest distance. Advantage should be taken of all local circumstances to render the artillery fire most effective, and at the same time shelter it from the fire of the enemy. The guns should be placed, if possible, under cover. This is easily effected upon heights, by keeping them so far back that the muzzles only are to be seen over them. Ravines, banks, ditches, &c., also offer facilities for the purpose. The perfection to which the *matériel* of field artillery has been brought, gives it comparatively great mobility of action; but large quantities of ammunition must be consumed to attain any positive result from its employment in battle. The transportation of this ammunition with an army involves serious economical considerations, constituting no small impediment to armies, from the number of horses, wagons, caissons, &c., required for each battery. The improvements made in the *matériel* of artillery will not, therefore, in all probability, cause a more frequent employment of light batteries; but on the contrary, the long range which has been given to the rifle and musket, and the facility with which the horses and gunners of field batteries may be picked off at 1,000 yards, will probably cause even the rifled field gun to become an arm of RESERVE, which brought up at a decisive moment may influence the result of a battle, defend entrenchments against attack, and be usefully employed against isolated field works.

Smooth-bore field pieces, fired at a distance of five or six hundred yards, will penetrate from one yard and a half to two yards in parapets recently constructed, and will traverse walls of ordinary construction; but a 12-pounder is necessary to make a breach in walls of good masonry four feet in thickness, and in this case the position of the battery must be favorable, and the operation is even then a slow one.

Moderate charges are employed in firing upon gates, block-houses, palisades, and in general upon all wooden structures. The heaviest siege pieces, by their great force of penetration, are best adapted for forming a breach in the walls of permanent fortifications. Their superior accuracy, and the mass of their projectiles, render them also very effective in ricochet firing. Balls of smaller calibre have not sufficient mass to destroy carriages offering such resistance as those employed in the defence of places. The force of penetration of balls in different substances increases with their calibre and velocity: at one hundred yards, a 24-pound ball fired with a cartridge of 12 pounds will be one yard in brick masonry, nearly two feet in rubble work, one yard and a half in oak wood, two yards in pine, two yards and a half in well rammed earth, and nearly five yards in a recent embankment. The ball of an 18-pounder, fired with a charge of nine pounds under the same circumstances, will give penetrations nearly six-sevenths of those indicated above.

Field guns, in general, may be employed to cannonade with force and perseverance; to reinforce the weakest points of positions, whether offensive or defensive; to secure a retreat by the occupation of points established as the base of defence of particular ground, or of any important object, as the defence of a village or defile, or the passage of a river, and to overthrow such obstacles as palisades, rampart walls, doors, &c., interposed by art; to prepare the way for an assault, and aid, at a decisive moment, to secure the victory by a united fire. A field cannon ball has sufficient force to disable seven or eight men at a distance of 900 yards. It is stated that a single cannon ball, at the battle of Zorndorf, disabled 42 men. Rifle projectiles, having more momentum, are effective at greater distances.

The following tables of Charges and Ranges for United States Field Guns, Howitzers, and Heavy Ordnance, are taken from Roberts' Handbook of Artillery.

CHARGES FOR A FLATTENED RICOCHET FOR SIEGE-GUNS.

DISTANCE.	ELEVATION.	CHARGE.
660 yards.	2° 45′	$1/12$ wt. of ball.
550 "	3°	$1/15$ " "
440 "	3° 15′	$1/20$ " "
330 "	3° 35′	$1/30$ " "

CHARGES FOR A FLATTENED RICOCHET FOR SIEGE-HOWITZERS.

DISTANCE.	ELEVATION.	CHARGE.
550 yards.	1° 45′	3 lbs.
440 "	2° 15′	2 lbs. 3 oz.
330 "	2° 15′	1 lb. 12 oz.
220 "	2° 45′	1 lb. 2 oz.

CHARGES FOR A CURVATED RICOCHET FOR SIEGE-HOWITZERS.

DISTANCE.	ELEVATION.	CHARGE.	REMARKS.
550 yards.	7° 30'	1 lb. 4 oz.	The height of the object above the level of the battery being supposed to be 20 feet.
440 "	"	1 lb. 1 oz.	
330 "	"	14 oz.	
220 "	"	10 oz.	

The charges vary with the elevation; or, if the elevation be fixed at any particular angle, they must be determined by the range.

CHARGES FOR FIELD-GUNS AND FIELD-HOWITZERS.

KIND.	FOR GUNS.		FOR HOWITZERS.			
	12-pdr.	6-pdr.	32-pdr.	24-pdr.	12-pdr.	Mountain.
	lbs.	lbs.	lbs.	lbs.	lbs.	lbs.
For shot............	2.5	1.25				
For spherical case or canister...........	1.5	1.	2.5	1.75	0.75	0.5
For shells, { small charge............			2.5	2.	1.	0.5
large charge............			3.25	2.50	1.	0.5

CHARGES FOR HEAVY GUNS, COLUMBIADS, AND HOWITZERS.

GUNS.					COLUMBIADS.		HOWITZERS.			
42-pdr.	32-pdr.	24-pdr.	18-pdr.	12-pdr.	10-inch.	8-inch.	Siege 8-in.	24-pdr. Garrison.	SEA-COAST.	
									10-in.	8-in.
lbs.	lbs.	lbs.	lbs.	lbs.	lbs.	lbs.	lbs.	lbs.	lbs.	lbs.
10.5	8.	8.	6.	4.	14.	8.	4.	2.	12.	8.

GREATEST CHARGES OF SEA-COAST, SIEGE, AND COEHORN MORTARS.

SEA-COAST.		SIEGE.		COEHORN.	STONE MORTAR.	
13-inch.	10-inch.	10-inch.	8-inch.		120 pounds of stones.	15 6-pdr. shells.
lbs.	lbs.	lbs.	lbs.	lbs.	lbs.	lb.
20.	10.	4.	2.	0.5	1.5	1

RANGES OF FIELD GUNS AND HOWITZERS.

KIND OF PIECE.	Powder.	Ball.	Elevation.	Range.	Remarks.
	lbs.		° ′	yards.	
6-Pounder Field Gun.	1.25	Shot.	0	318	
		"	1	647	P. B. Range.
		"	2	867	
		"	3	1138	
		"	4	1256	
		"	5	1523	
	1.	Sph. case.	2	650	Time of flight 2″
		"	2 30	840	do. 3″
		"	3	1050	do. 4″
12-Pounder Field Gun.	2.5	Shot.	0	347	
		"	1	662	P. B. Range.
		"	1 30	785	
		"	2	909	
		"	3	1269	
		"	4	1455	
		"	5	1663	
	1.5	Sph. case.	1	670	Time 2 seconds.
		"	1 45	950	" 3 "
		"	2 30	1250	" 4 "
12-Pounder Field Howitzer.	1.	Shell.	0	195	
		"	1	539	
		"	2	640	
		"	3	847	
		"	4	975	
		"	5	1072	
	0.75	Sph. case.	2 15	485	Time 2 seconds.
		"	3 15	715	" 3 "
		"	3 45	1050	" 4 "
24-Pounder Field Howitzer.	2.	Shell.	0	295	
		"	1	516	
		"	2	793	
		"	3	976	
		"	4	1272	
		"	5	1322	
	1.75	Sph. case.	2	600	Time 2 seconds.
		"	3	800	" 3 "
		"	5 30	1050	" 4 "
	2.	"	3 30	880	" 3 "
32-Pounder Field Howitzer.	2.5	Shell.	0	290	
		"	1	531	
		"	2	779	
		"	3	1029	
		"	4	1203	
		"	5	1504	
	2.5	Sph. case.	3	800	Time 2¾ seconds.
Mountain Howitzer.	0.5	Shell.	0	170	
		"	1	300	
		"	2	392	
		"	2 30	500	Time 2 seconds.
		"	3	637	
		"	4	785	Time 3 seconds.
		"	5	1005	

RANGES OF FIELD GUNS AND HOWITZERS—(Continued.)

KIND OF PIECE.	Powder.	Ball.	Elevation.	Range.	Remarks.
Mountain Howitzer—*Continued.*	lbs. 0.5	Sph. case. " " " "	° ' 0 2 30 3 4 4 30	yards. 150 450 500 700 800	 Time 2 seconds. Time 2¾ seconds. Time 3 seconds.
	0.5	Canister.	4 to 5'	250	

RANGES OF HEAVY ARTILLERY.

KIND OF PIECE.	Powder.	Ball.	Elevation.	Range.	Remarks.
18-Pdr. Siege and Garrison Gun on Barbette Carriage.	lbs. 4.5	Shot. " " " " "	° ' 1 1 30 2 3 4 5	yards. 641 800 950 1256 1450 1592	 Point Blank.
24-Pdr. Siege and Garrison Gun on Siege Carriage.	6. 8.	Shot. " " " " " " " " " " "	0 1 1 30 2 3 4 5 1 2 3 4 5	412 842 953 1147 1417 1666 1901 883 1170 1454 1639 1834	 Point Blank.
32-Pdr. Sea-Coast Gun on Barbette Carriage.	6. 8. 10.67	Shot. " " " " " " " " "	1 45 1 1 30 1 35 2 3 4 5 1 2 3	900 713 800 900 1100 1433 1684 1922 780 1155 1517	
42-Pdr. Sea-Coast Gun on Barbette Carriage.	10.5 14.	Shot. " " " " " " " " " "	1 1 30 2 3 4 5 1 2 3 4 5	775 800 1010 1300 1600 1955 770 1128 1380 1687 1915	

RANGES OF HEAVY ARTILLERY—(Continued.)

KIND OF PIECE.	Powder.	Ball.	Elevation.	Range.	Remarks.
	lbs.		° ′	yards.	
8-inch Siege Howitzer on Siege Carriage.	4.	45-lb. Shell.	0	251	Time ¾ seconds.
		"	1	435	" 1½ "
		"	2	618	" 2 "
		"	3	720	" 3 "
		"	4	992	" 4 "
		"	5	1241	" 5 "
		"	12 30	2280	
24-Pdr. Iron Howitzer on a Flank Casemate Carriage.	2.	17-lb. Shell.	0	295	
		"	1	516	
		"	5	1322	
	1¾	Sph. cases.	2	600	Time 2 seconds.
		"	5 30	1050	" 4 "
	2.	"	3 30	880	" 3 "
8-inch Sea-Coast Howitzer on a Barbette Carriage.	4.	45-lb. Shell.	1	405	
		"	2	652	
		"	3	875	
		"	4	1110	
		"	5	1300	
	6.	"	1	572	
		"	2	828	
		"	3	947	
		"	4	1168	
		"	5	1463	
	8.	"	1	646	
		"	2	909	
		"	3	1190	
		"	4	1532	
		"	5	1800	
10-inch Sea-Coast Howitzer on Barbette Carriage.	12.	90-lb. Shell.	1	580	
		"	2	891	Time 3 seconds.
		"	3	1185	" 4 "
		"	3 30	1300	
		"	4	1426	" 5¼ "
		"	5	1650	" 6 "
8-in. Columbiad on Barbette Carriage.	10.	65-lb. Shot.	1	932	Axis of gun 16 feet above the water.
		"	2	1116	
		"	3	1402	
		"	4	1608	
		"	5	1847	
		"	6	2010	
		"	8	2397	Shot ceased to ricochet on the water.
		"	10	2834	
		"	15	3583	
		"	20	4322	
		"	25	4875	
		"	27	4481	
	15.	"	27 30	4812	
	10.	50-lb. Shell.	1	919	
		"	2	1209	
		"	3	1409	
		"	4	1697	
		"	5	1813	
		"	6	1985	
		"	8	2203	

RANGES OF HEAVY ARTILLERY—(*Continued.*)

KIND OF PIECE.	Powder.	Ball.	Elevation.	Range.	Remarks.
	lbs.		° ′	yards.	
8-in. Columbiad on Barbette Carriage—*Continued.*	10.	50-lb. Shell.	10	2657	
		"	15	3556	
		"	20	3716	
		"	25	4387	
		"	27	4171	
	15.	"	27 30	4468	
10-inch Columbiad on Barbette Carriage.	18.	128-lb. Shot.	0	394	Axis of gun 16 feet above the water.
		"	1	752	
		"	2	1002	
		"	3	1230	
		"	4	1570	
		"	5	1814	
		"	6	2037	Shot ceased to ricochet on the water.
		"	8	2519	
		"	10	2777	
		"	15	3525	
		"	20	4020	
		"	25	4304	
		"	30	4761	
		"	35	5433	
	20.	"	39 15	5654	
	12.	100-lb Shell.	1	800	
		"	2	1012	
		"	3	1184	
		"	4	1443	
		"	5	1604	
	18.	"	0	448	
		"	1	747	
		"	2	1100	
		"	3	1239	
		"	4	1611	
		"	5	1865	
		"	6	2209	
		"	8	2489	
		"	10	2848	
		"	15	3200	
		"	20	3885	
		"	25	4150	
		"	30	4651	
		"	35	4828	Time 35 seconds.
13-in. Sea-Coast Mortar.	20.	200-lb. Shell.	45	4325	Time 40 seconds.
10-in. Sea-Coast Mortar.	10.	98-lb. Shell.	45	4250	Time 36 seconds.
10-inch Siege Mortar.	1.	90-lb. Shell.	45	300	Time 6.5 sec'ds.
	1.5	"	"	700	" 12. "
	2.	"	"	1000	" 14. "
	2.5	"	"	1300	" 16. "
	3.	"	"	1600	" 18. "
	3.5	"	"	1800	" 19. "
	4.	"	"	2100	" 21. "
8-inch Siege Mortar.	lbs. oz.	45-lb. Shell.	45	209	Time 6.75 sec'ds.
	0 8	"	"	376	" 9. "
	0 12	"	"	650	" 11.5 "
	1 0	"	"	943	" 14. "
	1 4	"	"	1318	" 16.5 "
	1 8	"	"	1522	" 18.5 "
	1 12	"	"	1837	" 20.5 "
	2 0				

RANGES OF HEAVY ARTILLERY—(*Continued.*)

KIND OF PIECE.	Powder.	Ball.	Elevation.	Range.	Remarks.
	oz.		° ′	yards.	
24-Pounder Coehorn Mortar.	0.5	17-lb. Shell.	45	25	
	1.	"	"	68	
	1.5	"	"	104	
	1.75	"	"	143	
	2.	"	"	165	
	2.75	"	"	260	
	4.	"	"	422	
	6.	"	"	900	
	8.	"	"	1200	
	lbs.	Stones.			
Stone Mortar.	1.5	120 lbs.	60	150 to 250	
	1	15 6-pdr. shells.	33	50 to 150	Fuze 15 seconds.

NOTE.—Fire-balls, according to their size, are fired from mortars of corresponding calibres. With a charge of ONE TWENTY-FIFTH its weight, the ball is thrown 600 to 700 yards.

Howitzers are used to drive the enemy from positions when he can only be reached by shells; against covered ground, and particularly forests and defiles; against strong cavalry attacks; to prepare the way for an attack of fortifications and posts, and to burn combustible objects of great extent. (Consult *Aide Memoire, par* GASSENDI; GIBBON; ROBERTS; BENTON; KINGSBURY; *Histoire et Tactique des Trois Armes, par* ILD. FAVÉ. *See* AMMUNITION; RIFLED ORDNANCE.)

ASSAULT. In any assault, it is necessary that the officer, commanding and responsible for the whole operation, should be in immediate communication with the troops during the assault, and be present with the reserve or supporting party; 2. The troops destined for this duty should be divided into two portions, *each* equal in strength to three-fourths of the garrison attacked: one portion being the attacking party, and the other half, the reserve or supporting party; 3. Each column of the attacking party will also be subdivided into advance, main body, and support, whatever may be the number of these columns; 4. The disposition of the attacking party, as it reaches the point of attack, will be regulated by the engineer officer, under the orders of the officer commanding—they having made the necessary reconnoissances; the party must be furnished with tools, ladders, and proper implements, adapted to the circumstances of the moment, and accompanied by a detachment of sappers; 5. The disposition of the reserve, equal, as before observed, to the whole attacking force, should be regulated by the officer intrusted with the execution of the assault; and this reserve should be accompanied or not, according to circumstances, by cavalry and field artillery. When these descriptions of force are present, the former should be placed under cover or out of gun shot, about 1,500 yards distant; the artillery should be kept in hand until the attacking party is engaged, when the guns should be spread out on the flanks, and open a vigorous fire upon the works; the infantry, brought immediately in rear of the leading attack, should be placed under cover, if possible, from fire of grape and musketry, and halted until the issue of the first assault is seen; 6. It is impossible to regulate an assault by any minute suggestions for the advance, except to observe that it is usual for each column to attack the salient points of the works, and least defended portions; to throw out skirmishers and firing parties under any cover available, and keep up a rapid and compact fire upon the defenders; to follow with the sappers and grenadiers to force all obstructions; and then to advance the main body, the supports of each column being judiciously planted in the rear. Eventually, as success occurs and the whole move on, points of security should be taken up, such as the reverse, or the exterior slope of the works; buildings, walls, as well as gorges and flanks, which frequently give cover. Men should be planted under an officer, with instructions to take no notice of the pell-mell, but to keep up a heavy firing in front; employing the sappers in entrenching the position taken up by the supporting party, or in collecting wagons, carts, carriages, &c., capable of being made into a barricade; 7. Either on the supposition that the success of the assault is doubtful, or that there is a check or repulse, the reserve, in case of doubtful success, to render the attack doubly sure, should move forward under the officer commanding the whole assaulting force, and relieve the assailants, who take their places as the reserve as soon as order can be restored; the artillery brought into position in the openings, between the advancing columns, would be directed upon the retreating or resisting forces; and if success is finally complete, the cavalry, in the event of their being employed, will move forward, either through the openings cleared, or by a detour, if a fortified town, in pursuit.

In the second case—that of a check—the reserve, on the reconnoissance of the officer commanding, will either march forward in support of the attack, or to cover the retreat, if further perseverance in the assault is deemed impracticable—the artillery and cavalry being warned as to the intention. In the event of the assault being repulsed, the reserve, which should be in echelon, having advanced guards in front, will allow the retreating party to move through the intervals, and the advanced guard will endeavor to check the pursuit; if overpowered, they will fall back on the reserve, and the whole may in that manner retreat until beyond gun shot, endeavoring to make a stand, repulse the garrison, and if possible convert failure into success, if the pursuit has been badly conducted and without due caution. As an important rule in all assaults, except in partial attacks, as an outwork, or any particular work in which a lodgement is to be made, the composition of the forces should be by regiments and corps, and not by detachments; and each non-commissioned officer should be provided with the means of spiking a gun, for which purpose even an old nail is sufficient. *Assaults*, if *feasible*, would seldom fail with these precautions, and there are few posts not open to assault, by taking the proper opportunity, an officer intrusted with the defence of a place should therefore exercise the most unremitting vigilance. (Consult DUFOUR, *Tactique des Trois Armes; Aide Memoire by British Officers.*)

ASSEMBLY. Drum beat to order troops to assemble; *assembly* for skirmishers, a bugle sound.

ASSIGNMENT. If, upon marches, guards, or in quarters, different corps of the army shall happen to join, or do duty together, the officer highest in rank of the line of the army, marine corps, or militia, by commission, there on duty or in quarters, shall command the whole, and give orders for what is needful to the service, unless otherwise specially directed by the President of the United States, according to the nature of the case; (ART. 62, *Rules and Articles of War*.)

It has been contended that the last clause of this article enables the President to make rank in the army vary at his pleasure, by an order of *assignment*. But inasmuch as the authority given to the President by the last clause of Article 62 is equally applicable to all commissions in the line of the army, marine corps, or militia, it would follow, under such a construction, that the laws creating rank did not fix a range of subordination; or, in other words, that Congress, after creating rank, or a range of subordination, and establishing rules of appointment and promotion, which require seniority or gallant and meritorious services, and the sanction of the Senate for the attainment of such promotion, have undone their whole work by giving to the President the power to deprive rank of the only quality which gives it consideration. The bare statement of this proposition is sufficient to show that such could never have been the meaning of the last clause of Article 62 of the Rules and Articles of War, and an attentive and candid examination of the article will, it is believed, convince all that its purpose was to declare that the officer highest in rank should command whenever different corps came together, "*unless otherwise specially directed by the President of the United States, according to the nature of the case.*" That is to say, unless the President, in any special case, should deem the highest officer inefficient or incompetent; then he might supersede him, by withdrawing him from the command. Or, in other cases, the President might desire to carve out of the general command particular trusts, or limit the discretion of the commanding officer in regard to what is needful for the service. This plain interpretation of the disputed passage in no case permits the violation of the rights of any officer, by placing a junior over a senior; but the authority which it gives the President is indispensable to a proper administration of his great office of commander-in-chief. And it may be here stated that, during the Mexican war, Mr. Polk's administration after much deliberation emphatically disavowed the possession of any legal authority to assign a junior major-general to command a senior. (See article RANK, for a statement of the case of Major-general Benton. See also BREVET; DETACHMENT; LINE; PRESIDENT.)

ASSIGNMENT OF PAY. No assignment of pay made by a non-commissioned officer or soldier, is valid; (*Act* of May 8, 1792.)

ASTRAGAL—Small convex moulding used in the ornamental work of ordnance, and usually connected with a *fillet* or flat moulding.

ASYLUM, (MILITARY.) The persons entitled to the benefits of the Asylum, or Soldier's Home, as it is now called, located in the District of Columbia, are: 1. All soldiers, and discharged soldiers of the army of the United States, who may have served honestly and faithfully for twenty years. 2. All soldiers, and discharged soldiers of the regular army, and of the volunteers, who served in the war with Mexico, and were disabled by disease or wounds contracted in that service and in the line of their duty, and who are, by their disability, incapable of further military service. This class includes the portion of the marine corps that served with the army in Mexico. 3. Every soldier, and discharged soldier, who may have contributed to the funds of the Soldier's Home since the passage of the act to found the same, March 3, 1851, according to the restrictions and provisions thereof, and who may have been disabled by disease or wounds incurred in the service and in the line of his duty, rendering him incapable of military service. 4. Every pensioner on account of wounds or disability incurred in the military service—though not a contributor to the funds of the Institution—who shall transfer his pension to the Soldier's Home during the period he voluntarily continues to receive its benefits. No provision is made for the wives and children of those admitted.

No mutineer, deserter, or habitual drunkard, or person convicted of felony or other disgraceful crime of a civil nature, while in the army or after his discharge, is admitted into the asylum without satisfactory evidence being shown to the Commissioners of the Soldier's Home of subsequent service, good conduct, and reformation of character. The Commissioners are: the adjutant-general, the commissary-general of subsistence, and the surgeon-general. The Soldier's Home has its governor, secretary, and treasurer, appointed from the army; (*Act* March 3, 1851.)

ATTACK AND DEFENCE. (*See* REDOUBT.) A redoubt may be either armed with cannon, or only defended by infantry. In the former case, it may be necessary to silence cannon by cannon; in the latter, we may march at once to the attack. Light infantry, principally riflemen, envelop the work, and even, at a distance of 1,000 yards, direct their fire upon the interior of the work and crest of the parapet, so as to prevent the defenders from showing themselves, or at least to cause them to fire hurriedly. Gradually approaching and converging their fire, the riflemen groove the parapet, and assert the superiority of their arm. Arrived at a short distance from the ditch, they run and leap into it, unless prevented by obstacles such as palisades, abatis, and trous-de-loup. In that event, they get rid of the obstacles by means of their axes, or fill the trous-de-loup with fascines, with which they have previously provided themselves. The whole number, however, do not throw themselves into the ditch, a portion remain upon the counterscarp, to fire upon any one daring to show himself behind the parapet. When the troops have taken breath at the bottom of the ditch, they *assault*, and to do this the soldiers aid each other in mounting upon the *berme*. From thence they mount together upon the parapet, leap into the redoubt, and force the defenders to ground their arms. If the redoubt is armed with cannon, and is of greater strength than has been supposed, it might be necessary at first to cannonade in such a manner as to break the palisades, dismount the pieces, and plough up the parapet. Favorable positions for the cannon used in the attack will be sought: these positions should command the work, or be on the prolongation of its faces, so as to give an enfilading fire. If the redoubt is pierced with embrasures, it is necessary to direct one or two pieces upon each embrasure so as to dismount the pieces, and to penetrate into the interior of the work, in order to demoralize the defenders. Some good riflemen will also approach towards the embrasures, shunning their direct range, and fire upon the artillerymen, who may attempt to reload their pieces.

It is only after the attacking artillery has produced its desired effect, that the light infantry envelop the work, and do what has been already indicated. When infantry of the line take part in the attack, it is formed in as many columns as there are salients of attack. Each of these columns is preceded by men armed with axes and carrying ladders. It is a wise precaution to give to front rank men, fascines, which not only serve as bucklers, but are also useful in filling up part of the ditch. The light infantry open to allow the passage of the columns, but redouble their fire to sustain the attack at the moment that the assailants begin to climb the parapet. *The essential thing in this decisive moment for the assailants is unity of effort, and to leap into the work from all sides at once.* It is necessary, then, that the troops stop a moment upon the berme, and await the concerted signal to clamber up the exterior slope, in order to mount upon the parapet. If the redoubt be not aided by other troops, or strengthened by works upon its flanks, it will be difficult to resist an attack thus directed when valiantly executed. Whatever may be the result, it is the first duty of the commandant of a post to sustain and invigorate the *morale* of his soldiers, by his own confident air, his valiant resolutions, and his activity in putting every thing in the best order. If the attack is not immediate, the commandant will surround the redoubt with abatis; he will provide heavy stones for the defence of the ditches; he will endeavor to procure bags of earth, to make embrasures upon the parapet. Wanting these he will supply himself with sods, making loopholes, through which the best marksmen will fire upon the enemy. A beam placed across these sods may, at the same time, serve as a protection to the marksmen, and a means of rolling down the assailants. Cannon begins the defence. As soon as the batteries of the enemy are discovered, the fire is opened. But when once the batteries have taken their positions, when their pieces are partly covered by the ground, and their fire begins to produce an effect, the struggle is no longer equal. It is then necessary to withdraw the cannon of the work into its interior, or to leave those pieces only which are covered by good traverses, throwing, however, from time to time, some canister among the light infantry, who may press too nearly. The artillery is at first only aided by a few good marksmen placed in the angles, behind traverses, or wherever the fire of the enemy is least felt. But when the work is so closely pressed that the artillery of the assailants cannot continue its fire without danger to their own men, the defenders mount upon the banquettes, the guns are brought back, and the warmest fire is directed upon the columns of attack, and upon the squads of light infantry, who seek to make a passage through the abatis to the counterscarp. This is the moment to explode such small mines as have been previously prepared under the glacis, or in the interior of the work.

If, notwithstanding such efforts, the enemy reaches the ditch, and collects his force for the assault, all is not yet lost. The defenders roll upon him shells, trunks of trees, and heavy stones, and then mounting upon the parapet, stand ready to receive him at the point of the bayonet, or to use the butt of the musket. History records the failure of more than one attack from such conduct on the part of the defenders; and if we reflect upon the disorder of the assailants, and the physical advantage which those standing upon the parapet must possess, it is necessary, for the success of the attacking force, that they should have a great moral superiority. This does often exist, but the commander of a work may infuse his own indomitable spirit into his men.

Temporary works may be attacked by SURPRISE or by OPEN FORCE. In all cases, the first thing to be done is for the commander of the attack to obtain the fullest possible information that circumstances will admit, of the character of the work, garrison, ground around it, defences, and probable aid at hand, &c. If an intrenched village is to be attacked, it should be ascertained by what means the streets and roads leading into it have been closed, whether by stockades or breastworks; how these obstacles are flanked; what obstructions are placed in front of them, &c., &c. If the post is an isolated building, such as a country house or church, attention should be directed to the mode in which the doors have been barricaded, or the windows blocked up; how the loopholes are arranged; what sort of flank defence has been provided; how it can best be approached; what internal preparations have been made for prolonging the defence, &c. Part of this knowledge may be obtained from spies, and reconnoissance must do the rest. In the attack of military posts, infantry are frequently thrown upon their own resources. They have no guns or howitzers for tearing up and destroying stockades, abatis, palisading, chevaux-de-frize, &c. Their reliance must therefore be their own activity and fertility of invention. Abatis may sometimes be fired by lighted fagots, or else passed by cutting away a few of the smaller branches. Small ditches may be filled up with fagots or bundles of hay; chevaux-de-frize may be displaced by main force with a rope, and a good pull together, or they may be cut up or blown to pieces by a box of powder. Stockade work or palisading may be escaladed with ladders brought up in a line under the protection of a firing party, and carried by two or four men according to their length; or a stockade, barricaded doors, gates, and windows may be breached by a bag of powder, &c. By such measures, decisively and boldly used, troops would be a match for any of the ordinary obstructions which might oppose their advance, whether the attack were made by night or day, by surprise or by open force. (Consult Dufour; *Aide Memoire*, &c.)

ATTACK AND DEFENCE OF PERMANENT FORTIFICATIONS. (*See* SIEGE.)

ATTENTION—Cautionary command addressed to troops, preparatory to a particular exercise or manœuvre.

ATTESTATION. A certificate, signed by the magistrate before whom a recruit is sworn in as a soldier.

AUDITORS. (*See* ACCOUNTABILITY for their duties.) They may administer oaths; (*Act* March 3, 1817.)

AUTHORITY, (CIVIL.) Any commissioned officer or soldier accused of a capital crime, or of having used violence, or committed any offence, against the person or property of any citizen of any of the United States, such as is punishable by the known laws of the land, must be delivered over upon application of the civil authority; and all officers and soldiers are required to use their utmost endeavors to deliver over such accused persons, and likewise to be aiding and assisting the officers of justice in apprehending and securing the persons so accused in order to bring them to trial. Any commanding officer or officers, wilfully neglecting or refusing upon application to deliver over such accused persons, or to be aiding and assisting the officers of justice in apprehending such persons, shall be cashiered; ART. 33. (*See* COMMAND; EXECUTION OF LAWS.)

AUXILIARY. Forces to aid.

AWARD. The decision or sentence of a court-martial.

B

BAGGAGE OF AN ARMY—Called by the Romans *impedimenta*, and by Bonaparte *embarras*. No question is more important in giving efficiency to an army, than the regulation of its baggage. Nothing so seriously impairs the mobility of an army in the field as its baggage-train, but this baggage is necessary to its existence; and the important question therefore arises, How shall the army be sustained with least baggage? Sufficient attention is not paid by Government to this subject in time of peace, and in war the commander of the troops finds himself therefore obliged to use the *unstudied* means which his Government hastily furnishes. In respect to artillery and artillery equipments, the minutest details are regulated. It should be the same with other supplies. In the United States Army, the quartermaster's department has charge of transports, and some steps have been taken to regulate the subject; but legislation is required for the necessary military organization of conductors and drivers of wagons, and perhaps, also, unless our arsenals may be so used, for the establishment of depots, where a studied examination of field transportation may be made; which will recommend rules, regulating the kinds of wagons or carts to be used in different circumstances; prescribing the construction of the wagon and its various parts in a uniform manner, so that the corresponding part of one wagon will answer for another, giving the greatest possible mobility to these wagons consistent with strength; prescribing the harness, equipment, valises of officers, blacksmith forges, tool chests, chests for uniforms, bales of clothing, packing of provisions, and, generally, the proportion, form, substance, and dimensions of articles of supply; what should be the maximum weight of packages; the means to be taken for preventing damage to the articles; the grade, duties and pay of the quartermasters, wagon masters, and drivers should be properly regulated; rules for loading should be given; and, finally, a complete system of marks, or modes of recognition should be systematized. With such rules, and the adoption of a *kitchen cart*, (*See* WAGON,) together with small cooking utensils for field service which may be carried by the men, an army would no longer always be tied to a baggage train, and great results might be accomplished by the disconnection. (*See* CONVOY; WAGON.)

BAKING. Troops bake their own bread, and the saving of $33\frac{1}{3}$ per cent. thus made in flour is carried to the credit of the Post Fund. (*See* OVENS.)

BALKS—are joist-shaped spars, which rest between the cleats upon the saddles of two pontoons, to support the chess or flooring.

BALL. (*See* CHAIN BALL; NAIL BALL; SOLID SHOT.)

BALLISTICS—is that branch of gunnery which treats of the Motion of Projectiles. The instruments used to determine the initial velocity of projectiles are the gun-pendulum, the ballistic pendulum, and the electro-ballistic machine. By the latter machine, the velocity of the projectile at any point of its trajectory is also determined. The initial velocity is determined by the gun pendulum, by suspending the piece itself as a pendulum, and measuring the recoil impressed on it by the discharge; the expression for the velocity is deduced from the fact, that the quantity of motion communicated to the pendulum is equal to that given to the projectile, charge of powder, and the air. The second apparatus is a pendulum, the bob of which is made strong and heavy to receive the impact of the projectile; and the expression for the velocity of the projectile is deduced from the fact, that the quantity of motion of the projectile before impact, is equal to that of the pendulum and projectile after impact. These machines have been brought to great perfection in France and in the United States. By the electro-ballistic machines wires are supported on target frames, placed in the path of the trajectory, which communicate with a delicate time-keeper. The successive ruptures of the wires mark on the time-keeper the instant that the projectile passes each wire, and knowing the distances of the wires apart, the mean velocities, or velocities of the middle points can be obtained by the relation velocity $= \frac{\text{space.}}{\text{time.}}$

The electro-ballistic machine of Capt. Navaez of the Belgian service, has been found too delicate and complicated for general service; that devised by Capt. J. G. Benton, Ordnance Department, is used at the United States Military Academy. (For description, &c., consult BENTON's *Ordnance and Gunnery*.)

BAND. Musicians, as Regimental Band, Post Band, &c. They are enlisted soldiers, and form a band of musicians under the direction of the adjutant, but are not permanently detached from their companies, and are instructed in all the duties of a soldier.

BANQUETTE—is the step of earth within the parapet, sufficiently high to enable the defenders, when standing upon it, to fire over the crest of the parapet with ease.

BARBETTE. Guns are said to be in barbette when they are elevated, by raising the earth behind the parapet, or by placing them on a high carriage, so that, instead of firing through embrasures, they can be fired over the crest of the parapet. In this position, the guns have a wide range, instead of being limited, as in firing through embrasures.

BARRACKS—from the Spanish *barraca*, are buildings erected by Government for lodging troops. Where the ground is sufficiently spacious, they are made to enclose a large area, for the purpose of exercising and drilling. Barracks should be very commodious, comprising mess-rooms, cooking-houses, guard-houses, magazines, &c. United States troops are generally badly quartered, sometimes in casemates of fortifications, and often in cantonments constructed by themselves. Officers and soldiers' quarters should be properly furnished by the Government; but in the United States, officers' quarters are bare of all conveniences when assigned to them for occupancy. The quarters of soldiers are provided with bunks, tables, &c. (Consult, for detailed information upon the proper construction of Barracks,

80 MILITARY DICTIONARY. [BAR.

and their necessary furniture, &c., BARDIN's *Dictionnaire de l'Armée de Terre; Spectateur Militaire, &c.; British Regulations.*)

BARRICADES. The following series of Barricades afford means of closing openings in various ways, most of them practicable under all circumstances:

1. Palisading; movable or fixed. ⎫ Loopholed; the bottom of the
2. Stockade of trees.   ⎬ loophole not less than 8 feet
3. Stockade of squared baulk.   ⎭ above ground outside.
4. Abatis; with or without parapet of earth and ditch behind.

(*See* PALISADES; STOCKADE; AND ABATIS.)

FIG. 64.

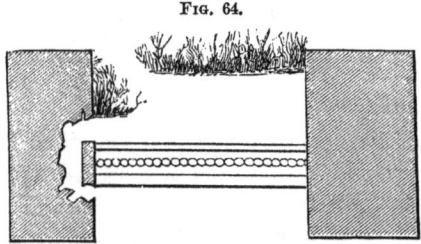

Fig. 64 represents a barricade in a street, with its means of communication.

FIG. 65.

Fig. 65. Barricade made in haste with tierces, boxes, wagon bodies, &c., and filled with earth or dung, avoiding parapets of paving stones.

Fig. 66. Barricades made with bales of merchandise, barrels of sugar, with the approaches also obstructed. Sand-bag parapets may also be used as barricades. (*See* REVETMENT.)

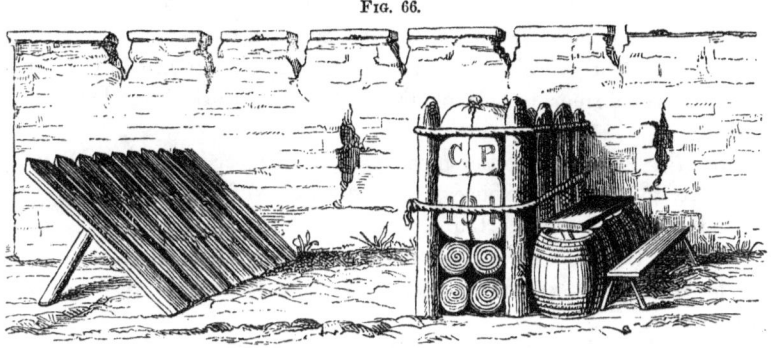

Fig. 66.

BARRIER. Carpentry obstructions in fortifications. The purpose regulates the construction. If the barrier is to be permanently defensible, it should be musket-proof, and then becomes a *Stockade*. If occasionally defensible, palisading will suffice, with a sand-bag or other temporary parapet when required, behind and near enough to fire between the palisades. The gates in both the above should, if possible, be of palisading, as the heavy stockade gate is unwieldy. Barrier gates should never be left unprotected.

BASE OF OPERATIONS. That secure line of frontier or fortresses occupied by troops, from which forward movements are made, supplies furnished, and upon which troops may retreat, if necessary.

BASTION. A work consisting of two faces and two flanks, all the angles being salient. Two bastions are connected by means of a CURTAIN, which is screened by the angle made by the prolongation of the corresponding faces of two bastions, and flanked by the line of defence. Bastions contain, sheltered by their parapets, marksmen, artillery, platforms, guards. They are protected by galleries of mines, and by demi-lunes and lunettes outside the ditch, and by palisades, if the ditch is inundated. Bastions should be large, and contain five or six hundred infantry, with the necessary artillery. The boyaux of the besiegers are directed towards the CAPITAL of the Bastion. The FACES of the BASTION are the parts exposed to being enfiladed by ricochet batteries, and also to being battered in breech. (*See* FORTIFICATION; SIEGES.)

Bastion (*Demi*)—is that which has only one face and one flank, cut off by the capital—like the extremities of horn and crown works.

Bastion (*Empty*). When the mass of rampart and parapet follows the windings of the faces and flanks, leaving an interior space in the centre of the bastion, on the level of the ground, it is called a hollow or empty bastion. In standing in a bastion, and looking towards the country, the face and flank on the right hand are called the right face and flank; and on the left hand, the left face and flank.

Bastion (*Flat*). When the demi-gorges and gorge are in the same line, and the former is half of the latter, the work is called a flat bastion.

Bastion (*Forts*)—are the most perfect of closed field works, with reference to flanking defences, as each side or front consists of two faces, two flanks, and a curtain.

Bastion (*Full*). When the interior space is filled up to the level of the terre plein of the rampart, the construction is called a full bastion.

BAT, BAT MEN, BAT HORSE, BAT AND FORAGE ALLOWANCE. Men who take charge of the baggage of officers and companies. Allowance given at the beginning of a campaign in the English army is called Bat and Forage allowance.

BATARDEAU—is a strong wall of masonry built across a ditch, to sustain the pressure of the water, when one part is dry and the other wet. To prevent this wall being used as a passage across the ditch, it is built up to an angle at top, and armed with iron spikes; and to render the attempt to cross still more difficult, a tower of masonry is built on it. In the batardeau is the sluice-gate, by the opening or closing of which the manœuvres of the water can be regulated. (*See* DITCH.)

BATTALION. An aggregation of from two to ten companies in the United States Service. Their instruction is regulated by Infantry and Light Infantry tactics.

BATTERY. A battery consists of two or more pieces of artillery in the field. The term Battery also implies the emplacement of ordnance destined to act offensively or defensively. It also refers to the company charged with a certain number of pieces of ordnance. The ordnance constitutes the Battery. Men serve the Battery. Horses drag it, and epaulments may shelter it. A battery may be with or without embrasures. In the latter case it is *en barbette*, and the height of the *genouillere* varies according to the description of the gun carriage used. The ordnance constituting the battery requires substantial bearings either of solid ground for field-pieces, or of timber, plank, or masonry platforms, for heavy artillery. Batteries are sometimes designated as follows: *Barbette battery,* one without embrasures, in which the guns are raised to fire over the parapet; *Ambulant battery*, heavy guns mounted on travelling carriages, and moved as occasion may require, either to positions on a coast, or in besieged places; *Covered battery*, intended for a vertical fire, and concealed from the enemy; *Breaching battery; Joint batteries*, uniting their fire against any object; *Counter battery*, one battery opposed against another; *Coast battery; Direct battery; Cross batteries*, forming a cross fire on an object; *Oblique battery* forms an angle of 20° or more, with the object against which it is directed, contradistinguished from direct battery; *Raised battery*, one whose terre plein is elevated considerably above the ground; *Sunken battery*, where the sole of the embrasures is on a level with the ground, and the platforms are consequently sunk below it; *Enfilading battery*, when the shot or shell sweeps the whole length of a line of troops or part of a work; *Horizontal battery*, when the terre plein is that of the natural level of the ground, consequently the parapet alone is raised and the ditch sunk; *Open battery*, without epaulment, or other covering wholly exposed; *Indented battery*, or battery à crémaillère, battery constructed with salient and re-entering angles for obtaining an oblique, as well as a direct fire, and to afford shelter from the enfilade fire of the enemy; *Reverse battery*, that which fires upon the rear of a work or line of troops; *Ricochet battery*, whose projectiles, being fired at low angles, graze and bound without being buried; *Masked battery*, artificially concealed until required to open upon the enemy.

Field Batteries, in sieges, are usually of two kinds, viz., Elevated Batteries and Sunken Batteries, and they are placed either in front of the parallel, in the parallel itself, or in rear of it. In an elevated battery, the platforms for the guns or mortars to stand upon, are laid on the natural level of the ground, and the whole of the covering mass, or parapet, is raised above that level, the earth for forming it being ob-

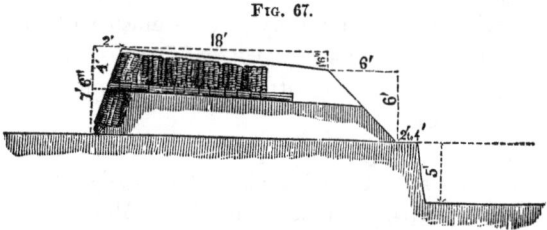

Fig. 67.

tained from a ditch in front; (Fig. 67.) In a sunken battery, the whole interior of the battery is excavated about three feet deep, and the platforms laid on the bottom, the earth is thrown to the front, and the parapet is formed out of it; (Fig. 68.) An inspection of these figures will show the difference; and it will be obvious that the whole of the parapet in the elevated battery has to be raised, and that in a sunken battery part of the cover is obtained by taking advantages of the excavation

Fig. 68.

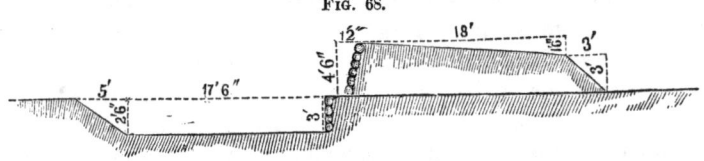

made for forming the mass. This construction is frequently used in turning the portion of a parallel into a battery, by increasing the width of the interior excavation of the trench so as to make room for the platforms of the guns. Great care must be taken that no rise in the ground before the battery obscures the view from the soles of the embrasures; for this purpose, the officer laying out the battery should lie down and look along the ground, in order to be sure that his guns can range freely from their embrasures, before he fixes his details for construction. When guns are fired with an elevation—when the soil is sandy or gravelly—when the weather is dry—or the ground elevated, this construction is approved. The depth of the excavation for the interior must depend on the height of the carriages upon which the guns are mounted: it should be deeper in rear than in front, that it may be drained. The interior slopes of these batteries, and the cheeks of the embrasures, must be supported by field revetments of gabions, fascines, sand-bags, casks, or sods. In batteries exposed to a heavy fire, especially of shells, it is necessary to provide as much cover as possible for the men serving in them; for this purpose, traverses are usually placed between every two guns; and as these masses serve to protect the men from the splinters of the bursting shells, they are generally called splinter-proof traverse. There is nearly twice as much work in the elevated as in the sunken battery. (Jebb's *Attack and Defence;* see Embrasure.)

BATTERY WAGON. A battery wagon accompanies each field-battery. (*See* Forge.)

BATTLE. Battles are either *parallel* or *oblique*, and they are *strategic* when, in consequence of a plan of campaign, they are fought upon a given and objective point, as the battles of Marengo or Austerlitz.

The following preparations for battle are usually made by great commanders: All disposable troops are held in hand; the readiness of the troops is ascertained by inspection of arms; proper nourishment is given to them before going into battle; the projects of the day are communicated from grade to grade; the points for the ambulances and caissons are indicated; the rendezvous for rallying or retreating are made known; measures are taken to secure the rear and communications, in order to retain the mastery of the base of operations; the army is ranged ordinarily in two lines, and the position of reserves given in the order of battle; the three arms are disposed according to the nature of the ground; decisive points are occupied; open or flanking batteries are established on proper elevations; the front and flanks of the army are furnished with artillery, in number, kind, and calibre according to circumstances. These are preparations for battle; the action commences ordinarily as follows: Marksmen are thrown forward, sometimes acting in conjunction with artillery. Either the enemy shows an equal disposition to attack, or else one party insults the other to bring on a combat. When the advanced guards have *felt* each other, the army disposed to make battle begins or increases its cannonade, to constrain the adversary to deploy his Masses, show his different arms, and thus make known the composition, number, importance, and the direction to be given to the adverse forces. The reserves remain stationary, while the cavalry, properly sheltered from fire, watch their opponents, and throw themselves upon weakened or staggered lines of infantry. When the affair has begun, and the position and dispositions of the enemy are known, and the proper effect has been produced by firing, the infantry may march to the charge, with the arms at a carry or on the right shoulder, leaving to the instinct of the soldier the determination of the proper moment of bringing the musket to the position of charge bayonet.

These details, however, constitute the mechanical parts of a battle. The art and science of battles consist, according to Professors of Strategy, in the subordination of tactical movements to the rule of attacking only with such Forces, as can overthrow those of the enemy, either by numbers, position, or vigor; in creating alarm upon many points to induce your adversary to take false steps; in surprising him in the midst of his bold movements, and punishing him in his irresolute ones; in penetrating his designs to neutralize their effects, or taking advantage of his faults; in occupying commanding positions; in avoiding masks or curtains, and in acting always, if possible, on the Offensive. When the action has seriously begun, the important business of the general is to follow it up to advantage. If he is skilful and valiant, he will preserve the Alignment and intervals of his battalions, by standing firm, or by marching; he will strengthen his flanks by enterprises against those of the enemy; by employing his fire so as not to stop the fire, at the same time, of all arms; by filling up, at the expense of the cavalry or second line, the holes made in the first line; by reinforcing or reanimating all corps which give way or falter; by leaving none in unfavorable positions; by sheltering the reserves from cannon shot; by bringing up, at opportune moments, fresh troops; by preserving the rear lines from being broken, while opening a free passage to repulsed troops; by exposing, when needed, his own person, securing united efforts in attacks, vigor in charges, and promptitude in rallying. Such is the theory of battles; but GENIUS and experience are necessary to apply the theory, and victory will be in vain sought from the mechanical application of any dogma whatever. Battles upon the same ground rarely occur, and never with soldiers of the same *morale*, the same arms, the same numbers, and the same relative proportions. It is by study of the campaigns of great commanders, by his own experience, and his own genius, that battles are properly initiated and won by a skilful general. (*See* MANŒUVRES IN COMBAT.)

BAYONET. At the battle of Spires, in 1703, charges of infantry were first made with fixed bayonet. From that time, however, until the wars of the French Revolution, the bayonet was more threatening than murderous. Since then it has changed, throughout, the whole system of the military art; cavalry has ceased to be the terror of foot; and the fire of lines of battle, even with new arms effective in range at 1,000 yards, does not impair the usefulness of the bayonet; and although Suwarof's maxim that "La balle est folle" cannot be admitted, yet it is true that "la bayonnette est sage." (Consult *Manual of Bayonet Exercise*, by CAPT. G. B. MCCLELLAN.)

BED. Straw and bedsacks are allowed to soldiers for bedding. The introduction of single iron bedsteads will make it necessary to increase the allowance of bed furniture. In Prussia and other countries, hammocks are used in place of bedsteads. *Bed* has also other applications, as mortar bed; camp bed; bed of a gun lock; bed of sand; bed of a river; to separate the beds of stone in a quarry, &c.

BELT. (*See* ACCOUTREMENTS.)

BERME. Narrow path round fortifications, between the parapet and the ditch, to prevent the earth from falling in.

BESIEGE. (*See* SIEGE.)

BILLET. No soldier shall, in time of peace be quartered in any house without the consent of the owner; nor in time of war, but in the manner to be prescribed by law; (ART. 3, *Amendments to the Constitution*.) The manner of quartering soldiers in time of war is usually by *Billets*, but no manner has been *prescribed by law in the United States*.

The constables and other persons duly authorized in England are required to billet the officers and soldiers of the army, and also the horses belonging to the cavalry, staff, and field-officers, in victualling and other houses specified in the mutiny act; and they must be received by the occupiers of these houses, and provided with proper accommodations. They are to be supplied with diet and small beer, and with stables, hay, and straw, for the horses; paying for the same the several rates prescribed by law. Troops, whether cavalry or infantry, are in no case to be billeted above one mile from the place mentioned in the route. Where cavalry are billeted, the men and their horses must be billeted in the same house, except in case of necessity. One man must always be billeted where there are one or two horses; and less than two men cannot be billeted where there are four horses; and so in proportion for a greater number. No more billets are at any time to be ordered than there are effective soldiers and horses present; and all billets are to be delivered into the hands of the commanding officer. Commanding officers may, for the benefit of the service, exchange any men or horses billeted in the same town, provided the number of men and horses so exchanged does not exceed the number at the time billeted on each house; and the constables are obliged to billet those men and horses accordingly. Any justice may, at the request of the officer or non-commissioned officer commanding any soldiers requiring billets, extend the routes or enlarge the district within which billets shall be required, in such manner as may be most convenient to the troops. In Scotland, officers and soldiers are billeted according to the provisions of the laws in force in that country at the time of its union with England; and no officer is obliged to pay for his lodging, where he shall be regularly billeted, except in the suburbs of Edinburgh.

BILL HOOK. An instrument for cutting twigs.

BIVOUAC. (*See* CAMP.)

BLACKING. (For SHOES.) Take three ounces of molasses, three ounces of ivory black, one ounce muriatic acid, one ounce sulphuric acid, and a spoonful of olive oil. Mix the ivory black and molasses, then add the muriatic acid, and subsequently the oil; when the paste is well formed, incorporate with it the sulphuric acid.

BLACKING, LIQUID. (For SHOES, &c.) Three parts of white wax, seven and a half parts essence of turpentine; one and a half parts of ivory black. The wax is cut into small pieces and put into a glazed vessel. Spread the turpentine over it, and leave it for 24 hours. Then mix it by degrees with ivory black. To use it, spread it with a rag in a thin layer on the leather, and afterwards rub with a soft brush.

BLACKING. (For Harness.) Yellow wax, four parts in weight, six parts essence of turpentine, one part of mutton suet, and one part of ivory black. Cut the wax into small pieces, and leave it to soak twenty-four hours in the essence of turpentine; grind in separately the ivory black and suet until there is a perfect mixture of the whole mass. When the leather has lost its color, it may be restored by the mud of ink, or by sulphate of iron in a thick solution, spread upon the edges.

BLACKSMITH AND FARRIER—Allowed to cavalry regiments. (*See* Forge; Army Organization.)

BLINDAGE. A siege work contrived, when defilement is impossible, as a shelter against a cross or ricochet fire of artillery. It is also used to guard against the effects of shells. The powder magazines, the hospitals, the cisterns, certain doors and windows are thus *blinded* by means of carpentry work, or shelters loaded with earth, dung, &c. Blindage of the trenches is also necessary, particularly when the besiegers begin the crowning of the covered way by means of the sap. Blindages are thus used to guard against stones or hand grenades thrown by the besieged. This blindage is entirely exposed to sorties, and also to the danger of being burned by the besieged.

BLOCK AND TACKLE. The power is equal to the weight divided by the number of ropes attached to the lower block, or by twice the number of raising pulleys.

BLOCK-HOUSE (Redoubt of wood.) A common defence against Indians—at two diagonal angles of a picket work. Figs. 69 and 70,

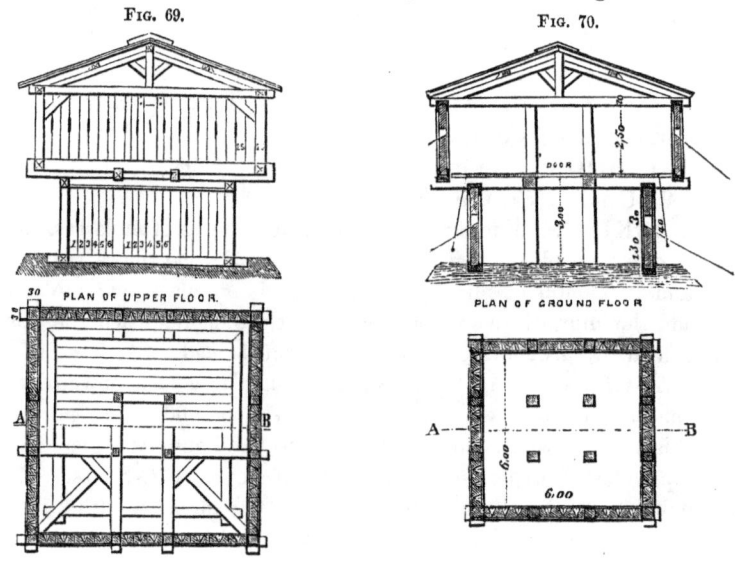

Fig. 69.   Fig. 70.

with dimensions in metres, show the construction used by the French in Algiers; or it may be built of logs 18 inches square on the ground floor, and 12 inches square in the upper story. Height of each story ten feet; loopholed; the upper story projecting all round, beyond the ground story, as machicoulis. Hatches should be made in the roof for the escape of smoke, and be grated.

BOARDS. A board composed of ordnance officers, designated by the Secretary of War, as the Ordnance Board, decides, with the approval of the secretary, on the models and patterns of all ordnance and ordnance stores for the land service of the United States.

Boards of Examination—are instituted to determine upon appointments in regiments, composed of army officers, and for appointments and promotion in the medical staff.

Boards of Survey—are to examine injured stores, &c., and to take an inventory of the public property in charge of a deceased officer.

Boards of Inspectors—determine upon the fitness of recruits for service.

BOAT. A boat has been invented by Colonel R. C. Buchanan, of the army, which has been used in several expeditions in Oregon and in Washington Territory, and has been highly commended by several experienced officers, who have had the opportunity of giving its merits a practical service test. It consists of an exceedingly light framework of thin and narrow boards, in lengths suitable for packing, connected by hinges, the different sections folding into so small a compass as to be conveniently carried upon mules. The frame is covered with a sheet of stout cotton canvas, or duck, secured to the gunwales with a cord running diagonally back and forth through eyelet-holes in the upper edge. When first placed in the water the boat leaks a little, but the canvas soon swells so as to make it sufficiently tight for all practical purposes. The great advantage to be derived from the use of this boat is, that it is so compact and portable as to be admirably adapted to the requirements of campaigning in a country where the streams are liable to rise above a fording stage, and where the allowance of transportation is small. It may be put together or taken apart and packed in a very few minutes, and one mule suffices to transport a boat with all its appurtenances, capable of sustaining ten men. Should the canvas become torn, it is easily repaired by putting on a patch, and it does not rot or crack like india-rubber or gutta-percha; moreover, it is not affected by changes of climate or temperature.—MARCY's *Prairie Traveller*. (*See* BRIDGE; PONTON.)

BOMB. The shell thrown by a mortar is called a bomb-shell; and the shelters made for magazines, &c., should be bomb-*proof*.

BOMBARDMENT. A shower of shells and other incendiary projectiles. Properly employed against fortifications, but not against open commercial cities.

BOOKS. *Regimental* books to be kept, are: 1. General order book; 2. Regimental order book; 3. Letter book; 4. Index of Letters; 5. Size or descriptive book; 6. Monthly returns. *Company* books required are: 1. Descriptive book; 2. Clothing book; and 3. Order book.

The following rules for keeping books at the head-quarters of the army and in the adjutant-general's office may, with modifications that will readily occur, be used with armies in the field, at the head-quarters of divisions, departments, regiments, &c.:

I. Letters Received.—(7 *quires, demy-Russia, with spring back.*) 1. All official communications received will be entered in this book, excepting only such letters of mere transmittal of orders, returns, certificates of disability, requisitions, &c., as need not be preserved. The orders, returns, certificates, requisitions, &c., themselves, will be appropriately entered in other books specially provided for the purpose.

2. Preliminary to being entered every letter will be folded and endorsed. Letter paper will be folded in three equal folds—*Cap* paper in four. The endorsement will give the place and date of letter, name, and rank of writer, and a summary of its contents, and if other papers accompany the letter, the number transmitted will also be noted on the back, in red ink. Each enclosure will be numbered and bear the same *office marks* as the letter transmitting it. Figures A, b, c, exemplify the manner of endorsing.

3. Every letter required to be preserved will be entered *alphabetically* and numbered—the series of numbers beginning and terminating with the year, and including all letters *dated* (whether received or not) within the year. Only one number will be given to each letter received with its enclosures, so that the sum of the numbers under each alphabetical entry in the book of " Letters Received," during any year, will show the number of letters received in that year.

4. As a general rule, every letter will be entered in the name of its writer; but there are cases where it is preferable, for convenience of reference, to enter it in the name of the person who forms the subject of the letter and not in that of the writer. Applications from citizens for the discharge of soldiers, &c., are of this nature. Usually, a single entry of each letter and its enclosures will suffice, but it may sometimes be necessary, in addition, to make entries in the names of one or more of the individuals to whom it relates. Such entries, however, will not be *numbered*, but merely contain the date of receipt, name of individual,

Fig. A.	fig. b.	fig. c.
G. 1 FORT ADAMS, R. I., *May* 8, 1849. Col. ———, 3d Artillery, Com'd'g. Relative to unhealthiness of quarters at the Post, and enclosing Surgeon ———'s report on the subject, dated Apr. 30, 1849; forwards also a copy of a report, dated Aug. 16, 1840, of a Board of Officers assembled to examine into the condition of the quarters. [Two enclosures.] Rec'd (Hd. Qrs.) May 11, 1849.	1. G. 1. (Hd. Qrs.) May 11, 1849.	2. G. 1. (Hd. Qrs.) May 11, 1849.

place and date of the letter concerning him, with a reference, in red ink, to the number of that letter. Fig. E is an illustration of an entry of this kind.

5. The book of "Letters Received" will contain a *side* index extending throughout, and will be divided among the several letters of the alphabet according to the probable space required for entries under each letter. The book will be paged, and each page divided into three columns, headed "When received," "Name," "Date and purport of letter," respectively, as shown by figure *D*, which also exhibits the entry in the book of the letter represented by figure *A*.

Fig. E.

S. LETTERS RECEIVED.
1849.

When received.	Name.	Date and purport of letter.
May 11th.	[Surgeon ———.]	Fort Adams, R. I., May 8, 1849. See No. 1, Letter *G*.

Fig. D.

LETTERS RECEIVED. G.
1849.

When received.	Name.	Date and purport of letter.
May 11th. 1	Col. ———, 3d Artillery, command'g.	FORT ADAMS, R. I., *May* 8, 1849. Relative to unhealthiness of quarters at the Post, and enclosing Surgeon ———'s report on the subject, dated April 30, 1849; forwards also copy of a report, dated Aug. 16, 1840, of a Board of Officers assembled to examine into the condition of the quarters.

6. Each entry will be separated from the one preceding it by a red ink line; and where two or more letters relate to the same subject they will be either filed together, or made to refer to each other by their *numbers*, and the filing or reference be noted in the book as well as on the letters themselves.

7. Letters from the Executive and Staff Departments and other public offices in Washington, will be entered alphabetically in the names of the *departments* or *offices* themselves, but the entry will always exhibit the writers' names likewise;—thus, communications from the War Department would be entered in the letter *W*, as follows: "*War*, Secretary of, (Hon. ———,) &c."

8. Communications from the President will be entered in the letter *P*—from State Department, in *S*—Treasury, *T*—War, *W*—Navy, and its bureaux, *N*—Post Office and its bureaux, *P*—Interior, *I*—Attorney-general, *A*—Adjutant-general's office, *A*—Quartermaster-general, *Q*—Subsistence, *S*—Surgeon-general, *S*—Paymaster-general, *P*—Engineer Department, *E*—Topographical Engineers, *E*—Ordnance, *O*—Recruiting service, Superintendent of, *R*—Pension Office, *P*—Comptrollers, (1st and 2d,) *C*—The several Auditors, *A*—Treasurer U. S., *T*—Commissioner Indian Affairs, *I*—General Land Office, *L*—Solicitor's Office, *S*—and Patent Office, *P*.

9. Communications from Governors of States will be entered in the names of the *States*, the entry showing likewise the Governors' names;—thus a letter from the Governor of New York would be entered in the letter *N*, as follows: "New York, Governor of, (His Excellency ———,)" &c.

10. Letters from *Staff Officers*, written by direction of their generals, will be entered in the names of the *Generals* themselves;—thus a communication from General K———'s Staff Officer would be entered in the letter K, as follows:

"Bvt. Major Gen'l ———, comd'g West'n Div'n,"
"(by Assist. Adjt. Gen'l ———.)"

11. Communications addressed to the War Department or Adjutant-general's office, and thence referred, without an accompanying letter, to head-quarters for report, or to be disposed of, will be entered, in the ordinary way, in the names of their writers, a note (in red ink) being simply made in the second column of the book, to show the fact of reference, thus—"(from A. G. O.)"

12. Where letters are referred from the office for report, &c., a note of the fact must be made (in red ink) in this book with a citation of the page, (or number of the letter,) in the "Endorsement" or "Letter

Book" where the reference is recorded, thus—Ref'd for report to Comd'g Offi'r Fort T., May 11—see Book of "Endorsements," p. 3,—(or, "see Letter No. 7, vol. 1st.") When the communication is returned, a memorandum to this effect will be made in the book—"Returned with report, May 25th."

13. Should the portion of this book appropriated to any particular letter of the alphabet prove insufficient for entries under that letter, they will be transferred to a few of the *last* leaves allotted to some other letter of the alphabet, where there is more space than will probably be required. The fact of transfer will be noted in large characters, (in red ink,) at the bottom of the page from which transferred, and at the top of the page to which carried, as follows:

"TRANSFERRED TO PAGE 250," and "BROUGHT FROM PAGE 60."

II. LETTER BOOK.—(7 *quires, demy-Russia, with spring back.*) 1. Every letter recorded in this book is numbered, (in red ink,) the numbers commencing and terminating with the year, and each letter is separated from the one which follows it by a red line.

2. The address of all letters should be at the top, the *surname* being written conspicuously in the margin, followed by the official title (if any) and Christian name, thus:

  Bvt. Maj. Gen'l ———.
  Comd'g, &c., &c., &c., or
  Esq. Samuel H.

3. Each letter should be *signed* in the record book by its *writer*.

4. Whenever copies of letters are furnished, the names of the persons to whom they are sent should be noted in red ink in the margin with the *date*, when the last differs from the date of the letter itself. In like manner, when a letter is addressed to one officer, under cover to his commander, &c., this fact should also be noted in red ink in the margin.

5. The name of every person to whom a letter is addressed is indexed alphabetically, in black ink, and the names of the individuals whom it principally concerns are indexed in red ink. A red ink line is drawn in the body of the letter under the names so indexed, to facilitate a reference to them. In the margin, immediately under the name of the person to whom a letter is addressed, there are two references, above and below a short red line, the one above (in red) indicates the last preceding letter to the same individual, and the one below (in black) the next following. A *detached* index is used until the record book is full, when the names are arranged under each letter as in City Directories, and thus classified they are transferred to the *permanent* index attached to the record book.

III. GENERAL ORDERS.—(7 *quires, demy-Russia, with spring back.*)
1. Every order recorded in this book should be *signed* by the staff officer whose signature was attached to the originals sent from the office, and each order should be separated from the one following by a red line.

2. The mode of numbering, distribution, and general form of orders are prescribed by the Regulations—(see paragraphs 904, 905, and 908, edition of 1847;) but the distribution in each particular case should be noted in red ink in the margin to show that the Regulations have been complied with; and where orders are sent to one officer, under cover to his commander, (which course ought always to be pursued,) or furnished at a date subsequent to that of their issue—these facts should likewise be added: where the order has been *printed*, it will be sufficient to write the word "*printed*" in red ink in the margin, to indicate that the widest circulation has been given to it.

3. There are *two* indexes attached to the book—one of *names*, the other of *subjects*—every order will be indexed in the *latter* immediately after being copied.

For *names*, a *detached* index will first be used until the record book is full, when they will be arranged under each letter as in City Directories, and thus classified, transferred to the *permanent* alphabetical index attached to the record book. Every proper name will be indexed and a red line drawn in the body of the order under it, to facilitate a reference to it.

IV. SPECIAL ORDERS.—(7 *quires, demy-Russia, with spring back.*)
1. Every order recorded in this book should be *signed* by the staff officer whose signature was attached to the originals sent from the office, and each order should be separated from the one following by a red line.

2. The mode of numbering, distribution, and general form of orders are prescribed by the Regulations—(see paragraphs 904, 905, and 908, edition of 1847;) but the distribution in each particular case should be noted in red ink in the margin, to show that the Regulations have been complied with; and where orders are sent to one officer, under cover to his commander, (which course ought always to be pursued,) or furnished at a date subsequent to that of their issue—these facts should likewise be added.

3. There are *two* indexes attached to the book—one of *names*, the other of *subjects*—every order will be indexed in the *latter* immediately after being copied.

For *names*, a *detached* index will first be used until the record book is full, when they will be arranged under each letter as in City Directories, and thus classified, transferred to the *permanent* alphabetical index attached to the record book. Every proper name will be indexed and a red line drawn in the body of the order under it, to facilitate a reference to it.

V. ENDORSEMENTS AND MEMORANDA.—(5 *quires, Cap—Russia, with spring back.*) 1. Every endorsement made on letters or other communications sent from the office will be copied in this book, and be *signed* by the staff officer whose signature was attached to the endorsement itself. A brief description of the communication sent out (the name of its writer, date, subject, and *office marks*) should precede the record of the endorsement, to render the latter intelligible; and where such communication has been entered in the book of "letters received," the disposition made of it should also be noted in that book, with a citation of the *page* where the endorsement is recorded. Should the communication be returned to head-quarters, a memorandum will be made to that effect, with the date when received back, in all the books where the fact of the reference from the office may have been noted.

2. In the case of such papers as proceedings of general courts-martial, certificates of disability for the discharge of soldiers, requisitions for ordnance, &c., which are not filed at head-quarters, but forwarded thence for deposit in other offices, it will generally suffice to make a brief memorandum of the general-in-chief's action upon them, instead of copying the endorsements. Where the endorsement, however, settles any rule or principle, it ought, of course, to be copied in full.

3. The name and address of every officer to whom a communication is referred will be written in the margin, and all *proper* names, no matter in what connection employed, must be indexed.

4. The name of the person to whom a communication is sent will be indexed in black ink, and the names mentioned in the description prefixed to the endorsement on the communication, as well as in the endorsement itself, will be indexed in red ink. To facilitate a reference to these last names, a red line will be drawn under them. In the margin, immediately under the name of the person to whom a communication is addressed, there are two references, above and below a short red line; the one above (in red) indicates the last preceding reference to the same individual, and the one below (in black) the next following.

VI. BOOK OF RETURNS.

Besides the foregoing blank books of appropriate size according to circumstances, the following books of reference are necessary: HETZEL'S

Military Laws; Army Regulations; Ordnance Manual; Artillery Manual; Prescribed Tactics for Infantry, Artillery, and Cavalry; McCLELLAND's Bayonet Exercise; Aide Memoire du Génie; Aide Memoire d'Etat Major; WHEATON's International Law; KENT's or STORY's Commentaries; MAHAN's Field Fortifications; Military Dictionary.

BOOM—is a chain of masts, or a large cable, or other obstacles stretched over a river for the protection of a military bridge which has been thrown across, or under the fire of fortifications to bar access within a harbor.

BOOTY. (SAXON, *bot, bote*, lawful profit, gain, advantage, distinguished from plunder or pillage.) Despoiling a people or city is barbarous and not tolerated in civilized warfare, but legitimate subjects of *booty* are well described in an act of the British Parliament (2 William IV., c. 53):—as arms, ammunition, stores of war, goods, merchandise, and treasure belonging to the *state* or any *public trading company of the enemy*, and found in any of the fortresses or possessions, and all ships and vessels in any road, river, haven, or creek belonging to any such fortress or possession. It should be the duty of commanding generals to cause an exact account of such captures to be kept, in order that the captors may be remunerated by the government for such stores as are reserved for the public service, and in order that *all* such prizes of war may be legally and equitably divided amongst the captors. Such is the practice in England. There land prizes are divided according to an established rule of division. In the Piedmontese army the administration of booty is intrusted to a special staff corps; the French laws (says Bardin, Dictionnaire de l'Armée de Terre) are silent on this subject, or else those which are in force announce nothing positive; and in their silence, there is inhumanity, hypocrisy, and mental reserve. In a memorial presented by the Duke of Wellington he claimed of his government for the English army, more than a million sterling which had been used in the king's service from captures made by the British army in Spain and France, and the English budget of 1823 shows that the amount so claimed was given to the army. The 58th article for the government of the armies of the United States provides, that " All public stores taken in the enemy's camp, towns, forts, or magazines, whether of artillery, ammunition, clothing, forage, or provisions, shall be secured for the service of the United States; for the neglect of which the commanding officer is to be answerable." This article of war is borrowed from a corresponding British article, which directs that the same stores shall be secured for the king's service. But by proclamation in Great Britain the money value of all captures is invariably divided amongst the captors. No practice can be more wise and just, for although it is necessary to proscribe *marauding* or *pillage*, it is impossible to extirpate the desire of gain from the human heart, and it is therefore necessary that the law should frankly provide for an equitable distribution of captures amongst the army. The absence of a law of division tends to introduce into an army the greatest evils: soldiers disband themselves in search of pillage, and their cupidity leads to the greatest horrors. These great evils are avoided by a legal division of booty, when all soldiers, animated by the hope of sharing the fruits of victory, are careful not to abandon to the greedy, the cowardly, and the wicked amongst themselves advantages properly belonging to the gallant victors. In the hope that Congress may yet do justice to our army in respect to captures made in the war with Mexico, the rules established in Great Britain are annexed in a series of prize proclamations taken from Prendergast's Law Relating to Officers of the Army:—

I.—*Prize Warrants.*

1.—SCINDE BOOTY.

Victoria R.

Victoria, by the Grace of God, of the United Kingdom of Great Britain and Ireland, Queen, Defender of the Faith, To all to whom these presents shall come, Greeting: Whereas the Commissioners of our Treasury have represented unto us, that certain hostilities were carried on in the year 1843 against the Ameers of Scinde by our land forces and the land forces raised and paid by the East India Company, in which a portion of the Indus Flotilla co-operated: and that during the said hostilities certain battles were fought, and a quantity of booty and plunder captured or taken possession of, consisting of gold and silver bars and coins, of ornaments, jewels, and ornamented arms, and of guns, cattle, and other property, of which the following schedule or account has been rendered to our said commissioners, (that is to say,)

	Rupees.
Paid in to the Public Treasury in Scinde on account of the articles sold, about	229,038
Realized at Kurrachie	17,743
Value of Silver	2,564,337
Gold sold	1,713,537
Gold remaining unsold, estimated at	123,273
Lead, valued at	15,000

to which are to be added the sum due from the Government for articles transferred to public departments, the sum due from individuals for articles sold in Scinde, and the sum which may be produced by the sale of the jewels, &c., which are at present in deposit at Bombay, but have been ordered to be sold;

And whereas it has been further represented unto us that the said booty and plunder do of right belong to us in virtue of our Royal prerogative, and that the said booty and plunder should be given and granted in such manner as to us may seem meet and just;

And whereas our said commissioners, under all the circumstances of this case, have recommended unto us to give and grant the said captured booty and plunder, or the produce or value thereof, as before stated, according to the following scheme, (that is to say :)

Such articles of personal use and ornament to be reserved for the Ameers as may be selected for that purpose by the Governor-general of India in council, with the approbation of the Commissioners of our Treasury;

The remaining property to be divided into sixths:

One-sixth to be given to all such of the troops stationed at, or between Shikarpoor, Seikkur, and Kurrachie, and all such of the Indus Flotilla stationed between Seikkur and Kurrachie on any day between the 17th of February and 24th of March, 1843, both included, as shall not be otherwise entitled to share in the booty;

The Major-general commanding in Scinde, and the officers of the general staff of the forces serving under his orders in the above-mentioned operations, to share in this portion as well as in the other portions hereinafter specified.

The remaining five-sixths (subject to the deductions hereinafter specified) to be divided in two equal parts, one moiety to be given to the troops who fought at Meanee, and the other to those who fought at Hyderabad; the troops who were in both battles receiving a share of each moiety; and from the share or shares accruing to each individual under the distribution to be made of this portion of the booty there should be deducted and repaid into the Company's Treasury the amount of the Donation of Batta, which the individual entitled to the said share or shares has received under the general order of the Government of India, dated 28th of February, 1844, as having been present at the battles of Meanee or Hyderabad;

And our said Commissioners likewise recommend that the troops under Lieutenant-colonel Outram, who were detached previously to the battle of Meanee, and directed to fire the Shikargah on upon the right flank of the army, as well as the detachment which so gallantly defended the British Residency on the 15th of February, and also such portion of the Indus Flotilla as was engaged in that defence, or co-operated with the detachment under Colonel Outram, or was in any other way in immediate connection with the army that achieved the victory of Meanee, should share as if they had all been actually present at the battle of Meanee; and in like manner the garrison of Hyderabad should be entitled to share in the sum alloted to those engaged in the second battle;

Now know ye that We, taking the premises into our Royal consideration, are graciously pleased to approve the said scheme, and do, with the advice and recommendation of our said Commissioners, by this our Royal Warrant, under our Royal sign-manual, give and grant the said captured booty and plunder, or the produce or value thereof as before stated, unto the Directors of the East India Company, or to such person or persons as they shall appoint to receive the same, upon the trust following, (that is to say,) upon trust, after making the reservations and deductions above stated, to distribute the remainder among our land forces, and the land forces of the said Company, and the officers and crews of the Indus Flotilla, engaged in the aforesaid hostilities in accordance with the scheme hereinbefore mentioned and set forth, and with the usage of the army of India;

And we are graciously pleased to order and direct that, in case any doubt shall arise respecting the claims to share in the distribution aforesaid, or respecting any demand upon the said captured booty or plunder, the same shall be determined by the Directors of the East India Company, or by such person or persons to whom they shall refer the same, which determination thereupon made shall, with all convenient speed, be notified in writing to the Commissioners of our Treasury, and the same shall be final and conclusive to all intents and purposes, unless, within three months after the receipt thereof at the office of the Commissioners of our Treasury, We shall be graciously pleased otherwise to order, hereby reserving to ourselves to make such order therein as to us shall seem meet.

Given at our Court at Windsor Castle, this 11th day of November, in the 9th year of our reign, and in the year of our Lord 1845.

By Her Majesty's Command,

(Signed) HENRY GOULBURN,
J. MILNES GASKELL,
WILLIAM CRIPPS.

2.—TARRAGONA BOOTY.

(*Conjunct Expedition of British Land and Sea forces.*)

GEORGE R.

Whereas ordnance arms, stores, magazines, and other booty have been captured from the enemy during the year 1813, at Tarragona, by that part of the British army under Field-marshal the Duke of Wellington, in Spain, which was under the immediate orders of Lieutenant-general Lord William Bentinck, and by H.M.S. *Malta, Fame, Invincible, Merope, Buzzard* and *Volcano,* forming part of the fleet under Admiral Lord Exmouth, then under the immediate orders of Admiral Sir Benjamin Hallowell, and appropriated to the public service; And whereas an Act passed in the 54th year of the reign of our late Royal Father, entitled an Act for regulating the payment of Army prize-money, and to provide for the payment of unclaimed and forfeited shares to Chelsea Hospital; And whereas application hath been made to us by the said F.M. the Duke of Wellington and Admiral Lord Exmouth to grant the sum of £31,531 18s. (being the estimated value of such ordnance and stores) in trust, to be distributed as booty to the officers, non-commissioned officers, and privates serving in that part of the British army under his command in Spain, which was under the immediate orders of Lieutenant-general Lord William Bentinck, and to the officers, non-commissioned officers, seamen, and marines, on board H.M.S. *Malta, Fame, Invincible, Merope, Buzzard* and *Volcano,* placed by Admiral Lord Exmouth under the immediate orders of Admiral Sir Benjamin Hallowell, at Tarragona; And whereas the said Field-marshal the Duke of Wellington, having expressed his wish not to participate in the distribution of the booty as Commander-in-chief of the British army serving in Spain; We, taking the same into our Royal consideration, are graciously pleased to give and grant, and do hereby give and grant, to the said Lieutenant-general Lord William Bentinck and Admiral Lord Viscount Exmouth the said sum of £31,531 18s.; and that the said sum be issued and paid without any fee or other deduction whatsoever, in trust, for the benefit of the said Lord William Bentinck and the officers, non-commissioned officers, and privates serving under him, and of Admiral Lord Viscount Exmouth, and the officers, non-commissioned officers, seamen, and marines actually on board of our before-mentioned ships employed in that service, as booty and prize, or bounty money in the nature of prize-money, under the provisions of the said Act passed in the 54th year of the reign of our late Royal Father, to be distributed under the provisions of the said Act of Parliament, and agreeably to our Proclamation for the distribution of prize, in force at the time of the said expedition, and this our Royal grant, in manner and in the several proportions following, (that is to say,) such sums being divided into eight equal parts:

To the said Lieut.-general Lord Wm. Bentinck, Admiral, Lord Viscount Exmouth, and such General Officers and Admirals under their command, who were actually present at the capture of the said booty, so that the said Lieut.-gen. Lord Wm. Bentinck and Admiral Lord Viscount Exmouth shall take one moiety, and the other General Officers and Admirals who were actually present at the capture of the said booty, the other moiety in equal proportions—*One-eighth.*

To the Colonels, Lieut.-colonels, and Majors in the army, and Captains and Commanders in the navy, who were actually present at the capture of the said booty, to be equally distributed among them, and the persons entitled by the usage of our army to share with them—*Two-eighths.*

To the Captains in the army and Lieutenants in the navy, and other description of persons entitled by the usage of our army and navy respectively to share with them—*One-eighth.*

To the Lieutenants, Cornets, Ensigns, and Quartermasters in the army, and Warrant and other Officers in the navy, and other description of persons entitled by the usage of our army and navy to share with them—*One-eighth.*

To the Sergeants in the army and Petty Officers in the navy, and other description of persons entitled by the usage of our army and navy respectively to share with them—*One-eighth.*

To the Trumpeters and Soldiers, Seamen, and Marines, and other description of persons entitled by the usage of our army and navy respectively to share with them.—*Two-eighths*

And we are further pleased to direct that all such respective sums of money shall be distributed as prize or bounty money, or money in the nature of prize-money, according to the provisions of the said Act of Parliament of the 54th year of the reign of our Royal Father, and the several Acts relating to the distribution of prize-money in our navy, and our said Proclamation, and this our grant, and the rules and customs heretofore used and observed in our army and navy respectively in that behalf, and the agents intrusted with the distribution thereof by the said Lieutenant-general Lord William Bentinck and Admiral Lord Viscount Exmouth shall give all such notices, and make such notifications of such distribution, as are required by the said Act of Parliament and the several Acts of Parliament in force relating to the distribution of prize-money in our army, and our said Proclamation, and pay over all unclaimed shares to Chelsea and Greenwich Hospitals respectively, to be hereafter paid to the persons entitled thereto, or remain for the benefit of the said respective Hospitals according to the provisions and regulations of the said Act of Parliament and the several Bills in force relating to the distribution of prize-money in our navy; And We are further graciously pleased to order and direct that in case any doubt shall arise respecting the said distribution, or with respect to any other matter or thing relating thereto, the same shall be determined by the said commanders of the said land and sea forces, Lieutenant-general Lord William Bentinck and Admiral Lord Viscount Exmouth, or by such person or persons to whom the said commanders of the said land and sea forces shall refer the same; and such determination shall be final and conclusive upon all persons concerned, and as to all matters and things relating to the said distribution.

Given at our Court, at Carlton House, this 7th day of June, 1820, in the first year of our reign.

By his Majesty's command,

(Signed)   BATHURST.

3.—GENOA BOOTY.
(Conjunct Expedition of British and Allied Forces.)

In the name and on behalf of His Majesty,

GEORGE P. R.

Whereas it has been represented to us that, at the capture of the Territory and City of Genoa and its dependencies, on the 18th of April, 1814, a quantity of ordnance, military and naval stores, ships and vessels, and other booty, being public property belonging to the enemies of the Crown of Great Britain, was seized and taken possession of by our sea and land forces, under the command of Vice-admiral Sir Edward Pellew, Bart. (now Lord Exmouth,) and Lieutenant-general Lord William Cavendish Bentinck, Knight of the Bath, commanding our naval and military forces in and upon the coasts of the Mediterranean, assisted by certain Sicilian and Italian troops, and troops in British pay, and has been condemned to us as good and lawful prize taken in the said conjunct expedition; And whereas no instructions were given by us for the division or distribution of the booty to be captured on the said conjunct expedition; *And whereas* application hath been made to us that we would be graciously pleased to order and direct that the same ordnance, military and naval stores, ships, vessels and other booty may be distributed between the officers and crews of our ships, and those of our

Ally the King of the Two Sicilies, and the officers and men of our land forces, and those of our Ally the King of the Two Sicilies, according to any plan of distribution We shall be graciously pleased to approve: We, taking the premises into our Royal consideration, are graciously pleased to give and grant, and do hereby give and grant, to the said Vice-admiral Sir Edward Pellew (now Lord Exmouth), Commander-in-chief of our fleet and vessels employed on the said expedition, and Lieutenant-general Lord William Cavendish Bentinck, Knight of the Bath, Commander-in-chief of our land forces employed on the said expedition, the said ordnance, military and naval stores, ships, vessels, and other booty, so as aforesaid taken and condemned to us, in trust, to distribute the same amongst the commanders-in-chief, general and flag officers, and all other officers serving on the said expedition in the following manner, (that is to say), that the division of the booty between the army and navy and the said Sicilian and Italian ships and troops serving in the said expedition, shall be made according to the following scheme or schemes: the whole being first divided into equal parts:

1 To the Commanders-in-chief and to the Flag and General Officers serving in the said expedition, one-eighth, to be distributed amongst them, so that each Commander-in-chief shall take double that share which each General and Flag Officer (not being Commander-in-chief) shall take; but if the number of Flag and General Officers, exclusive of the two Commanders-in-chief, shall exceed four, in that case a moiety of the said one-eighth shall be divided between the two Commanders-in-chief, and the other moiety amongst the other Flag and General Officers—*One-eighth.*

2 To the Colonels, Lieutenant-colonels, and Majors in the army, and Post Captains, and Masters and Commanders in the navy, *and to the persons of like rank belonging to the said Sicilian and Italian ships and troops*, to be equally distributed amongst them—*One-eighth.*

3 To the Captains of Marines and land forces, and the sea Lieutenants, and other description of persons entitled by our Proclamation for the distribution of prize of the 11th November, 1807, or by the usage of our army, to share with them, *and to the persons in like rank belonging to the said Sicilian and Italian ships and troops—One-eighth.*

4 To the Lieutenants and Quartermasters of marines, and Lieutenants, Ensigns, and Quartermasters of land forces, and the Boatswains, Gunners, Pursers in the navy, and other description of persons entitled by our said Proclamation or by the usage of our army, to share with them, *and to the persons in like rank belonging to the said Sicilian and Italian ships and troops—One-eighth.*

5 To the Midshipmen, Captains' Clerks, Sergeants of marines and land forces, and the other description of persons entitled by our said Proclamation or by the usage of our army, to share with them, *and to the persons in like rank belonging to the said Sicilian and Italian ships and troops—One-eighth.*

6 To the Trumpeters, Quarter-gunners, Seamen, Marines, and Soldiers, and the other description of persons entitled by our said Proclamation, or by the usage of our army, to share with them, *and to the persons in like rank belonging to the said Sicilian and Italian ships and troops—One-eighth.*

And that the portion of the said booty, so belonging to our said land forces employed on the said expedition, *and the persons belonging to the said Sicilian and Italian troops*, shall be distributed between the Commanders-in-chief, officers, and privates composing the same, according to the rule heretofore used and observed by the army, under the above scheme or schedule;

And that the portion of the said booty so as aforesaid belonging to our naval forces employed in the said expedition, *and the persons belonging to the said Sicilian and Italian ships*, be distributed amongst the Commander-in-chief, flag and other officers, and men belonging to our navy employed on the said expedition, *and the persons belonging to the said Sicilian and Italian ships*, agreeably to our Proclamation for the distribution of prize in force at the time of the said expedition.

And we are graciously pleased to order and direct that, in case any doubt shall arise respecting the said distribution, or respecting any charge or demand upon the said captured property, the same shall be determined by the Commanders-in-chief, and flag and general officers, or such of them as can conveniently be assembled, or by such person or persons to whom they, or a majority of them, shall agree to refer the same; which determination so thereupon made, shall, with all convenient speed, be notified in writing to the Clerks of our Council, and the same shall be final and conclusive to all intents and purposes, unless within three months after the receipt thereof at our Council Office, we shall be pleased otherwise to order; hereby reserving to ourself to make such orders therein as to us shall seem fit. Given at our Court at Carlton House, this second day of August, 1815, in the 55th year of our reign.

By command of H.R.H. the Prince Regent, in the name, and on the behalf of, His Majesty. (Signed) BATHURST.

II.—*India Prize-Money.*

The following is the present standing scale of distribution of prize-money in India, to European commissioned and non-commissioned officers, privates, &c.

	SHARES.
Commander-in-chief	⅛ of the whole.
General Officers	1,500
Colonels	600
Lieut.-colonels, Adjutant-gen. and Quartermaster-general of Her Majesty's and the Hon. Company's troops, Commissary-general, Members of the Medical Board, Inspector of Hospitals of Her Majesty's Troops	360
Majors, Deputy Adjutant-general, and Deputy Quartermaster-general of Her Majesty's and the Hon. Company's Troops, Deputy Commissary-general, and Superintending Surgeons	240
Captains, Surgeons, Assistant Adjt.-general, and Assistant Quartermaster-general of Her Majesty's and the Hon. Company's Troops, Assistant Commissary-general, Deputy Assistant Adjutant-general, Quartermaster-general and Commissary-gen., Paymaster, Surgeon to His Excellency the Commander-in-chief, Brigade-majors, Aides-de-camp to His Excellency the Commander-in-chief and General Officers, and Commissaries of Ordnance	120
Lieutenants, Assistant-surgeons, Cornets, Ensigns, Adjutants and Quartermasters of Her Majesty's Dragoons and Infantry, Veterinary Surgeons, Deputy Commissaries, and Deputy Assistant Commissaries of Ordnance	60
Conductors, Riding Masters, Apothecaries, Stewards, Sub-assistant and Veterinary Surgeons and Provost Martial	15
Sub-conductors, Assistant-apothecaries, Assistant-stewards, Regimental Sergeant-majors, Staff-brigade and Farrier-sergeants of Horse Artillery, Park Sergeant, Armorer, and Sergeants of Artillery	3
Trumpet-majors, Paymaster-sergeants, Saddler-sergeants, Schoolmaster-sergeants, Hospital-sergeants, Drill-sergeants, Color-sergeants, Armorer-sergeants, Drum-majors, Brigade and Staff-sergeants of Foot Artillery, Magazine-sergeants, Laboratory-sergeants, and Sergeants	2

	SHARES.
Fife-majors, Corporals, Bombardiers, Trumpeters, Farriers, Rough Riders, Gunners, Drummers, and Privates	1
Volunteers	1

The following scale of distribution of prize-money, for the several classes and ranks of native troops, has been adopted at all the Presidencies of India.

	SHARES.
Subedar, Syrang	6
Woordee, Major, Russaldar	
Jemedar, Tindal	2
Naib Russaldar	
Havildar, Native Doctor	1
Naik, Drummer	
Trumpeter, Gun Lascar	
Private, Puckallie	
Native Farrier, Duffadar	
Nishan Burder, Nuggurchee	$\frac{2}{3}$
Vakell and Hirkarrah	
Gun-driver, Bheestie	
Nakeeb	

For the Royal Army there is no standing scale of distribution, though, by the foregoing Prize Warrants, it will be seen that a uniform practice is generally observed.

III.—*Prize Proclamation for the Russian War of* 1854.

VICTORIA R.

Whereas by our Royal Proclamation, bearing date the Twenty-ninth day of March, One thousand eight hundred and fifty-four, We have ordered and directed that the net proceeds of all prizes taken during the present War with Russia, by any of our ships or vessels of war, after the same shall have been to us finally adjudged lawful prize, shall be for the entire benefit of the officers and crews of such ships and vessels of war (save as therein excepted), in which Proclamation We have directed in what proportion the land forces, doing duty as Marines, shall be entitled to share: And whereas in the said Proclamation We have reserved to ourselves the division and distribution of all prize and booty taken on any conjunct expedition of our ships and vessels of war with our army; and it is desirable that We should provide for the division and distribution of all prize and booty taken on such conjunct expedition, as also by our army alone: We therefore hereby order and direct, that in such cases the net proceeds of the share which shall be assigned by us to our army, under our Royal Sign Manual, shall be divided and distributed in the following manner and proportions, viz.:—

Commander of the Forces } *One-fourth of One-tenth part of the net proceeds.*

General Officers:

    1st Class.—General Officers commanding Divisions, and other Officers, &c., holding equivalent Staff Appointments . . . . . .
    2d Class.—Other General Officers, and all other Officers, &c., holding equivalent Staff Appointments . .
} *The remaining Three fourths of One-tenth part of the net proceeds; the same to be so divided that a General Officer, &c., of the 1st Class shall receive One-half more in amount than a General Officer, &c., of the 2d Class.*

Field Officers:

    1st Class.—Colonels, Lieutenant-colonels, and Brevet Lieutenant-colonels, and other Officers holding Staff Appointments equivalent thereto . . . . . .
    2d Class.—Brevet Lieutenant-colonels not holding an Appointment qualifying them to share in the preceding Class of Field Officers, and all Majors, Regimental or Brevet, and all other Officers holding Appointments equivalent thereto . . . . . .
} *One-eighth of the remainder of the net proceeds; the same to be so divided that a Field Officer, &c., of the 1st Class shall receive One-half more in amount than a Field Officer, &c., of the 2d Class.*

The remainder of the net proceeds shall be distributed in the following Classes, so that every Officer, Non-commissioned Officer, &c., shall receive shares or a share according to his Class, as set forth in the following scale:

    1st Class.—Captains, and all other Officers entitled according to the usage of our army to share in that rank . . } *Thirty-five Shares each.*
    2d Class.—Subalterns, and all other Officers entitled according to the usage of our army to share in that rank . } *Twenty Shares each.*

3d Class.—Sergeant majors, Quartermaster Sergeants, and all other Staff Sergeants, and others holding equivalent rank . . . . . . } *Ten Shares each.*

4th Class.—Sergeants, and others holding equivalent rank . . . . } *Eight Shares each.*

5th Class.—Corporals . . . . *Four Shares each.*

6th Class.—Private Soldiers, Trumpeters, Drummers, &c. . . . . } *Three Shares each.*

And in the event of any difficulty arising with respect to the Class in which any Officer, &c., shall be entitled to share, our will and pleasure is, that the same shall be determined and adjusted by the Commander-in-chief of our land forces for the time being.

Given at our Court at Buckingham Palace, this Eleventh day of August, in the year of our Lord One thousand eight hundred and fifty-four, and in the eighteenth year of our reign.

God Save the Queen.

BOUNTY. "Every able-bodied musician or soldier, *re-enlisting* in his company or regiment within two months before, or one month after the expiration of his term of service, shall receive two months' extra pay, besides the pay and allowances due him on account of the unexpired period of his enlistment;" (*Act* March 2, 1833.) *Bounty* lands have also been given by Congress for military service. The principal characteristic of those acts has been to reward alike all grades, and to make no distinction of service, except by granting forty acres for the *minimum* degree of service, and one hundred and sixty acres for the *maximum* of service. A very marked and utterly indefensible departure from the principle upon which such rewards of merit and services were made by the several States immediately after the Revolutionary War.

BOYAU—is a small trench, or a branch of a trench, leading to a magazine, or to any particular point. They are generally called boyaus of communication.

BREACH. Rupture made in a fortification to facilitate the assault. The best mode of doing this is by dividing the wall up into detached parts by making one horizontal and several vertical cuts, and battering each part down. The easiest way to make the cut is to direct the shots upon the same line, and form a series of holes a little greater than a diameter apart, and then fire at the intervals until the desired cut is made. The horizontal cut is finished first. The vertical cuts are then commenced at the horizontal cut, and raised until the wall sinks, overturns, and breaks into pieces. The effective breaching power of rifle cannon has been shown by recent successful experiments in England, against a martello tower 30 feet high and 48 feet diameter, the walls being of good solid brick masonry, from 7 to 10 feet thick. Armstrong guns with 40 and 80-pounder solid shot, and 100-pounder percussion shells were used at a distance of 1,032 yards, more than twenty times the usual breaching distance. The 80-pounder shot passed completely through the masonry, (7 feet 3 inches,) and the 40-pounder shot and 100-pounder percussion shells lodged in the brickwork, at a depth of five feet. After firing 170 projectiles, a small portion of which were loaded shells, the entire land side of the tower was thrown down, and the interior space was filled with the *debris* of the vaulted roof, forming a pile which alone saved the opposite side from destruction. The superior breaching power of rifle projectiles depends not only on penetration, but on accuracy of flight and consequent concentration on any desired point; (BENTON.)

BREACH OF ARREST. Any arrested officer who shall leave his confinement, before he shall be set at liberty by his commanding officer, or by a superior officer, shall be cashiered; (ART. 77, *Rules and Articles of War*.)

BREAK GROUND—is to commence the siege of a place by opening trenches, &c.

BREASTWORK—is a hastily constructed parapet, not high enough to require a banquette, or at least generally without one; (*See* FIELD WORKS.)

BREECH. The mass of solid metal behind the bottom of the bore of a gun extending to the rear of the base ring. The *base of the breech* is a frustum of a cone or spherical segment in rear of the breech.

Breech of a musket; Breech screw; Breech pin. (For breech-loading arms, *See* CARBINES; PISTOL.)

BREVET. (*French.*) It is derived from Latin, *breve, brevia*, which signify a brief; a parchment containing an annotation or notification; (BARDIN, *Dictionnaire de l'Armée de Terre.*) So also, according to Ainsworth, *To issue out a writ*, Mandatum, *vel* BREVE emittere. This Latin word *breve, brevia*, is also still preserved in English law, as signifying a writ, or mandatory precept issued by the authority, and in the name of the sovereign or state.—See *Breve*, a writ, *Breve de Recto*, a writ of right, *Brevia Formata*, the register of writs; (BOUVIER's *Law Dictionary*.) So also in *Scots Law*, *Breve Testatum* (*Lat.*) an acknowledgment in writing, which, by the ancient practice, was made out on the land at the time of giving possession to the vassal, and signed by the superior; (OGILVIE.)

The word brevet in French signifies, when applied to officers in the army or navy, *commission*; (SPIERS and SURENNE.) Brevet was taken by the English from the French with this meaning. As used in the United States army, brevet was borrowed with our Articles of War from England, and in the British service it means a commission in the army at large, distinctive of a commission in a particular regiment or corps. But, as both in the British service and our own, *payments* are made for the authorized number of officers of the various grades in the several corps composing an army, ordinary English lexicographers have set down the meaning of brevet as a *commission* which gives an officer title and rank in the army above his pay; (WEBSTER, WORCESTER, and OGILVIE.) This would be the true meaning of brevet, if there was no legislation on the subject of rank by brevet other than that authorizing such rank to be conferred. But as rank by brevet is given in the army of the United States, by and with the advice and consent of the Senate, for " gallant actions or meritorious services," the laws have justly provided that, whenever an officer is on duty, and exercises a command according to his brevet, he shall be entitled to the pay of such grade; (*Acts of 1812 and 1818.*) Brevets, however, being commissions in the army at large, it would also follow, if there was no further legislation, that such commissions would be exercised in the particular regiment in which an officer was mustered. To avoid this, and also to give efficacy to commissions in particular corps where different corps come together, the 61st and 62d Articles of War have regulated the whole subject. The 61st Article provides that within a regiment or corps officers shall take rank and do duty according to the commissions by which they are mustered in their regiments or corps, but brevets or former commissions may take effect in detachments and courts-martial composed of different regiments or corps. As rank, however, means *a range of subordination in the body in which it is held*, it is manifest that rank in any particular body, as a regiment, corps, or the army at large, would not of itself give the right to command out of that particular body, without being *enabled* by further legislation. Hence the necessity of the 62d Article of War, which provides that, when *different* corps come together, the officer highest in rank of the line of the army, marine corps, or militia, by commission there on duty or in quarters shall command the whole, and give orders for what is needful for the service, unless otherwise specially directed by the President of the United States, according to the nature of the case; (*See* COMMAND; DETACHMENT; LINE; PRESIDENT; RANK.)

BRIBE AT MUSTER. Art. 16 of the Rules and Articles of

War provides that any officer convicted of taking any bribe on mustering, or on signing muster rolls, shall be displaced from his office, and be utterly disabled from ever after holding any office or employment in the service of the United States.

BRICOLE. Men's harness for dragging guns, length 18 feet—used for harnessing men to guns when horses cannot be used.

BRIDGE. If you are at the side of a narrow but deep and rapid river, on the banks of which trees grow long enough to reach across, one or more should be felled, confining the trunk to its own bank, and letting the current force the head round to the opposite side; but if " the river be too wide to be spanned by one tree—and if two or three men can in any manner be got across—let a large tree be felled into the water on each side, and placed close to the banks opposite to each other, with their heads lying up-streamwards. Fasten a rope to the head of each tree, confine the trunks, shove the heads off to receive the force of the current, and ease off the ropes, so that the branches may meet in the middle of the river, at an angle pointing upwards. The branches of the trees will be jammed together by the force of the current, and so be sufficiently united as to form a tolerable communication, especially when a few of the upper branches have been cleared away. If insufficient, towards the middle of the river, to bear the weight of men crossing, a few stakes, with forks left near their heads, may be thrust down through the branches of the trees to support them;" (SIR H. DOUGLAS.)

When a river, which cannot be forded, must be crossed by animals and carriages, a bridge becomes necessary; and in all cases it is better, if possible, to cross by a bridge than by a ford, unless the latter be exceedingly shallow. Military bridges may be of three kinds: 1st. Fixed structures of timber. 2d. Floating-bridges. 3d. Flying-bridges. Timber bridges may be either supported on piles or on trestles. Pile-bridges are the most secure, and where bridges are required to remain in use for a considerable period, as those which may be constructed on the lines of communication of an army, with its base of operations, this form of bridge will generally be adopted. To construct a good pile-bridge over a considerable river, much skilled labor is necessary, and an ample supply of materials essential. When the bottom of the channel is firm, and the river not subject to floods, a pile-bridge may be constructed without dfficulty, and will be very durable. The piles must be driven by an engine, which may be constructed of an 8-inch or 10-inch shell run full of lead, suspended by a rope over a pulley. This may be worked by hand, and will drive piles to a depth sufficient to allow of the passage of the heaviest artillery over the bridge. The pulley of the pile engine should be supported on a framework, some 16 feet high, which may be made to act as a guide to the shell during its fall, and also for the pile while it is being driven. This framework should be erected upon a large flat-bottomed boat. If such a boat is not to be procured, a raft must be made to answer the purpose. When timber of a considerable length can be procured for the joists of the bridge, it will be advisable to make the intervals between the piers or rows of piles, as great as the length of the joists will allow, so that the current of the river may be impeded as little as possible, and its action on the bridge be reduced to a minimum. By this arrangement, too, as much space as possible is given for the passage of floating bodies, and the danger of their damaging the bridge is proportionately diminished. When all the piles have been driven as far as the power of the engine can accomplish, they must be sawn off to the same level, and the superstructure of timber be strongly and carefully fitted. With bays of 20 feet, and a roadway 14 feet wide, there must be at least five or six beams not less than 7 inches by 8. With wider bays, timbers of larger dimensions will be necessary. The planking should not be less than 2 inches thick laid transversely. Bridges on piles, for the passage of infantry over shallow rivers only, may be expeditiously constructed, as the piles may be slight, 6 inches in diameter would suffice, and they can be driven by hand by heavy mauls, or by two men using a beetle. See diagram, Fig. 71.

Fig. 71.

Here the pile is set and kept in its place by means of two spars of planks resting their extremities upon a stool placed on the bank. A plank is then laid across, on which one or two men may stand to drive the pile. The weight of the men may be increased, if necessary, by stones placed on the platform assisting to force the piles into the ground. When one row of piles is placed, and the floor laid to a cross beam fixed upon them, another row may be set and driven in the same manner, fixing the stool on that part of the floor which will thus have been completed. Piles driven in this way may be safely depended upon to bear infantry with a front of two or three files in open ranks, not keeping step.

Bridges on Trestles.—When rivers are shallow, and not liable to sudden floods, and when their channels are firm and even, very useful bridges may be constructed on trestles. Trestles for this purpose should each consist of a stout transom or ridge piece some 8 inches square and 16 feet long; to this should be fitted four legs adapted to the depth of the river slanting outwards from the vertical, and strengthened by diagonal bracing, (Fig. 72.) For large bridges it will be found advantageous to add an additional pair of legs to each trestle. These, from the difficulty of fitting six legs to the uneven surface of the bottom of the river, should not be attached until the trestle is placed in position; they should then be driven into the bed of the river, and their upper extremities should be firmly nailed to the ridge piece. When the different parts of the trestles are all prepared beforehand, they can be speedily put together and the bridge completed with great expedition. Fascines may be used for flooring, where plank cannot be obtained. When the intervals or bays are ten feet, the dimensions of the trestle and beams may be as follows:—

Fig. 72.

		Length.	Breadth.	Thickness.
Trestles.	1 Head beam....................	16	8	8
	4 Legs.............................	*	$4\frac{1}{2}$	$4\frac{1}{2}$
	6 Braces..........................	—	—	—
	Balks.............................	12	$4\frac{1}{2}$	$4\frac{1}{2}$
	Planks for floor...............	12	12	2

If there be a strong current, a cable should be stretched across the river on each side of the bridge, and the trestles be firmly lashed to them. It may, moreover, sometimes be necessary to load the trestles with shot or stones, to keep them in their position until the flooring is laid upon them.

Floating-Bridges are those generally adopted for the passage of troops over rivers. They may be very expeditiously constructed, and can be made strong enough to carry the heaviest artillery. During the last century boats were generally used for this purpose; and, although on navigable rivers, boats are readily found, it was frequently a work of time and difficulty to collect a sufficient number, particularly if the enemy had had the opportunity of removing or destroying them previously. The inconveniences and delays resulting from this cause, always hazardous and often fatal to the success of an expedition, led to the introduction of regular bridge equipages or pontoon trains, duly organized to accompany the march of armies. An efficient pontoon train renders an army independent of the rivers which may intersect its route. By its aid rivers of very considerable magnitude may be bridged in a few hours, and a march of a given distance may thus be with certainty completed in a given time—a matter often of momentous importance to the success of military operations.

Bridges of Boats.—Boats of almost any kind will make a serviceable bridge. For wide rivers the boats should be large. The boats of which a bridge is constructed should, if possible, be nearly of the same size, unless they are all very large, and then variations in dimensions will be of little consequence. Should some be large and some small, the passage of large bodies of troops, of heavy guns and ammunition wagons will depress them unequally, causing the flooring of the bridge to assume an irregular line, straining and injuring, and in some cases fracturing, the timber and destroying the bridge. When boats, all of the same size, cannot be obtained, the larger boats should be placed at wider intervals, so that they may sustain a heavier weight, proportioned to their greater capacity, during the passage of troops, and be depressed to an equal distance with the smaller. The superstructure will consist of balks of timber laid across the gunwales of the boats, and securely fastened, and the flooring of planks laid transversely over. A certain rigidity results from this arrangement, by which, if the boats were subject to much motion, the bridge would be speedily destroyed. In tidal rivers, where a considerable swell must generally be encountered, this manner of securing the timbers will not answer. In this case, it will be found advantageous to erect a trestle or support in the centre of each boat, over which the timbers may be bolted to each other: thus each boat will be allowed independent motion, and this will not endanger the fracture of the bridge.

The boats should be moored head and stern, and should be kept at their relative distances by timbers fixed at the head and at the stern,

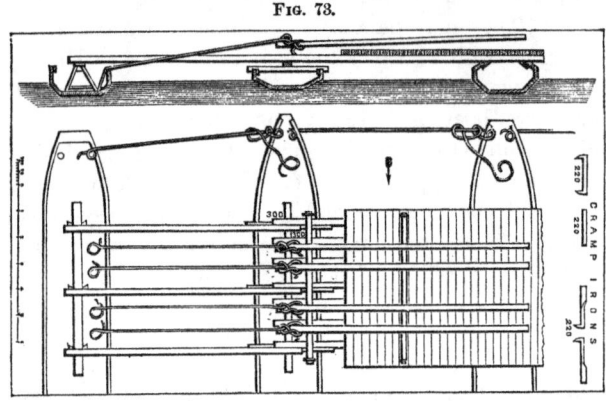

Fig. 73.

stretching across the bays, so as to remove unnecessary strain from the timbers of the bridge. The timbers should be as nearly as possible square, and of dimensions proportioned to the space of the intervals. With good timbers, 8 inches by 6, twenty feet may be allowed from trestle to trestle. The width of the bridge should also be proportioned to the dimensions of the timbers. With five balks of 7 inches by 8, the bridge should not exceed 14 feet in width. If too wide there will be danger of the beams being broken by the overcrowding of troops on the bridge.

When there is no regular pontoon train, and boats cannot be procured, rafts may be used in place of boats. These rafts may be made of *casks*, which, if properly arranged and securely lashed, will answer all the purposes of pontoons. Eight or ten casks, all of the same size, should be placed side by side on a level piece of ground, touching each other, bung-holes uppermost. Two stout balks, $4\frac{1}{2}$ inches square, and about 2 feet longer than the sum of the diameters of the casks which are to form the pier, must then be prepared and laid along the upper surface of the casks, parallel to each other, and each about a foot distant from the line of the bung-holes. A piece of 3-inch rope should then be attached to one end of each of these balks, passed under all the casks, and secured to the other end of the same balk.

These ropes are then drawn up towards the balks and tightly lashed by small ropes between every pair of casks, and the smaller ropes of the one side are again lashed across to those of the other side (Fig. 74.) The whole pier thus becomes so compact that it may be rolled and launched and rowed with as little danger of breaking up as though it were a single pontoon. Piers of casks constructed in this way may be used exactly like pontoons, and will form a most efficient bridge.

Fig. 74.

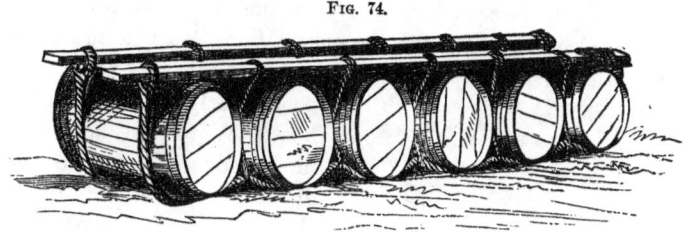

Pontoons are vessels of various forms and dimensions, and are made of various materials. They are generally boat-shaped, of wood, of copper, or of tin, sometimes with decks, and sometimes without. Each boat, or pontoon, is carried on a suitable wagon, which also conveys the portion of superstructure necessary for one bay or interval.

Flying-Bridges.—A flying-bridge is an arrangement by which a stream with a good current may be crossed, when, from a want of time or a deficiency of materials, it may not be possible to form a bridge. It consists of a large boat or raft firmly attached by a long cable to a mooring in the centre of the stream, if the channel be straight, or on the bank if the channel be curved. By hauling the boat or raft into proper positions, it will be driven across the stream in either direction as may be desired.

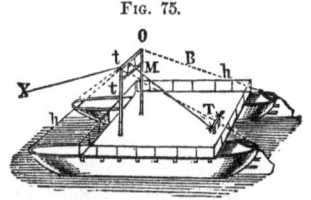

Fig. 75.

The bridge is made usually of two, (Fig. 75,) three, and sometimes six boats, connected together, and very solidly floored over, the beams being fastened to the gunwales of the boats with iron bolts or bands, and the flooring planks nailed down upon them. The floor is sometimes surrounded with a guard-rail. The most suitable boats are long, narrow, and deep, with their sides nearly vertical, in order to offer greater resistance to the action of the current. At the end of the rope is fixed an anchor X, which is moored in the channel, if this is in the middle of the stream. If the channel is not in the middle, the anchor is placed a little on one side of it toward the most distant shore. By means of the rudder, the bridge is turned in such a direction that it is struck obliquely by the current, and the force resulting from the decomposition of the action of the current makes it describe an arc of a circle around the anchor as a centre, and this force acquires its maximum effect when the sides of the boats make an angle of about 55° with the direction of the current.

Suppose M N (Fig. 76) to represent the side of the boat, and A B the resultant of the forces of the current against it. The force A B will be decomposed into two forces; the one, A C, will act in the direction M N as friction, and may be neglected, and the other, A D, will act perpendicularly to the side of the boat.

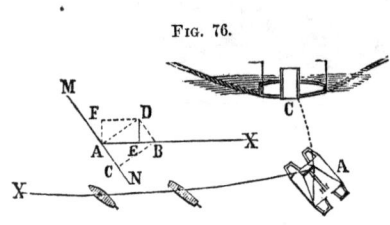

Fig. 76.

Were the boat free to move, and headed in the same direction, it would descend the river, at the same time crossing it. A D is then decomposed into two other forces, the one A E, in the direction of the current, causing the boat to drift, the other A F, perpendicular to this, which pushes the boat across. If the boat is now attached to a fixed point by the rope A X, the force A E will be neutralized, and all the effort of the current will be reduced to the force A F, which makes the boat revolve around the point X. The length of rope used should be once and a half or twice the width of the river. With a shorter rope the arc described by the bridge is too great, and it performs the ascending branch with difficulty; with a longer one, the rope becomes too heavy, sinks in the water, and fetters the movement. Generally, the arc described by the bridge should not be more than 90°. To prevent the rope from dragging over the deck, which would interfere with the load, it is held up by an arrangement such as is indicated in Fig. 76, and buoyed out of the water nearly to the anchor by skiffs, empty casks, or other floating bodies. When the stream to be crossed is not very wide, a flying-bridge may be made with two ropes, one fastened on each shore, the ropes being used alternately. If the stream, on the contrary, is very wide, several boats are fastened together, floored over, and anchored in the middle, and communication kept up with each shore by a flying-bridge, like the one already described. In about one hour 36 men can construct a flying bridge composed of 6 bridge-boats, and capable of carrying 250 infantry, or 2 pieces of artillery and 12 horses. At least one spare anchor should always be carried on the bridge, to anchor it in case the rope should break or become detached; and oars, a small boat, and a long rope, should also be provided. A flying-bridge may, in case of emergency, be made of any kind of boats with the means of fixing rudders to them. For want of an anchor, a large stone, mill-stone, or a bag or box of sand may be made use of. A flying-bridge may be made of a raft, the best form being lozenge-shaped, with the front angle about 55°. It is attached to a rope stretched across the stream by three others with pulleys, which slide along the first rope, this being tightly stretched across and not allowed to hang in the water. Buttresses constructed on boats or trestles, according to the means at hand, are formed on both sides of the river, at the points where the flying-bridge lands. Wagons impermeable to water may, by means of a rope attached to the wagon body, be used to pass a company with its baggage.

Where large bodies are to be crossed, a common contrivance is the RAFT of logs, but it is the last expedient to be adopted from its want of buoyancy and general manageability, and is inapplicable when the passage of a river is likely to be contested with animation. Its merits are that, at the expense of time, it can be constructed with less experienced workmen; it saves carriage, as it can only be made of materials near the spot. It is, however, an indifferent substitute for boats, pontoons, or casks. An independent raft will require two rows of trees, at least, to float as many men as can stand upon it, and the logs are best bound together by withes, or ropes, and stiffened with cross and diagonal traces.

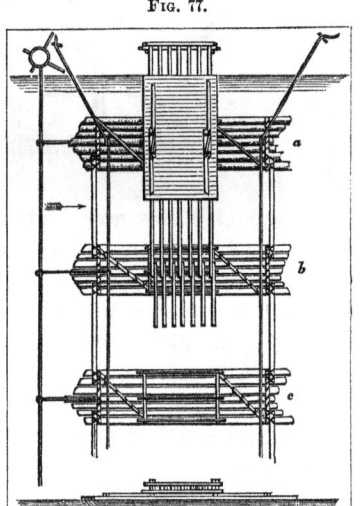

Fig. 77.

Timber Bridges.—The rudest form of arch is very strong, easy of construction, and of frequent occurrence; the timbers being roughly notched into each other as in log-houses, and gradually jutting over

Fig. 78.

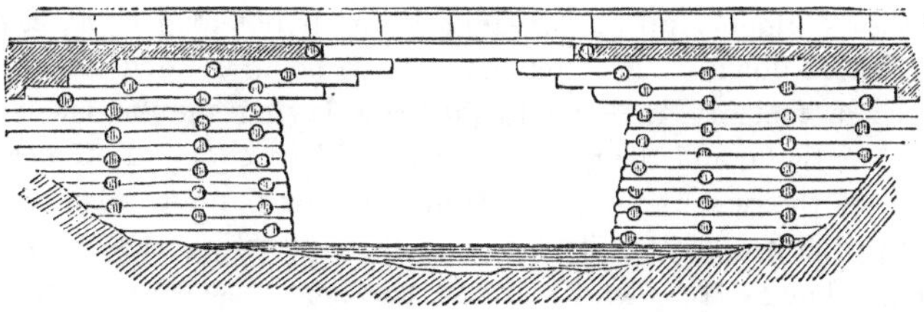

the pier or abutment near each other. A few of the upper courses may be trenailed down. Figure 79 shows the manner of construction with hewn or rough timber.

FIG. 79.

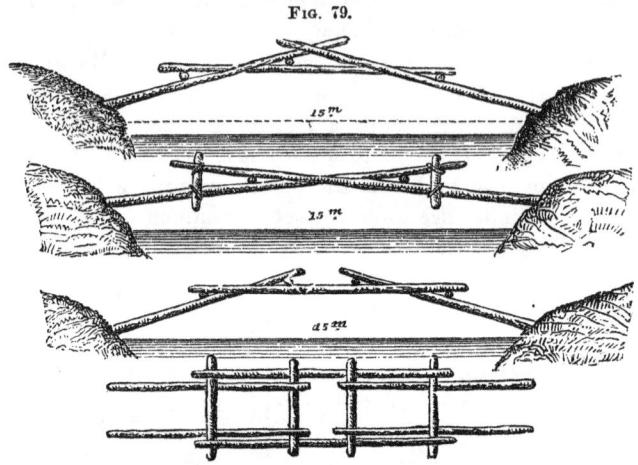

The wagon bodies now made for the United States army are galvanized or zincked iron; the lower and upper rails are of oakwood, covered with sheet iron; wooden supporters are framed into the lower rails like the usual wagon body, the tail piece is hung upon hinges. An important application of these iron wagon bodies, (suggested by Lieutenant-colonel Crossman, United States army,) would be their employment as boats in bridging rivers. If they are so perfected as to render them water-tight, they might be readily converted into a system of pontoons, each one carrying a portion of the string pieces and planks necessary to construct a bridge, without materially interfering with the usual load. Arranged and lashed together in double rows, they would afford a sufficient breadth of roadway for the passage of both cavalry and artillery with facility.

Large trees may be felled to enable infantry to cross narrow streams, placing them so that their butts may rest upon the banks with the top directed obliquely up the stream; if one is not long enough, others may be floated down so as to extend across, being guided and secured by ropes: a footway may be formed by laying planks, fascines, or hurdles over them, and their branches should be chopped off nearly to the level of the water and intertwined below; poles also may be driven into the bed of the river, to aid in supporting the trees by attaching the boughs to them. *Wheel carriages* used to form a foot bridge may be connected by beams; or a single pair of wheels with an axle-tree to admit two strong posts may be attached and placed in the centre of the stream if it is not too wide. Poles reaching from each bank may be secured to the posts, and the wheels would act as a trestle. With a flooring over the poles, a slight bridge could be

FIG. 80.

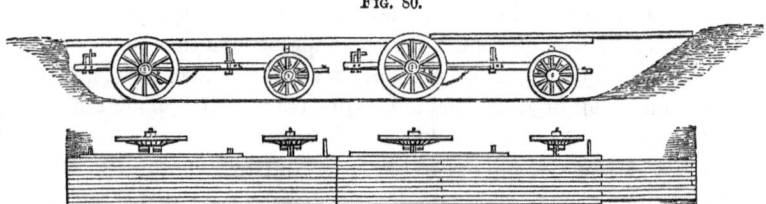

rapidly constructed for an advanced guard. *Hide boats* are made of four buffalo hides strongly sewed together with buffalo sinew, and stretched over a basket work of willow 8 feet long and 5 feet broad, with a rounded bow, the seams then being covered with ashes and tallow. Exposed to the sun for some hours, the skins contract and tighten the whole work. Such a boat with four men in it draws only four inches of water. *Inflated skins* have been used since the earliest times for crossing, and if four or more are secured together by a frame, they form a very buoyant raft. *Canvas* (rendered water-proof by a composition of pitch 8 lbs., beeswax 1 lb., and tallow 1 lb., boiled together and laid on quite hot) will serve as a raft or pontoon, if placed over framework or wicker work; (Consult *Memorial des Officiers d'Infanterie et Cavalerie; Aide Memoire of the Military Sciences;* DOUGLAS's *Principles and Construction of Military Bridges;* HYDE's *Fortifications;* GIBBON's *Manual;* HAILLOT, *Instruction sur le Passage des Rivieres et la Construction des Ponts Militaires.*)

BRIDGE-HEAD (*la tête du pont*)—is a work consisting of one or more redans or bastions, constructed on the bank of a river, to cover a bridge, to protect a retiring army in crossing the river, and to check an enemy when pressing upon it. (*See* REDAN.)

BRIDOON. The snaffle and rein of a military bridle, which acts independently of the bit, at the pleasure of the rider.

BRIGADE. Two regiments of infantry or cavalry constitute a brigade. (*Act* March 3, 1799.)

BRIGADIER-GENERAL. Rank next below major-general. The commander of a brigade. Entitled to one aide-de-camp.

BRIGADE-INSPECTOR. (*See* MILITIA.)

BRIGADE-MAJOR. An officer appointed to assist the general commanding a brigade in all his duties. (*See* MILITIA.)

BUILDING. (*See* BRIDGES; CARPENTRY.)

BUILDINGS, DEFENCE OF. The objects now under consideration are churches, country-houses, factories, prisons, or other substantial buildings; and as there is but little difference in the mode to be pursued for placing any of them in a state of defence, an explanation of the details applied to a single house will perhaps be sufficient to convey an idea on the subject. A building proper for defensive purposes, should possess some or all of the following requisites: 1. It should COMMAND all that surrounds it. 2. Should be SUBSTANTIAL, and of a nature to furnish materials useful for placing it in a state of defence. 3. Should be of an EXTENT PROPORTIONED TO THE NUMBER OF DEFENDERS, and only require the TIME AND MEANS which can be devoted to completing it. 4. Should have walls and projectings that mutually FLANK each other. 5. Should be DIFFICULT OF ACCESS on the side exposed to attack, and yet have a SAFE RETREAT for the defenders. 6. And be in a situation proper for fulfilling the object for which the detachment is to be posted. A church will be found usually to unite all these good properties more than any other building. It may be remarked that though good strong walls are an advantage, yet their thickness should be limited to 2 or 3 feet, from the difficulty there would be in piercing loopholes; unless when they are likely to be battered by artillery, in which case the musketry must be confined to the windows, and the more solid the walls are, the better. It should also be remembered that brick houses and walls are preferable, on several accounts, to those built of stone; for when exposed to artillery, a round shot merely makes a small hole in the former, but stone is broken up in large masses, and dangerous splinters fly from it in all directions. It is much easier also to make loopholes through brickwork than through masonry. Wooden houses, or those made of plaster, are to be avoided, from the facility with which an enemy can set fire to them, and they are frequently not even musket-proof. Thatched houses are equally objectionable, on account of fire, unless there is time to unroof them; and after all it must not be forgotten, that earthen works, when exposed to artillery, are to be preferred to houses, as far as affording security to the defenders is concerned. In seeking this security, however, it should be borne in mind that they are not so *defensible*—for troops cannot be run into a house; but they are not exempt from such an intrusion in an earthen work of the nature under discussion. The two together can be made to form a more respectable post than *either* can be made into singly, for the merits of both will be enhanced, and the defects be modified, by the union. A building is therefore at all times a capital base to go to work upon. The walls may be partially protected from cannon shot by throwing up earthen parapets round it, and the house may "reciprocate" by acting the part of a keep, and afford the garrison a place of refuge, in which they may either defend themselves with advantage, or if it "suits their book," resume the offensive and drive the assailants out again.

An officer will be able to make his selection at first sight, with reference to most of these points, but it requires a little more consideration to determine whether a building and its appliances are convertible into a post, of a size proportioned to the force under his command. The average number of men, however, proper for the defence of a house, may be roughly estimated on some such data as the following:—That in a lower story it might generally be proper to tell off one man for every 4 feet that the walls measured round the interior. In the second story one man for every 6 feet, and in an attic or roof one man for every 8 feet. For example, if a house of three stories high were found, on pacing it, to measure 140 feet round the interior walls, the number of men for its defence on the above data would be determined thus:—

Feet.

$\frac{140}{4}$ Would give 35; which would be the number of men for the lower story.

$\frac{140}{6}$ Would be about 23 men for the second floor.

$\frac{140}{8}$ Would be 18 men for the attic.

making a total of 76 men for the three stories; to which about one-sixth of the whole, say 14 men, should be added as a reserve, making altogether a garrison of 90 men. If there were out-buildings or walls in addition, the number of men required for their defence, would be determined in a similar manner, by assuming certain data adapted to the circumstances as a guide in the calculation. These numbers are not to be considered *definitive*, but merely to convey an idea on the subject; for if a detachment were much weaker in proportion to the extent, a vigorous defence might still be made. The force might be concentrated where most required, as it is not a matter of course that a place will be attacked on all sides at once; or if a building were found so large that the disposable force would be too much disseminated, or if there were a want of materials and time for putting the *whole* of it in a state of defence, a *part* of it only might be occupied. Should there exist any doubt about having sufficient time to complete all that might be wished, it would become matter for consideration what were the points which it would be of the greatest importance to secure first, so as to be in a condition to repel an *immediate* attack, because such points would naturally claim attention to the exclusion of all others. In such a case, it might be well to employ as many men as could work without hindering each other by being too crowded. 1. To collect materials and barricade the doors and windows on the ground floor, to make loopholes in them, and level any obstruction outside that would give cover to the enemy, or materially facilitate the attack. 2. To sink ditches opposite the doors on the outside, and arrange loopholes in the windows of the upper story. 3. To make loopholes through the walls generally, attending first to the most exposed parts, and to break communications through all the party-walls and partitions. 4. To place abatis or any feasible obstructions on the outside, and to improve the defence of the post by the construction of tambours, &c. 5. To place out-buildings and garden walls in a state of defence, and establish communications between them. To make arrangements in the lower story especially, for defending one room or portion after another, so that partial possession only could be obtained on a sudden rush being made. These different works to be undertaken *in the order of their relative importance*, according to circumstances; and after securing the immediate object for which they were designed, they might remain to be improved upon if opportunity offered. An endeavor will now be made to explain the mode of executing these works in the order in which they are mentioned.

Collecting Materials.—The materials that will be found most useful in barricading the passages, doors, and windows, are boxes, casks, cart bodies, bricks, stones, cinders, dung, &c., and timber of any sort that comes to hand; if they cannot be found elsewhere on the premises, the roof and floors must be stripped to furnish what is required.

Barricading Doors.—In the application of these materials, the boxes and casks filled with cinders or dung, and placed against the doors to a height of 6 feet, will prevent their being forced open, and loopholes may be made through the upper portions, which can be rendered musket-proof to protect the men's heads; short lengths of timber piled one upon another to the same height, leaving a space between any two of them in a convenient situation for firing through, and their ends being secured in the side walls of a passage, or propped with upright pieces on the inside, will effect the same object; or a door may be loosely bricked up, leaving loopholes, &c. If it is probable that artillery will be brought up for knocking away these barricades, and so forcing an entrance, a passage may be partially filled with dung or rubbish to the thickness of 8 or 10 feet, or thick beams of timber may be reared up on the outside of a door, and the interval filled with the same, or with earth if more convenient. A hole, about 3 feet square, may be left through an ordinary barricade for keeping up a communication with the exterior; but for effecting a retreat, or making sorties, it will be necessary to make a door musket-proof, by nailing on several additional thicknesses of plank, and arrange it so as to open as usual, or contrive something on the spot which shall equally protect the men when firing through the loopholes, and yet be removable at pleasure.

Barricading Windows.—Windows do not require to be barricaded so strongly as doors, unless from their situation an entrance may easily be effected, or an escalade be attempted. The principal object is to screen and protect the defenders whilst giving their fire; any thing, therefore, that will fill up the window to a height of 6 feet from the floor, and that is musket-proof, will answer the purpose. Thus two or three rows of filled sand-bags, laid in the sill of a window, Fig. 81, or

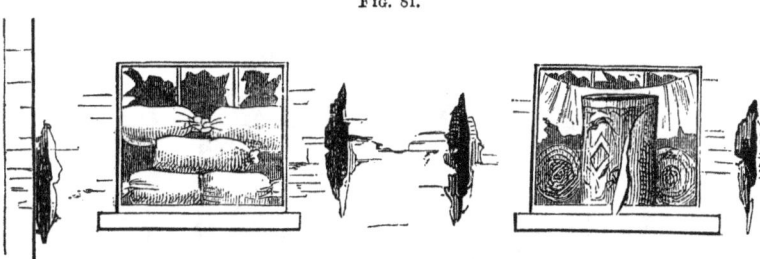

Fig. 81.

short lengths of timber would do; or a carpet, a mattrass, or blankets rolled up, would be ready expedients. Loopholes would, in all cases, be arranged whatever materials were used. If time presses, and windows could not be blocked up, one means of obtaining concealment, which is the next best thing to security, would be to hang a great coat or blanket across the lower part of them as a screen, and make the men fire beneath it, kneeling on the floor. The glass should be removed from windows before an attack commences, as it is liable to injure the defenders, when broken by musketry.

Levelling Obstructions outside.—Any shrubberies, fences, or outbuildings, within musket-shot, which would favor an attack by affording cover to an enemy, and allowing him to approach unperceived, should be got rid of as soon as possible. The trees should be felled, leaving the stumps of different heights, so as to encumber the ground, and the materials of walls, &c., should be spread about with the same view; but whatever is convertible for barricades should be carried to the house. The thatch from roofs, and any combustibles, should also be removed or destroyed.

Ditches in Front of the Doors, &c.—As a means of preventing a door being forced, a ditch may be dug in front of it, about 7 feet wide and 5 feet deep; such a ditch is also necessary in front of the lower windows, if the loopholes cannot be conveniently made high enough from the outside to prevent an enemy reaching them. These partial ditches may afterwards be converted into a continued ditch all round a house if opportunity offers, as it would contribute to the defence of the post. The floors may also be taken up on the inside, opposite the doors or windows open to attack.

Loopholes.—If the walls are not too thick, they may be pierced for loopholes, at every 3 feet, in the spaces between the windows, &c. (Fig. 82.)

FIG. 82.

Two tiers of these loopholes may be made if opportunity offers, and a temporary scaffolding of furniture, benches, casks, or ladders, &c., erected for firing from the upper ones: on the lower story a row of loopholes may be made close to the ground. The floor must, in this case, be partly removed, and a small excavation made between the beams for the convenience of making use of them. Just under the eaves of a roof there is generally a place where loopholes can be made with great facility, and a tile or slate knocked out here and there with a musket, will give other openings, from which an assailant may be well plied as he comes up.

Communications.—A clear communication must be made round the whole interior of the building, by breaking through all partitions that interfere with it: and for the same purpose, if houses stand in a row or street, the party walls must be opened, so as to have free access from one end to the other. Means should likewise be at hand for closing these openings against an enemy, who may have obtained any partial possession. Holes may also be made in the upper floors to fire on the assailants, if they force the lower ones, and arrangements made for blocking up the staircases, with some such expedient as a tree, prepared in the same manner as for an abatis, or by having a rough palisade gate placed across. Balconies may be covered or filled up in front with timber or sand-bags and made use of to fire from downwards. (Fig. 83.)

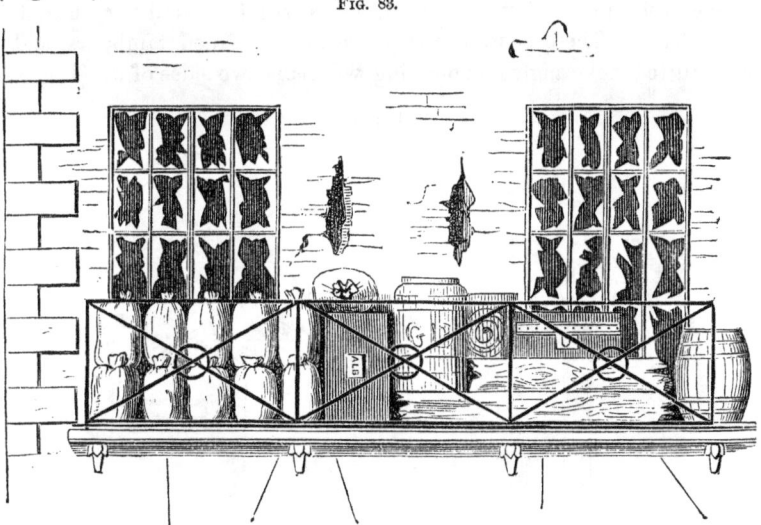

Fig. 83.

Abatis.—The partial levelling of any object on the outside, that would give concealment to an enemy, and favor an attack, is supposed to have been already attended to: but if time admits, after loopholes, &c. are completed, this system must be extended and perfected, and the formation of a more regular abatis should be commenced, and any other obstruction added that opportunity permits. The best distance for such obstructions, if they are continuous and cannot be turned, is within 20 or 30 yards of a work, or even less, so that every shot may tell whilst the assailants are detained in forcing a passage through them; within such a distance also of defenders securely posted, it would not be pleasant for a hostile force in confusion, to "*Fall in,*" or "*Re-form Column.*" If hand-grenades are to play their part in the defence of a post, the obstruction, whatever it may be, should be placed within their influence. A man will easily throw them 20 yards, but a trial on the spot will best determine the distance at which they can be used with effect.

Tambours.—If the building that has been selected has no porches, wings, or projecting portions from which flank defence can be obtained, it will be advisable to construct something of a temporary nature to afford it. Stockade work offers a ready means of effecting this object; it may be disposed in the form of a triangle, projecting 8 or 10 feet in front of a door or window, planted as described in Article STOCKADE, and with the precautions of having the loopholes high enough. A small hole should be left in the barricade of the door or window to communicate with the interior. Three or four loopholes on each face of the projection cut between the timbers will be found very useful in the defence. These contrivances are usually termed tambours, and if constructed at the angle of a building, will flank two sides of it. (Fig. 84.)

FIG. 84.

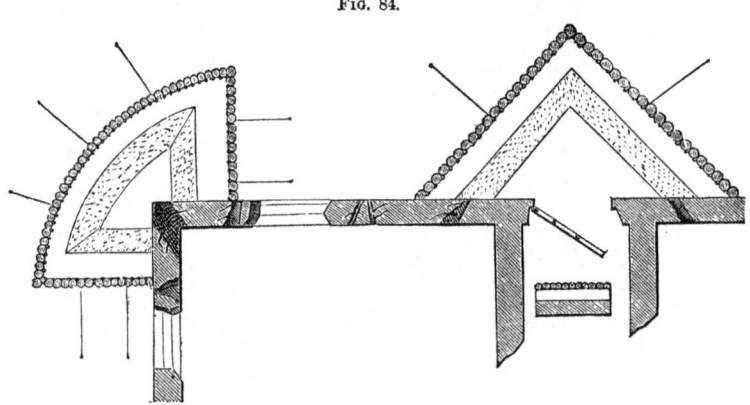

Out-buildings and Walls.—When the defences of the main building are in a state of forwardness, any out-buildings or walls which have been found too solid to be levelled at the moment, or which have been preserved for the chance of having time to fortify them, and thus to increase the strength of the post, must be looked to. They may be placed in a state of defence by the means already described, and separate communications should be established between them and the principal building by a trench, or a line of stockade work, and by breaking through the walls when necessary. In this way a post may be enlarged in any required proportion, by turning all objects that present themselves, such as out-buildings, sheds, walls, hedges, ponds, &c., to the best account; first taking the precaution to secure what is absolutely necessary for *immediate* protection, and for placing it in a state to be defended on the shortest notice. An exterior wall or fence, tolerably close to a house and parallel to it, may be retained for the purposes of defence, without the danger of affording cover, and thus facilitating an attack, by throwing up a slope of earth on the outside of it, or planting an abatis in the same situation; (Fig. 85.) An enemy would thus remain completely exposed, and it would be worse than useless to him. If a post of the description under consideration were composed of two or more buildings, and it were to be left to itself, and were open to attack on all sides, the stockades or trenches, forming the communications between them, would obviously require to be so arranged as to afford cover, and the means of resistance *on both sides*. This would be effected by merely making them *double*, as shown in Fig. 82; but for greater security, the exterior of such communications should be laid under fire from the buildings at their extremities. If cover cannot from circumstances be obtained, screens should be contrived that will conceal the movements that may be necessary. In arranging the defences of such posts, it is an essential point to make each portion of them so far independent of the others, that if any one part, such as a building for instance, be taken, it shall not compromise the safety of the remainder, nor materially impair the defence they will make by themselves; so that whilst free communications are essential in most cases to a vigorous defence, the means must be at hand for instantly cutting them off by some such expedients as would be afforded by a loopholed, musket-proof door, or rough gates, or by letting fall a tree, prepared as for an abatis, and which till wanted might be reared on its end in the situation required, the means of bringing a close fire upon it having been previously secured; (JEBB's *Attack and Defence*.)

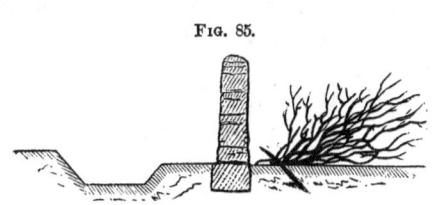

FIG. 85.

BULLET. (*See* AMMUNITION; ARMS; PERCUSSION BULLET; PROJECTILES; RIFLED ORDNANCE.)

BUNK. A word used in the army, a place for bedding.

BUREAU—of the War Department. During the absence of the quartermaster-general, or the chief of any military bureau of the War Department, his duties in the bureau, prescribed by law or regulations, devolve on the officer of his department empowered by the President to perform them in his absence; (*Act* July 4, 1836.)

BURIAL. The funeral honors paid to deceased officers and soldiers are prescribed by orders from the President contained in the Army Regulations. The coffin is furnished by the quartermaster's department.

BUSHING A GUN—is drilling a hole into the piece where the vent is usually placed, about one inch in diameter, and screwing therein a piece of metal which had previously a vent; the metal used in bushing is pure copper for brass pieces.

C

CADET. A warrant officer; students at the West Point Military Academy are cadets of the Engineer Corps. The number of cadets by appointments hereafter to be made shall be limited to the number of representatives and delegates in Congress and one for the District of Columbia; and each Congressional District, Territory, and District of Columbia shall be entitled to have one cadet at said Academy; nothing in this section shall prevent the appointment of an additional number of cadets, not exceeding ten, to be appointed at large, without being confined to a selection by Congressional Districts; (*Act* March 1, 1843, *Sec.* 2). Pay $30 per month. (*See* ACADEMY.)

CAISSON. The number of rounds of ammunition carried by each caisson and its limber are for 6-pounder guns 150 rounds; 12-pounder guns, 96 rounds; 12-pounder howitzers, 117 rounds; 24-pounder howitzer 69 rounds, and 32-pounder howitzers 45 rounds. The number of caissons with field-batteries are: with a battery of 12-pounders, 8 caissons for guns, and 4 for howitzers; and with a battery of 6-pounders, 4 for guns, and 2 for howitzers.

CALIBRE. The calibre of bullets is determined by the number required to weigh a pound. The calibre of guns is designated by the weight of the shot; siege and sea-coast howitzers, columbiads, mortars by the number of inches of their respective diameters. (Consult ORDNANCE MANUAL.)

CALLING FORTH MILITIA. Congress shall have power to provide for calling forth the militia to execute the laws of the Union, suppress insurrections, and repel invasions; (*Constitution, Art.* 1, *Sec.* 8, *Clause* 15.) By Act of Congress, Feb. 28, 1795, the President is authorized to call forth the militia whenever: 1.—"the United States shall be invaded or be in his judgment in imminent danger of invasion, (from any foreign nation or Indian tribe;) and to issue his orders for that purpose to such officer or officers of militia as he may think proper. 2.—In case of an insurrection in any State against the government thereof, on application of the Legislature of such State, or of the Executive, (when the Legislature cannot be convened.) 3.—Whenever the laws of the United States shall be opposed, or the execution thereof obstructed in any State, by combinations too powerful to be suppressed by the ordinary course of judicial proceedings, or by the powers vested in the marshals; but whenever it may be necessary, in the judgment of the President, to *use* the military force hereby directed to be called forth in case of insurrection or obstruction to the laws, the President shall forthwith, by proclamation, command such insurgents to disperse, and retire peaceably to their respective abodes within a limited time;" (*Act* Feb. 28, 1795.) In cases where it is lawful for the President to call forth the militia, it shall be lawful for him to employ for the same purposes, such part of the land or naval forces of the United States as shall be judged necessary, *having first observed all the pre-requisites of the law in that respect*; (*Act* March 3, 1807.) (*See* INVASION; MARSHAL; OBSTRUCTION; EXECUTION OF LAWS; INSURRECTION.)

CAMEL. The camel is used in the East as a beast of burthen from 3 to about 16 years of age, and in hot sandy plains, where water and food are scarce, is invaluable. With an army, however, generally speaking, it is not so valuable as the mule or horse. The camel under a burthen is very slow-going, about half the pace of a mule, or from $1\frac{1}{2}$ to 2 miles per hour; he can, however, travel 22 out of the 24 hours, and only requires food once a day. His load varies exceedingly in different countries. In Egypt it is as high as 10 cwt.; and for the short distance from Cairo to Boulac, even 15 cwt. is, it is said, sometimes carried. But in Syria it rarely exceeds 500 lbs., and the heaviest load in the engineer equipment for the British army of the Indus is stated to be 4 cwt. 48 lbs., independent of the pack-saddle. About 400 lbs. is a sufficient load on the march. The pack-saddle or pad is secured in its place by the hump on the back, a hole being made in the pad to let it come through, also by a breast-plate and breeching; no dependence is placed on the girth, which is not kept tight. From the great size of the camel, averaging about 7 feet to the top of the hump, and 8 feet from his nose to his tail, when standing in a natural position, he is capable of carrying light field artillery, and the 12-pounder mountain howitzer, which, with its side arms, weighs from 330 to 350 lbs. The bed or carriage is carried by a second, and the ammunition by a third camel. In rocky or slippery ground the camel is apt to slip, and his fore feet then are frequently spread out right and left: when this is the case, he splits up inside the arms, and dies, or becomes useless. Though patient and obedient to his keeper, at whose command he lies down to be loaded, he is frequently very savage with strangers, and his bite is very severe. The camels introduced into the service of the United States on our Western frontiers, carry from 300 to 600 lbs. on continuous journeys, depending on the kind of camel employed. These weights they will carry from 18 to 30 miles a day, according to the character of the country. With lighter loads they travel a little faster. The saddle dromedary will travel 50 miles in 8 or 10 hours; and on an emergency they make 70 or 90 miles a day, but only for a day or two, on a level road. Their use in the United States is still an experiment.

CAMOUFLET—is a small mine, of about 10 lbs. of powder, sufficient to compress the earth all around it, without disturbing the surface of the ground. It is sometimes formed in the wall or side of an enemy's gallery, in order to blow in the earth, and to cut off the retreat of the miner.

CAMP—is the temporary place of repose for troops, whether for one night or a longer time, and whether in tents, in *bivouac*, or with any such shelter as they may hastily construct, as sheds, bowers, &c. Troops are *cantoned* when distributed at any time among villages, or when placed in huts at the end of campaign. *Barracks* are permanent military quarters. *Tents* (says Napoleon) are not wholesome. It is better for the soldier to bivouac, because he can sleep with his feet towards the fire, and he may shelter himself from the wind by means of sheds, bowers, &c. In woods there is great facility in making warm encampments, even in the most bitter weather. A young tree, when felled, yields poles to support branches as shields against weather, and flooring above the snow or damp. A common arrangement is as follows:—A cross-bar is supported by two uprights; against this cross bar a number of poles are made to lean; on the back of the poles abundance of fir branches are laid horizontally; and, lastly, on the back of the fir branches are

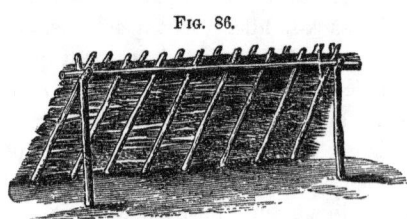
Fig. 86.

another set of leaning poles, in order to make all secure by their weight. A cloth of any kind is made to give shelter by an arrangement of this kind. The corners of the cloth should be secured by a simple hitch in the rope and not by a knot. The former is sufficient for all purposes of security, but the latter will jam, and you may have to injure both cloth and string to get it loose again. It is convenient to pin a skewer in the middle of the sides of the cloth, round the ropes.

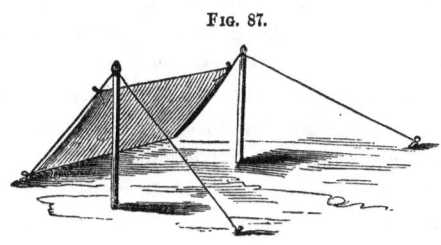

Fig. 87.

Good water within a convenient distance is essential in the selection of a camp, as is also the proximity of woods for firewood, material for shelter, &c. Good roads, canals, or navigable streams are important to furnish the troops with the necessaries of life if troops are encamped for long periods. The ground should not be near swamps or stagnant water. This requirement is essential to health. The ground, to be suitable for defence, must admit the manœuvres of troops. The front of the camp of each battalion of infantry or squadron of cavalry must, therefore, be equal to the front of the battalion or squadron. And as far as possible camps for cavalry and infantry should be established on a single line—the cavalry upon the wings, the infantry in the centre. The shelters or huts are alligned, *as well as the nature of the ground admits*, from one extremity of the camp to the other, and arranged by companies in streets, perpendicular to the front. The general thus has the whole extent of his camp in view, and order can be better preserved. When the army is formed upon two lines, there are two camps—one in front of the other. The reserve has also its particular camp. Artillery usually encamps behind the infantry, and thus forms a little separate camp or camps of its own. In establishing a camp, however, no universal rule can be laid down; but it is necessary (says Napoleon) that the genius of the commander should, according to circumstances, decide whether an army ought to be confined to one single encampment or to form as many as it has corps or divisions; where the vanguard and flanks should be posted; where the cavalry, artillery, and wagons should be placed, and whether the army should occupy one or more lines; what should be the distance between the lines; and whether the cavalry should be in reserve behind the infantry, or should be placed on the wings.

Baron Larrey suggests the following sanitary considerations in relation to camps: A camp, especially if permanent, should be selected so as to be accessible to the troops by easy marches; it should occupy a spacious plain, in a province exempt from both epidemical and endemical diseases; the soil should be dry, but not too hard, so that it may quickly imbibe the rain; because it then becomes fit for military operations a few hours after the most violent shower. This prompt absorption, moreover, preserves the troops from the baneful influence of dampness without exposing them to the inconveniences of want of water, since in such a soil wells may be easily dug and water found at an inconsiderable depth, as is the case at Chalons. A good camp should not be intersected by streams or ditches, nor enclosed by large forests. The tents should not be too closely packed, in order to insure good ventilation throughout, and diminish the probability of epidemics. When a river is too near a camp, and its banks are somewhat marshy, the breaking out of intermittent fever should be prevented by deepening the bed of the river, cleansing it as much as possible of all putrefying vegetable and animal substances, raising the banks and giving them at the same time a greater inclination, making channels for carrying off the water, and establishing tents and barracks at a sufficient distance, and as much as possible on rising ground. When the supply of water to a camp is derived from a river, the latter ought to be divided into three sections: the first and upper one to be exclusively used for drink by the men, the second to be reserved for the horses, and the third and lowermost for washing the linen of the troops. These demarcations should be strictly guarded by sentinels stationed at the proper places. To drive off dampness, bivouac-fires ought to be lighted in the evening; each tent, moreover, should be surrounded with a gutter communicating with a main ditch to carry off rain-water; the space occupied by certain corps should also be sanded over, to facilitate the absorption of humidity by the soil. In pitching tents care should be taken to maintain between them a distance of at least two metres; those of the general officers should be situated in the healthiest quarter. Tents made of white stuff are prejudicial to the eyesight in summer, and should be therefore discarded. A tent being liable to infection like a room, it ought not to be hermetically closed, as is the custom with soldiers, but, on the contrary, well aired; and the ground ought not only to be scraped and swept, but should also be well rammed. The men ought not to sleep in the tents with their heads near the centre and their feet towards the circumference, but in the contrary position, else they breathe a vitiated instead of a pure air. A tent, generally calculated for 16 men, ought never to contain more than 12 or 13 infantry, and 8 or 10 cavalry. Of the different kinds of tents the conical Turkish tent is the best; for ambulances the marquee is preferable. The *tente-d'abri*, which is made by joining two camp-sacks together by means of a wooden pole, and keeping them stretched by small stakes stuck into the ground, is a most precious invention. Four men can find shelter under it, and the weight it adds to their kit is trifling, but it can only be used in provisional encampments. The tents of the cavalry ought to be freed from the encumbrance of saddles and accoutrements, which vitiate the air, and should be placed under small sheds in front of the tents, or, better still, in the stable-barracks. The men should be encouraged to cultivate little patches of ground around their tents as gardens; it is both an amusement and a means of purifying the air, only they must not be allowed to manure the soil. As regards sleeping, each soldier should fill a camp-sack with straw and lie down on it as on a mattress, with his blanket to cover him; or, better still, he should get into the sack filled with straw—a much better plan than allowing the men to sleep together in couples on two sacks spread out on the straw, and with the same blanket to cover them. The ground on which the men sleep ought to be swept daily and sanded over, for it easily gets infected; in which case it becomes necessary to shift the tents—a measure which is often sufficient to stop an epidemic at its outbreak. A reserve of planks and trestles ought to be kept in store for extempore bedsteads when the ground has become too damp; or water-proof canvas may be spread over to protect the straw from humidity. In autumn a single blanket is not sufficient, each man should be provided with two.

The guards of camps are: 1. The *Camp-guard*, which serves to keep good order and discipline, prevent desertions and give the alarm; 2. Detachments of infantry and cavalry, denominated *pickets*, in front and on the flanks, which intercept reconnoitring parties of the enemy, and give timely notice of the approach of an enemy; and 3. *Grand-guards*, or out-posts, which are large detachments posted in surrounding villages, farm-houses or small field-works, from which they can watch the movements of the enemy. They should not be so far from the camp as to be beyond succor in case of attack, and not so near as to prevent timely notice being given to the main body of the army on the approach of an enemy. If the camp is to present the same front as the troops in order of battle, 400 military paces will be necessary per regiment of 500 files front. Immediately after arriving on the ground, the number of men to be furnished for guards and pickets are detailed; the posts to be occupied by them are designated; the places of distribution of provisions are mentioned, and, in general, all arrangements made concerning the interior and exterior police and service of the camp.

The *tente-d'abri* has been introduced in the French service since 1837, when first used at the camp of Compiegne. These tents consist of a tissue of cotton cloth impregnated with caoutchouc, and thus made water-proof. Every man carries a square of this cloth, with buttons and button-holes around it, by which it is attached to the squares carried by his comrades, and an excellent shelter for six soldiers is made as follows:—Three tent-sticks are fixed into the ground, whose tops are notched; a light cord is then passed round their tops, and fastened into the ground with a peg at each end; (Fig. 88.) Two sheets, A and B, are buttoned together and thrown over the cord, and then two other sheets, C and D; and C is buttoned to A, and D to B. Lastly, another sheet is thrown over each of the slanting cords, the one buttoned to A and B, and the other to C and D; (Fig. 89.) The sides of the tent are of course pegged to the ground.

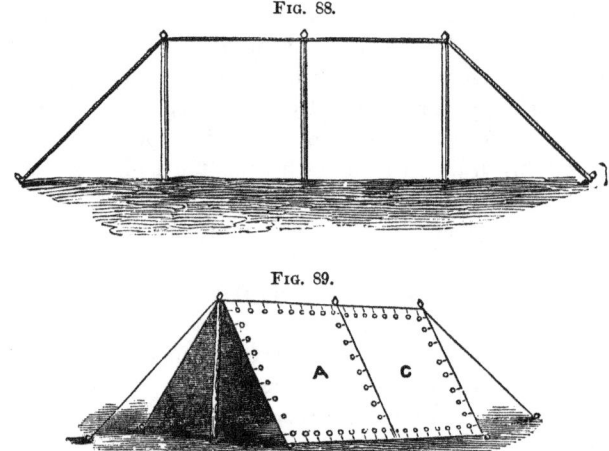

Fig. 88.

Fig. 89.

There are many modifications in the way of pitching these tents. For want of sticks, muskets can be used.

Preparations for a Storm.—Before a storm, dig a ditch as deep as you can, round the outside of the tent, to turn aside the rain-water, and to drain the ground on which the tent is standing—even a furrow scratched with a tent-peg is better than nothing at all. Fasten guy-ropes to the spike of the tent-pole; and be careful that the tent is not too much on the strain, else the further shrinking of the materials, under the influence of the rain, will certainly tear up the pegs. Earth, banked up round the bottom of the tent, will prevent gusts of wind from finding their way beneath. The accompanying sketch shows a tent pitched for a lengthened habitation. It has a deep drain, a seat and table dug out, and a fireplace. (Fig. 90.)

Tent Furniture.—A portable bedstead, with musquito-curtains, is a very great luxury, raising the sleeper above the damp soil, and the attacks of most creatures that creep on it; where a few luxuries can be carried, it is a very proper article of baggage. It is essential where white ants are numerous. Hammocks and cots have but few advocates, as it is rare to find places adapted for swinging them; they are quite out of place in a small tent.

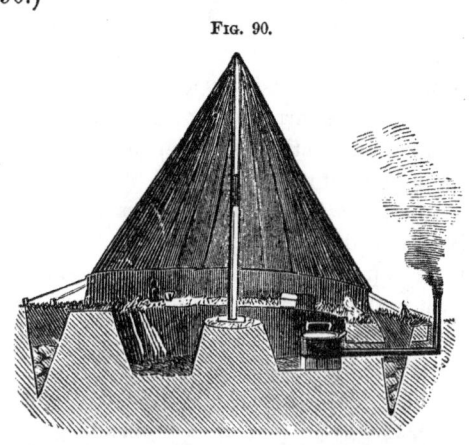

Fig. 90.

Chairs and Tables.—It is advisable to take very *low* strong and *roomy* camp-stools, with tables to correspond in height, as a chamber is much less choked up when the seats are low, or when people sit, as in the East, on the ground. The seats should not be more than 1 foot high, though as wide and deep as an ordinary footstool; but without a seat, a man can never write, draw, nor calculate as well as if he has one. The stool represented in Fig. 91 is a good one; it has a full-sized seat made of leather or canvas, or else of strips of dressed hide. For want of a chair, it is convenient to dig a hole or a trench in the ground, and to sit on one side of it, with the feet resting on its bottom; the opposite side of the trench serves as a table, for putting things on, within easy reach.

Fig. 91.   Fig. 92.

To tie clothes, or any thing, up to a smooth tent-pole, a strap with hooks in it, to buckle round the pole, is very convenient. The method shown in Fig. 92 suffices, if the pole is notched, or jointed, or in any way slightly uneven. Bags, &c., are hung upon the bit of wood that is secured to the loose end. The luxuries and elegancies practicable in tent life are only limited by the means of transport. The articles that make the most show are handsome rugs, and skins, and pillows; canteens of dinner and coffee services, &c.; and candles, with screens of glass, or other arrangements to prevent them from flickering. The art of luxurious tenting is better understood in Persia than in any other country, even than in India.

Losing things.—Small things are constantly mislaid and trampled in the sand: to search for them, the ground should be disturbed as little as possible—it is a usual plan to score its surface in parallel lines with a thin wand. It would be well worth while to make and use a small light rake for this purpose.

Huts.—In making a depot, it is usual to build a house; often the men have to pass weeks in inactivity, and they may as well spend them in making their quarters comfortable, as in idleness. Whatever huts the natives live in are sure, if made with extra care, to be sufficient for travellers.

Walls.—The materials whence the walls of huts may be constructed, are very numerous, and there is hardly any place which does not furnish one or other of them. Those principally in use are as follows: Skins, canvas, felt, tarpauling, bark, reed mats, reed walls, straw walls, wattle-and-dab, log-huts, fascines or fagots, boards, &c., fastened by Malay-hitch, brick, sunburnt or baked, turf, stones, gabions, bags or mats filled with sand or shingle, snow huts, underground huts, tents over holes in earth.

Roofs.—Many of the above list would be perfectly suitable for roofs: in addition may be mentioned slating with flat stones, thatch, sea-weed, and wood shingles.

Floors.—Cowdung and ashes make a hard, dry, and clean floor, such as is used for a threshing-floor. Ox-blood and fine clay, kneaded together, are excellent; both these compositions are used in all hot, dry countries.

Tarpaulings, made in the sailors' way, are much superior to others in softness and durability. As soon as the canvas is sewn together, it is thoroughly wetted with sea-water; and, while still wet, is done over on one side with tar and grease boiled together—about two parts tar and one of grease. Being hung up till dry, it is turned; and the other side, being a second time well wetted, is at once painted over with the tar and grease just as the first side had been done before. The sailors say that " the tar dries in as the water dries out."

Bark.—It is an art to strip it quickly—the Australians understand it well. Two rings are cut round the tree; the one as high as can be reached, the other low down. A vertical slit is then made, and the whole piece forced off with axes, &c. In spring the bark comes off readiest from the sunny side of the tree. A large sheet of bark is exceedingly heavy. It is flattened, as it lies on the ground, by weighting it with large stones, and allowing it to dry, partially at least, in that position.

Straw Walls of the following kind are very effective, and they have the advantage of requiring a minimum of string (or substitute for string) in their manufacture. The straw, or herbage of almost any description, is simply nipped between two pair of long sticks, which are respectively tied together at the two ends, and at a sufficient number of intermediate places. The whole is neatly squared and trimmed; (Fig. 93.) A few of these would help in finishing the roof or walls of a house. They can be made movable, so as to suit the wind, shade, and aspect. Even the hut door can be made on this principle.

Log-huts.—In building log-huts, four poles are planted in the ground to correspond to the four corners: against these, logs are piled one above another, as in Fig. 94; they are so deeply notched where they

Fig. 93. Fig. 94.

cross one another, that the adjacent sides are firmly dovetailed together. When the walls are entirely completed, the doors and windows are chopped out, and the spaces between the logs must be well caulked with moss, &c., or the log-cabin will be little better than a log-cage. It of course requires a great many trees to make a log-hut; for, supposing the walls to be 8 feet high, and the trees to average 8 inches in diameter, it would require 12 trees to build up one side, or 48 to make all four walls.

Malay hitch.—I know no better name for the following wonderfully simple way of attaching together wisps of straw, rods, laths, reeds, planks, poles, or any thing of the kind, into a secure and flexible mat;

the sails used in the far East are made in this way, and the movable decks are made of bamboos joined together with a similar but rather more complicated stitch; (Fig. 95.) Soldiers might be trained to a great deal of hutting practice in a very inexpensive way if they were drilled at putting together huts whose roofs and walls were made of planks lashed together by this simple hitch, and whose supports were short scaffolding-poles planted in deep holes dug without spades or any thing but the hand and a small stick. The poles, planks, and cords might be used over and over again for an indefinite time. Further, bedsteads could be made in a similar way by short cross planks lashed together, and resting on a framework of horizontal poles lashed to uprights planted in the ground. The soldier's bedding would not be injured by being used on these bedsteads, in the way it would be if laid on the bare gound. Many kinds of designs and experiments in hutting could be practised without expense in this simple way.

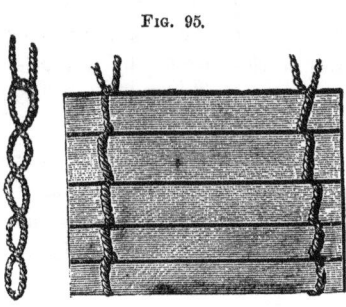

FIG. 95.

Snow-houses.—Few travellers have habitually made snow-houses, except Sir J. Franklin's party, and that of Dr. Rae. Great praises are bestowed on the comfort of them by all travellers, but skill and practice are required in building them. The mode of erection of these dome-shaped buildings is as follows:—It is to be understood that the hard, compact, underlying snow is necessary for the bottom of the hut; and that the looser textured, upper layer of snow is used to build the house. First, select and mark out the circular plot on which the hut is to be raised. Then, cut out with knives deep slices of snow, six inches wide, three feet long, and of a depth equal to that of the layer of loose snow, say one or two feet. These slices are curved, so as to form a circular ring when placed on their edges, and of a size to make the first row of snow-bricks for the house. Other slices are cut for the succeeding rows; and, when the roof has to be made, the snow-bricks are cut with the necessary double curvature. A conical plug fills up the centre. Loose snow is then heaped over the house, to fill up crevices. Lastly, a doorway is cut out with knives; also a window, which is glazed with a sheet of the purest ice at hand. For the inside accommodation, there is a pillar or two, to support lamps.

Underground Huts are used in all quarters of the globe. The experience of the British troops encamped before Sebastopol tells strongly in their favor, as habitations during an inclement season. The timely adoption of them was the salvation of the British army. They are, essentially, nothing else than holes in the ground, roofed over. The shape and size of the hole correspond to that of the roof it may be possible to procure for it; its depth is no greater than requisite. If the roof have a pitch of 2 feet in the middle, the depth of the hole need not exceed 4½ feet. In the Crimea, the holes were rectangular, and roofed like huts; (Fig. 96.) Where there is a steep hill side, a, a, an underground hut, b, is easily contrived; because branches laid over its top have sufficient pitch to throw off the rain, without having recourse to any uprights, &c. Of course the earth is removed from a, at the doorway.

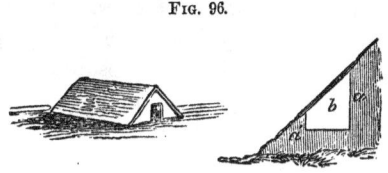

FIG. 96.

Tents pitched over excavations.—A hole may be dug deeply beneath the tent floor, partly as a store-room, and partly as a living-room when the weather is very inclement. This, also, was done before Sebastopol in the manner shown in the engraving.

Thatching.—After the framework of the roof has been made, the thatcher begins at the bottom, and ties a row of bundles of straw, side by side, on to the framework. Then he begins a second row, allowing the ends of the bundles composing it to overlap the heads of those in the first row.

Wood Shingles are tile-shaped slices of wood, easily cut from fir-trees, and used for roofing on the same principle as tiles or slates.

Fix hooked sticks, and cow or goat horns, round the walls, as pegs to hang things on; and if you want a luxurious bed, make a framework of wood, with strips of raw hide lashed across it from end to end, and from side to side; (Fig. 97.) If you collect bed feathers, recollect that if

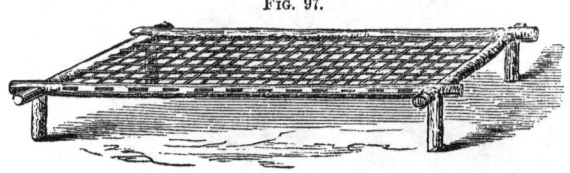

FIG. 97.

cleanly plucked they require no dressing of any kind, save drying and beating. Concrete for floors is made of 80 parts large pebbles, 40 river sand, 10 lime; lime is made by burning limestone, chalk, shells, or coral, in a simple furnace, and whitewash is lime and water. Bark makes a good roof. The substitutes for glass are—waxed or oiled paper or cloth, bladder, fish-membranes, talc, and horn. Glass cannot be cut with any certainty without a diamond; but it may be shaped and reduced to any size by gradually chipping, or rather biting, away at its edges with a key, if the slit between its wards be just large enough to admit the pane of glass easily. A window, or rather a hole in the wall, may be rudely shuttered by a stick run through loops made out of wisps of grass. In hot weather the windows of the hut may be loosely filled with grass, which, when well-watered, makes the hut much cooler. A mosquito-curtain may be taken and suspended over the bed, or place where you sit. It is very pleasant, in hot, mosquito-plagued countries, to take the glass sash entirely out of the window frame, and replace it with one of gauze. Broad network, if of fluffy thread, keeps wasps out. The darker a house is kept, the less willing are flies, &c., to flock in. If sheep and other cattle be near the house, the nuisance of flies, &c., becomes almost intolerable; (GALTON's *Art of Travel.*)

Major H. H. Sibley, 2d Dragoons, has invented a tent in which a fire can be made in its centre, and all soldiers sleep with their feet to the fire. Major Sibley's tent is conical, light, easily pitched, erected on a tripod holding a single pole, and will comfortably accommodate twelve soldiers with their accoutrements. Where means of transportation admit of tents being used, Major Sibley's will probably supersede all others. (Fig. 98.)

FIG. 98.

A commander of troops usually sends in advance to prepare the camp. The camping party of a regiment may be the regimental quartermaster, and quartermaster-sergeant, and a corporal and two men per company. The camp of a larger detachment is prepared by the chief quartermaster or some officer of the general's staff, designated by the commander of the troops assisted by the company camping parties of regiments. With camp colors the direction of the front line of the camp is marked, and the extent of the front of each corps, the intervals between corps, and the beginning, breadth, and direction of streets designated. When the encampment is on two lines, let there be 450 paces between their respective fronts. Behind intrenchments there ought to be about 300 paces between the entrenchments and the front of the camp. The posts of the police guard will be designated, and the necessary works to secure communication between the parts of the camp will also be determined. Fig. 99 gives details for the camp of a regiment of infantry.

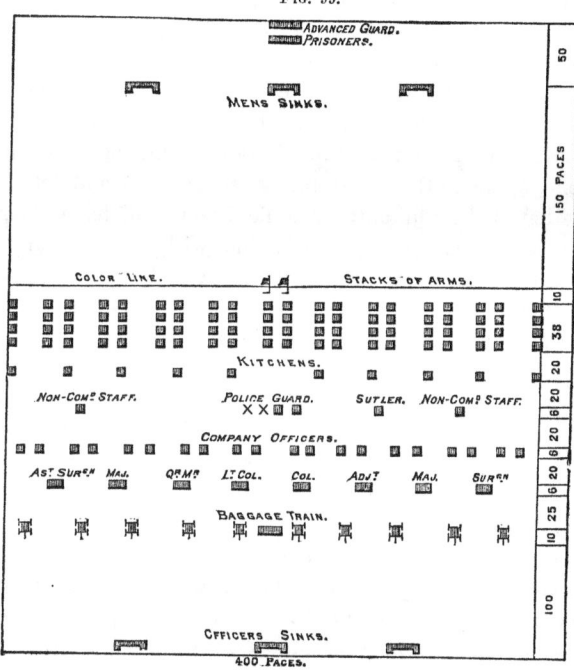

Fig. 99.

Camp of Cavalry.—In the cavalry, each company has one file of tents—the tents opening on the street facing the left of the camp. The horses of each company are placed in a single file, facing the opening of the tents, and are fastened to pickets planted firmly in the ground, from 3 to 6 paces from the tents of the troops. The interval between the file of tents should be such that, the regiment being broken into columns of companies, each company should be on the extension of the line on which the horses are to be picketed. The streets separating the squadrons are wider than those between the companies by the interval separating squadrons in line; these intervals are kept free from any obstruction throughout the camp. The horses of the rear rank are placed on the left of those of their file-leaders. The horses of the lieutenants are placed on the right of their platoons; those of the captains on the right of the company. Each horse occupies a space of about 2 paces. The number of horses in the company fixes the depth of the camp, and the distance between the files of tents; the forage is placed between the tents. The kitchens are 20 paces in front of each file of tents. The non-commissioned officers are in the tents of the front rank. Camp-followers, teamsters, &c., are in the rear rank. The police guard in the rear rank, near the centre of the regiment. The tents of the lieutenants are 30 paces in rear of the file of their company; the tents of the captains 30 paces in rear of the lieutenants. The colonel's tent 30 paces in rear of the captains', near the centre of the regiment; the lieutenant-colonel on his right; the adjutant on his left; the majors on the same line, opposite the 2d company on the right and left; the surgeon on the left of the adjutant. The field and staff have their horses on the left of their tents, on the same line with the company horses; sick horses are placed in one line on the right or left of the camp. The men who attend them have a separate file of tents; the forges and wagons in rear of this file. The horses of the train and of camp-followers are in one or more files extending to the rear, behind the right or left squadron. The advanced post of the police guard is 200 paces in front, opposite the centre of the regiment; the horses in one or two files. The sinks for the men are 150 paces in front—those for officers 100 paces in rear of the camp.

Camp of Artillery.—The artillery is encamped near the troops to which it is attached, so as to be protected from attack, and to contribute to the defence of the camp. Sentinels for the park are furnished by the artillery, and when necessary, by the other troops. For a battery of six pieces the tents are in three files—one for each section; distance between the ranks of tents 15 paces; tents opening to the front. The horses of each section are picketed in one file, 10 paces to the left of the file of tents. In the horse artillery, or if the number of horses make it necessary, the horses are in two files on the right and left of the file of tents. The kitchens are 25 paces in front of the front rank of tents. The tents of the officers are in the outside files of company tents, 25 paces in rear of the rear rank—the captain on the right, the lieutenants on the left. The park is opposite the centre of the camp, 40 paces in rear of the officers' tents. The carriages in files 4 paces apart; distance between ranks of carriages sufficient for the horses when harnessed to them; the park guard is 25 paces in rear of the park. The sinks for the men 150 paces in front; for the officers 100 paces in rear. The harness is in the tents of the men. (Consult BARDIN; *Memorial des Officiers d'Infanterie et de Cavalerie;* GALTON's *Art of Travel.*)

CAMP AND GARRISON EQUIPAGE. (*See* CAMP; CLOTHING; TOOLS; UTENSILS; QUARTERMASTER'S DEPARTMENT.)

CAMPAIGN. The period of a year that an army keeps the field from the opening of a campaign until the return to quarters or cantonments at the end of the campaign. A series of continuous field operations. An ordinary campaign, in respect to recompense for length of service, is counted as two years of effective service in the French army. In all services excepting our own, additional allowances in campaign are made to troops beyond those given at other periods. (*See* ALLOWANCES.)

CANISTER—for field service, consists of a tin cylinder attached to a sabot, and filled with cast-iron shot. For siege and garrison guns the bottom is of cast iron, and the cover of sheet iron with a handle made of iron wire. (*See* SABOT.)

CANNON. (*See* CALIBRE; ORDNANCE.)

CANTEEN. A small tin caoutchouc or circular wooden vessel, used by soldiers on active service to carry liquor, &c. A small trunk or chest, containing culinary and other utensils for the use of officers. A kind of suttling house, kept in garrisons, &c., for the convenience of the troops.

CANTONMENTS. Troops are said to be in cantonments when detached and quartered in the different towns and villages, lying as near as possible to each other. (*See* CAMP.)

CAPITAL. The line drawn bisecting the salient angle of a work.

CAPITULATION. Articles of agreement, by which besieged troops surrender at discretion, or with the honors of war. The terms granted depend upon circumstances of time, place, &c. Any surrender in the open field without fighting was stigmatized by Napoleon as dishonorable, as was also the surrender of a besieged place without the advice of a majority of a council of defence, before the enemy had been forced to resort to successive siege-works, and had been once repulsed from an assault through a practicable breach in the body of the place, and the besieged were without means to sustain a second assault; or else the besieged were without provisions or munitions of war.

CAPONNIERE. Passage from the place to an outwork; it is either single or double, sometimes bomb-proof and loopholed. (*See* FORTIFICATION.)

CAPS. Percussion caps for small arms are formed by a machine which cuts a star or *blank* from the sheet of copper, and transfers it to a die in which the cap is shaped by means of a punch. The powder with which caps are charged consists of fulminate of mercury, mixed with half its weight of saltpetre.

CAPTAIN. Rank in the army between major and 1st lieutenant, charged with the arms, accoutrements, ammunition, clothing, or other warlike stores belonging to the troops or company under his command; (Art. 40.)

CAPTURE. (*See* Prize; Booty.)

CARBINE. A cavalry weapon intermediate in weight and length between rifle and pistol, and usually breech-loading. (For Pistol-carbine, *see* Arms.) Carbines for the United States' service have been obtained from the following manufactories:—Samuel Colt's, Hartford, Conn.—Colt's Revolving Pistols, Rifles, and Carbines; Sharpe's Arms-Manufacturing Company, Hartford, Conn., for Sharpe's Carbines and Rifles; Charles Jackson, Providence, R. I., for Burnside's Carbines; and Maynard's Arms Company, Washington, D. C., for Maynard's Rifles and Carbines. The breech-loading arms of the foregoing manufactories have been tried more or less in service, and favorably reported upon by boards of officers. They are considered good cavalry arms, but neither have yet been pronounced the best by the ordnance department. (*See* Ordnance Department.)

The distinguishing feature of a breech-loading arm is the method of closing the breech. One of the most serious defects of these arms was the escape of gas through the joint. This defect has been removed by closing the joint at the moment of discharge by the action of the gas itself. This operation, called *packing the joint*, is accomplished: 1st. By the use of cartridge cases of sheet brass, India rubber, or other material; or, 2d. By the use of a thin, elastic ring of steel, which overlies the joint. By the first method the case is permanently distended, (but may be safely used for several fires,) and some arrangement is required to remove it from the chamber. In the second method, the ring or *gas check* is a part of the arm; and its elasticity causes it to return to its original form after the discharge.

Burnside's Carbine is an example of the first method; it has a movable chamber which opens by turning on a hinge. A brass cartridge case is used which packs the joint and cuts off the escape of the gas. The advantages of this arm are: its strength, water-proof cartridges, perfectly tight joint, and working machinery. Its disadvantages are the cost, and difficulty of getting the cartridges.

Car.] MILITARY DICTIONARY. 147

Sharpe's Carbine has a fixed chamber, and the breech is closed by a slide which moves nearly at right angles to the axis of the barrel. By boring a recess into the face of the slide, opposite to the chamber, and inserting a tightly-fitting ring into it, so that the inner rim is pressed against the end of the barrel at the instant of discharge, the escape of gas is prevented.

Maynard's Carbine has a fixed chambered piece, with the joint closed by a metallic cartridge case. (*Consult* BENTON.)

CARCASS. Combustible composition enclosed in globes, formed with iron hoops, canvas, and cord, generally of an oblong shape, and thrown from mortars or stone mortars; it is used in bombardments, firing shipping, &c.

CARPENTRY. An assemblage of pieces of timber connected by framing or letting them into each other, as are the pieces of a roof, floor, centre of a bridge, &c. It is distinguished from joiners' work, by being put together without using other tools than the axe, adze, saw, and chisel. Troops frequently are obliged to hut themselves, make bridges, &c., and some knowledge of rough carpentry is essential in roofing and centring. The obvious mode of covering a building is to place two sloping rafters upon two walls, meeting in the apex, where we will suppose them connected. (Fig. 100.) It is plain that the weight of this rafter will tend to thrust the walls from its vertical line. This is prevented by tying together the feet of the rafters, by means of another beam called a *tie beam*. Beyond certain lengths or spans, however, it is apparent that the *tie beam* will itself have a tendency to bend or sag in the middle, and accordingly it becomes necessary to resort to another contrivance called a *king post*, but more properly a king piece, as it performs the office of tying up the tie beam to prevent it from bending. If the rafters be so long as to be liable to bend, two pieces called *struts* are introduced, which have their footing against the sides of the king post, and act as posts to *strut* up the rafters at their weakest point. This piece of framing thus contrived is called a *truss*. It is obvious that, by means of the upper joints of the struts, we can obtain more points of support or rather suspension.

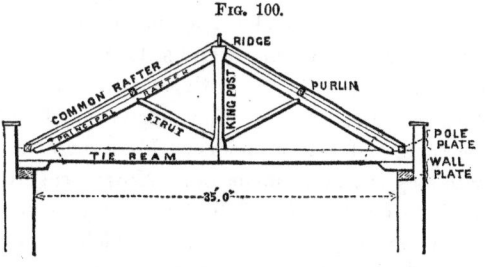

FIG. 100.

It is not, however, necessary to truss

all, but only the principal rafters of a building. These principal rafters must never be more than ten feet apart, and by the intervention of a *purline* they are made to bear the smaller rafters, the latter being notched down on the purline. These common rafters are received by or pitch upon a plate called a *pole plate*, and the principal rafters which pitch upon the tie beam, are ultimately borne by a wall plate. When beams in either roofs or floors are so long that they cannot be procured in one piece, two pieces to form the required length are *scarfed* together, by indenting them at their joints, and bolting them together thus: (Fig. 101.)

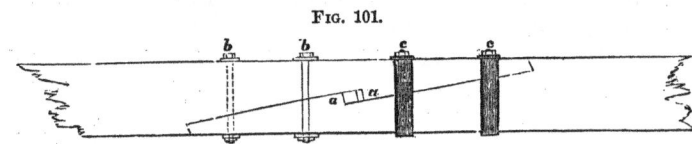

Fig. 101.

The following simple manner of putting up balloon frames, that is, frames without tenons or mortises, is given in the language of a builder in our western country: The best size for a small house is 16 by 32 feet, divided into three rooms and only one story high, unless roofing is very expensive. For such a building six pieces of scantling are required, cut 2 by 8, or 2½ by 10 inches, 16 feet long for sills, and seventeen pieces for sleepers, with seventeen pieces of same size, 18 feet long, for upper floor joists. The studs must be 2 by 4, or 2½ by 5 inches, and 8, 9 or 10 feet long, as you wish the height of your ceiling. The end studs may be longer, so as to run up to the rafters; but this is not important, since studs may be spliced anywhere by simply butting the ends together and nailing strips of boards upon each side, or the timbers may lap by each other and be held in place by a few nails till the siding is nailed on. But to begin at the foundation: Lay down two of the sixteen feet timbers flatwise upon blocks or stones, if you can get them, and make them level all around. Nail on strips where the ends of the sills butt together, and halve on the end sills and nail them together at the corners, and put on the sleepers, with a stout nail toed-in upon each side to hold them in place. Cut all your side studs of an exact length and square at each end, and set up one at each corner exactly plumb and fasten them with stay-laths on the inside. Now measure off for your doors and windows on the sides of the house, and set up studs for them. You are now ready to put on the plates, which are nothing but strips of inch board, just the width of your studs, spliced in length just as directed for splicing studs. The next step is to put up the rest of the studs, nailing through the plate into their tops, and toeing nails through the bottoms into the sills. Hands may now commence at once to nail the sheathing-boards upon the sides, while others are putting up the joists, which should be 18 feet long and either 2 by 8 or 2½ by 10 inches, according to the strength of the timber. Pine and poplar should always be of the larger dimensions. Cut notches one inch deep in the lower edge of the joists, so that they will lock on to the plate, and project over the sides one foot at each end. Nail up through the plate into the joists with stout nails, having just as many joists as pairs of rafters, the feet of which are to stand on and be nailed to the joists, which project the eaves a foot beyond the sides. This, however, may be dispensed with, if short eaves are preferred, or if timber cannot be got long enough. The end studs will be nailed both to the sill and end sleeper and to the end joists, and to the rafter if long enough to reach up, and if not splice them as before directed. Finish sheathing the sides and ends before you put on the roof. The siding may be afterward put on at your leisure. Boards three-fourths of an inch thick make good sheathing; and the best plan is to put them on without any regard to fitting the edges, and batten all the cracks on the inside with waste pieces of boards or shingles. When shingles are inexpensive they make a better siding and cheaper than sawed clapboards. You will find it a great saving of labor to lay the upper floor before you put on the roof. If you wish to make your house one and a half or two stories high, the following is the way the chamber floor joists are supported: Take a strip of board one inch thick and five inches wide, and let it into the face of the studs on the inside and nail it fast and set your joists on this and nail them to the studs, and also notch your floor boards in between all the studs and nail fast; and you will find, when done, that no old-fashioned frame with its heavy oak timbers and months of mortising, with all its braces, was ever stiffer than your "balloon," which two men can frame and raise, and cover and lay the floors, and get ready to move into in one week's time. There is no difficulty in making a balloon frame-house of any other size desired, by putting in the partitions before you put on the upper joists, so as to rest them upon the caps in the same way as upon the sides. For a house, say thirty-two feet wide, the upper joists would be the same length as for a house sixteeen feet, the inner ends resting upon the cap of a centre partition, where they would be strongly spliced, as we have directed, by nailing strips upon each side. The rafters of such a wide roof should be stayed in the middle by strips nailed upon the sides of rafters and joists, to prevent sagging; as it is always to be borne in mind that all the timbers of such a building are to be as light as possible; the strength being obtained by nailing all fast together.

CARRIAGES. A gun carriage is designed to support its piece when fired, and also to transport cannon from one point to another. Field, mountain, and siege artillery have also limbers, which form when united with the carriage a four-wheeled vehicle. Sea-coast carriages are divided into *barbette, casemate,* and flank defence carriages, depending upon the part of the work in which they are mounted. They are now made of wrought iron and found to possess lightness, great strength, and stiffness. The sea-coast carriages are made in a similar manner, and one carriage can be altered to fit another piece by changing the trunnion-plates and transom straps. The carriage consists of two cheeks of thick sheet-iron, each one of which is strengthened by three flanged iron-plates bolted to the cheeks. Along the bottom of each cheek, an iron shoe is fixed with the end bent upwards. In front, this bent end is bolted to the flange of the front strengthening plate. In rear the bent portion is longer, and terminated at top by another bend, which serves as a point of application for a lever on a wheel, when running to and from battery. The trunnion-plates fit over the top ends of the strengthening plates, which meet around the bed, and are fastened to the flanges of the latter by movable bolts and nuts. The *cheeks* are joined together by transoms made of bar-iron. The front of the carriage is mounted on an axle-tree, with truck wheels similar to the wooden casemate carriages. The elevating screws are of two kinds: one for low angles of elevation, and the second for columbiads where great angles of elevation are required. The elevating arc is made of brass and attached to the upper edge of the right cheek, and may be folded down. It is employed to measure the elevation of the piece.—ROBERTS & BENTON. (*See* CHASSIS; COLUMBIAD.)

CARTE BLANCHE. A blank paper sent to a person, to fill up with such conditions as he may think proper to insert. In the general acceptation of the term, it implies an authority to act at discretion.

CARTEL. An agreement between two hostile powers for a mutual exchange of prisoners. (*See* WAR.)

CARTRIDGE. Bullets for small arms are made by pressure. To prepare the lead for the press, it is cast into cylinders or drawn out into wires somewhat less in diameter than the bullet. One press can make 3,000 bullets in an hour. Bullets may also be cast in moulds and afterwards *swaged* in a die to proper size and shape.

Table of dimensions for formers for making cartridges with elongated expanding bullets. (The dimensions are referred to the plate by means of the letters placed opposite to them.)

	Altered musket.	New rifle musket.	Pistol carbine.	
	Inches.	Inches.	Inches.	
a	3.5	3.5	3.5	
d	2.5	2.25	2.25	Outer wrapper.
c	5.25	4.25	4.25	
a	1.1	1.	.8	Cylinder case.
e	2.75	2.	2.	
f	1.5	1.3	1.1	
g	2.75	2.2	2.2	Cylinder wrapper.
h	3.75	3.	3.	

The diameters of the round sticks on which the powder cases are formed should be .69 inch for the old, and .58 inch for the new calibre. This will make the exterior diameter of the case somewhat larger than the bullet, and will prevent the outer wrapping from binding around its base when the cartridge is broken. The outer wrapper should not be made of too strong paper: that prescribed in the Ordnance Manual for blank cartridges, and designated as No. 3, will answer a better purpose for these cartridges than that designated as No. 1. The cylinder case should be made of stiff rocket paper, No. 4; and its wrapper may be made of paper No. 1, 2, or 3. Before enveloping the bullets in the cartridges, their cylindrical parts should be covered with a melted composition of one part beeswax and three parts tallow. It should be applied hot, in which case the superfluous part would run off; care should be taken to remove all of the grease from the bottom of the bullet, lest by coming in contact with the bottom of the case it penetrate the paper and injure the powder. The bullets being thus prepared, and the grease allowed to cool, the cartridges are made up as follows, viz.: place the rectangular piece of rocket paper, called the cylinder case, on the trapezoidal piece, called the cylinder wrapper, as shown by the broken lines of Fig. 102, and roll them tightly round the former stick, allowing a portion of the wrapper to project beyond both case and stick. Close the end of the case by folding in this projecting part of the wrapper. To prevent the powder from sifting through the bottom, paste the folds, and press them on to the end of the stick, which is made slightly concave to give the bottom a form of greater strength and stiffness. After the paste is allowed to dry, the former stick is inserted in the case, and laid upon the outer wrapper, (the oblique edge from the operative, the longer vertical edge towards his left hand,) and snugly rolled up.

The bullet is then inserted in the open end of the cartridge, the base resting on the cylinder case, the paper neatly choked around the point of the bullet, and fastened by two half hitches of cartridge thread. The former stick is then withdrawn, the powder is poured into the case, and the mouth of the cartridge is "pinched" or folded in the usual way. To use this cartridge, tear the fold and pour out the powder; then seize the bullet end firmly between the thumb and fore finger of the right hand and strike the cylinder a smart blow across the muzzle of the piece; this breaks the cartridge and exposes the bottom of the bullet; a slight pressure of the thumb and forefingers forces the bullet into the bore clear of all cartridge paper. In striking the cartridge the cylinder should be held square across, or at right angles to the muzzle; otherwise, a blow given in an oblique direction would only bend the cartridge without rupturing it. Cartridges constructed on these principles present a neat and convenient form for carrying the powder and bullet attached to each other, and they obviate two important defects of the elongated bullet cartridges in common use, viz.: the reversed position of the bullet in the cartridge, and the use of the paper wrapper as a patch. (Fig. 103.)

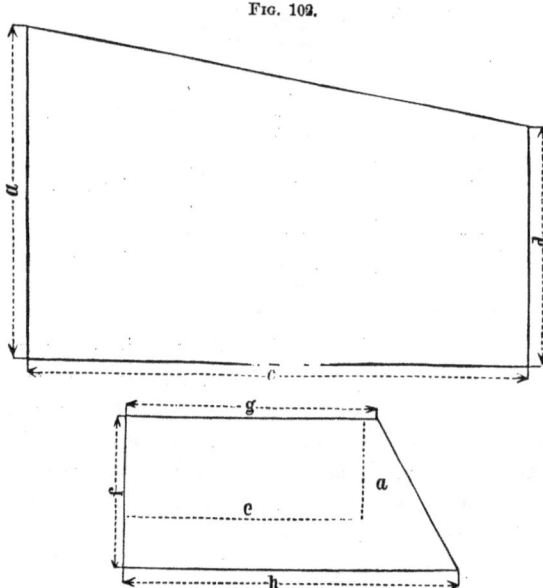

Fig. 102.

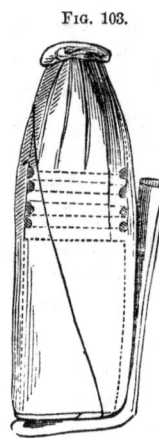

Fig. 103.

Cartridge-bags for field-pieces should be made of wild-bore, merino or bombazette, composed entirely of wool, free from any mixture of thread or cotton, which would be apt to retain fire in the piece. The texture and sewing should be close enough to prevent the powder sifting through. Untwilled stuff is to be preferred. Flannel may be used when other materials cannot be obtained. The bag is of two rectangular pieces, which forms the cylinder, and a circular piece for the bottom. As the stuff does not stretch in the direction of its length, the long side of the rectangle should be taken in that direction, otherwise the cartridge might become too large for convenient use.

Blank-cartridge Bags, or those intended for immediate use, may be made of two rectangular pieces with semicircular ends sewed together. The pieces are marked out with stamps made of one-inch board with a handle in the middle of one side, and on the other two projecting rims of copper or tin, parallel to each other and half an inch apart.

Siege and Garrison Cartridges consist of the charge of powder in a bag, and the projectile always separate from the cartridge.

The Cartridge-bags are usually made of woollen stuff. They are made of two pieces, in the form of a rectangle with semicircular end, which are marked out with stamps and sewed together as described for making blank-cartridge bags for the field service, and are filled, preserved, and packed in the same way.

Paper Bags.—Bags for heavy ordnance may be made entirely of paper. The bottom is circular, and one end of the cylindrical part is cut into slips about one inch long, which are pasted over the paper bottom on a cylindrical former. When a paper bag is filled, the open end is folded down about three-fourths of an inch wide, and this fold is rolled on itself down to the powder, and the part which projects beyond the cylinder is turned in on the top of it. The bags are apt to leave paper burning in the gun, for which reason those made of woollen stuff are preferable. Bags are sometimes made of both paper and woollen stuff, by forming the cylindrical part of paper, and sewing to it a bottom of woollen stuff made of two semicircular pieces.

CARTS AND KITCHEN CART. A system of army transportation proposed by Colonel Cavalli. (*See* AMBULANCE; WAGON.)

CASCABLE—is the part of the gun in rear of the base ring; it is composed generally of the following parts: the *knob*, the *neck*, the *fillet*, and the *base of the breech*.

CASEMATE. Vaulted chamber with embrasures for guns. It is necessary that they should be bomb-proof and distributed along the faces and flanks of the bastion, to serve as quarters and hospital to the garrison in war; but such subterranean barracks are always unwholesome.

CASE SHOT—are small balls enclosed in a case or envelope, which, when broken by the shock of the discharge in the piece, or by a charge of powder within the case, exploding during the flight of the case, scatters the balls. The kinds of case shot in use are GRAPE, CANISTER, and SPHERICAL CASE.

CASHIERED. When an officer is sentenced by a court-martial, to be dismissed the service, he is said to be cashiered.

CASTING AWAY—*Arms and Ammunition.* Punishable with death or other punishment, according to the nature of the offence, by the sentence of a general court-martial; (ART. 52.)

CASTRAMETATION. The art of encampment. (*See* CAMP.)

CASUALTIES. A word comprehending all men who die, desert, or are discharged.

CAVALIER—is a term applied to a work of more than ordinary height. It is sometimes constructed upon the terre-plein of the bastion, with faces and flanks parallel to those of the bastion which it commands. Cavaliers are not confined to bastions, but are placed wherever a great command of fire is required, and are sometimes traced straight, on other occasions curved.

CAVALRY. There are two regiments of dragoons, one of mounted riflemen, and two styled cavalry in our army. It has been recommended that these regiments should all be called regiments of cavalry. (*See* ARMY for their organization.) Cavalry is usually divided into heavy and light cavalry. Heavy cavalry acts in heavy masses. Its essential condition is united ranks. It finds its true type in the mailed chivalry of the middle ages, but it is believed that the general introduction into service of rifled muskets will render heavy cavalry entirely useless in war. Formerly cavalry could move against infantry in columns of squadrons first at a trot, then at a gallop, and finally at full speed from a position taken up within 400 yards of infantry. But now that the cavalry comes within range of the rifle at 1,000 yards, the infantry must be greatly demoralized before cavalry can have the least chance of success in a charge. Accordingly at the camp of Chalons, where all arms of the service were supposed to be represented, heavy cavalry were not seen. Light cavalry on the contrary is intended rather to envelop an enemy. Quickness and agility are its primary conditions. Indefatigable and careless of repose it ought to occupy an enemy during entire hours, harass and fatigue him. If he lays himself open pierce him with the quickness of lightning, and cut him to pieces with the sabre. The cavalry soldier must consider his horse as part of himself, and the perfect management of the horse cannot be learned either in schools, or in a few weeks of practice. If daily exercises are dispensed with, both horse and man return to their natural state, and such mounted men cease to be efficient. The main body in all campaigns against Indians should be infantry. But a small mounted force, kept in high condition, would add much to the efficiency of such a main body. The horses should be well fed; and upon long marches in uninhabited districts this is impossible. The idea of employing such a force as a main body, in order to make rapid marches, is also untenable; for upon long marches of many days, infantry will improve every day, accomplish a greater distance in many successive days, and have at the opportune moment greater vigor than a large cavalry force, necessarily with broken-down horses from want of food; whereas a small cavalry force might be held in hand and maintained in the highest state of efficiency. Cavalry is indispensable in time of war. It will always take a leading part in pursuing a retreating enemy; it is the proper arm in ordinary reconnoissances; it will always serve as eclaireurs, and as escorts, and should, in the present state of the art of war, carry carbines and be prepared for service on foot. It is weakened and destroyed when in a country without forage. Its first cost, its constant maintenance, the defects of its employment, and the system of providing horses make it expensive; but it ought nevertheless to be maintained in a complete state, for its art can only be exercised by men and horses that are properly instructed.

Cavalry Tactics.—The individual instruction of men and horses should be regarded as the most important point of the whole system, and should be as simple as possible; the man should be taught to manage his horse with ease and address over all kinds of ground and at all gaits, to swim rivers, to go through certain gymnastic exercises—such as vaulting, cutting heads, to fence, to fire very frequently at a mark, and to handle his weapon with accuracy and effect at all gaits, and in all situations. Individual instruction has been recently made a supplementary instruction in France. Every thing in reference to heavy cavalry, lancers, hussars, &c., should be omitted. Insist upon the sabre being kept sharp in the field, provide the men with means of doing so, and lay it down as a rule that the strength of cavalry is in the " spurs and sabre." The instruction on foot should be carried no further than its true object requires —that is, to bring the men under discipline, improve their carriage, and enable them to comprehend the movements they are to execute mounted. The formation for review, parade, inspection, &c., to be: the companies deployed in one line, with intervals of 12 paces, or else in a line of columns of companies by platoons, according to the ground. It should be laid down as a fixed rule that no cavalry force should ever charge without leaving a reserve behind it, and that against civilized antagonists the compact charge in line should be used in preference to that as foragers.

Columns to be formed with wheeling distance, and closed in mass; when closed in mass, the file-closers close up to 1 pace from the rank, and the distance between the subdivisions to be just enough to permit each company to wheel by fours. Marching columns to be by file, twos, fours, or platoons; by fours and platoons in preference when the ground permits. Columns of manœuvre to be by fours, platoons, companies, or in double column; the latter always a regimental column, and to be formed on the two central companies, or platoons, without closing the interval between them. Deployments to be made habitually at a gallop, and the individual oblique to be used as much as possible. The instruction in two lines to be provided for. The Russian tactics give a good basis for the system of skirmishers, and charging as foragers. For the use of the mounted rifles, and cavalry acting as such, a thorough system for dismounting rapidly, and fighting on foot, has already been submitted by Captain Maury, and adopted. (*Consult* McClellan.)

CENTRE OF THE BASTION—is the intersection made by the two demi-gorges.

CERTIFICATE. (*See* Muster.)

CHAIN-BALL. It has been proposed to attach a light body by means of a chain to the rear of an oblong projectile, when thrown under high angles with a moderate velocity, so as to cause it to move with its point foremost.

CHAIN-SHOT—consist of two hemispheres, or two spheres connected together by a chain. The motion of rotation of these projectiles in flight would render them useful in cutting the masts and riggings of vessels, if their flight was not so inaccurate. When the mode of connection is a *bar* of iron instead of a chain, they are called Bar-shot.

CHALLENGE. No officer or soldier shall send a challenge to another officer or soldier to fight a duel, or accept a challenge if sent, upon pain if a commissioned officer of being cashiered; if a non-commissioned officer or soldier, of suffering corporeal punishment at the discretion of a court-martial; (Art. 25.) If any commissioned or non-commissioned officer commanding a guard shall knowingly or willingly suffer any person whatsoever to go forth to fight a duel, he shall be punished as a challenger; and all seconds, promoters, and carriers of challenges, in order to duels, shall be deemed principals, and be punished accordingly. And it shall be the duty of every officer commanding an army, regiment, company, post or detachment, who is knowing to a challenge being given, or accepted, by any officer, non-commissioned officer or soldier under his command, or has reason to believe the same to be the case, immediately to arrest and bring to trial such offenders;

(ART. 26.) Any officer or soldier who shall upbraid another for refusing a challenge shall himself be punished as a challenger; and all officers and soldiers are hereby discharged from any disgrace, or opinion of disadvantage, which might arise from their having refused to accept challenges, as they will only have acted in obedience to the laws, and done their duty as good soldiers, who subject themselves to discipline; (ART. 28.)

CHALLENGE OF MEMBERS OF COURT-MARTIAL. When a member shall be challenged by a prisoner, he must state his cause of challenge, of which the court shall, after due deliberation, determine the relevancy or validity, and decide accordingly; and no challenge to more than one member at a time shall be received by the court; (ART. 71.) Challenges of members are made in writing. The member withdraws and the court is cleared for deliberation. If the challenge is disallowed the member resumes his seat. Blackstone says: A *principal* challenge is where the cause assigned carries *prima facie* evidence of malice or favor; as that a juror is of kin to either party within the 9th degree; that he has been arbitrator on either side; that he has formerly been a juror in the same cause; that he is the party's master, servant, &c. These grounds of challenge, if true, cannot be overruled. Challenges to the favor are, where the party hath no principal challenge, but objects only on probable circumstances of suspicion, as acquaintance and the like; the validity of which is left to the triers; (HOUGH.)

CHALLENGE OF A SENTINEL. Who goes there?

CHAMADE—is a signal made for parley by beat of drum.

CHAMBER OF A MINE—is a cell of a cubical form, made to receive the powder.

CHAMBER of howitzers, columbiads, and mortars, is the smallest part of the bore, and contains the charge of powder. In the howitzers and columbiads the chamber is cylindrical, and is united with a large cylinder of the bore by a conical surface; the angles of intersection of this conical surface with the cylinders of the bore and chamber, are rounded (in profile) by arcs of circles. In the 8-inch siege howitzer, the chamber is united with the cylinder of the bore by a spherical surface, in order that the shell may, when necessary, be inserted without a sabot.

CHAPLAIN. Punished by a court-martial for undue absence; (ART. 4.) One allowed to Military Academy who shall be professor of geography, history, and ethics—with pay of professor of mathematics. Chaplains allowed to military posts, not exceeding twenty, are selected by the council of administration of the post, and are also to be schoolmasters, with $70 per month, 4 rations per day, and quarters and fuel; (*Acts* July 5, 1838; and Feb. 21, 1857.)

CHARACTER. Where a witness is introduced by a prisoner to prove character, the court may ask how long he has known the prisoner, and whether he has known him from that time to the present without interruption, and whether he speaks from his own knowledge or from general report.—*Cross-examination* by the prosecutor, of witnesses introduced by the prisoner to prove character, is not allowed. (*Consult* Phillips' *Law of Evidence.*)

CHARGE. Cavalry charges have been sometimes made silently. Those of Frederick the Great always began the Hurrah at fifty paces from the enemy. If at the moment of the shock the infantry is not disturbed, but their bayonets and fire have on the contrary saved them from the impulsive force of the charge, the fall of the front ranks of the cavalry will have interposed a rampart behind which infantry cannot fail to be victorious. But if the cavalry has practised the stratagem of beginning operations by drawing the fire of infantry upon skirmishers, and the commander of the cavalry ready for the charge has pushed forward curtains of light cavalry in a single rank, who succeed, by means of clouds of dust, in making an unskilful infantry believe that to be an attack which in reality is only a feint, the infantry may fire its balls at random—the thinness of the curtain of light cavalry will render the infantry's fire of little effect—the infantry will be eager to reload, and this may be done in agitation and disorder. The proper moment is then at hand, and the heavy cavalry in mass, concealed by the dust of their skirmishers, may charge, break, and sabre the infantry. The light cavalry finish the fugitives. The passage of defiles in retreat ought to be secured by a charge of cavalry. Coolness, silence, immobility, contempt of hurrahs, and a reserved fire until within suitable range, are the principal means of resisting a charge of cavalry. The file-closers must prevent firing, not ordered; watch the execution of the fire by ranks; see that it does not commence at too great a distance, then enjoin upon the soldiers to aim at the breast; to act only upon signals of the drum, or at the command of officers on horseback, who occupy the centre of the square, and who from that height alone can judge whether the charge of cavalry is a mere feint or a real attack. This necessary impassibility of infantry is obtained by discipline and experience, and is only perfected upon battle-fields. Without *sang froid*, and also promptness in manœuvring upon any ground, infantry will not be able to exhibit the whole strength of its arm against the best cavalry. Charges by infantry are made in order of battle, in column of attack, and in close columns in mass. Charges in order of battle are executed as follows: If the combat is between infantry and infantry, the troops receiving the charge, fire at the moment at which it is almost joined with the enemy. The troops making the charge, fire at one hundred or one hundred and twenty paces from the enemy; without waiting to reload, they march forward at the quick step; at two-thirds the distance take charging step, and if the ground permits they subsequently take a running step, keeping up the touch of the elbow, and throw themselves upon the enemy with HURRAHS. Frederick the Great says that it is "better for a line to falter in a charge than to lose the touch of the elbow," so necessary is it that the charge should be *en muraille*.

In modern wars the charge in column has been used but not exclusively, and sometimes with fatal results. But whatever may be the form of the charge, success must not make the victor at once pursue his enemy. He must, on the contrary, halt, rally his men, form line if the charge was made in column, reload, fire upon the fugitives, and continue thus to gain ground, by a regulated fire, until at last the cavalry which seconds him comes to his aid. It must be considered that there may be a second line of the enemy, fresh troops, masked batteries, flank fires, or squadrons of cavalry ready to oppose an unforeseen resistance. It may be, that the attacking party has experienced some disadvantage, not far from the point where the infantry has just triumphed in the charge. Such circumstances may cause the infantry to pay dearly for its temporary success, a temporary success sometimes owing to stratagem on the part of the enemy. These precepts are given by the best writers on charges of infantry. (*Consult* DECKER; BARDIN, &c., &c.)

CHARGER. The horse rode by an officer in the field or in action.

CHARGES AND SPECIFICATIONS. The form of indictments tried by courts-martial. (*See* COURT-MARTIAL; EVIDENCE.) As to the perspicuity and precision of charges: If the description of the offence is sufficiently clear to inform the accused of the military offence for which he is to be tried, and to enable him to prepare his defence, it is sufficient; (*Opinions of Attorney-general*, p. 189.)

A copy of charges, as well as a list of witnesses for the prosecution, should be given to the prisoner in all cases as soon as possible. Antecedent to arraignment, charges may be framed and altered by the party who brings forward the prosecution, or by the officer ordering the court, both in regard to substance and in other respects; but the court, where the deviation was material, would probably deem it sufficient cause for delaying proceedings upon application of the prisoner. As the witnesses of an officer may be at a distance, the sooner a copy is given the better; (Hough's *Law Authorities*.)

CHASE. The conical part of a piece of ordnance in front of the reinforce.

CHASSIS. A traversing carriage. The barbette and casemate carriages consist of gun carriages and chassis. The wrought-iron chassis now made consists of two rails of wrought iron, the cross-section of each being in form of a T, the flat surface on top being for the reception of the shoe-rail of the gun carriage. The rails are parallel to each other, and connected by iron transoms and braces. The chassis is supported on traverse wheels. A prop is placed under the middle transom of the chassis to provide against sagging. The pintle is the fixed centre around which the chassis traverses. In the ordinary barbette, the pintle is placed under the centre of the front transom; but in the columbiad carriage, it is placed under the centre of the middle transom. (*See* COLUMBIAD.)

CHEMIN DES RONDES—is a berme from four to twelve feet broad, at the foot of the exterior slope of the parapet. It is sometimes protected by a quickset hedge, but in more modern works by a low wall, built on the top of the revetment, over which the defenders can fire, and throw hand grenades into the ditch.

CHESSES—are the platforms which form the flooring of military bridges. They consist of two or more planks, ledged together at the edges, by dowels or pegs.

CHEVAUX-DE-FRISE. The principal uses of chevaux-de-frise are to obstruct a passage, stop a breach, or form an impediment to cavalry. Those of the modern pattern are made of iron, whose barrel is six feet in length, and four inches in diameter, each carrying twelve spears, five feet nine inches long, the whole weighing sixty-five pounds. (*See* OBSTACLES.)

CHOLERA. (*See* SANITARY PRECAUTIONS.)

CIRCUMVALLATION. Works made by besiegers around a besieged place facing outwards, to protect their camp from enterprises of the enemy.

CITADEL. A citadel is a small strong fort, constructed either within the place, or on the most inaccessible part of its general outline, or very near to it; it is intended as a refuge for the garrison, in which to prolong the defence, after the place has fallen.

CIVIL AUTHORITY. (*See* AUTHORITY; CONTRACTS; EXECUTION OF LAWS; INJURIES; REMEDY.)

CLERKS. Whenever suitable non-commissioned officers or privates cannot be procured from the line of the army, paymasters, with the approbation of the Secretary of War, may employ citizens to perform the duties of clerks at $700 per year; (*Acts* July 5, 1838; and Aug. 12, 1848.) One ration per day allowed when on duty at their station; (*Act* Aug. 31, 1852.)

CLOTHING. The President of the United States is authorized to prescribe the kind and quality of clothing to be issued annually to the troops of the United States. The manner of issuing and accounting for clothing shall be established by general regulations of the War Department. But whenever more than the authorized quantity is required, the value of the extra articles shall be deducted from the soldiers' pay; and, in like manner, the soldiers shall receive pay according to the annual estimated value for such authorized articles of uniform as shall not have been issued to them in each year. And when a soldier is discharged, it is the duty of the paymaster-general to pay him for clothing not drawn; (*Act* April 24, 1816.) The quartermaster's department distributes to the army the clothing, camp and garrison equipage required for the use of the troops. Every commander of a company, detachment, or recruiting station, or other officer receiving clothing, &c., renders quarterly returns of clothing according to prescribed forms to the quartermaster-general. All officers charged with the issue of clothing to make good any loss or damage, unless they can show to the satisfaction of the Secretary of War, by one or more depositions, that the deficiency was occasioned by unavoidable accident, or was lost in actual service, without any fault on their part; or, in case of damage, that it did not result from neglect; (*Act* May 18, 1826.) Purchasing clothing from a soldier prohibited under penalty of three hundred dollars, and imprisonment not exceeding one year; (*Act* March 16, 1802, and Jan. 11, 1812.)

The French system of making up clothing is as follows: Officers commanding regiments make their requisitions for the regulated quantities of cloth and other materials necessary for the clothing of the number of men under their command. The intendant having checked this demand gives an order for the issue, and the materials are made up by soldiers in the regimental workshops under the direction of the clothing captain, an officer holding an appointment in some respects analogous to that of our quartermasters; a fixed rate being paid for each article. Organized as the European armies are, those troops have always a large proportion of skilled workmen undergoing their term of military service; but it is not so with us. Still there are many points in the European system of clothing the troops which might, with advantage to the soldier and with economy to the public, be adapted to the wants of our service.

STATEMENT *of the cost of Clothing, Camp and Garrison Equipage for the Army of the United States, furnished by the Quartermaster's Department, during the year commencing July 1, 1859, with the allowance of clothing to each soldier during his enlistment, and his proportion for each year respectively.*

CLOTHING.	Engineer Troops.	Hospital Stewards.	Ordnance Sergeants.	Ordnance Mechanics.	Dragoons.	Cavalry.	Mounted Riflemen.	Light Artillery.	Artillery.	Infantry.	First.	Second.	Third.	Fourth.	Fifth.	Allowance during enlistm't
	$ c.	$ c.	$ c.	$ c.	$ c.	$ c.	$ c.	$ c.	$ c.	$ c.						
Uniform Hat.	2 35	2 35	2 35	2 35	2 35	2 35	2 35	2 35	2 35	2 35	1	1	1	1	1	5
" " Feather.	11	11	11	11	11	11	11	11	11	11	1	1	1	1	1	5
" " Cord and tassels.	13	13	13	13	13	13	13	13	13	13	1	1	1	1	1	5
" " Eagle.	3	3	3	3	3	3	3	3	3	3	..	..	..	..	..	..
" " Castle.	14															
" " Shell and flame.			4	4												
" " Crossed sabres.					4	4										
" " Trumpet.							4									
" " Crossed cannon.								3	3							
" " Bugle.										3						
" " Letter.	2				2	2	2	2	2	2						
" " Number.					2	2	2	2	2	2						
" Cap (old pattern).								1 13			1	1	1	1	1	5
" " Tulip.								7								
" " Cord and tassel.								56								
" " Plate.								4								
" " Rings, pairs of.								5								
" " Hair plume.								62								
Forage Cap.	57	57	57	57	57	57	57	57	57	57	1	1	1	1	1	5
Uniform Coats, Musicians.	6 89							6 89	6 89		1	1	1	1	1	5
" Privates.	6 56	6 56	6 56	6 56					6 56	6 56	1	1	1	1	1	5
" Jackets, Musicians.					5 52	5 52	5 52	5 52			1	1	1	1	1	5
" Privates.					5 17	5 17	5 17	5 17			1	1	1	1	1	5
Chevrons, N. C. S., pairs of.			1 24		1 24	1 24	1 24		1 24	1 24	1	1	1	1	1	5
" 1st Sergeants, pairs of.	37				37	37	37	37	37	37	1	1	1	1	1	5
" Sergeants, "	25				25	25	25	25	25	25	1	1	1	1	1	5
" Corporals, "	19				19	19	19	19	19	19	1	1	1	1	1	5
Caduceus.		95									1	1	1	1	1	
Shoulder Scales, brass, pr of N. C. S.		95	95		95	95			95	95						
Do. do. Sergeants.	80				80	80			80	80						
Do. do. Privates.	50			50	50	50			50	50						
Do. bronze, N. C. S.							1 15									
Do. do. Sergeants.							90									
Do. do. Privates.							60									
Trowsers, Sergeants.	3 00	3 00	3 00		4 05	4 05	4 05	4 05	3 00	3 00	3	2	3	2	3	13
" Corporals.	2 87				3 93	3 93	3 93	3 93	2 87	2 87	3	2	3	2	3	13
" Privates.	2 82			2 82	3 87	3 87	3 87	3 87	2 82	2 82	3	2	3	2	3	13
Sash.	3 00	3 00	3 00		3 00	3 00	3 00	3 00	3 00	3 00						
Blue flannel Sack Coats.	2 10	2 10	2 10	2 10	2 10	2 10	2 10	2 10	2 10	2 10	} 2	2	2	2	2	10
Do. do. lined, for Recruits.	2 56				2 56	2 56	2 56	2 56	2 56	2 56						
Flannel Shirts.	90	90	90	90	90	90	90	90	90	90	3	3	3	3	3	15
Drawers.	71	71	71	71	71	71	71	71	71	71	3	2	2	2	2	11
*Bootees, pairs.	2 20	2 20	2 20	2 20	2 20	2 20	2 20	2 20	2 20	2 20	4	4	4	4	4	20
*Boots, pairs.					3 60	3 60	3 60	3 60			1	1	1	1	1	5
Stockings, pairs.	24	24	24	24	24	24	24	24	24	24	4	4	4	4	4	20
Great Coats.	6 40	6 40	6 40	6 40	7 63	7 63	7 63	7 63	6 40	6 40	1	0	0	0	0	1
" " straps, sets.	24	24	24	24					24	24						
Blankets.	2 44	2 44	2 44	2 44	2 44	2 44	2 44	2 44	2 44	2 44	1	0	1	0	0	2
Leather Stocks.	17	17	17	17	17	17	17	17	17	17	1	0	1	0	0	2
Knapsacks and straps.	2 78	2 78	2 78	2 78	2 78	2 78	2 78	2 78	2 78	2 78						
Havresacks.	39	39	39	39	39	39	39	39	39	39						
Canteens.	32	32	32	32	32	32	32	32	32	32						
Canteen Strap.	14	14	14	14	14	14	14	14	14	14						
Fatigue Overalls.	71	71	71	71							1	1	1	1	1	5
Stable Frock.					62	62	62	62			1	0	1	0	0	2
Talma.						5 00										

* Mounted men may, at their option, receive *one* pair of "boots" and *two* pairs of "bootees," instead of *four* pairs of Bootees.

NOTE.—Metallic Eagles, Castles, Shell and flame, Crossed Sabres, Trumpets, Crossed Cannon, Bugles, Letters, Numbers, Tulips, Plates, Shoulder Scales, Rings, the Cap cord and tassels, and the hair Plume of the Light Artillery, the Sashes, Knapsacks and Straps, Havresacks, Canteens, Straps of all kinds, and the Talmas, will not be issued to the soldiers, but will be borne on the Return as company property while fit for service. They will be charged on the Muster Rolls against the person in whose use they were when lost or destroyed by his fault.

CAMP AND GARRISON EQUIPAGE.

Bedsack, single	$1 02	Drum case		$ 20
" double	1 13	Wall tent	$17 86	
Mosquito bars	1 13	" " fly	5 04	
Axe	85	" " poles, sets	1 18	
" helve	10	" " pins, sets	72	
" sling	70			24 80
Hatchet	29	Sibley tent	$32 30	
" helve	03	" " poles and tripod	4 72	
" sling	40	" " sets	48	
Spade	58			37 50
Pickaxe	56	" " stove		4 00
" helve	10	Hospital tent	$64 13	
Camp kettle	50	" " fly	23 50	
Mess pan	18	" " poles, sets	5 60	
Iron pot	1 23	" " pins, sets	1 28	
Garrison flag	36 66			94 51
" " halliard	3 00	Servant's tent	$6 62	
Storm flag	12 35	" " poles, sets	1 10	
Recruiting flag	3 77	" " pins, sets	28	
" " halliard	20			8 00
Guidon	5 28	Tent pin, large size, hospital		05
Camp color	1 82	" " " wall		04
National color, Artillery	35 48	" small size, common		02
" " Infantry	35 48	Regimental book, order	$2 25	
Regimental color, Artillery	42 60	" " general order	2 25	
" " Infantry	47 60	" " letter	3 50	
Standard for Mounted Regiments	20 87	" " index	1 75	
Trumpet	3 88	" " descriptive	2 25	
Bugle, with extra mouth-piece	3 12			12 00
Cord and tassels for Trumpets and Bugles	75	Post book, morning report	$2 00	
Fife, B	47	" " guard	2 00	
" C	41	" " order	1 15	
Drum, complete, Artillery or Infantry	5 90	" " letter	1 15	
Drum head batter	60			6 30
" " snare	19	Company book, clothing	$2 50	
" sling	45	" " descriptive	1 80	
" sticks, pairs	23	" " order	1 70	
" " carriage	64	" " morning report	2 00	
" cord	20			8 00
" snares, sets	17	Record book, for target practice		60

The tunic of the French infantry soldier lasts three years and a half, the shell jacket two years, the great coat three years, and the trowsers one year. In the Sardinian and Belgian armies the great coat is intended to last eight years. Those governments credit every man on his enlistment with about eight dollars as outfit money, which is about the annual cost of the clothing of each soldier, and a daily allowance of 10 centimes is given for repairs. Regimental master-tailors are required to make all repairs at a fixed annual contribution from the soldiers' pay. This does not often exceed 80 centimes; and the surplus, after the soldier has paid the cost of his clothing, is handed to him at the end of the year. By this means the soldier is taught economy, but if at any time an article of dress is found to be unfit for use, captains of companies may order it to be renewed at the cost of the soldier. The great durability of the clothing of European armies is attributable to the precautions taken to insure good materials from the manufacturers by whom the cloth is supplied. Not only is every yard of cloth, when delivered into store, subjected to several distinct and minute examinations by boards of officers assisted by *experts*, who weigh it, shrink it, and view it inch by inch against a strong light, so that the slightest flaw may be detected; but they likewise apply chemical tests to detect the quality of the dye, and the manufactories are at all times open to inspectors, who watch the fabrication at every stage. When clothing has once been manufactured, it is hardly possible with any degree of accuracy to ascertain the quality of the material.

COEHORN MORTAR. Brass 24-pdr. mortar, weighing 164 lbs.

COLONEL. Rank in the army between brigadier-general and lieutenant-colonel.

COLORS. Each regiment of artillery and infantry has two silken colors, but only one is borne or displayed at the same time, and on actual service that is usually the regimental one.

COLUMBIAD. An American cannon invented by Colonel Bomford, of very large calibre, used for throwing solid shot or shells, which, when mounted in barbette, has a vertical field of fire from 5° depression to 39° elevation, and a horizontal field of fire of 360°. Those of the old pattern were chambered, but they are now cast without, and otherwise greatly improved. The 10-inch weighs 15,400 lbs., and is 126 inches long. The 8-inch columbiad is 124 inches long and weighs 9,240 lbs. Rodman's 15-inch columbiad, represented in Fig. 104, was cast at Pittsburgh, Pennsylvania, by Knapp, Rudd & Co., under the directions of Captain T. J. Rodman, of the Ordnance Corps, who conceived the design, which he has happily executed, of casting guns of large size *hollow*, and by means of a current of water introduced into the *core*, which forms the mould of the bore, cooling it from the interior, and thus making the metal about the bore the hardest and densest, and giving the whole thickness of metal subjected to internal strain its maximum strength. The gun has the following dimensions:

Total length	190 inches.
Length of calibre of bore,	156 "
Length of ellipsoidal chamber,	9 "
Total length of bore,	165 "
Maximum exterior diameter,	48 "
Distance between rimbases,	48 "
Diameter at muzzle,	25 "
Thickness of metal behind the chamber,	25 "
Thickness at junction of bore with chamber,	16½ "
Thickness at muzzle,	5 "
Diameter of shell,	14.9 "
Weight of gun,	49,100 lbs.
Weight of shell,	320 "
Bursting charge,	17 "

The gun is mounted upon the new iron centre pintle carriage, (Fig. 104,) which with requisite lightness has great strength and stiffness; and to facilitate the pointing from 5° depression to 39° elevation, a slot is cut in the knob of the cascable, and a ratchet is formed on the base of the breech to receive a "pawl" attached to the elevating screw. If the distance be greater than the length of a single notch of the ratchet, the piece is rapidly moved by a lever which passes through an opening in the pawl. If the distance is less, then the elevating screw is used. The piece was fired and manœuvred during the trials at Fort Monroe, with great facility, being manned by 1 sergeant and 6 negroes; the times of loading were 1′ 15″ and 1′ 3″. Time in traversing 90° 2′ 20″, and in turning back 45° 1′. Time of loading, including depression and elevation, 4′ and 3′ 18″.

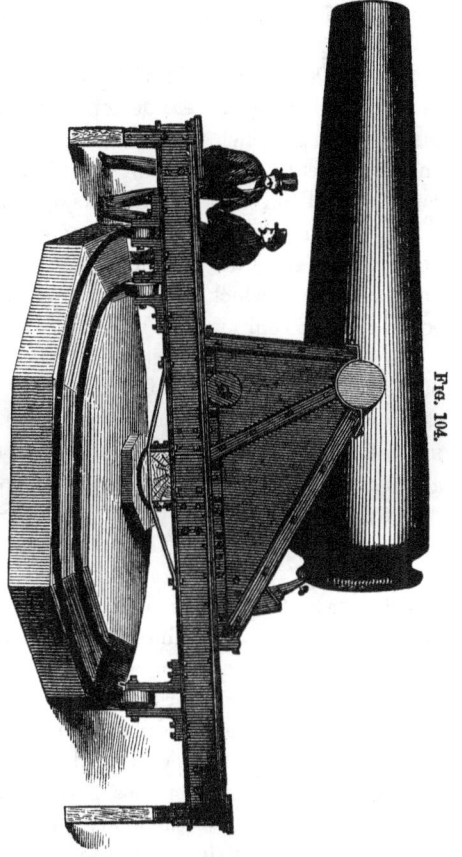

Fig. 104.

The mean ranges at 6° elevation, of ten shots, was 1,936 yards, and the mean lateral deviation 2.2 yards; 35 lbs. of .6-inch grain powder being the charge and 7″ the time of flight. At 10° elevation and 40 lbs. of powder, large grain, the range was 2,700 yards, and time of flight 11″.48. At 28° 35′ elevation the range was 5,730 yards; time of flight 27″, and the lateral deviation, as observed with a telescope attached to one of the trunnions, very slight. (*See* ARTILLERY; GUNPOWDER; ORDNANCE AND ORDNANCE STORES; RANGES.)

COLUMN—of attack; in route; close column; column of divisions; column at half distance; open column. (*See* MANŒUVRES IN BATTLE; TACTICS.)

COMMAND. An officer may be said to command at a separate post, when he is out of the reach of the orders of the commander-in-chief, or of a superior officer, in command in the neighborhood. He must then issue the necessary orders to the troops under his command, it being impossible to receive them from a superior officer; (PETER's *Digest of Decisions of Federal Courts*, vol. 1. p. 179.)

Officers having brevets or commissions of a prior date to those of the regiment in which they serve, may take place in courts-martial and on detachments, when composed of different corps, according to the ranks given them in their brevets, or dates of their former commissions; but in the regiment, troop, or company, to which such officers belong, they shall do duty and take rank, both in courts-martial and on detachments, which shall be composed only of their own corps, according to the commissions by which they are mustered in said corps; (ART. 61.) If, upon marches, guards, or in quarters, different corps of the army shall happen to join and do duty together, the officer highest in rank of the line of the army, marine corps, or militia, by commission there, on duty or in quarters, shall command the whole, and give orders for what is needful to the service, unless otherwise specially directed by the President of the United States, according to the nature of the case; (ART. 62.) The great principle that rank, when an officer is on duty, and military command, are ideas only to be separated by positive law, has always been recognized in legislation. The 61st Article of War, for instance, forbids the exercise of brevet rank within the regiment, troop, or company, to which such officers belong. The 63d forbids engineers to assume, and declares they are not subject to be ordered on any duty beyond the line of their immediate profession, except by the special order of the President of the United States. The acts of Congress giving rank to officers of the medical and pay departments of the army, provide that they shall not, in virtue of such rank, be entitled to command in the line or other staff departments of the army; and so, if any other legal restrictions on rank exist, they must be found in some positive statute. This necessity is made plain by the consideration that *military rank means a range of military subordination*. Higher rank therefore, created by law, cannot be made subordinate to lower rank, except by positive law; or, in other words, a junior cannot command a senior, unless the law shall otherwise decree. The 61st Article of War declares that officers holding commissions of a prior date to the regiment in which they serve, shall nevertheless take rank " both in courts-martial and on detachments composed only of their own corps, according to the commissions by which they are mustered in said corps." The 98th Article declares that militia officers, when serving in conjunction with the regular forces, shall take rank next after all officers of the like grade in said regular forces, notwithstanding the commissions of such militia officers may be older than the commissions of the officers of the regular forces of the United States. The 27th Article declares that all officers have power to part and quell all quarrels, &c., and to order officers into arrest, and whosoever shall refuse to obey such officer (though of inferior rank) shall be punished, &c. Here are cases in which Congress has decreed that seniors in commission may be commanded by juniors; and if any other cases exist, they likewise must be found in some positive statute. The 62d Article of War is ambiguous, from the use of the words " line of the army ; " our legislation having applied those words to contradistinguish regular troops from militia, and also, in many cases, the same words are correlative and contradistinctive of staff of the army. "But," says President Fillmore, after a careful examination on his part, to determine this question, "I find but one act of Congress in which the words 'line of the army' have been employed to designate the regular army in contradistinction to the militia, and none in which they have manifestly been used as contradistinctive of brevet." Whatever ambiguity, therefore, may exist under the 62d Article, in respect to the right of command on the part of officers of staff corps and departments, the article does not decree any restriction on brevet rank; and hence the great principle that rank on duty confers military command has its full force in respect to commissions by brevet, and all other commissions not restricted by law. The President, as commander-in-chief under the 62d Article of War, may relieve any officer from duty with a particular command, or he may assign some officer of superior rank to duty with a command; but the laws have not authorized him to place a junior in command of a senior, and that power which creates rank, viz., Congress, is alone authorized to place restrictions on its meaning. (*See* ASSIGNMENT; BREVET; LINE; RANK.)

The word *command*, when applied to ground, is synonymous with overlook; and any place thus commanded by heights within range of cannon is difficult to defend, if the enemy have been able to seize the heights. (*See* BREVET; OATH; OBEDIENCE; RANK.)

COMMAND OF FIRE. When a work has a sufficient elevation over the work before it, to enable the defensive weapons to act in both works at the same time upon an advancing enemy, even to the foot of the glacis, then the inner work is said to have a command of fire over the other.

COMMAND OF OBSERVATION. When the interior work has only sufficient elevation to look into or even over the work before it, but not sufficient to fire clear of it, then it is said to have only a command of observation.

COMMANDER-IN-CHIEF. The President shall be commander-in-chief of the Army and Navy of the United States, and of the militia of the several States, when called into the actual service of the United States; (*See* CONSTITUTIONAL RELATION OF CONGRESS AND THE PRESIDENT TO THE LAND FORCES.)

COMMANDER OF THE ARMY. That whenever the President shall deem it expedient, he is hereby empowered to appoint, by and with the advice and consent of the Senate, a commander of the army which may be raised by virtue of this act, and who, being commissioned as lieutenant-general, may be authorized to command the armies of the United States; (*Sec.* 5, *Act* May 28, 1798.)

COMMISSARY OF SUBSISTENCE. An officer of the subsistence department. (*See* SUBSISTENCE.)

COMMISSION. The President shall commission all officers of the United States; (*Sec.* 3 *Constitution.*) Officers of the United States army may hold their commissions through rules of appointment prescribed by Congress under its authority to raise armies and make rules for their government and regulation, but their commissions must be signed by the President. The words introduced into every officer's parchment :—" this commission to continue in force during the pleasure of the President of the United States for the time being "—have been inserted without authority of law. There has been no legislation on the subject of the form of an officer's commission. The form adopted was borrowed originally from British commissions, and was " probably the pen work of some clerk, or at the most, the hasty direction of the Secretary of War, without reflecting that the chief magistrate in a republic is not the fountain of all honor and power," and that Congress alone has the power to raise armies, and to make rules for their government and regulation.

COMPANY. Companies are commanded by captains having under their orders lieutenants, sergeants, corporals, musicians, and privates. (*See* ARMY ORGANIZATION.)

COMPTROLLER. (*See* ACCOUNTABILITY.)

CONDUCT UNBECOMING AN OFFICER AND A GENTLEMAN—punished with dismission by sentence of general court-martial. What constitutes the offence is not defined, but it is left to the moral sense of the court-martial to determine.

CONFINEMENT. Non-commissioned officers and soldiers charged with crimes shall be confined until tried by a court-martial, or released by proper authority; (Art. 78.) No officer, or soldier who shall be put in arrest, shall continue in confinement more than eight days, or until such time as a court-martial can be assembled; (Art. 79.) (*See* Arrest.)

CONGRESS. (*See* Constitutional Relation of Congress.)

CONNIVING AT HIRING OF DUTY. If a non-commissioned officer, shall be reduced. If a commissioned officer, punished by the judgment of a general court-martial; (Art. 48.)

CONSCRIPTION. The only means of raising a National Army. The system of voluntary enlistments will always divide an army into two castes—officers and soldiers, and the latter will hardly ever be found qualified for promotion. The system of conscription is, too, the only means of raising large armies. This was made plain during the last war with England. Even with the largest bounties in land and money, soldiers could not be procured, and the President and Secretary of War (Messrs. Madison and Monroe) recommended in strong terms a system of conscription. The legislature of New York passed an act at the same time, for raising 12,000 troops by conscription. (*See* Defence, National; Raise.)

CONSTITUTION. The following provisions of the constitution relate to the land and naval forces: *Preamble*—We, the people of the United States, in order to * * provide for the common defence * * do ordain and establish this constitution for the United States of America.

Art. I. Sec. 1. All legislative powers herein granted, shall be vested in a Congress of the United States, which shall consist of a Senate and House of Representatives.

Art. I. Sec. 8. The Congress shall have power :—

Clause 1. * * To pay the debts and provide for the common defence and general welfare of the United States; * *

Clause 9. * * To define and punish offences against the law of nations; * *

Clause 10. To declare war, grant letters of marque and reprisal, and make rules concerning captures on land and water;

Clause 11. To raise and support armies; but no appropriation of money to that use, shall be for a longer term than two years;

Clause 12. To provide and maintain a navy;

Clause 13. To make rules for the government and regulation of the land and naval forces;

Clause 14. To provide for calling forth the militia to execute the laws of the Union, suppress insurrections, and repel invasions;

Clause 15. To provide for organizing, arming, and disciplining the militia, and for governing such part of them as may be employed in the service of the United States, reserving to the States, respectively, the appointment of the officers, and the authority of training the militia according to the discipline prescribed by Congress.

Clause 16. To exercise exclusive legislation * * over all places purchased, by consent of the legislature of the State in which the same shall be, for the erection of forts, magazines, arsenals, dock-yards, and other needful buildings—and

Clause 17. To make all laws which shall be necessary and proper for carrying into execution the foregoing powers, and all other powers vested by this constitution in the Government of the United States, or in any department or officer thereof.

Sec. 9. *Clause* 2. * * The privilege of the writ of habeas shall not be suspended, unless when, in cases of rebellion or invasion, the public safety may require it. * *

Sec. 10. *Clause* 2. * * No State shall, without the consent of Congress * * keep troops or ships of war in time of peace * * or engage in war, unless actually invaded, or in such imminent danger as will not admit of delay.

Art. II. Sec. 1. *Clause* 1. The executive power shall be vested in a President of the United States of America. * *

Sec. 2. *Clause* 1. The President shall be commander-in-chief of the army and navy of the United States, and of the militia of the several States, when called into the actual service of the United States. * *

Sec. 3. *Clause* 1. * * He shall take care that the laws be faithfully executed; and shall commission all officers of the United States.

Art. III. Sec. 3. *Clause* 1. Treason against the United States shall consist only in levying war against them, or in adhering to their enemies, giving them aid and comfort. No person shall be convicted of treason, unless on the testimony of two witnesses to the same overt act, or on confession in open court.

Clause 2. The Congress shall have power to declare the punishment of treason; but no attainder of treason shall work corruption of blood, or forfeiture, except during the life of the person attainted.

Art. IV. Sec. 4. *Clause* 1. The United States shall guarantee to every State in this Union a republican form of government; and shall protect each of them against invasion, and on the application of the legislature, or of the executive, (when the legislature cannot be convened,) against domestic violence.

Amendments to the Constitution:—1. Congress shall make no law respecting an establishment of religion, or prohibiting the free exercise thereof; abridging the freedom of speech, of the press; or the right of the people peaceably to assembly, and to petition the Government for redress of grievances.

Art. II. A well-regulated militia being necessary to the security of a free State, the right of the people to keep and bear arms shall not be infringed.

Art. III. No soldier shall, in time of peace, be quartered in any house, without the consent of the owner; nor in time of war, but in a manner to be prescribed by law.

Art. V. No person shall be held to answer for a capital or otherwise infamous crime, unless on a presentment or indictment by a grand jury, except in cases arising in the land or naval forces, or in the militia, when in actual service, in time of war, or public danger; nor shall any person be subject for the same offence to be twice put in jeopardy of life or limb; nor shall be compelled, in any criminal case, to be a witness against himself, nor be deprived of life, liberty, or property, without due process of law; nor shall private property be taken for public use without just compensation.

CONSTITUTIONAL RELATION OF CONGRESS AND THE PRESIDENT TO THE LAND AND NAVAL FORCES OF THE UNITED STATES. The power of making rules for the government and regulation of armies, as well as the power of raising armies, having in express terms been conferred on *Congress*, it is manifest that the President as commander-in-chief is limited by the constitution to the simple command of such armies as Congress may raise, under such rules for their government and regulation as Congress may appoint: "The authorities, (says Alexander Hamilton, *Federalist*, No. 23,) essential to the care of the common defence are these: To raise armies; to build and equip fleets; to prescribe rules for the government of both; to direct their operations; to provide for their support. These powers ought to exist without limitation; because it is impossible to foresee or to define the extent and variety of national exigencies, and the correspondent extent and variety of the *means* which may be necessary to satisfy them."

. . "Defective as the present (old) Confederation has been proved to be, this principle appears to have been fully recognized by the framers of it; although they have not made proper or adequate provision for its exercise. Congress have an unlimited discretion to make requisitions of men and money; to govern the army and navy; to direct their operations." "The government of the military is that branch of the code, (says BARDIN, *Dictionnaire de l'Armée de Terre,*) which embraces the military *Hierarchy,* or the gradual distribution of inferior authority." From this principle proceeds the localization of troops, their discipline, remuneration for important services, the repression of all infractions of the laws, and every thing in fine which the legislature may judge necessary either by rules of appointment or promotion, penalties or rewards, to maintain an efficient and well-disciplined army. But, as if to avoid all misconstruction on this point, the constitution not only declares that Congress shall make rules for the *government,* but also for the *regulation* of the army; and *regulation* signifies precise determination of functions; method, forms and restrictions, not to be departed from. It is evident, therefore, that the design of the framers of the constitution, was not to invest the President with powers over the army in any degree parallel with powers possessed by the king of Great Britain over the British army, whose prerogative embraces the *command* and *government* of all forces raised and maintained by him with the consent of parliament, (BLACKSTONE;) but their purpose, on the contrary, was to guard in all possible ways against executive usurpation by leaving with *Congress* the control of the Federal forces which it possessed under the articles of the Confederation, and at the same time to strengthen the powers of Congress by giving that body an unrestricted right to *raise* armies, provided appropriations for their support should not extend beyond two years. The command of the army and navy and militia called into service, subject to such rules for their government and regulation as Congress may make, was given by the constitution to the President; but the power of making rules of government and regulation is in reality that of SUPREME COMMAND, and hence the President, to use the language of the *Federalist,* in his relation to the army and navy, is nothing more than the "*first General and Admiral of the Confederacy;*" or the first officer of the military hierarchy with functions assigned by Congress. A curious example of this contemporaneous construction of the constitution is found in a letter from Sedgwick to Hamilton (vol. 6, Hamilton's Works, p. 394.) Congress, in raising a provisional army in 1798, created the office of commander of the army with the title of *Lieutenant-general.* A year subsequently a provision was made by law for changing this title to that of *General.* This last provision gave great offence to Mr. Adams, then President, who considered it as an evidence of the desire of Congress to make "*a general over the President.*" So strangely was he possessed with this idea that he never commissioned Washington as General, but the latter died in his office of *Lieutenant-*general; the President evidently thinking that the title of General conveyed a significancy which belonged to the President alone, although the commander of the army might in his opinion very properly take the title of *Lieutenant-general*, and thus have his subordination to the commander-in-chief of the army and navy and militia clearly indicated. It is plain therefore no less from the appointment by the constitution of the President as commander-in-chief, than from all contemporaneous construction, that his functions in respect to the army are those of First General of the U. S., and in no degree derived from his powers as first civil magistrate of the Union. The advocates of executive discretion over the army must therefore seek for the President's authority in his military capacity, restrained as that is by the powers granted to Congress, which embrace the raising, support, government, and regulation of armies; or, to use the language of the *Federalist*, No. 23, " there can be no limitation of that authority, which is to provide for the defence and protection of the community, in any matter essential to its efficacy; that is, in any matter essential to the *formation, direction,* or *support* of the NATIONAL FORCES." After the foregoing investigation of the unrestricted power of Congress in respect to the army, *save only in the appointment of the head of all the national forces, naval and military*, it will be plain that the 2d Section of the constitution, in giving to the President the nomination and appointment, by and with the advice and consent of the Senate, of *all other officers* of the United States, *whose appointments are not herein otherwise provided for, excludes* officers of the army and navy. The power of raising armies and making rules for their government and regulation, necessarily involves the power of making rules of appointment, promotion, reward, and punishment, and is therefore a provision in the constitution otherwise providing for the appointment of officers of the land and naval forces. So true is this that the principle has been acted on from the foundation of the Government. Laws have been passed giving to general and other officers the appointment of certain inferior officers. In other cases the President has been confined by Congress, in his selection for certain offices in the army, to particular classes. Again, rules have been made by Congress for the promotion of officers, another form of appointment; and in 1846, an army of volunteers was raised by Congress, the officers of which the acts of Congress directed should be appointed according to the laws of the States in which the troops were raised, excepting the general officers for those troops, who were to be appointed by the President and Senate (*Act* June 26, 1846)—a clear recognition that the troops thus raised by Congress were United States troops, and not militia. It is certainly true that the military legislation of the country has for long years vested a large discretion in the President in respect to appointments and other matters concerning the army; but it may well be asked whether fixed rules of appointments and promotion which would prevent the exercise of favoritism by the executive might not, with the greatest advantage to the army and the country, be adopted by Congress? "Military prejudices (says Gen. Hamilton) are not only inseparable from, but they are essential to the military profession. The government which desires to have a satisfied and useful army must consult them. They cannot be moulded at its pleasure; it is vain to aim at it." These are maxims which should lead Congress to the adoption of rules of appointment and promotion in the army which would prevent all outrages to the just pride of officers of the army. The organization of every new regiment, where the appointment of the officers has been left to executive discretion, shows that, if the desire has been felt in that quarter to cherish or cultivate pride of profession among the officers of the army, the feeling has been repressed by other considerations. All pride of rank has been so far crushed by this system of executive discretion that it is apparent, if Congress cannot provide a better rule for the government and regulation of the army, a generous rivalry in distinguished services must be superseded by political activity. Rules of appointment and promotion limiting the discretion of the President, and at the same time giving effect to opinions in the army, might easily be devised; or borrowed from existing rules in the French army, which, without ignoring the important principle of seniority, would at the same time afford scope and verge for rewards for distinguished services. (*See* PROMOTION.) No army can be kept in war in the highest vigor and efficiency without rewards for distinguished activity, and the appointment of Totleben at the siege of Sevastopol shows how far almost superhuman efforts may be prompted by investing a commander in the field with the power of selecting his immediate assistants. Colonels of regiments with us now exercise this authority in selecting regimental adjutants and quartermasters. Why should not the same trust be reposed in commanding generals of departments, brigades, divisions, and armies? And why should not all necessary restrictions (such as those in operation in the French armies) be put upon the President in making promotions for distinguished services, and also in original appointments, in order to secure justice to the army, and thereby promote the best interests of the country? (*Consult Federalist;* HAMILTON's *Works;* MADISON's *Works; Acts of Congress; Report of Committee of the Senate,* April 25, 1822. See PRESIDENT; RAISE; VICE-PRESIDENT; PROMOTION.)

CONTEMPT. Any officer or soldier who shall use contemptuous or disrespectful words against the President of the United States, the Vice-President, against the Congress of the United States, or against the chief magistrate or legislature of any of the United States in which he may be quartered, shall be punished as a court-martial shall direct. Any officer or soldier who shall behave himself with contempt or disrespect towards his commanding officer, shall be punished by the judgment of a court-martial; (ARTS. 5 and 6.)

No person whatsoever shall use any menacing words, signs, or gestures, in presence of a court-martial, or shall cause any riot or disorder, or disturb their proceedings, on the penalty of being punished at the discretion of the said court-martial; (ART. 76.) *Contempts* thus rendered summarily punishable by courts-martial are of public and self-evident kind, not depending on any interpretation of law admitting explanation, or requiring further investigation. Courts-martial sometimes act on this power. At other times individuals so offending are placed in arrest, and charges are preferred for trial. A regimental court-martial may punish summarily, but are not competent to award punishment to commissioned officers. A regimental court-martial in such cases would impose arrest. Citizens, not soldiers, would be removed from court; (HOUGH's *Military Law Authorities*.)

CONTRACTS. Supplies for the army, unless in particular and urgent cases the Secretary of War should otherwise direct, shall be purchased by contract, to be made by the commissary-general on public notice, to be delivered on inspection in bulk, and at such places as shall be stipulated; which contract shall be made under such regulations as the Secretary of War may direct; (*Act* April 14, 1818, *Sec.* 7.) No contract shall hereafter be made by the Secretary of State, or of the Treasury, or of the Department of War, or of the Navy, except under a law authorizing the same, or under an appropriation adequate to its fulfilment; and excepting also contracts for the subsistence and clothing of the army and navy, and contracts by the quartermaster's department which may be made by the secretaries of those departments; (*Act* May 1, 1820.) Members of Congress cannot be interested in any contract, and a special provision must be inserted in every contract that no member of Congress is interested in it. Penalty—forfeiture of three thousand dollars for making contracts with members of Congress; (*Act* April 21, 1808.)

Liability of Contracts.—By analogy to the rule which protects an officer from the treatment of a trespasser or malefactor, in regard to acts done by him in the execution of the orders of his own government, a similar immunity is extended to him in respect to contracts which he enters into for public purposes within the sphere of his authority. No private means or resources would otherwise be adequate to the responsibilities which, under any other rule, would effectually deter the best citizens of a state from rendering their services to the government. On high grounds, therefore, of public policy, it has long been established, that no action will lie against any government officer upon contracts made by him in his official character for public purposes, and within the legitimate scope of his duties.

"Great inconveniences (says Mr. Justice Ashurst) would result from considering a governor or commander as personally responsible in such cases. For no man would accept of any office of trust under government upon such conditions. And indeed it has been frequently determined that no individual is answerable for any engagements which he enters into on their behalf." . . . . . "In any case (says Mr. Justice Buller) where a man acts as agent for the public, and treats in that capacity, there is no pretence to say that he is personally liable." This doctrine applies in full force to military officers in the exercise of their professional duties. One of the earliest cases of this nature was Macheath v. Haldimand, in which it appeared that General Haldimand, being commander-in-chief and governor of Quebec, had, in those capacities, appointed Captain Sinclair to the command of a fort upon Lake Huron, with instructions to employ one Macheath in furnishing supplies for the service of the Crown. In pursuance of these orders, Macheath had furnished various articles for the use of the fort; and Captain Sinclair, according to his instructions from General Haldimand, drew bills upon him for the amount. Macheath also remitted his accounts to General Haldimand at Quebec, with the following words prefixed: "Government debtor to George Macheath for sundries paid by order of Lieutenant-governor Sinclair." General Haldimand objected to several of the charges, and refused payment of the amount; but ultimately made a partial payment on account, without prejudice to Macheath's right to the remainder, to recover which he brought the present action. At the trial it appeared so clearly that Macheath had dealt with General Haldimand solely in the character of commander-in-chief, and as an agent of government, that Mr. Justice Buller told the jury they were bound to find for the defendant in point of law. The jury gave their verdict accordingly; and upon the express ground of General Haldimand's freedom from personal liability in such a case, the Court of King's Bench were unanimous in refusing a new trial.

In a case which was tried before Lord Mansfield, one Savage brought an action against Lord North, as First Lord of the Treasury, for the expenses which he (Savage) had incurred in raising a regiment for the service of government; and Lord Mansfield held that the action did not lie. So in another case of Lutterlop v. Halsey, an action was brought against a commissary for the price of forage, supplied to the army by the plaintiff, at the request of the defendant, in his official character; and the commissary was held not to be liable. On another occasion, a suit was instituted in chancery against General Burgoyne, for a specific performance of a contract for the supply of artillery carriages in America. But Lord Chancellor Thurlow said there was no color for the demand as against General Burgoyne, who acted only as an agent for government; and his lordship dismissed the suit with costs. In 1818 an action was brought against Hall, the late purser of H. M. S. *La Belle Poule*, by the purser's steward of the same ship, to recover the amount of pay due to the latter for his services on board. It appeared that the purser's steward could not be appointed without the consent of the commander, and that he was entitled to the pay of an able seaman, but usually received pay under a private contract with the purser. The chief justice, Lord Ellenborough, at first felt some difficulty in the case; but considering how very extensive the operation of the principle might be, if such an action could be supported, and if a person, receiving a specific salary from the Crown in respect of his situation, could recover remuneration for his services from the officer under whose immediate authority he acted, and that the purser had no fund allowed him out of which such services were to be paid, his lordship was of opinion that the plaintiff had no right of action against the purser.

It is quite immaterial also, whether the officer gives the orders in person, or through a subordinate agent appointed by himself. The creditor cannot, in the latter case, charge the officer with a personal liability. In Myrtle v. Beaver, the plaintiff, a butcher at Brighton, brought an action against Major Beaver, the captain of a troop in the Hampshire Fencible Cavalry, for the price of meat supplied to the troop when quartered at Brighton, in January and February, 1800. One Bedford, a sergeant in the troop, had been employed by Major Beaver, according to his duty as captain, to provide for the subsistence of the men; and so long as Major Beaver remained with the troop, he regularly settled the butcher's bill monthly, up to the 24th January, 1800. At that date Major Beaver was detached with a small party to command at Arundel, the greater part of the regiment remaining at Brighton under the command of the colonel; and the command of

Major Beaver's troop, with the duties of providing for its subsistence, devolved on Lieutenant Hunt, who continued to employ Sergeant Bedford in providing supplies for the men, and gave him money for that purpose. The plaintiff furnished meat as before, under Sergeant Bedford's orders, but it did not appear that he had been apprised of the change of the authority, under which the sergeant gave those orders. On the 20th February, and before the usual monthly period of settling the butcher's bill, Lieutenant Hunt, who was also paymaster of the regiment, absconded with the regimental moneys, and left the plaintiff's demand and the regimental accounts unsettled. As Sergeant Bedford had, in the first instance, been accredited by Major Beaver, as his agent for ordering the supplies, the plaintiff Myrtle contended that until he had been informed of the discontinuance of that authority, he had a right to presume its continuance, and to look to Major Beaver for payment as before. But the Court of King's Bench held, that although the sergeant acted by Major Beaver's orders, he was not to be considered as the agent of a private individual, as it was plain that he acted as agent for whatever officer happened to have the command of the troop. There was, therefore, no ground for fixing Major Beaver with any personal liability in the matter.

An agent of government may, however, render himself personally liable upon contracts made by himself in the execution of his office. On this principle an action was brought against General Burgoyne, to recover a sum of money due to the plaintiff as provost-marshal of the British army in America; the general having promised that the plaintiff should be paid at the same rate as the provost-marshal under General Howe had been. At the trial, an objection was taken to the legality of the action; but Lord Mansfield refused to stop the case, and the plaintiff thereupon went into his evidence. It appeared, however, in the course of the inquiry, that the plaintiff's demand had been satisfied; and, therefore, the verdict was in favor of General Burgoyne. But it is evident from Lord Mansfield's suffering the trial to go on, that his lordship thought a commanding officer might so act as to make himself personally liable in such a case; and the question, whether he had so acted or not, was for the determination of a jury. In the next case it was accordingly sought to fix a naval officer with a personal liability for supplies furnished to his crew, on the ground of the language used by him on the occasion of ordering the supplies. Lieutenant Temple was first lieutenant of H. M. S. *Boyne*, and on her arrival at Portsmouth from the West Indies, he inquired for a slop-seller to supply the crew with new clothes, saying, "He will run no risk; I will see him paid." One Keate being accordingly recommended for this purpose, Lieutenant Temple called upon him and used these words, "I will see you paid at the pay-table; are you satisfied?" Keate answered, "Perfectly so." The clothes were delivered on the quarter-deck of the *Boyne*, though the case states that slops are usually sold on the main-deck. Lieutenant Temple produced samples to ascertain whether his directions were followed. Some of the men said that they were not in want of any clothes, but were told by the lieutenant that if they did not take them he would punish them; and others, who stated that they were only in want of part of a suit, were obliged to take a whole one, with anchor buttons to the jacket, such as were then worn by petty officers only. The former clothing of the crew was very light, and adapted to the climate of the West Indies, where the *Boyne* had been last stationed. Soon after the delivery of the slops, the *Boyne* was destroyed by fire, and the crew dispersed into different ships. On that occasion Keate, the slop-seller, expressed some apprehension for himself, but was thus answered by Lieutenant Temple:—"Captain Grey (Captain of the *Boyne*) and I will see you paid; you need not make yourself uneasy." After this the commissioner came on board the *Commerce de Marseilles* to pay the crew of the *Boyne*, at which time Lieutenant Temple stood at the pay-table, and took some money out of the hat of the first man who was paid, and gave it to the slop-seller. The next man, however, refused to part with his pay, and was immediately put in irons. Lieutenant Temple then asked the commissioner to stop the pay of the crew, but he answered that it could not be done. It was in evidence that though the crew were pretty well clothed, yet from the lightness of their clothing they were not properly equipped for the service in which they were engaged; and the compulsory purchases were not improperly ordered by the officer. Under these circumstances, Keate, the slop-seller, being unable to obtain the payment to which he was entitled, brought his action against Lieutenant Temple for the price of the clothing; and Mr. Justice Lawrence told the jury that if they were satisfied that the goods were advanced on the credit of the lieutenant as immediately responsible, Keate was entitled to recover the amount; but if they believed that Keate, on supplying the goods, relied merely on the lieutenant's assistance to get the money from the crew, the verdict ought to be in favor of the lieutenant. The jury found a verdict against Lieutenant Temple, but the Court of Common Pleas set it aside. Eyre, C. J.: "The sum recovered is 576*l*. 7*s*. 8*d*., and this against a lieutenant in the navy, a sum so large that it goes a great way towards satisfying my mind that it never could have been in contemplation of the defendant to make himself liable, *or of the slop-seller to furnish the goods on his credit.* I can hardly think that had the *Boyne* not been burnt, and the plaintiff been asked whether he would have the lieutenant or the crew for his paymaster, but that he would have given preference to the latter. . . . From the nature of the case it is apparent, that the men were to pay in the first instance; the defendant's words were, 'I will see you paid *at the pay-table;* are you satisfied?' and the answer was, 'Perfectly so;' the meaning of which was, that however unwilling the men might be to pay of themselves, the officer would take care that they should pay. . . . I think this a proper case to be sent to a new trial." The verdict found against Lieutenant Temple was accordingly set aside. But where an officer, acting in his private capacity and for his own private purposes, enters into any contract with another officer or a private individual, the ordinary rules and principles of law apply to such cases in the same manner as between civilians. (Consult PRENDERGAST.)

CONVOYS—have for their object the transportation of munitions of war, money, subsistence, clothing, arms, sick, &c. If convoys to an army do not come from the rear, through a country which has been mastered, and consequently far from the principal forces of the enemy, they will be undoubtedly attacked and broken up, if not carried off. There is no more difficult operation than to defend a large convoy against a serious attack. Ordinarily, convoys are only exposed to the attacks of partisan corps or light troops which, in consequence of their insignificant size, have thrown themselves in rear of the army. It is to guard against such attacks, that escorts are usually given to convoys. These escorts are principally infantry, because infantry fights in all varieties of ground, and in case of need may be placed in the intervals between the wagons, or even inside the wagons, when too warmly pressed. Cavalry is, however, also necessary to spy out an enemy at great distances, and give prompt information of his movements, as well as to participate in the defence of the convoy against cavalry. An enemy's cavalry being able rapidly to pass from the front to the rear of the train, would easily find some part of it without defence, if the escort were composed only of infantry. To give an idea of the facility of such attacks, it may be stated that a wagon drawn by four horses occupies ten yards. Two hundred wagons marching in single file and closed as much as possible form a train more than 2,000 yards in extent. In a long line of wagons, therefore, it would be impossible for infantry to meet the feints of cavalry and repulse real attacks.

The escort should then be composed of an advance guard entirely of cavalry preceding the train, some two or three miles, searching the route on the right and on the left; but as it may happen that the enemy, eluding the vigilance of the advance guard, have made ambuscades between the advance and the head of the column, it is necessary to place another body immediately in front of the train, with a small party in advance and flankers on the right and left. The longer the train the greater the danger of surprise, and consequently the greater the precautions to be used. A convoy is almost as much exposed to attack in rear as in front; it is therefore necessary to have, with a rear guard, some horsemen, who may be despatched to give information of what passes in rear. When the troops constituting the body of the escort are principally composed of infantry, they are divided into three bodies. Workmen will march with the advanced party, and the wagons loaded with tools of all kinds, rope, small beams, thick plank and every thing necessary for the repair of bridges and roads, will lead the convoy. The second detachment will be placed in the middle of the column of wagons, and the third in rear. Care is taken not to disseminate the troops along the whole extent of the train. A few men only are detached from the three bodies mentioned, to march abreast of the wagons, and to force the drivers to keep in their prescribed order, without opening the distance between the wagons. If a wagon breaks down on the route its load is promptly distributed among other wagons. A signal is made if it is necessary for the column to halt, but for slight repairs the train is not halted. The wagon leaves the column, is repaired on one side of the road, and afterwards takes its place in rear. Soldiers should never be permitted to place their knapsacks in the wagons, for a soldier should never be separated from knapsack or haversack, and the wagons would also become too much loaded. Whenever the breadth of the road permits, the wagons should be doubled and march in two files. The column is thus shortened one half, and if circumstances require it, the *defensive park* is more promptly formed. This is done by wheeling the wagons round to the right and left so as to bring the opposite horses' heads together and facing each other—turning towards the exterior the hind wagon wheels. This movement requires ground and time. It ought not to be ordered then except when absolutely necessary. It is much better to hold the enemy in check, by manœuvres of the escort when that can be done, and let the convoy move on. When the park has been formed, however, it constitutes an excellent means of defence, under shelter of which infantry can fight with advantage even when they have been compelled to take such refuge. A convoy usually halts for the night near a village, but it should always pass beyond it, because on commencing its march in the morning it is better to have the defile behind than before it, in order to avoid ambuscades of the enemy. Places for parking the wagons are sought where there are hedges or walls, as those obstructions offer greater security than any others. The troops, with the exception of the park guard, bivouac at a short distance from the park, in some position which offers the best military advantages. An advance guard and a sufficient number of sentinels for the safety and police of the park and bivouac are then posted. The park is ordinarily a hollow square, but locality will dictate its form. It should furnish an enclosed space for the horses and drivers, and at the same time be an intrenchment in case of attack. The wagons are ranged either lengthwise or side by side—the rule being that the poles are turned in the same direction and towards the place of destination. The wagons laid lengthwise may be doubled, so that the intervals of ranks may be closed by pushing forward the wagon of another rank. When the space for the park is small and the number of wagons great, the wagons are placed upon many lines, and streets sufficiently broad to receive the horses, &c., are made parallel to each other. The important principle in defending convoys on the march is, that the escort should not consider itself tied to wagons, but should repulse the enemy by marching to meet him. It is only after the escort has been repulsed, that it should fall back on the wagons and use them as an intrenchment. Even then a very long resistance may be ill judged if the enemy be greatly superior. It is better to abandon a part of the convoy to save the rest, or else try to destroy it, by cutting the traces, breaking the wheels, overthrowing the wagons, and even setting fire to the most inflammable parts. An attack upon a flank is most dangerous because the convoy then presents a larger mark. The three detachments in this case should be united on the side attacked and pushed forward sufficiently to compel the enemy to describe a great circle, in order to put himself out of reach when he wishes to attack the front or rear of the convoy. The best position to take is that of three echelons, the centre in advance. The convoy, which has doubled its wagons, continues to move forward, regulating its march by the position of the troops which cover it. If the attack be in front, as soon as the enemy has been announced by the first advance guard, which falls back at a gallop for the purpose, the wagons are closed or formed in two files if the road permits; the centre detachment joins the first, either in echelon or according to locality, to prevent a movement upon the flank of the convoy. The third detachment should be held in reserve immediately at the head of the wagons. If however this position be too near that taken by the first and second detachments united, the reserve must then take some position on the flank of the convoy. The defence against an attack upon the rear will be conducted on the same principles. It may be concluded that the attack of a convoy is an operation in which little is to be lost and much gained; for if the enemy be deficient in numbers or skill, a part of his convoy is easily destroyed or brought off. If the attack fail, nothing is to be feared upon retiring. The corps which attacks should be half cavalry and half infantry. It is clear, that if the attacking party has been concealed behind a wood, a height, a corn field, &c., and has been able to surprise the front or rear of the convoy, and enveloped it before aid arrives, full success will be obtained. But this negligence will not often occur on the part of the commander of the escort. If his troops then be in good order and united at the moment of the attack, it is necessary to divide his attention by directing against him many little columns and skirmishers, who seek to open a way to the wagons by killing the horses, and thus encumbering the road. The cavalry making a circuit throw themselves rapidly upon parts badly protected. If they reach some of the wagons they content themselves with driving off the conductors and cutting the traces of the wagons because all the wagons in rear are thus stopped. If we are at liberty to choose the time and place of attack, it is clear that the best time is when the convoy is passing a defile and we can envelop the front or the rear. Success is then certain; the inevitable encumbrance of the defile preventing one part of the troops from coming to the aid of another part. When the whole or part of a convoy has been seized, the prize must be brought to a safe place, before the enemy is in sufficient force to make us abandon it. But sooner than do this, the most precious articles should be placed on horses, the wagons should be destroyed, and the horses put to their speed. The attacking force should avoid further combat, for its object has been accomplished. (*Consult* DUFOUR; BARDIN; *Ordonnance sur le Service des Armées en Campagne*).

COOKING. Bread and soup are the great items of a soldier's diet: to make them well is, therefore, an essential part of his instruction. Scurvy and diarrhœa more frequently result from *bad* cooking than any other cause whatever. Camp ovens may be made in twenty-four hours. One hundred and ninety-six pounds when in dough hold about 11 gallons or 90 pounds of water, 2 gallons yeast, and 3 pounds salt, making a mass of 305 pounds, which evaporates in kneading, baking, and cooling about 40 pounds, leaving in bread weighed when stale about 265 pounds. Bread ought not to be burnt, but baked to an equal brown color. The troops ought not to be allowed to eat soft bread fresh from the oven without first toasting it. Fresh meat ought not to be cooked before it has had time to bleed and to cool; and meats will generally be *boiled*, with a view to soup; and sometimes roasted or baked. Meat may be kept in hot weather by half boiling it; or by exposing it for a few minutes to a thick smoke. To make soup, put into the vessel at the rate of five pints of water to a pound of fresh meat; apply a quick heat, to make it boil promptly; skim off the foam, and then moderate the fire; put in salt according to palate. Add the vegetables of the season one or two hours, and sliced bread some minutes before the simmering is ended. When the broth is sensibly reduced in quantity, that is, after five or six hours' cooking, the process will be complete. If a part of the meat be withdrawn before the soup is fully made, the quantity of water must be proportionally less. Hard or dry vegetables, as the bean ration, will be put in the camp kettle much earlier than fresh vegetables. The following receipts for army cooking are taken from Soyer's *Culinary Campaign*:

SOYER'S HOSPITAL DIETS.

THE IMPORTANCE OF WEIGHTS AND MEASURES IN THE ACCOMPANYING RECEIPTS IS FULLY RECOGNIZED; IT IS THEREFORE NECESSARY THAT TROOPS SHOULD BE SUPPLIED WITH SCALES, AND WITH MEASURES FOR LIQUIDS.

No. 1.—SEMI-STEWED MUTTON AND BARLEY. SOUP FOR 100 MEN. Put in a convenient-sized caldron 130 pints of cold water, 70 lbs. of meat, or about that quantity, 12 lbs. of plain mixed vegetables, (the best that can be obtained,) 9 lbs. 6 oz. of barley, 1 lb. 7 oz. of salt, 1 lb. 4 oz. of flour, 1 lb. 4 oz. of sugar, 1 oz. of pepper. Put all the ingredients into the pan at once, except the flour; set it on the fire, and when beginning to boil, diminish the heat, and simmer gently for two hours and a half; take the joints of meat out, and keep them warm in the orderly's pan; add to the soup your flour, which you have mixed with enough water to form a light batter; stir well together with a large spoon; boil another half-hour, skim off the fat, and serve the soup and meat separate. The meat may be put back into the soup for a few minutes to warm again prior to serving. The soup should be stirred now and then while making, to prevent burning or sticking to the bottom of the caldron. The joints are cooked whole, and afterwards cut up in different messes; being cooked this way, in a rather thick stock, the meat becomes more nutritious.

Note.—The word "about" is applied to the half and full diet, which varies the weight of the meat; but ½ lb. of mutton will always make a pint of good soup: 3 lbs. of mixed preserved vegetables must be used when fresh are not to be obtained, and put in one hour and a half prior to serving, instead of at first; they will then show better in the soup, and still be well done. All the following receipts may be increased to large quantities, but by all means closely follow the weight and measure.

No. 2.—BEEF SOUP. Proceed the same as for mutton, only leave the meat in till serving, as it will take longer than mutton. The pieces are not to be above 4 or 5 lbs. weight; and for a change, half rice may be introduced; the addition of 2 lbs more will make it thicker and more nutritive; ¼ lb. of curry powder will make an excellent change also. To vary the same, half a pint of burnt sugar water may be added— it will give the soup a very rich brown color.

No. 3—BEEF TEA. RECEIPT FOR SIX PINTS. Cut 3 lbs. of beef into pieces the size of walnuts, and chop up the bones, if any; put it into a convenient-sized kettle, with ½ lb. of mixed vegetables, such as onions, leeks, celery, turnips, carrots, (or one or two of these, if all are not to be obtained,) 1 oz. of salt, a little pepper, 1 teaspoonful of sugar, 2 oz. of butter, half a pint of water. Set it on a sharp fire for ten minutes or a quarter of an hour, stirring now and then with a spoon, till it forms a rather thick gravy at bottom, but not brown: then add 7 pints of hot or cold water, but hot is preferable; when boiling, let it simmer gently for an hour; skim off all the fat, strain it through a sieve, and serve.

No. 3A.—ESSENCE OF BEEF TEA. For camp hospitals.—" Quarter pound tin case of essence." If in winter set it near the fire to melt; pour the contents in a stewpan and twelve times the case full of water over it, hot or cold; add to it two or three slices of onion, a sprig or two of parsley, a leaf or two of celery, if handy, two teaspoonfuls of salt, one of sugar; pass through a colander and serve. If required stronger, eight cases of water will suffice, decreasing the seasoning in proportion. In case you have no vegetables, sugar, or pepper, salt alone will do, but the broth will not be so succulent.

No. 4.—THICK BEEF TEA. Dissolve a good teaspoonful of arrowroot in a gill of water, and pour it into the beef tea twenty minutes before passing through the sieve—it is then ready.

No. 5.—STRENGTHENING BEEF TEA WITH CALVES-FOOT JELLY, OR ISINGLASS. Add ¼ oz. calves-foot gelatine to the above quantity of beef tea previous to serving, when cooking.

No. 6.—MUTTON AND VEAL TEA. Mutton and veal will make good tea by proceeding precisely the same as above. The addition of a little aromatic herbs is always desirable. If no fresh vegetables are at hand, use 2 oz. of mixed preserved vegetables to any of the above receipts.

No. 7.—CHICKEN BROTH. Put in a stewpan a fowl, 3 pints of water, 2 teaspoonfuls of rice, 1 teaspoonful of salt, a middle-sized onion, or 2 oz. of mixed vegetables; boil the whole gently for three-quarters of an hour: if an old fowl, simmer from one hour and a half to two hours, adding 1 pint more water; skim off the fat and serve. A small fowl will do.

Note.—A light mutton broth may be made precisely the same, by using a pound and a half of scrag of mutton instead of fowl. For thick mutton broth proceed as for thick beef tea, omitting the rice; a tablespoonful of burnt sugar water will give a rich color to the broth.

No. 8.—PLAIN BOILED RICE. Put two quarts of water in a stewpan, with a teaspoonful of salt; when boiling, add to it $\frac{1}{2}$ lb. of rice, well washed; boil for ten minutes, or till each grain becomes rather soft; drain it into a colander, slightly grease the pot with butter, and put the rice back into it; let it swell slowly for about twenty minutes near the fire, or in a slow oven; each grain will then swell up, and be well separated; it is then ready for use.

No. 9.—SWEET RICE. Add to the plain boiled rice 1 oz. of butter, 2 tablespoonfuls of sugar, a little cinnamon, a quarter of a pint of milk; stir it with a fork, and serve; a little currant jelly or jam may be added to the rice.

No. 10.—RICE WITH GRAVY. Add to the rice 4 tablespoonfuls of the essence of beef, a little butter, if fresh, half a teaspoonful of salt; stir together with a fork, and serve. A teaspoonful of Soyer's Sultana Sauce, or relish, will make it very wholesome and palatable, as well as invigorating to a fatigued stomach.

No. 11.—PLAIN OATMEAL. Put in a pan $\frac{1}{4}$ lb. of oatmeal, $1\frac{1}{2}$ oz. of sugar, half a teaspoonful of salt, and 3 pints of water; boil slowly for twenty minutes, "stirring continually," and serve. A quarter of a pint of boiled milk, an ounce of butter, and a little pounded cinnamon or spice added previous to serving is a good variation. This receipt has been found most useful at the commencement of dysentery by the medical authorities.

No. 12.—CALVES-FOOT JELLY. Put in a proper-sized stewpan $2\frac{1}{4}$ oz. of calves-foot gelatine, 4 oz. of white sugar, 4 whites of eggs and shells, the peel of a lemon, the juice of three middle-sized lemons, half a pint of Marsala wine; beat all well together with the egg-beater for a few minutes, then add $4\frac{1}{2}$ pints of cold water; set it on a slow fire, and keep whipping it till boiling. Set it on the corner of the stove, partly covered with the lid, upon which you place a few pieces of burning charcoal; let it simmer gently for ten minutes, and strain it through a jelly-bag. It is then ready to put in the ice or some cool place. Sherry will do if Marsala is not at hand. For orange jelly use only 1 lemon and 2 oranges. Any delicate flavor may be introduced.

JELLY STOCK, made from calves' feet, requires to be made the day previous to being used, requiring to be very hard to extract the fat. Take two calf's feet, cut them up, and boil in three quarts of water; as soon as it boils remove it to the corner of the fire, and simmer for five hours, keeping it skimmed, pass through a hair sieve into a basin, and let it remain until quite hard, then remove the oil and fat, and wipe the top dry. Place in a stewpan half a pint of water, one of sherry, half a pound of lump sugar, the juice of four lemons, the rinds of two, and the whites and shells of five eggs; whisk until the sugar is melted, then add the jelly, place it on the fire, and whisk until boiling, pass it through a jelly-bag, pouring that back again which comes through first until quite clear; it is then ready for use, by putting it in moulds or glasses. Vary the flavor according to fancy.

No. 13.—SAGO JELLY. Put into a pan 3 oz. of sago, 1½ oz. of sugar, half a lemon-peel cut very thin, ¼ teaspoonful of ground cinnamon, or a small stick of the same; put to it 3 pints of water and a little salt; boil ten minutes, or rather longer, stirring continually, until rather thick, then add a little port, sherry, or Marsala wine; mix well, and serve hot or cold.

No. 14.—ARROWROOT MILK. Put into a pan 4 oz. of arrowroot, 3 oz. of sugar, the peel of half a lemon, ¼ teaspoonful of salt, 2½ pints of milk; set it on the fire, stir round gently, boil for ten minutes, and serve. If no lemons at hand, a little essence of any kind will do. When short of milk, use half water; half an ounce of fresh butter is an improvement before serving. If required thicker, put a little milk.

No. 15.—THICK ARROWROOT PANADA. Put in a pan 5 oz. of arrowroot, 2¼ oz. of white sugar, the peel of half a lemon, a quarter of a teaspoonful of salt, 4 pints of water; mix all well, set on the fire, boil for ten minutes; it is then ready. The juice of a lemon is an improvement; a gill of wine may also be introduced, and ½ oz. of calves-foot gelatine previously dissolved in water will be strengthening. Milk, however, is preferable, if at hand.

No. 16.—ARROWROOT WATER. Put into a pan 3 oz. of arrowroot, 2 oz. of white sugar, the peel of a lemon, ¼ teaspoonful of salt, 4 pints of water; mix well, set on the fire, boil for ten minutes. It is then ready to serve either hot or cold.

No. 17.—RICE WATER. Put 7 pints of water to boil, add to it 2 ounces of rice washed, 2 oz. of sugar, the peel of two-thirds of a lemon; boil gently for three-quarters of an hour; it will reduce to 5 pints; strain through a colander; it is then ready. The rice may be left in the beverage or made into a pudding, or by the addition of a little sugar or jam, will be found very good for either children or invalids.

No. 18.—BARLEY WATER. Put in a saucepan 7 pints of water, 2 oz. of barley, which stir now and then while boiling; add 2 oz. of white sugar, the rind of half a lemon, thinly peeled; let it boil gently for about two hours, without covering it; pass it through a sieve or colander; it is then ready. The barley and lemon may be left in it.

No. 19.—SOYER'S PLAIN LEMONADE. Thinly peel the third part of a lemon, which put into a basin with 2 tablespoonfuls of sugar; roll the lemon with your hand upon the table to soften it; cut it into two, lengthwise, squeeze the juice over the peel, &c., stir round for a minute with a spoon to form a sort of syrup; pour over a pint of water, mix well, and remove the pips; it is then ready for use. If a very large lemon, and full of juice, and very fresh, you may make a pint and a half to a quart, adding sugar and peel in proportion to the increase of water. The juice only of the lemon and sugar will make lemonade, but will then be deprived of the aroma which the rind contains, the said rind being generally thrown away.

No. 20.—SEMI-CITRIC LEMONADE. RECEIPT FOR 50 PINTS. Put 1 oz. of citric acid to dissolve in a pint of water, peel 20 lemons thinly, and put the peel in a large vessel, with 3 lbs. 2 oz. of white sugar well broken; roll each lemon on the table to soften it, which will facilitate the extraction of the juice; cut them into two, and press out the juice into a colander or sieve, over the peel and sugar, then pour half a pint of water through the colander, so as to leave no juice remaining; triturate the sugar, juice, and peel together for a minute or two with a spoon, so as to form a sort of syrup, and extract the aroma from the peel and the dissolved citric acid; mix all well together, pour on 50 pints of cold water, stir well together; it is then ready. A little ice in summer is a great addition.

No. 21.—SOYER'S CHEAP CRIMEAN LEMONADE. Put into a basin 2 tablespoonfuls of white or brown sugar, ½ a tablespoonful of lime juice, mix well together for one minute, add 1 pint of water, and the beverage is ready. A drop of rum will make a good variation, as lime juice and rum are daily issued to the soldiers.

No. 22.—TARTARIC LEMONADE. Dissolve 1 oz. of crystallized tartaric acid in a pint of cold water, which put in a large vessel; when dissolved, add 1 lb. 9 oz. of white or brown sugar—the former is preferable; mix well to form a thick syrup; add to it 24 pints of cold water, slowly mixing well; it is then ready. It may be strained through either a colander or a jelly-bag; if required very light, add 5 pints more water, and sugar in proportion; if citric acid be used, put only 20 pints of water to each ounce.

No. 23.—CHEAP PLAIN RICE PUDDING, FOR CAMPAIGNING, in which no eggs or milk are required: important in the field. Put on the fire, in a moderate-sized saucepan, 12 pints of water; when boiling, add to it 1 lb. of rice or 16 tablespoonfuls, 4 oz. of brown sugar or 4 tablespoonfuls, 1 large teaspoonful of salt, and the rind of a lemon thinly peeled; boil gently for half an hour, then strain all the water from the rice, keeping it as dry as possible. The rice water is then ready for drinking, either warm or cold. The juice of a lemon may be introduced, which will make it more palatable and refreshing.

THE PUDDING. Add to the rice 3 oz. of sugar, 4 tablespoonfuls of flour, half a teaspoonful of pounded cinnamon; stir it on the fire carefully for five or ten minutes; put it in a tin or pie-dish, and bake. By boiling the rice a quarter of an hour longer, it will be very good to eat without baking. Cinnamon may be omitted.

No. 23A.—BATTER PUDDING. Break two fresh eggs in a basin, beat them well, add one tablespoonful and a half of flour, which beat up with your eggs with a fork until no lumps remain; add a gill of milk, a teaspoonful of salt, butter a teacup or a basin, pour in your mixture, put some water in a stewpan, enough to immerge half way up the cup or basin in water; when boiling, put in your cup or basin and boil twenty minutes, or till your pudding is well set; pass a knife to loosen it, turn out on a plate, pour pounded sugar and a pat of fresh butter over, and serve. A little lemon, cinnamon, or a drop of any essence may be introduced. A little light melted butter, sherry, and sugar may be poured over. If required more delicate, add a little less flour. It may be served plain

No. 24.—BREAD AND BUTTER PUDDING. Butter a tart-dish well, and sprinkle some currants all round it, then lay in a few slices of bread and butter; boil one pint of milk, pour it on two eggs well whipped, and then on the bread and butter; bake it in a hot oven for half an hour. Currants may be omitted.

No. 25.—BREAD PUDDING. Boil one pint of milk, with a piece of cinnamon and lemon-peel; pour it on two ounces of bread crumbs; then add two eggs, half an ounce of currants, and a little sugar: steam it in a buttered mould for one hour.

No. 26.—CUSTARD PUDDING. Boil one pint of milk, with a small piece of lemon-peel and half a bay-leaf, for three minutes; then pour these on to three eggs, mix it with one ounce of sugar well together, and pour it into a buttered mould: steam it twenty-five minutes in a stewpan with some water, turn out on a plate and serve.

No. 27.—RICH RICE PUDDING. Put in ½ lb. of rice in a stewpan, washed, 3 pints of milk, 1 pint of water, 3 oz. of sugar, 1 lemon peel, 1 oz. of fresh butter; boil gently half an hour, or until the rice is tender; add 4 eggs, well beaten, mix well, and bake quickly for half an hour, and serve: it may be steamed if preferred.

No. 28.—STEWED MACARONI. Put in a stewpan 2 quarts of water, half a tablespoonful of salt, 2 oz. of butter; set on the fire; when boiling, add 1 lb. of macaroni, broken up rather small; when boiled very soft, throw off the water; mix well into the macaroni a tablespoonful of flour, add enough milk to make it of the consistency of thin melted butter; boil gently twenty minutes; add in a tablespoonful of either brown or white sugar, or honey, and serve. A little cinnamon, nutmeg, lemon-peel, or orange-flower water may be introduced to impart a flavor; stir quick. A gill of milk or cream may now be thrown in three minutes before serving. Nothing can be more light and nutritious than macaroni done this way. If no milk, use water.

No. 29.—MACARONI PUDDING. Put 2 pints of water to boil, add to it 2 oz. of macaroni, broken in small pieces; boil till tender, drain off the water and add half a tablespoonful of flour, 2 oz. of white sugar, a quarter of a pint of milk, and boil together for ten minutes; beat an egg up, pour it to the other ingredients, a nut of butter; mix well and bake, or steam. It can be served plain, and may be flavored with either cinnamon, lemon, or other essences, as orange-flower water, vanilla, &c.

No. 30.—SAGO PUDDING. Put in a pan 4 oz. of sago, 2 oz. of sugar, half a lemon-peel or a little cinnamon, a small pat of fresh butter, if handy, half a pint of milk; boil for a few minutes, or until rather thick, stirring all the while; beat up 2 eggs and mix quickly with the same; it is then ready for either baking or steaming, or may be served plain.

No. 31.—TAPIOCA PUDDING. Put in a pan 2 oz. of tapioca, 1½ pint of milk, 1 oz. of white or brown sugar, a little salt, set on the fire, boil gently for fifteen minutes, or until the tapioca is tender, stirring now and then to prevent its sticking to the bottom, or burning; then add two eggs well beaten; steam or bake, and serve. It will take about twenty minutes steaming, or a quarter of an hour baking slightly. Flavor with either lemon, cinnamon, or any other essence.

No. 32.—BOILED RICE SEMI-CURRIED, FOR THE PREMONITORY SYMPTOMS OF DIARRHŒA. Put 1 quart of water in a pot or saucepan; when boiling, wash ½ a lb. of rice and throw it into the water; boil fast for ten minutes; drain your rice in a colander, put it back in the saucepan, which you have slightly greased with butter; let it swell slowly near the fire, or in a slow oven till tender; each grain will then be light and well separated. Add to the above a small tablespoonful of aromatic sauce, called "Soyer's Relish or Sultana Sauce," with a quarter of a teaspoonful of curry powder; mix together with a fork lightly, and serve. This quantity will be sufficient for two or three people, according to the prescriptions of the attending physician.

No. 33.—FIGS AND APPLE BEVERAGE. Have 2 quarts of water boiling, into which throw 6 dry figs previously opened, and 2 apples, cut into six or eight slices each; let the whole boil together twenty minutes; then pour them into a basin to cool; pass through a sieve; drain the figs, which will be good to eat with a little sugar or jam.

No. 34.—STEWED FRENCH PLUMS. Put 12 large or 18 small-size French plums, soak them for half an hour, put in a stewpan with a spoonful of brown sugar, a gill of water, a little cinnamon, and some thin rind of lemon; let them stew gently twenty minutes, then put them in a basin till cold with a little of the juice. A small glass of either port, sherry, or claret is a very good addition. The syrup is excellent.

No. 35.—FRENCH HERB BROTH. This is a very favorite beverage in France, as well with people in health as with invalids, especially in spring, when the herbs are young and green. Put a quart of water to boil, having previously prepared about 40 leaves of sorrel, a cabbage lettuce, and 10 sprigs of chervil, the whole well washed; when the water is boiling, throw in the herbs, with the addition of a teaspoonful of salt, and ½ oz. of fresh butter; cover the saucepan close, and let simmer a few minutes, then strain it through a sieve or colander. This is to be drunk cold, especially in the spring of the year, after the change from winter. I generally drink about a quart per day for a week at that time; but if for sick people, it must be made less strong of herbs, and taken a little warm. To prove that it is wholesome, we have only to refer to the instinct which teaches dogs to eat grass at that season of the year. I do not pretend to say that it would suit persons in every malady, because the doctors are to decide upon the food and beverage of their patients, and study its changes as well as change their medicines; but I repeat that this is most useful and refreshing for the blood.

No. 36.—BROWNING FOR SOUPS, &c. Put ½ lb. of moist sugar

into an iron pan and melt it over a moderate fire till quite black, stirring it continually, which will take about twenty-five minutes: it must color by degrees, as too sudden a heat will make it bitter; then add 2 quarts of water, and in ten minutes the sugar will be dissolved. You may then bottle it for use. It will keep good for a month, and will always be found very useful.

No. 37.—TOAST-AND-WATER. Cut a piece of crusty bread, about a ¼ lb. in weight, place it upon a toasting-fork, and hold it about six inches from the fire; turn it often, and keep moving it gently until of a light-yellow color, then place it nearer the fire, and when of a good brown chocolate color, put it in a jug and pour over 3 pints of boiling water; cover the jug until cold, then strain it into a clean jug, and it is ready for use. Never leave the toast in it, for in summer it would cause fermentation in a short time.

Baked Apple Toast-and-Water.—A piece of apple, slowly toasted till it gets quite black and added to the above, makes a very nice and refreshing drink for invalids.

Apple Rice Water.—Half a pound of rice, boiled in the above until in pulp, passed through a colander, and drunk when cold. All kinds of fruit may be done the same way. Figs and French plums are excellent; also raisins. A little ginger, if approved of, may be used.

Apple Barley Water.—A quarter of a pound of pearl barley instead of toast added to the above, and boil for one hour, is also a very nice drink.

Citronade.—Put a gallon of water on to boil, cut up one pound of apples, each one into quarters, two lemons in thin slices, put them in the water, and boil them until they can be pulped, pass the liquor through a colander, boil it up again with half a pound of brown sugar, skim, and bottle for use, taking care not to cork the bottle, and keep it in a cool place.

For Spring Drink.—Rhubarb, in the same quantities, and done in the same way as apples, adding more sugar, is very cooling. Also green gooseberries.

For Summer Drink.—One pound of red currants, bruised with some raspberry, half a pound of sugar added to a gallon of cold water, well stirred, and allowed to settle. The juice of a lemon.

Mulberry.—The same, adding a little lemon-peel. A little cream of tartar or citric acid added to these renders them more cooling in summer and spring.

Plain Lemonade.—Cut in very thin slices three lemons, put them in a basin, add half a pound of sugar, either white or brown; bruise all together, add a gallon of water, and stir well. It is then ready.

French Plum Water.—Boil 3 pints of water; add in 6 or 8 dried plums previously split, 2 or 3 slices of lemon, a spoonful of honey or sugar; boil half an hour, and serve.

For *Fig, Date, and Raisin Water*, proceed as above, adding the juice of half a lemon to any of the above. If for fig water, use 6 figs. Any quantity of the above fruits may be used with advantage in rice, barley, or arrowroot water.

EFFERVESCENT BEVERAGES. *Raspberry Water.*—Put 2 tablespoonfuls of vinegar into a large glass, pour in half a pint of water; mix well.

Pine-Apple Syrup.—Three tablespoonfuls to a pint.

Currant Syrup.—Proceed the same.

Syrup of Orgeat.—The same.

FIELD AND BARRACK COOKERY FOR THE ARMY, BY THE USE OF SOYER'S NEW FIELD STOVE, NOW ADOPTED BY THE MILITARY AUTHORITIES.—Each stove will consume not more than from 12 to 15 lbs. of fuel, and allowing 20 stoves to a regiment, the consumption would be 300 lbs. per thousand men. Coal will burn with the same advantage. Salt beef, pork, Irish stew, stewed beef, tea, coffee, cocoa, &c., can be prepared in these stoves, and with the same economy. They can also be fitted with an apparatus for baking, roasting, and steaming.

FIG. 105.

NO. 1.—RECEIPT TO COOK SALT MEAT FOR FIFTY MEN. 1. Put 50 lbs. of meat in the boiler. 2. Fill with water, and let soak all night. 3. Next morning wash the meat well 4. Fill with fresh water, and boil gently three hours, and serve. Skim off the fat, which, when cold, is an excellent substitute for butter. For salt pork proceed as above or boil half beef and half pork—the pieces of beef may be smaller than the pork, requiring a little longer time doing.

Dumplings, No. 21, may be added to either pork, or beef in proportion; and when pork is properly soaked, the liquor will make a very good soup. The large yellow peas, as used by the navy, may be introduced; it is important to have them, as they are a great improvement. When properly soaked, French haricot beans and lentils may also be used to advantage. By the addition of 5 pounds of split peas, half a pound of brown sugar, 2 tablespoonfuls of pepper, 10 onions; simmer gently till in pulp, remove the fat and serve; broken biscuit may be introduced. This will make an excellent mess.

No. 1A.—How to soak and plain-boil the Rations of Salt Beef and Pork, on Land or at Sea. To each pound of meat allow about a pint of water. Do not have the pieces above 3 or 4 lbs. in weight. Let it soak for 7 or 8 hours, or all night if possible. Wash each piece well with your hand in order to extract as much salt as possible. It is then ready for cooking. If less time be allowed, cut the pieces smaller and proceed the same, or parboil the meat for 20 minutes in the above quantity of water, which throw off and add fresh. Meat may be soaked in sea water, but by all means boiled in fresh when possible. I should advise, at sea, to have a perforated iron box made, large enough to contain half a ton or more of meat, which box will ascend and descend by pulleys; have also a frame made on which the box might rest when lowered overboard, the meat being placed outside the ship on a level with the water, the night before using; the water beating against the meat through the perforations will extract all the salt. Meat may be soaked in sea water, but by all means washed.

No 2.—Soyer's Army Soup for Fifty Men. 1. Put in the boiler 60 pints, $7\frac{1}{2}$ gallons, or $5\frac{1}{2}$ camp kettles of water. 2. Add to it 50 lbs. of meat, either beef or mutton. 3. The rations of preserved or fresh vegetables. 4. Ten small tablespoonfuls of salt. 5. Simmer three hours and serve. When rice is issued, put it in when boiling. Three pounds will be sufficient. About eight pounds of fresh vegetables. Or four squares from a cake of preserved vegetables. A tablespoonful of pepper, if handy. Skim off the fat, which, when cold, is an excellent substitute for butter.

No. 2A.—Salt Pork with Mashed Peas, for One Hundred Men. Put in two stoves 50 lbs. of pork each, divide 24 lbs. in four pudding-cloths, rather loosely tied; putting to boil at the same time as your pork, let all boil gently till done, say about two hours; take out the pudding and peas, put all the meat in one caldron, remove the liquor from the other pan, turning back the peas in it, add two teaspoonfuls of pepper, a pound of the fat, and with the wooden spatula smash the peas and serve both. The addition of about half a pound of flour, and two quarts of liquor, boiled ten minutes, makes a great improvement. Six sliced onions, fried and added to it, make it very delicate.

No. 3.—STEWED SALT BEEF AND PORK. For a company of one hundred men, or a regiment of one thousand men. Put in a boiler, of well soaked-beef 30 lbs., cut in pieces of a quarter of a pound each, 20 lbs. of pork, 1½ lb. of sugar, 8 lbs. of onions, sliced, 25 quarts of water, 4 lbs. of rice. Simmer gently for three hours, skim the fat off the top, and serve.

Note.—How to soak the meat for the above mess:—Put 50 lbs. of meat in each boiler, having filled them with water, and let soak all night; and prior to using it, wash it and squeeze with your hands, to extract the salt. In case the meat is still too salt, boil it for twenty minutes, throw away the water, and put fresh to your stew. By closely following the above receipt you will have an excellent dish.

No. 4.—SOYER'S FOOD FOR ONE HUNDRED MEN, USING TWO STOVES. Cut or chop 50 lbs. of fresh beef in pieces of about a ¼ lb. each; put in the boiler, with 10 tablespoonfuls of salt, two tablespoonfuls of pepper, four tablespoonfuls of sugar, onions 7 lbs. cut in slices: light the fire now, and then stir the meat with a spatula, let it stew from 20 to 30 minutes, or till it forms a thick gravy, then add a pound and a half of flour; mix well together, put in the boiler 18 quarts of water, stir well for a minute or two, regulate the stove to a moderate heat, and let simmer for about two hours. Mutton, pork, or veal can be stewed in a similar manner, but will take half an hour less cooking.

Note.—A pound of rice may be added with great advantage, ditto plain dumplings, ditto potatoes, as well as mixed vegetables. For a regiment of 1,000 men use 20 stoves.

No. 5.—PLAIN IRISH STEW FOR FIFTY MEN. Cut 50 lbs. of mutton into pieces of a quarter of a pound each, put them in the pan, add 8 lbs. of large onions, 12 lbs. of whole potatoes, 8 tablespoonfuls of salt, 3 tablespoonfuls of pepper; cover all with water, giving about half a pint to each pound; then light the fire; one hour and a half of gentle ebullition will make a most excellent stew; mash some of the potatoes to thicken the gravy, and serve. Fresh beef, veal, or pork will also make a good stew. Beef takes two hours doing. Dumplings may be added half an hour before done.

No. 6.—TO COOK FOR A REGIMENT OF A THOUSAND MEN. Place twenty stoves in a row, in the open air or under cover. Put 30 quarts of water in each boiler, 50 lbs. of ration meat, 4 squares from a cake of dried vegetables—or, if fresh mixed vegetables are issued, 12 lbs. weight—10 small tablespoonfuls of salt, 1 ditto of pepper; light the fire, simmer gently from two hours to two hours and a half, skim the fat from the top, and serve. It will require only four cooks per regiment, the provisions and water being carried to the kitchen by fatigue parties; the kitchen being central, instead of the kitchen going to each company, each company sends two men to the kitchen with a pole to carry the meat.

No. 7.—SALT PORK AND PUDDINGS WITH CABBAGE AND POTATOES. Put 25 lbs. of salt pork in each boiler, with 50 lbs. from which you have extracted the large bones, cut in dice, and made into puddings; when on the boil, put five puddings in each, boil rather fast for two hours. You have peeled 12 lbs. of potatoes and put in a net in each caldron; put also 2 winter cabbages in nets, three-quarters of an hour before your pudding is done; divide the pork, pudding, and cabbage, in proportion, or let fifty of the men have pudding that day and meat the other; remove the fat, and serve. The liquor will make very good soup by adding peas or rice, as No. 1. For the pudding-paste put one-quarter of a pound of dripping, or beef or mutton suet, to every pound of flour you use; roll your paste for each half an inch thick, put a pudding-cloth in a basin, flour round, lay in your paste, add your meat in proportion; season with pepper and a minced onion; close your pudding in a cloth, and boil. This receipt is more applicable to barrack and public institutions than a camp. Fresh meat of any kind may be done the same, and boiled with either salt pork or beef.

No. 8.—TURKISH PILAFF FOR ONE HUNDRED MEN. Put in the caldron 2 lbs. of fat, which you have saved from salt pork, add to it 4 lbs. of peeled and sliced onions; let them fry in the fat for about ten minutes; add in then 12 lbs. of rice, cover the rice over with water, the rice being submerged two inches, add to it 7 tablespoonfuls of salt, and 1 of pepper; let simmer gently for about an hour, stirring it with a spatula occasionally to prevent it burning, but when commencing to boil, a very little fire ought to be kept under. Each grain ought to be swollen to the full size of rice, and separate. In the other stove put fat and onions the same quantity with the same seasoning; cut the flesh of the mutton, veal, pork, or beef from the bone, cut in dice of about 2 oz. each, put in the pan with the fat and onions, set it going with a very sharp fire, having put in 2 quarts of water; steam gently, stirring occasionally for about half an hour, till forming rather a rich thick gravy. When both the rice and meat are done, take half the rice and mix with the meat, and then the remainder of the meat and rice, and serve. Save the bones for soup for the following day. Salt pork or beef, well soaked, may be used—omitting the salt. Any kind of vegetables may be frizzled with the onions.

No. 9.—BAKING AND ROASTING WITH THE FIELD STOVE. By the removal of the caldron, and the application of a false bottom put over the fire, bread bakes extremely well in the oven, as well as meat, potatoes, puddings, &c. Bread might be baked in oven at every available opportunity at a trifling cost of fuel. The last experiment I made with one was a piece of beef weighing about 25 lbs., a large Yorkshire pudding, and about 10 lbs. of potatoes, the whole doing at considerably under one pennyworth of fuel, being a mixture of coal and coke; the whole was done to perfection, and of a nice brown color. Any kind of meat would, of course, roast the same.

Baking in fixed Oven.—In barracks, or large institutions, where an oven is handy, I would recommend that a long iron trough be made, four feet in length, with a two-story movable grating in it, the meat on the top of the upper one giving a nice elevation to get the heat from the roof, and the potatoes on the grating under, and a Yorkshire pudding at the bottom. Four or five pieces of meat may be done on one trough. If no pudding is made, add a quart more water.

No. 10.—FRENCH BEEF SOUP, OR POT-AU-FEU, CAMP FASHION. FOR THE ORDINARY CANTEEN-PAN. Put in the canteen saucepan 6 lbs. of beef, cut in two or three pieces, bones included, $\frac{3}{4}$ lb. of plain mixed vegetables, as onions, carrots, turnips, celery, leeks, or such of these as can be obtained, or 3 oz. of preserved in cakes, as now given to the troops; 3 teaspoonfuls of salt, 1 teaspoonful of pepper, 1 teaspoonful of sugar, if handy; 8 pints of water, let it boil gently three hours, remove some of the fat, and serve. The addition of $1\frac{1}{2}$ lb. of bread cut into slices, or 1 lb. of broken biscuit, well soaked in the broth, will make a very nutritious soup; skimming is not required.

No. 11.—SEMI-FRYING, CAMP FASHION, CHOPS, STEAKS, AND ALL KINDS OF MEAT. If it is difficult to broil to perfection, it is considerably more so to cook meat of any kind in a frying-pan. Place your pan on the fire for a minute or so, wipe it very clean; when the pan is very hot, add in it either fat or butter, but the fat from salt and ration meat is preferable; the fat will immediately get very hot; then add the meat you are going to cook, turn it several times to have it equally done; season to each pound a small teaspoonful of salt, quarter that of pepper, and serve. Any sauce or maître-d'hôtel butter may be added. A few fried onions in the remaining fat, with the addition of a little flour to the onion, a quarter of a pint of water, two tablespoonfuls of vinegar, a few chopped pickles or picalilly, will be very relishin r.

No. 11A.—TEA FOR EIGHTY MEN, which often constitutes a whole company. One boiler will, with ease, make tea for eighty men, allowing a pint each man. Put forty quarts of water to boil, place the rations of tea in a fine net, very loose, or in a large perforated ball; give one minute to boil, take out the fire, if too much, shut down the cover; in ten minutes it is ready to serve.

No. 12.—COFFEE A LA ZOUAVE FOR A MESS OF TEN SOLDIERS, as made in the camp, with the canteen saucepan holding 10 pints. Put 9 pints of water into a canteen saucepan on the fire; when boiling add 7½ oz. of coffee, which forms the ration, mix them well together with a spoon or a piece of wood, leave on the fire for a few minutes longer, or until just beginning to boil. Take it off and pour in 1 pint of cold water, let the whole remain for ten minutes or a little longer. The dregs of the coffee will fall to the bottom, and your coffee will be clear. Pour it from one vessel to the other, leaving the dregs at the bottom, add your ration sugar or 2 teaspoonfuls to the pint; if any milk is to be had, make 2 pints of coffee less; add that quantity of milk to your coffee, the former may be boiled previously, and serve. This is a very good way for making coffee even in any family, especially a numerous one, using 1 oz. to the quart if required stronger. For a company of eighty men use the field-stove and four times the quantity of ingredients.

No. 13.—COFFEE, TURKISH FASHION. When the water is about to boil add the coffee and sugar, mix well as above, let it boil, and serve. The grounds of coffee will in a few seconds fall to the bottom of the cups. The Turks wisely leave it there, I would advise every one in camp to do the same.

No 14.—COCOA FOR EIGHTY MEN. Break eighty portions of ration cocoa in rather small pieces, put them in the boiler, with five or six pints of water, light the fire, stir the cocoa round till melted, and forming a pulp not too thick, preventing any lumps forming, add to it the remaining water, hot or cold; add the ration sugar, and when just boiling, it is ready for serving. If short of cocoa in campaigning, put about sixty rations, and when in pulp, add half a pound of flour or arrowroot.

EASY AND EXCELLENT WAY OF COOKING IN EARTHEN PANS. A very favorite and plain dish amongst the convalescent and orderlies at Scutari was the following:—Cut any part of either beef (cheek or tail), veal, mutton, or pork, in fact any hard part of the animal, in 4-oz. slices; have ready for each 4 or 5 onions and 4 or 5 pounds of potatoes cut in slices; put a layer of potatoes at the bottom of the pan, then a layer of meat, season to each pound 1 teaspoonful of salt, quarter one of pepper, and some onion you have already minced;

then lay in layers of meat and potatoes alternately till full; put in 2 pints of water, lay on the lid, close the bar, lock the pot, bake two hours, and serve. Remove some of the fat from the top, if too much; a few dumplings, as No. 21, in it will also be found excellent. By adding over each layer a little flour it makes a rich thick sauce. Half fresh meat and salt ditto will also be found excellent.

SERIES OF SMALL RECEIPTS FOR A SQUAD, OUTPOST, OR PICKET OF MEN, which may be increased in proportion of companies. *No.* 15. *Camp Soup.*—Put half a pound of salt pork in a saucepan, two ounces of rice, two pints and a half of cold water, and, when boiling, let simmer another hour, stirring once or twice; break in six ounces of biscuit, let soak ten minutes; it is then ready, adding one teaspoonful of sugar, and a quarter one of pepper, if handy.

No. 16. *Beef Soup.*—Proceed as above, boil an hour longer, adding a pint more water.

Note.—Those who can obtain any of the following vegetables will find them a great improvement to the above soups :—Add four ounces of either onions, carrots, celery, turnips, leeks, greens, cabbage, or potatoes, previously well washed or peeled, or any of these mixed to make up four ounces, putting them in the pot with the meat. I have used the green tops of leeks and the leaf of celery as well as the stem, and found that for stewing they are preferable to the white part for flavor. The meat being generally salted with rock salt, it ought to be well scraped and washed, or even soaked in water a few hours if convenient; but if the last cannot be done, and the meat is therefore too salt, which would spoil the broth, parboil it for twenty minutes in water, before using for soup, taking care to throw this water away.

No. 17.—For fresh beef proceed, as far as the cooking goes, as for salt beef, adding a teaspoonful of salt to the water.

No. 18. *Pea Soup.*—Put in your pot half a pound of salt pork, half a pint of peas, three pints of water, one teaspoonful of sugar, half one of pepper, four ounces of vegetables, cut in slices, if to be had; boil gently two hours, or until the peas are tender, as some require boiling longer than others—and serve.

No 19. *Stewed Fresh Beef and Rice.*—Put an ounce of fat in a pot, cut half a pound of meat in large dice, add a teaspoonful of salt, half one of sugar, an onion sliced; put on the fire to stew for fifteen minutes, stirring occasionally, then add two ounces of rice, a pint of water; stew gently till done, and serve. Any savory herb will improve the flavor. Fresh pork, veal, or mutton may be done the same way, and half a pound of potatoes used instead of the rice, and as rations are served out for three days, the whole of the provisions may be cooked at once.

No. 20.—Receipts for the Frying-pan. Those who are fortunate enough to possess a frying-pan will find the following receipts very useful:—Cut in small dice half a pound of solid meat, keeping the bones for soup; put your pan, which should be quite clean, on the fire; when hot through, add an ounce of fat, melt it and put in the meat, season with half a teaspoonful of salt; fry for ten minutes, stirring now and then; add a teaspoonful of flour, mix all well, put in half a pint of water, let simmer for fifteen minutes, pour over a biscuit previously soaked, and serve. The addition of a little pepper and sugar, if handy, is an improvement, as is also a pinch of cayenne, curry-powder or spice; sauces and pickles used in small quantities would be very relishing; these are articles which will keep for any length of time. As fresh meat is not easily obtained, any of the cold salt meat may be dressed as above, omitting the salt, and only requires warming; or, for a change, boil the meat plainly, or with greens, or cabbage, or dumplings, as for beef; then the next day cut what is left in small dice —say four ounces—put in a pan an ounce of fat; when very hot pour in the following:—Mix in a basin a tablespoonful of flour, moisten with water to form the consistency of thick melted butter, then pour it in the pan, letting it remain for one or two minutes, or until set; put in the meat, shake the pan to loosen it, turn it over, let it remain a few minutes longer, and serve. To cook bacon, chops, steaks, slices of any kind of meat, salt or fresh sausages, black puddings, &c.: Make the pan very hot, having wiped it clean, add in fat, dripping, butter, or oil, about an ounce of either; put in the meat, turn three or four times, and season with salt and pepper. A few minutes will do it. If the meat is salt, it must be well soaked previously.

No. 21.—Suet Dumplings. Take half a pound of flour, half a teaspoonful of salt, a quarter teaspoonful of pepper, a quarter of a pound of chopped fat pork or beef suet, eight tablespoonfuls of water, mixed well together. It will form a thick paste, and when formed, divide it into six or eight pieces, which roll in flour, and boil with the meat for twenty minutes to half an hour. Little chopped onion or aromatic herbs will give it a flavor.

A plainer way, when Fat is not to be obtained.—Put the same quantity of flour and seasoning in a little more water, and make it softer, and divide it into sixteen pieces; boil about ten minutes. Serve round the meat. One plain pudding may be made of the above, also peas and rice pudding thus:—One pound of peas well tied in a cloth, or rice ditto with the beef. It will form a good pudding. The following ingredients may be added: a little salt, sugar, pepper, chopped onions, aromatic herbs, and two ounces of chopped fat will make these puddings palatable and delicate.

CORDON—is the coping of the escarp or inner wall of the ditch, sometimes called the magistral line; as from it the works in permanent fortification are traced. It is usually rounded in front, and projects about one foot over the masonry: while it protects the top of the revetment from being saturated with water, it also offers, from projection, an obstacle to an enemy in escalading the wall.

CORPORAL. Grade between private and sergeant.

CORPOREAL PUNISHMENT, BY STRIPES AND LASHES. Prohibited excepting for the crime of desertion; (*Act* May 16, 1812 and *Act* March 2, 1833.)

CORPS. The Articles of War use the word *corps* in the sense of a portion of the army organized by law with a head and members; or any other military body having such organization, as the marine corps. A regiment is a corps; an independent company is a corps—a body of officers with one head is a corps, as the Topographical Engineers. Detachments of parts of regiments, or of whole regiments, united for a particular object, whether for a campaign or a part of a campaign, are not corps in the sense of the Rules and Articles of War, for such bodies have neither head nor members commissioned in the particular body temporarily so united; but the officers with such detachment hold commissions either in the corps composing the detachment, in the army at large, in the marine corps, or militia.

CORRECTING PROOFS. (*See* PRINTING.)

CORRESPONDENCE WITH THE ENEMY. Whoever shall be convicted of holding correspondence with or giving intelligence to the enemy, directly or indirectly, shall suffer death or such other punishment as shall be ordered by sentence of a court-martial; (ART. 57.)

COSINE. The complement of the sine.

COUNCIL OF ADMINISTRATION. Under the act of Congress of July 5, 1838, the council of administration may, from time to time, employ such person as they think proper to officiate as chaplain; who shall also perform the duties of schoolmaster at such post. The chaplain is paid on the certificate of the commanding officer, not exceeding forty dollars per month, as may be determined by the said council of administration with the approval of the Secretary of War. Councils of administration fix a tariff to the prices of sutler's goods—regulate the sutler in other matters, and make appropriations for specific objects determined by regulations from the post and regimental funds. Those funds are collected in great part by savings of flour, in making bread by troops.

COUNCIL OF WAR. An assemblage of the chief officers in the army, summoned by the general to concert measures of importance.

COUNSEL. All writers admit it to be the custom to allow a prisoner to have counsel.

COUNTER-BATTERY. When a number of guns are placed behind a parapet, for the purpose of dismounting or silencing by direct fire the guns in an enemy's work, it is called a counter-battery.

COUNTERFORTS—are the buttresses by which the revetment walls are backed and strengthened interiorly.

COUNTERGUARD—is a work composed of two faces, forming a salient angle, sometimes placed before a bastion, sometimes before a ravelin, and sometimes before both, to protect them from being breached.

COUNTERMINES—are galleries excavated by the defenders of a fortress, to intercept the mines, and to destroy the works of the besiegers.

COUNTERSCARP. The outer boundary of the ditch—revetted with masonry in permanent fortification to make the ditch as steep as possible.

COUNTERSIGN. A particular word given out by the highest in command, intrusted to those employed on duty in camp and garrison, and exchanged between guards and sentinels.

COUNTERSLOPE. In the case of a revetment, the slope is within instead of on the outside; and is usually formed in steps. In the case of a parapet, the slope is upwards instead of downwards.

COUP D'ŒIL. The art of distinguishing by a rapid glance the weak points of an enemy's position, and of discerning the advantages and disadvantages offered by any given space of country, or selecting with judgment the most advantageous position for a camp or battle-field. Experience is a great aid in the acquisition of this necessary military faculty, but experience and science alone will not give it.

COUP DE MAIN. A sudden and vigorous attack.

COUPURES—are short retrenchments made across the face of any work, having a terre-plein. The ditch of the coupure is carried quite across the terre-plein, and through the parapet of the work in which it is formed, but not through the revetment.

COURT-MARTIAL. Any general officer commanding an army, or colonel commanding a separate department, may appoint general court-martials whenever necessary; (Art. 65.) General courts-martial may consist of any number of commissioned officers, from five to thirteen, but they shall not consist of less than thirteen, where that number can be convened without manifest injury to the service; (Art. 64.) But no sentence of a court-martial shall be carried into execution until after the whole proceedings shall have been laid before the officer ordering the same, or the officer commanding the troops for the time being; neither shall any sentence of a general court-martial, in time of peace, extending to the loss of life, or the dismission of a commissioned officer, or which shall, either in time of peace or war, respect a general officer, be carried into execution, until after the whole proceedings shall have been transmitted to the Secretary of War, to be laid before the President of the United States for his confirmation or disapproval, and orders in the case. All other sentences may be confirmed and executed by the officer ordering the court to assemble, or the commanding officer for the time being, as the case may be; (Art. 65.) Whenever a general officer commanding an army, or a colonel commanding a separate department, shall be the accuser or prosecutor of any officer of the army under his command, the general court-martial for the trial of such officer shall be appointed by the President of the United States, and the proceedings and sentence of the said court shall be sent directly to the Secretary of War to be laid by him before the President for his confirmation or approval or orders in the case; (*Act* May 29, 1830.) Every officer commanding a regiment or corps may appoint, for his own regiment or corps, courts-martial to consist of three commissioned officers, for the trial and punishment of offences not capital, and decide upon their sentences. For the same purpose, all officers commanding any of the garrisons, forts, barracks, or other places where troops consist of different corps, may assemble courts-martial, to consist of three commissioned officers, and decide upon their sentences; (Art. 66.) No garrison or regimental court-martial shall have the power to try capital cases, or commissioned officers; neither shall they inflict a fine exceeding one month's pay, nor imprison, nor put to hard labor, any non-commissioned officer or soldier, for a longer time than one month; (Art. 67.) The judge-advocate, or some person deputed by him, or by the general, or officer commanding the army, detachment, or garrison, shall prosecute in the name of the United States, but shall so far consider himself as counsel for the prisoner, after the said prisoner shall have made his plea, as to object to any leading question to any witness, or any question to the prisoner, the answer to which might tend to criminate himself; and administer to each member of the court, before they proceed upon any trial, the oath prescribed in the Articles of War for General, Regimental and Garrison Courts-martial. The president of the court then administers an oath to the judge-advocate; (ART. 69.) If a prisoner when arraigned stands mute, the trial goes on as if he pleaded not guilty; (ART. 70.) If a member be challenged by a prisoner the court judges of the relevancy of the challenge. Only one member can be challenged at a time; (ART. 71.) All members are to behave with decency and calmness, and in giving their votes to begin with the youngest; (ART. 72.) All persons who give evidence are examined on oath or affirmation; (ART. 73.) On trials of cases not capital before courts-martial, the deposition of witnesses, not in the line or staff of the army, may be taken before some justice of the peace and read in evidence; provided the prosecutor and person accused are present at the taking of the same, or are duly notified thereof; (ART. 74.) No officer shall be tried but by a general court-martial, nor by officers of inferior rank, if it can be avoided. Nor shall trials be carried on except between 8 in the morning and 3 in the afternoon, excepting in cases requiring immediate example in the opinion of the officer ordering the court; (ART. 75.) No person to use menacing words, signs, or gestures before a court-martial, or cause any disorder or riot, or disturb their proceeding, on the penalty of being punished at the discretion of the said court-martial; (ART. 76.) (*Consult* DE HART, KENNEDY, and SIMMONS; *See* ADDRESS; ALIBI; AMICUS CURIÆ; APPEAL; ARREST; CHALLENGE OF MEMBERS; CHARACTER; CHARGES; CONTEMPT; COUNSEL; CRIMES; CUSTOM OF WAR; DEATH; DECISIONS; DEFENCE; DISMISSION; EVIDENCE; FALSEHOOD; FINDING; JUDGE-ADVOCATE; JURISDICTION; MISNOMER; NEW MATTER; NOTES; OATH; PLEA; PRESIDENT; PRISONERS; PROCEEDINGS; PROSECUTORS; QUESTIONS; RECOMMENDATION; REJOINDER; REPLY; REVISION; SENTENCE; SUMMING UP; SUSPENDED; TRIAL; VERDICT; VOTES; WITNESSES; and References under the heading ARTICLES OF WAR.)

FORM No. 1.

FORM of a General Order appointing a General Court-martial.

General Orders, }   Head-quarters of the Army,
    No.   }   March   , 18—.

A General Court-martial, to consist of thirteen members, will convene at Fort Monroe, in the State of Virginia, on Monday the 2d of April, 18—, at 11 o'clock, A. M., or as soon thereafter as practicable, for the trial of Captain A. B., of the 1st Regiment of Artillery, and such other prisoners as may be brought before it.

The following Officers are detailed as members of the Court:

   1. Colonel A. B.   1st Regiment of ———
   2. Colonel C. D.   3d Regiment of ———
   3. Lieut.-col. E. F.   1st Regiment of ———
   4. Lieut.-col. F. G.   2d Regiment of ———
   5. Major W. T.   3d Regiment of ———
   6. Major N. M.   1st Regiment of ———
   7. Captain A. N.   3d Regiment of ———
   8. Captain B. N.   1st Regiment of ———
   9. Captain C. N.   2d Regiment of ———
10. Captain D. M.   3d Regiment of ———
11. Captain E. L.   1st Regiment of ———
12. Captain F. H.   1st Regiment of ———
13. Captain G. W.   1st Regiment of ———

And the following Officers are detailed as supernumeraries:

   Captain N. P.   2d Regiment of Infantry.
   Captain D. B.   1st Regiment of Infantry.
   Captain N. O.   1st Regiment of Artillery.

Captain S. R., of the 4th Regiment of ———, is hereby appointed Judge-advocate.

            By command of
                Lieut.-general ———.
                ———, Adjutant-general.

FORM No. 2.

General Orders, }   Head-quarters.
    No.

A General Court-martial is hereby appointed to meet at ———, on the ——— day of ———, or as soon thereafter as practicable, for the trial of ———, and such other prisoners as may be brought before it.

            Detail for the Court.

   1. ———   5. ———   9. ———   13. ———
   2. ———   6. ———   10. ———
   3. ———   7. ———   11. ———
   4. ———   8. ———   12. ———

                               ———, Judge-advocate.

No other officers than those named can be assembled without manifest injury to the service.

              By order of ———,
                  ———, Asst. Adjt.-gen.

FORM No. 3.

General Orders, }   Head-quarters of the Army,
No.   }   April   , 18—.

A General Court-martial, to consist of as many members [within the prescribed limits] as can be assembled without manifest injury to the service, will convene at ———, in the State of ———, on Tuesday the 23d of April, 18—, at 10 o'clock, A. M., or as soon thereafter as practicable, for the trial of Lieutenant C. D., of the 1st Regiment, and such other prisoners as may be brought before it.

The Commanding Officer, at ———, will cause the members of the Court to be detailed from the officers of his command. First Lieutenant B. M., 2d Regiment of Artillery, is hereby appointed the Judge-advocate of the Court.

                 By order of ———,
                         Major-general Commanding in Chief,
                     R. J.
                                    Adjutant-general.

The above form delegating authority for the detail of members of a Court-martial to a distant commander, although not latterly used, is of the greatest practical importance. It conforms to the custom of war in other services, was long used in our own without question of its legality, and might with great benefit to the service be revived.

FORM No. 4.

Mode of recording the proceedings of a General [or other] Court-martial.

Proceedings of a General Court-martial, held at Fort Monroe, in the State of Virginia, by virtue of the following Orders, viz.:

     [Here insert a copy of the Order convening the Court.]
                    Fort Monroe, Virginia,
                          Monday, April —, 18—.

The Court met pursuant to the above Orders.

PRESENT.

    1. Colonel A.B. 1st Regt. of ———, *President.*

2. Colonel C. D.	3. Lieut.-col. E. F.
4. Lieut.-col. F. G.	5. Major W. T.
6. Major N. M.	7. Capt. A. N.
8. Capt. B. N.	9. Capt. C. N.
10. Capt. D. M.	11. Capt. E. L.
12. Capt. F. H.	13. Capt. W. G.

Members.

       Captain S. R., Judge-advocate.

The Court then proceeded to the trial of Captain A. B., of the ———— Regiment of ————, who, being called into Court, and having heard the General Order read, was asked if he had any objection to any of the members named in the General Order, to which he replied in the negative.

The Court was then duly sworn, in his presence, and Captain A. B. was arraigned on the following charge and specifications, viz. :

[Here insert the charge and specifications.]

To which the prisoner pleaded as follows :

   Not Guilty, to the 1st specification,
   Not Guilty, to the 2d specification,
   Not Guilty, to the charge.

All persons required to give evidence were directed to withdraw, and remain in waiting until called for.

Lieut. A. B. of the 2d Regiment of Infantry, a witness for the prosecution, being duly sworn, says : that on the ———— day of ————, &c. ———— &c. ————.

Question by the Judge-advocate. ————?
Answer. ————.
Question by the prisoner. ————?
Answer. ————.
Question by the Court. ————?
Answer. ————.

The prosecution was here closed, and the prisoner produced the following evidence :

Capt. C. D. of the Corps of ————, a witness for the defence, being duly sworn, says : that on the ———— day of ————, &c. &c. ————

Question by the prisoner. ————?
Answer. ————.
Question by the Judge-advocate. ————?
Answer. ————.
Question by the Court. ————?
Answer. ————.

The prisoner, having no further testimony to offer, requested to be indulged with ———— days to prepare for his final defence. The Court granted his request, and adjourned at ———— o'clock, P. M., to meet again at ———— o'clock, A. M., on Wednesday, the ———— day of ————.

SECOND DAY.

Wednesday, ———, 18—.

The Court met pursuant to adjournment: present all the members.

The proceedings having been read over to the Court by the Judge-advocate, the prisoner, Captain A. B., made the following address in his defence:

[Here insert the defence, or if it be too long, it may be marked, and annexed.]

The Court then closed, and proceeded to deliberate on the testimony adduced, and pronounced the following

SENTENCE.

The Court, having maturely weighed and considered the evidence adduced in support of it, is of opinion that &c. ——— &c. ———, and does therefore ——— &c. ——— &c.

            A. B. Col. 1st Regt. of ———,
S. R. Capt. —— Regt. of ———,   President.
   Judge-advocate.

FORM No. 5.

Form of an Order appointing a Garrison or Regimental Court-martial.

Orders, }   Head-quarters,
No.   Fort Columbus, N. Y.
         April , 18—.

A Garrison, [or Regimental Court-martial,] to consist of Captain C. D. ———, 1st Lieutenant D. F. ———, and 2d Lieutenant G. H. ———, will convene at the President's quarters to-morrow morning, at 11 o'clock, for the trial of Sergeant D. E. of ——— Company, ——— Regiment of Artillery, and such other prisoners as may be brought before it.

      By order of Colonel A. B.,
        Commanding,
      J. A.,
       Adjutant.

FORM No. 6.

Form of charges and specifications against a prisoner.

Charges and specifications preferred against Capt. C. D., of the 1st Regiment of Infantry.

CHARGE 1st.

DISOBEDIENCE OF ORDERS.

Specification 1*st.* . . . In this, that he, the said Captain C. D., of the 1st Regiment of Infantry, being ordered, on the 30th day of September, 18—, at the Recruiting Dépôt, in the town of Newport, Kentucky, by Colonel A. B., of the 1st Regiment of Infantry, the commanding officer of said Dépôt, to take command of and march with a detachment of recruits, to Jefferson Barracks, in the State of Missouri, did at said town of Newport, at the time aforesaid, refuse to take command of and march with said detachment of recruits, thereby disobeying the lawful commands and orders of his superior and commanding officer, the said Colonel A. B.

Specification 2*d.* . . . In this, that he the said Captain C. D., &c. &c.

E. F.
Major 1st Regiment of Infantry.

FORM No. 7.

Form of a General Order approving or disapproving the proceedings of a General Court-martial.

General Order, No.   Head-quarters of the Army,
January —, 18—.

I. . . At a General Court-martial, which convened at ——— on the ——— of ———, 18—, pursuant to General Orders, No. ——— of January 18—, and of which Brevet Brigadier-general ——— is President, was tried Captain ———, of the ——— Regiment of Artillery, on the following chargers and specifications preferred by Major ———, of the ——— Artillery, to wit:

CHARGE.

[Here insert charge. See Form No. 5.]

To which charge and specification the prisoner pleaded as follows:

To the 1st specification—[plea.]
To the 2d specification—[plea.]
And guilty [or not guilty] to the charge.

FINDINGS AND SENTENCE

The Court, after mature deliberation on the testimony adduced, find the prisoner, Capt. ———, of ——— Regiment of Artillery, as follows:

Of the 1st specification—[finding.]
Of the 2d specification—[finding.]
And guilty [or not guilty] of the charge.

And the Court do therefore sentence him, Captain ———, of ——— Regiment of Artillery, to [here insert sentence.]

II. . . The proceedings, findings, and sentence are approved, [or disapproved,] &c., &c., &c.

(Here the authority which constituted the Court will add such remarks as he may think proper.)

III. . . The General Court-martial, of which Brevet Brigadier-general ——— is President, is hereby dissolved.

By Command of
Major-general ———,
———, Adjutant-general.

COURT OF INQUIRY. In cases where the general or commanding officer may order a court of inquiry to examine into the nature of any transaction, accusation, or imputation, against any officer or soldier, the said court shall consist of one or more officers, not exceeding three, and a judge-advocate, or other suitable person as a recorder, to reduce the proceedings and evidence to writing, all of whom shall be sworn to the faithful performance of duty. This court shall have the same power to summon witnesses as a court-martial, and to examine them on oath. But they shall not give their opinion on the merits of the case, excepting they shall be thereto specially required. The parties accused shall also be permitted to cross-examine and interrogate the witnesses, so as to investigate fully the circumstances in the question; (ART. 91.) The proceedings of a court of inquiry must be authenticated by the signature of the recorder and the president, and delivered to the commanding officer, and the said proceedings may be admitted as evidence by a court-martial, in cases not capital, or extending to the dismission of an officer, provided that the circumstances are such that oral testimony cannot be obtained. But courts of inquiry are prohibited, unless directed by the President of the United States, or demanded by the accused; (ART. 92.)

The court may be ordered to report the *facts* of the case, with or without an opinion thereon. Such an order will not be complied with, by merely reporting the evidence or testimony; facts being the result, or conclusion established by weighing all the testimony, oral and documentary, before the court.

When a court of inquiry is directed to be assembled, the order should state whether the court is to report the facts or not, and also whether or not it is to give an opinion on the merits. The court should also be instructed, whether its attention is to be extended to a general investigation, or to be confined to the examination of particular points only, as the case may seem to require, in the judgment of the officer under whose authority it is assembled. Where the subject is multifarious, the court should be instructed to state its opinion on each point separately, that the proper authority may be able to form his judgment.

The court may sit with open or closed doors, according to the nature of the transaction to be investigated. The court generally sits with open doors; but there may be delicate matters to be examined into, that might render it proper to sit with doors closed.

The form of proceeding, in courts of inquiry, is nearly the same as that in courts-martial: the members being assembled, and the parties interested called into court, the judge-advocate, or recorder, by direction of the president, reads the order by which the court is constituted, and then administers to the members the following oath: "You shall well and truly examine and inquire, according to your evidence, into the matter now before you, without partiality, favor, affection, prejudice, or hope of reward: so help you God;" (ART. 93.)

The accusation is then read, and the witnesses are examined by the court; and the parties accused are also permitted to cross-examine and interrogate the witnesses, so as to investigate fully the circumstances in question; (ART. 91.)

The examination of witnesses being finished, the parties before the court may address the court, should they see fit to do so; after which the president orders the court to be cleared. The recorder then reads over the whole of the proceedings, as well for the purpose of correcting the record, as for aiding the memory of the members of the court. After mature deliberation on the evidence adduced, they proceed to find a state of facts, if so directed by the order constituting the court, and to declare whether or not the grounds of accusation are sufficient to bring the matter before a general court-martial; and also to give their opinion of the merits of the case, if so required.

The court should be careful to examine the order by which it is constituted, and be particular in conforming to the directions contained therein, either by giving a general opinion on the whole matter, a statement of facts only, or an opinion on such facts. The proceedings of courts of inquiry have been returned to be reconsidered, when the court has been unmindful of these points.

It has been settled that a member of a court of inquiry may be objected to, for cause.

The proceedings must be authenticated by the signatures of the president and recorder, and delivered to the commanding officer or authority which ordered the court; and the said proceedings may be admitted in evidence by a court-martial, in cases not capital, nor extending to the dismission of an officer, provided oral testimony cannot be obtained; (Art. 92.)

Transactions may become the subject of investigation by courts of inquiry after the lapse of any number of years, on the application of the party accused, or by order of the President of the United States; the limitation mentioned in the 88th Article of War, being applicable only to general courts-martial.

It is not necessary to publish the proceedings or opinion of the court, although it is usually done in general orders.

The court is dissolved by the authority that ordered it to convene.

COVERED WAY. A space between the counterscarp and the crest of the glacis in permanent works, and within the palisades, over which the garrison can run without being seen or subjected to the fire of the enemy. The *crowning* of the covered way by the besiegers is a difficult operation, and often costs them dearly.

COWARDICE. In all cases where a commissioned officer is cashiered for cowardice or fraud, it shall be added in the sentence, that the crime, name, and place of abode and punishment of the delinquent be published in the newspapers, after which it shall be deemed scandalous for an officer to associate with him; (Art. 85.)

CRATER OF A MINE—is the excavation or cavity formed in the ground, by the explosion of the powder.

CREMAILLERE—is an indented or zigzag outline.

CRENELLATED—loop-holed.

CRIMES, Disorders, and Neglects. All crimes not capital, and all disorders and neglects which officers and soldiers may be guilty of, to the prejudice of good order and military discipline, though not mentioned in the Articles of War, are to be taken cognizance of by a general or regimental court-martial, according to the nature and degree of the offence, and be punished at their discretion; (Art. 99.) (*See* Authority, Civil.)

CRIMINATE. (*See* Evidence.)

CROTCHETS—are openings cut into the glacis at the heads of traverses, to enable the defenders to circulate round them. These passages are closed by a gate when necessary.

CROWNING. A lodgment prepared by besiegers upon the crest of the glacis to make themselves masters of the *covered way*. It is effected usually by means of the SAP—a method apparently slow, but which, advancing night and day without intermission, accomplishes great objects. The work is done by sappers rolling before them a very large gabion stuffed with wool or cotton, or fascines, to shelter themselves from musketry. They fill thus one gabion after another, and do not push forward until the portion of the trench already made has been well consolidated.

CROWN-WORK—is a similar work to horn-work, but consisting of two fronts instead of one. It is connected to the main works in a similar way, and is used for the same purposes as the horn-work.

CROWS' FEET—are iron-pointed stars, or stout nails, so fixed as to radiate, that in any position they may have a point uppermost. They are strewed on the ground over which cavalry may be expected to pass. (*See* OBSTACLES.)

CUNETTE—is a narrow ditch in the middle of a dry ditch, to keep it drained, as well as to form, especially when filled with water, an obstacle to an enemy.

CURTAIN. The curtain is that part of the rampart of the body of the place, which lies between two bastions, and which joins their two flanks together.

CURTAIN ANGLE—is that formed by the meeting of the flank and the curtain.

CUSTOM OF WAR. The custom of war in like cases is the common law of the army recognized by Congress in the 69th Article of War, as a rule for the government of the army whenever any doubt shall arise not explained by the rules and articles established by Congress for the government and regulation of the army. To render a custom valid the following qualities are requisite:—1. Antiquity; 2. Continuance without interruption; 3. Have been acquiesced in without dispute; 4. It must be reasonable; 5. Certain; 6. *Compulsory*, that is, not left to the option of every man whether he will use it or not; 7. Customs must be consistent with each other.

D

DAM. An impediment formed of stones, gravel, and earth, by which a stream of water is made to overflow and inundate the adjacent ground.

DAMAGE. The costs of repairs of *damage* done to arms, equipments, or implements, in the use of the armies of the United States, shall be deducted from the pay of any officer or soldier in whose care or use the said arms, equipments or implements were when the said damages occurred: Provided, the damage was occasioned by the abuse or negligence of said officer or soldier. Every officer commanding a regiment, corps, garrison, or detachment, to make once every two months, or oftener if required, a written report to the colonel of ordnance stating all damages to arms so belonging to his command, and naming the officers and soldiers by whose negligence or abuse the damages were occasioned; (*Act* Feb. 8, 1815.)

DEAD ANGLE or (DEAD GROUND)—is any angle or piece of ground which cannot be seen, and which therefore cannot be defended from behind the parapet of the fortification.

DEATH. Sentence of death may be rendered by a general court-martial for the following crimes only: 1. Beginning, exciting, causing or joining in, any mutiny or sedition in any troop or company in the service of the United States, or in any party, post, detachment, or guard; (ART. 7.) 2. Being present at any mutiny or sedition and not using the utmost endeavors to suppress the same, or coming to the knowledge of any intended mutiny and not giving without delay information to the commanding officer; (ART. 8.) 3. Striking his superior officer, or drawing or lifting up any weapon, or offering any violence against him, he being in the execution of his office, on any pretence whatsoever; or disobeying any lawful command of his superior officer; (ART. 9.) 4. Desertion in time of war; (ART. 20 modified by *Act* May 28, 1830.) 5. Advising or persuading an officer or soldier to desert the service; (ART. 23.) 6. Any sentinel found sleeping on his post, or leaving it before being regularly relieved; (ART. 46.) 7. Any officer occasioning false alarms in camp, garrison, or quarters, by discharging fire-arms, drawing of swords, beating of drums, or by any other means whatsoever; (ART. 49.) 8. Doing violence to any person who brings provisions or other necessaries to the camp, garrison, or quarters of the forces of the United States employed in any parts out of the said States; (ART. 51.) 9. Misbehavior before the enemy, running away or shameful abandonment of any fort, post, or guard, which he may be commanded to defend, or speaking words inducing others to do the like; or casting away arms and ammunition, or quitting his post or colors to plunder and pillage; (ART. 52.) 10. Making known the watch-word to any person not entitled to receive it, or giving a parole or watch-word different from that received; (ART. 53.) 11. Forcing a safe-guard in foreign parts; (ART. 55.) 12. Relieving the enemy with money, victuals or ammunition; or knowingly harboring or protecting an enemy; (ART.

56.) 13. Holding correspondence with, or giving intelligence to the enemy, either directly or indirectly; (Art. 57.) 14. Compelling their commanding officer to give up to the enemy or abandon any garrison, fortress, or post; (Art. 59.) Every sentence of death in time of peace (in time of war it may be carried into execution by the officer ordering the court, or by the commanding officer) must, before being carried into execution, be laid before the President of the United States for his confirmation or disapproval and orders in the case; and no one can be sentenced to suffer death, except by the concurrence of two-thirds of the members of the court-martial, nor except in cases expressly mentioned; (Arts. 65 and 87.)

DEBLAI—is the quantity of earth excavated from the ditch to form the remblai. Under ordinary circumstances the one is equal to the other, but not always; as, from the nature of the soil, earth may have to be brought to supply the remblai.

DEBT. All non-commissioned officers, artificers, privates, and musicians enlisted in the actual service of the United States are exempted, during their term of service, from all personal arrests for any debt or contract; (*Act* March 3, 1799.) No non-commissioned officer, musician, or private shall be arrested or subject to arrest, or be taken in execution for any debt under the sum of twenty dollars, contracted before enlistment, nor for any debt contracted after enlistment; (*Act* March 16, 1802.)

DECEASED OFFICERS AND SOLDIERS. The major of the regiment or, in his absence, the second in command, secures the effects of an officer, and transmits an inventory to the department of war, that his executor or administrators may receive the same; (Art. 94.) In the case of a soldier, the commanding officer of the troop or company, in presence of two other officers, takes an account of the effects he died possessed of, and transmits the same to the department of war, which said effects are to be accounted for and paid to the representatives of such deceased non-commissioned officer or soldier; (Art. 95.)

DECISIONS. On courts-martial the majority of votes decides all questions as to the admission or rejection of evidence, and on other points involving law or custom. If equally divided, the doubt is in favor of the prisoner; (Hough's *Military Law Authorities*.)

DEFAULTERS. If any officer employed or who has heretofore been employed in the civil, military, or naval departments of the Government, to disburse the public money appropriated for the service of those departments respectively, shall fail to render his account or pay over, in the manner and in the times required by law, or the regulations of the department to which he is accountable, any sum of money remaining in the hands of such officer, the 1st or 2d comptroller of the treasury, as the case may be, shall cause to be stated and certify the account of such delinquent officer to the solicitor of the treasury, who shall immediately proceed to issue a warrant of distress against such delinquent officer and his sureties, directed to the marshal or marshals of the district or districts where they reside; and the marshal shall proceed to levy and collect the sum remaining due by distress and sale of goods and chattels of such delinquent officer; and, if the goods are not sufficient, the same may be levied upon the person of such officer, who may be committed to prison, there to remain until discharged by due course of law. But the solicitor of the treasury, with the approbation of the secretary of the treasury, may postpone for a reasonable time such proceedings where, in his opinion, the public interest will sustain no injury by such postponement. If any person shall consider himself aggrieved by any warrant issued as above, he may prefer a bill of complaint to any district judge of the United States, and thereupon the judge may, if in his opinion the case requires it, grant an injunction to stay proceedings. If any person shall consider himself aggrieved by the decision of such judge either in refusing to issue the injunction, or, if granted, on its dissolution, such person may lay a copy of the proceedings had before the district judge, before a judge of the supreme court, who may either grant the injunction, or permit an appeal, as the case may be, if, in his opinion, the equity of the case requires it; (*Act* May 15, 1820.) The judgment on a warrant of distress under this act, and the proceedings under the judgment, are a bar to any subsequent action for the same cause. U. S. *v.* Nourse, 9 Peters 8. (*See* DELINQUENT.) No money hereafter appropriated shall be paid to any person for his compensation, who is in arrears to the U. S., until such person shall have accounted for and paid into the treasury, all sums for which he may be liable; provided, that nothing herein contained shall be construed to extend to balances arising solely from depreciation of treasury notes received by such person, to be expended in the public service; but in all cases where the pay or salary of any person is withheld, in pursuance of this act, it shall be the duty of the accounting officers, if demanded by the party, his agent or attorney, to report, forthwith, to the agent of the treasury department the balance due; and it shall be the duty of the said agent, within sixty days thereafter, to order suit to be commenced against such delinquent and his sureties; (*Act* January 25, 1828.) (*See* REMEDY; STOPPAGE OF PAY.)

DEFENCE (COAST). Possible causes and objects of attack may be conquest or the destruction of commercial ports of more or less value; the possession of depots; the destruction of naval docks; or taking advantage of the weakness or absence of troops, to levy contributions. The parapets of all coast and harbor defences should be constructed of earth, where favorable sites can be found; but for low sites that can be approached within grape-shot range, such batteries must give place to masonry defences, and where masonry-casemated castles with three tiers of guns in casemates, and with guns and mortars on the roofs are resorted to, embrasures of wrought iron, like the model embrasures of Fort Richmond, New York harbor, will be found applicable. With such batteries well constructed, the direct fire of ships has little effect. Movable columns of troops in numbers, depending on the probable object of the enemy, must be held in some central position. If railroads are to convey the troops, a central point within a radius of sixty miles will be within good supporting distance. If railroads are not relied on, the distance should not be greater than fifteen miles. The columns should be at least seven-tenths infantry, one-tenth cavalry, and two-tenths field artillery. The latter being useful to oppose the debarcation of troops. The French charge both the fleet and the army with the movable defence of coasts. Steamers and flotillas, armed with howitzers, are particularly suited to that object. Corps of troops assembled at some central position are held ready to be thrown upon a threatened point. Batteries of howitzers give their aid to these corps. Concerted signals are arranged.

The ordinance of Jan. 3, 1843, directs that in military ports the naval forces shall be specially charged, under the orders of the commanding officer of the land forces, with the armament, service, and guard of the batteries looking *directly* upon the harbors, and upon interior roadsteads adjacent to these harbors, as well as upon the passes conducting to these interior roadsteads. Whenever the works to which those batteries belong do not form a principal part of the system of defence on the land side of the place and its dependencies, the *personnel* of the permanent batteries intrusted to the land forces is furnished from the artillery, by other troops, by the national guard, by revenue service men, or by ancient cannoneers taken from the coast population, at the rate of five men to a gun, one of whom must be an experienced gunner. The permanent works for defence are divided into three classes, according to their importance: 1*st Class.* Works for the defence of military harbors, large commercial harbors, and the principal points of islands. These fortifications are composed of exterior forts, capable of resisting regular attacks, obstructing bombardments, &c. *2d Class.* Works which protect anchorages and channels suited to ships of war. They consist of a system of forts or batteries tying them to the place. *3d Class.* Works defending small commercial ports, anchorages suited to merchantmen, places of refuge for coasting vessels. These consist of batteries with redoubts.

This classification regulates the supply of the batteries, but does not determine absolutely their armament. This must be regulated by various circumstances, as must also the relative strength of the redoubts. The armament of batteries is regulated by the strength of the ships they may have to repel, and the latter depend upon the nature of the coast, and principally upon the depth of water. 32-pounder guns and 8-inch howitzers are employed against ships at a distance of 2,600 yards. Guns begin the fire with round shot; the fire is continued with hollow shot. 13-inch mortars, whose range extends to 4,300 yards, are reserved for the ships at anchor. Experience has proved that a battery of four pieces of heavy calibre has the advantage of a ship of 120 guns. Projectiles *ricochet* better upon the water than upon the land, and lose less of their force; they can, after having *ricoched* at 1,300 yards, pass through the sides of a three-decked ship. Hollow projectiles penetrate the sides underneath the water line, and open large water holes by their explosion.

The number of 24 and 32-pound shot that timber ships have received in their sides without being disabled, ought perhaps to have caused their relinquishment in the armament of coast batteries in Europe. With James' projectile (*See* RIFLED ORDNANCE) such guns, when rifled, will again play an important part in defence. In the United States, such guns have been replaced by larger guns. Even the 42-pounder, retained of late years only as a hot-shot gun, may soon give way to 8 and 10-inch columbiads capable of being used as shell or shot guns; adding also, when necessary, Rodman's 15-inch columbiad, which, with shells of from 305 to 410 lbs., might with a single missile disable, if not entirely destroy the vessel at which it was directed with 6° elevation, when 2,000 yards distant. In many trials at that distance the lateral deviations were only from 1 to 5 yards, and the time of flight $6\frac{1}{2}$ to 7 seconds. With 28° 35′ elevation, and a charge of 40 lbs., the range of the shell is from 5,435 to 5,730 yards, and time of flight 27 seconds.

The height to be given the battery above the level of the sea is from 11 to 16 yards. To fire at point blank: if the aim is a little lower the ricochet brings it upon the ship. Red-hot shot may be fired from columbiads. If engaged with many ships, direct all the pieces of the battery upon that one most in range. Learn exactly the distances of all the most remarkable points, and post the information in the store-room and guard-room, in order that the distance of vessels may be easily determined. Observe the ricochets upon the water. Fire round shot upon disembarkations. Guard carefully against surprises. Observe every thing going on at sea and on land. Be attentive to all signals. Watch over the preservation of material with care; air the magazine in dry weather; move the gun carriages every day. It is important that a battery should have the elevation above given. With that elevation it will not be exposed to ricochet shot from ships, but the ricochets from the battery, losing but little of their force upon the water, will enable even 24-pounder shots, fired under four degrees, to pierce the side of a vessel, however strong it may be, at a distance of 640 yards and more. It is important to direct a heavy fire on ships before anchoring, especially upon the rigging, as the loss of a spar and a few ropes may oblige them to anchor where it is not intended, and thus derange the other ships. In the formation of batteries, regard should be had to the probable number of men that may be obtained to serve them. In the defence of coasts, booms are essential either to bar access to a harbor or river, or to cut off the retreat of the enemy if an entrance has been effected by surprise. Booms should be immediately under the fire of a battery, and are usually made of heavy chains floated by logs. It is unsafe to trust to a single line of booms in the main channel. Booms need not extend entirely across an entrance. Shallow or otherwise inaccessible parts may be omitted, and in order not to impede navigation unnecessarily, 100 yards of boom may be withdrawn from the channel, but always kept ready for replacing; (*Aide Memoire a l'Usage d'Artillerie, &c.*)

DEFENCE, BEFORE A COURT-MARTIAL. In point both of law and reason, a court-martial has as much power over the evidence introduced by the prisoner as over that of the prosecutor, and can reject the witnesses of the one as well as the other, or any part of such witnesses' testimony. Courts-martial are particularly guarded in adhering to the custom which obtains, of resisting every attempt on the part of counsel to address them; but cases have occurred, in which professional gentlemen in attendance have been permitted to read the defence prepared for the prisoner. A court will prevent a prisoner from adverting to parties not before the court, or only alluded to in evidence, further than may be actually necessary. All coarse and insulting language should be avoided, in any part of the defence; (HOUGH's *Law Authorities.*)

DEFENCE, (NATIONAL.) This subject is much associated, in

the popular mind, with ships, forts, and the preparation and proper distribution of all munitions of war; but important as they are, it is not here proposed to discuss those questions. It is not necessary to combat an idea which all history controverts, that a large naval force will ever be able, by cruising in front of our extended coast, to prevent a hostile expedition from landing on our shores.* The reluctant admission of the historian Alison may be accepted, that in the face of greatly superior maritime forces, Ireland was, for sixteen days, in 1796, at the mercy of Hoche's expedition of 25,000 men, and neither the skill of English sailors, nor the valor of English armies, but the fury of the elements, saved them from the danger. "While these considerations," continues Alison, "are fitted to abate confidence in invasion, they are, at the same time, calculated to weaken an overweening confidence in naval superiority, and to demonstrate that the only base on which certain reliance can be placed, even by an insular power, *is a well-disciplined army and the patriotism of its own subjects.*

Nor is it necessary to waste argument on the exploded idea that ships can contend with forts.† The results of such contests in our country, at Fort Moultrie, Mobile Point, Stonington, and Fort M'Henry, abundantly show that our sea-board defences, if completed under the supervision of our able engineers, *and properly garrisoned*, will resist, successfully, any merely naval aggressions, and it has been well said that in the British and French naval attack on Sebastopol, (Oct. 17, 1854,) the final experiment of wooden ships against granite and earthen walls was made, never we believe again to be repeated, until iron clad-ships range up in line of battle; (*See* IRON PLATES.) But the Crimean war did show with what facility large armies are transported by water, and it conclusively proves that the great maritime powers will look to their armies to accomplish in future wars what it would be idle to expect from a navy alone, and that by the organization of forces "fitted to bring into action the physical strength of the country with a competent knowledge of their duty and just ideas of discipline and subordination,"‡ such armies must be met. The means here proposed to accomplish this great object will leave unchanged the present militia laws of the Union, but an effort will be made to show in what manner

---

* For a sketch of the principal maritime expeditions, see Jomini's Art of War, translated by Major Winship and Lieut. McLain. See also the report of a board of officers submitted at the first session of the 26th Congress (Doc. 45), containing numerous illustrations from history, showing the impracticability of covering even a small extent of coast by cruising in front of it.

† The subject is ably discussed in "Halleck's Military Art and Science," under the head of "Sea Coast Defences."

‡ Report of Gen. Cass, while Secretary of War, on National Defence.

existing institutions may be applied to the great purpose in view, by a simple enactment granting to the States, in the words of the Constitution, the consent of Congress "*to keep troops.*"

Francis Lord Bacon has wisely said that "the principal point of greatness in any state is to have a race of military men;" and elsewhere, in his enumeration of the elements of true greatness in a state, he writes: "that it consisteth also in the value and military disposition of the people it breedeth, and in this that they make profession of arms. And it consisteth also in the commandment of the sea." But he writes: "In the measuring or balancing of greatness, there is commonly too much ascribed to largeness of territory, to treasure or riches, to the fruitfulness of the soil or affluence of commodities, and to the strength and fortification of towns and holds." What was made evident to Bacon by the lore of ages is equally true now. If we, as a people, neglect our military resources, do not foster the military spirit of the people, but on the contrary disregard military merit, and even neglect to honor and reward great military services rendered to the state, we cannot breed a race of military men, and are in danger of verifying the assertion of de Tocqueville, in his Observations upon Democracy in America, that "the military career was little honored and badly followed in time of peace." * * * That "this public disfavor is a very heavy burden, which bows down all military spirit," and that if such a people should undertake "a war after a long peace, they would run a much greater risk than any other people of being beaten."

The existing institutions which may be used as aids in organizing a system of National Defence are the Military Academy, the army of the United States, and the militia of the States. The Military Academy is already in successful operation. The first step, then, towards proper State organizations should be to give attention to the regular army—to make it, in fact, an aid or staff for the perfect development of the physical strength of the country. To do this, a system of recruiting is needed in harmony with our institutions and the manner in which all militia force must be collected. It is the several States which furnish the militia force; let the regular army, therefore, be recruited by States. Let every regiment have its depot in a particular district of country, and, with the present rate of pay given to the non-commissioned officers and privates, with the reward of promotion from the ranks bestowed whenever merited, we should soon have an army, in the different parts of which the various sections of the country would take a lively interest. In an army thus collected, which offered a career worthy of being sought, an *esprit-de-corps* would soon be developed which we may in vain seek in our present establishment, and such an army, instead of being regarded by their countrymen as strangers in sympathy and pursuit, might be made the nucleus of science and strength, around which the mental and physical force of the country could be concentrated in war. To accomplish this great object, other changes are also necessary, but much lies within the discretion of the President, and upon his recommendation it is not doubted that Congress will legislate where legislation is required.

If the idea be just that the skeleton regular establishment is maintained in peace, as a *nucleus* to be expanded in war, to meet the wants of the country, the President should be careful not so to distribute that force as to make this great purpose unattainable or difficult when war may impend. If it be possible so to locate the troops as to give them all possible instruction, and, at the same time, not neglect our Indian frontiers, the latter object should not be suffered to override that other most paramount consideration.

Look at any map of the United States, and attempt for a moment to realize the vast extent of our possessions. Bring your mind back to the period when railroads did not afford those facilities which we now have, in a portion of our country, for quickly passing over hundreds of miles, and you may no longer consider that military posts in Texas, New Mexico, California, Oregon, &c., and on the routes to those distant States and Territories, have such means of communication as would enable us to bring together any respectable force in a short period. Bear in mind that the whole army of the United States consists of but one hundred and ninety-eight companies, and that these companies are scattered in posts which dot our immense territory. Realize this, and then answer, is it possible for the small number of troops thus stationed to prevent marauding parties of Indians from passing between these posts and committing depredations either in Mexico or upon our own people? No candid inquirer will assert the possibility! What, then, is remedy? Settlers upon our Indian frontiers must be provided with arms; and the United States Government, besides encouraging Indians to engage in agriculture and other arts of peace, must hold tribes responsible for the acts of individuals. Where predatory bands of Indians have been known to proceed against Mexico or our own people, the tribe must be made answerable, and no vain pursuit be made after the marauding party. We must severely chastise such tribes, and make them understand that the United States require head men to govern and control their young men. That, for the acts of any individuals of the tribe, we will not fail, in any instance, to punish the tribe for such predatory acts. An occasional campaign made against Indians to punish them for misdeeds, produces lasting effects, and will always prove far more efficacious in guarding the lives and property of our citizens, than the present system of small posts, which, by the impunity they afford, only encourages a spirit of adventure in Indian tribes. Another advantage in breaking up the present vicious system of small posts, would be the establishment of schools of instruction for cavalry, artillery, engineers, and infantry. We now have a preparatory school for the cultivation of military science, at West Point; but, if officers of the army, after graduating there, are left without means or motives for improvement, and on remote stations suffer their minds to degenerate from want of exercise and competition, the Military Academy will have accomplished but very partially the great object of its institution. If the army is to be made the rallying point and instructor of our countrymen in war, it should keep pace with the improvements made in Europe, and this can only be done by assembling the engineers, and the three arms of the service, together, in schools of *practice*. Let those schools of practice be properly located: and, besides, the great results thus to be obtained by embodying the troops, detachments could at any time be sent to strike and punish tribes of Indians that failed to keep the peace. With one large detachment on the Atlantic coast; another at Jefferson barracks; a third in New Mexico, and a fourth on the Pacific, the army might be kept in a high state of discipline and efficiency, and soon made, by legislation, all that it should be. With an army so established, it would be apparent that all officers should be active, intelligent, and progressive. A retired list should provide for veterans, and proper legislation would enable commanding officers to appoint their own staff officers, in recognition of the established principle that such officers are the assistants of commanders of troops. Such a change would be necessary to insure the just responsibility of commanding officers, as well as proper instruction by alternation of duty in the line and staff; and by instituting a rigid system of inspection, which would inform the general-in-chief and Secretary of War of the legitimacy of the acts of all commanders, defects of organization, errors of administration, and pernicious customs of service would be made known and corrected by the Executive and Congress.

General Orders, No. 17, of 1854, contain very well-considered regulations for carrying into effect the 5th section of the Act of Congress of August 4, 1854, relative to the promotion of non-commissioned officers. Let us now abandon a system of recruiting, which burdens the army with the scum of cities, and promotion from the ranks would follow as regularly as from a lower to a higher grade of commissions. In a republican army *caste* should not exist, and it will help to break down that distinction now dividing officers and solders, leaving only the necessary difference in grades from private to general, if the army should be recruited by means of regimental recruiting depots so located, that different States shall consider different regiments as raised within their respective limits.

Our army organized and collected, as herein recommended, could easily, on the approach of war, by the addition to each regiment of two battalions, and by increasing the number of privates in a company, be made fifty thousand strong, and this federal force, organized, as it would be, in harmony with State troops, would constantly have kept pace with the advance of professional knowledge in Europe, and be capable of diffusing that knowledge throughout the country by means of the respective State organizations to be now considered.

If the first French revolution did not inaugurate the ideas of liberty and equality, it at least first inculcated by practice the correlative duty of every citizen to defend his country. Accustomed as Americans are to borrow ideas from the English press, it is not remarkable that the outcry made by that aristocratic community against French conscription should have been echoed in our own country. But in the language of General Knox, " It is the wisdom of political establishments to make the wealth of individuals subservient to the general good, and not to suffer it to corrupt or attain undue indulgence. Every State possesses not only the right of personal service from its members, but the right to regulate the service on principles of equality for the general defence. If people, solicitous to be exonerated from their proportion of public duty, exclaim against the only reliable means of defence, as an intolerable hardship, it cannot be too strongly impressed upon them, that while society has its charms, it also has its indispensable obligations. That to attempt such a degree of refinement as to exonerate the members of the community from all personal service, is to render them incapable of the exercise and unworthy of the characters of freemen."

Let us, then, no longer permit the marvels of industry in which our countrymen have been eminently successful, so far to dazzle us as to make us forget the lessons of past history. The Italian republics of the Middle Ages had made great strides in industry and the arts. The republic of the United Netherlands was enriched by commerce in the time of De Witt. But it has been well said, that in bending their whole energies to the attainment of riches, and neglecting their military resources, Italy became the prey of foreigners, and Holland only secured national independence by the sacrifice of political liberty.

The history of modern tactics proves " that preparation in peace gives victory upon fields of battle." The mobility of troops, as now organized, armed, and instructed; the quantity, and still more the kind of artillery used, render a passive resistance, such as that formerly made, impossible. The impossibility of resisting attacks by such means causes the defence to seize the moment in which the attacking party uncovers himself to resort to the offensive, and hence the issue is now more quickly decided, and conquest more rapid than it was a hundred years ago. The ease with which large bodies of men are now transported, the rapidity of all preparatory manœuvres, as well as the greatly increased mobility in action of instructed troops, admits of the ready concentration of great numbers of such men, without the machine becoming too heavy or unmanageable, or its component parts losing the sentiment of order. It therefore follows that the loss of a battle, in consequence of the numbers engaged, is now much more important than it formerly was, and that such loss resulting from incapacity to manœuvre, or want of discipline, may involve the most disastrous consequences. If the people of the United States suppose that the facilities which our railroads offer enable us to concentrate larger masses of men in a short period, the answer must be made that DISCIPLINE is the soul of an army, and that without the habit of obedience, an assemblage of men in battle can never be more than a panic-stricken mob. Instances in our own history are not rare to verify this truth. The fields of Princeton, Savannah River, Camden, Guilford Court-House, &c., during our Revolutionary War, not to speak of later disasters, amply sustain the declaration of Washington, that such undisciplined forces are nothing more than a "*destructive, expensive, and disorderly mob.*" "When danger is a little removed from them, they will not turn out at all. When it comes home to them, the well-affected, instead of flying to arms to defend themselves, are busily employed in removing their families and effects; while the disaffected are concerting measures to make their submission, and spread terror and dismay all around, to induce others to follow their example. Daily experience and abundant proofs warrant this information. Short enlistments and a mistaken dependence upon our militia, have been the origin of all our misfortunes, and the great accumulation of our debt. The militia come in, you cannot tell how; go, you cannot tell when; and act, you cannot tell where; consume your provisions, exhaust your stores, and leave you at last at a critical moment." Such facts, bringing fearfully home to us the contrast between indiscipline and discipline, it is hoped, may yet cause our countrymen to heed the admonition of the Father of his country, that " In peace we must prepare for war." Let us not deceive ourselves by supposing that, when danger becomes imminent, Congress will take the necessary measures to meet it. The steps which are necessary call for sacrifices from the people, and unless public opinion sanctions the means, Congress, in the day of trial, will always be found to represent misdirected popular opinions.

The veteran, Mr. Gales, in the *National Intelligencer* on the occasion of the death of Mrs. Madison, gave a picture of the inertness of the last session of the War Congress of 1814-15. His recollections of the past furnish instructive lessons of what we may expect in the future, if the attention of the people of the United States be not fixed on the necessary sacrifices which love of country demands. So believing, extracts from his historical sketch are here quoted in the firm persuasion that the measures, then recommended, are essential to the safety of our cities and towns, if some organization by States, at least, as efficient as the militia scheme recommended by General Knox, with the sanction of General Washington, be not adopted in time of peace when a matured scheme may be well digested. Mr. Gales writes: " Congress had assembled on the 19th of September preceding—not, as might be supposed from the date, in consequence of the then recent capture of the city [of Washington] by the enemy, but in pursuance of a requisition by the President anterior to that event, calling Congress together (as the President informed the two Houses, in his message at the opening of that session) for the purpose of supplying the inadequacy of the finances to the existing wants of the Treasury, and of making further and more effectual provisions for prosecuting the war. During the recess of Congress, the honor of the arms of the United States had been gallantly sustained in every conflict by land and sea; politically considered, the capture of Washingon itself, and the destruction of the Capitol and the other public buildings, so far from being a misfortune, was for the administration a fortunate event, by its effect in exciting indignant feelings throughout the country, uniting the people in support of the common cause, and preparing their minds for the additional burden of taxation which it had become obvious that they must be called upon to bear. All that was wanting to the vigorous prosecution of the war, was the provision of men and money for the purpose. The progress of recruiting for filling the ranks of the regular army had already proved entirely too slow, if not total failure, as had the resource of loans for the support of the Government, as well as for carrying on the war. The army, whose organization was, on paper, more than 62,000 men, comprised an actual force of only 32,000, exclusive of officers, of which force probably not more than one-half could be relied on for effective service; and the credit of the Government had sunk so low that plummet could hardly sound the depth of its degradation.

"At the opening of the session, the President, in his communication to the two Houses of Congress, with eloquent persuasion, endeavored to impress upon them the necessity of making *immediate* provision for filling the ranks of the army, and replenishing the treasury. In this purpose he was earnestly seconded by Secretary Monroe, of the War Department, and the new Secretary (Mr. Dallas) of the Treasury Department.

"Towards the first of these objects, a bill was soon matured, and afterwards received the assent of Congress, extending the age at which recruits might be enlisted to fifty years, doubling the bounty in land to each, and removing the interdiction upon recruiting minors and apprentices. This measure was a mere experiment, of no practical value, as the event showed. The plan for filling the ranks of the army upon which the Executive relied, and which was placed before the Senate in a bold and energetic report from the War Secretary, was to form into classes of 100 each, all the population of the United States fit for militia duty, out of every class of which four men for the war were to be furnished within thirty days after the classification, by choice or by draught, and delivered over to the recruiting officer of each district, to be marched to such places of general rendezvous as might be directed by the Secretary of War. This plan, which, as the reader will perceive, comprised all the essential features of the French conscription, though, perhaps, the only one which at the time promised effective results, found from the first no favor, especially in the House of Representatives; and became more and more obnoxious, the more the administration seemed to have it at heart. Hardly any one in Congress had the courage to allude to it. Mr. Troup did indeed prevail upon the Military Committee, of which he was chairman, to allow him to report a bill, conformable to the Executive recommendation, by the pregnant title of 'An Act making provision for filling the ranks of the regular army, by classing the free male population of the United States;' and the bill was referred to a Committee of the whole House, and never after heard of. In the course of the session some acts had passed, looking to the employment of volunteers and detachments of militia, under the old plan, for short terms; and one of more importance, " to authorize the President of the United States to accept the service of State troops and volunteers.' This last was not only the most effective measure which had passed towards the supply of men for carrying on the war, but it was the most so that was likely to pass.

"The truth to say, indeed, notwithstanding the nature of the emergency, a dogged inertness seemed to paralyze the action of Congress during the latter part of that session. The recommendation to recruit the army by drafts from the militia was not only unwelcome, as we have said, but revolting to the inclination of the popular branch of Congress; so much so, that a great proportion of the members of that body (and among them some of the leading and most conspicuous members of the republican party) shrunk from it as from the plague; and, as though the leprous influence of that proposition contaminated every other part of the plans of the administration, it was with almost equal reluctance that the House approached the consideration of adequate measures (such as Mr. Secretary Dallas frankly and fearlessly recommended) for the support of the public credit, and for strengthening the sinews of war." *

From the foregoing sketch of the past, it is evident that, unless the opinions and prejudices of the people of the United States be greatly changed, any attempt to raise large armies in the most critical emergencies, without the agency of States, must prove a failure. In order, therefore, to provide for the "common defence," the aid of State organizations will be necessary, and several plans, more or less efficient, have consequently been proposed to better the organization of the militia. All such attempts have, however, met with no favor from the people; and, indeed, it is much to be doubted whether the constitutional reservation to the States " of training the militia according to the discipline prescribed by Congress," and governing them, except when called forth "to execute the laws of the Union, suppress insurrections and repel invasions," will admit of any "good, energetic, general, uniform, and national system of organization." The division of authority made by the constitution between the United States and the several States, in regard to the militia, until called forth by the Federal Government, has left with Congress only the right to provide for "organizing, arming, and disciplining the militia;" but discipline, in that restricted sense, without power to regulate the appointment of officers

---

* In striking contrast with this inertness of Congress, the Legislature of New York assembled on the 26th of September, 1814, passed by the 24th of October a bill giving additional pay to the militia from the State treasury, an act to encourage privateering and an act to raise twelve thousand State troops by conscription or classification. See Hammond's Political History of New York, vol. 1. pp. 380–1.

or otherwise to govern, means little more than prescribing a system of tactics, and such discipline can never make soldiers.

There is, however, another suggestion in the Constitution of the United States, for providing for the common defence, which is obnoxious to none of the objections made against large standing armies, and which commends itself to favorable consideration, as being in harmony with the Federal Government, and capable of furnishing any number of disciplined soldiers which the exigency of our foreign relations may require, without outrage to the instincts of the people of the States. The tendency of the multiplication of States in our confederacy is to restrict the authority of the general Government over the internal affairs of the people of the States. This has been shown by breaking down the Bank of the United States, establishing the independent treasury, refusing appropriations for internal improvements, and, lastly, leaving to the people of Territories the regulation of their own institutions. The maxim " that the world is governed too much," has been sturdily preached, and it may become necessary not to shrink from maintaining our doctrine in the face of foreign powers. To do this we must arm for defence, and the consistent mode of doing so, is for Congress to give its consent for the several States to " *keep troops;* " — more particularly as the history of our country has shown that public opinion will not admit any other efficient military organization. States now have authority to keep troops in time of war, but for such troops to be useful in war, they must be prepared in peace; but as the Constitution of the United States forbids States " to keep troops in time of peace without the consent of Congress," that consent could be given with conditions attached, and those conditions, besides providing for the common defence in war, should require the organization and instruction of State troops to conform with that of the army of the United States, or rather with the cavalry, harnessed batteries of artillery, and infantry of the army.

To encourage States in such organizations, let Congress provide for the annual distribution of ———— dollars among the several States and Territories in proportion to their enrolled militia force, upon satisfactory evidence being furnished to the Secretary of War, that such States have organized camps of instruction during two months in the year, containing a number of troops not less than one-twentieth of the enrolled militia force of the State. Direct the President to furnish to the several State governors, upon their requisition, such army officers as they may desire to aid the commanders of the camps of instruction, and the information collected and kept up in the army will thus be diffused throughout the country. The different States will take pride in their respective organizations, and would recruit their respective armies according to the genius of their people. Their military codes would react upon each other, and upon that of the United States. An interest in military affairs would take the place of present derision, and more than all, the United States might laugh to scorn the efforts of any invader.

The Prussian Landwehr of the first ban, to which the proposed organization is assimilated, is considered a reserved army, remaining by their firesides in times of peace, except during their annual seasons of manœuvring, but ready to appear in case of war upon the first call, organized, equipped, and armed to serve like the line of the army, either at home or abroad. The Prussian territory is divided into as many districts as there are battalions of the Landwehr of the first ban. Each district furnishes a battalion of infantry, a squadron of cavalry, a company of artillery, and some other detachments. The battalions and squadrons are named from the principal town of their district, and depots of arms, clothing, camp and garrison equipage, and cavalry and artillery equipments, are there located. The districts of the Landwehr are also the recruiting districts of the line of the army; and, as troops from the same district serve together, there naturally exist between those corps ties of consanguinity, which dispel all feelings of superiority, and cause them mutually to sustain each other in time of danger.

In each district of the Landwehr, the following small list of officers are permanently paid. For the infantry: one major commanding, one adjutant, who is also accountant, four first sergeants, and four second sergeants, (one per company,) eight corporals, (two per company,) and one armorer. For the cavalry: one captain, or first lieutenant, one quartermaster-sergeant, and three corporals. The paid commanders of battalions are charged with the assistance of their staff, with the *personnel* and *materiel* of the Landwehr, and are accountable for the ordnance and military stores in depot in their districts. The first sergeants keep the list of names belonging to their companies, and no man can absent himself without notifying them.

If all the States of the Union did not deem it better under this system to keep up a small permanent force, it is supposed that they would all find it necessary to maintain a small skeleton organization of officers and non-commissioned officers, similar to that of the Prussian Landwehr of the first ban. And if such officers and non-commissioned officers were appointed by the States from officers and non-commissioned officers who have honorably retired from the army, a new link would be established between the army and State troops which would prove mutually beneficial.

To resume, then: the system of national defence or military organization herein suggested, as suitable for the United States is: 1. The promotion of the most thorough organization and instruction of the United States army, by concentrating troops at strategic points; changing the system of recruiting; creating a retired list for officers of the army, and providing for alternation of duty in the line and the staff, so that the whole army may be made really an aid or staff for the perfect development of the physical strength of the whole country. 2. An act of Congress authorizing the several States to keep troops in time of peace, provided their respective regimental organizations of cavalry and infantry shall conform to the regimental organization of those arms instituted by Congress. 3. An annual appropriation by Congress to be distributed among the several States in proportion to the enrolled militia force of the State, provided satisfactory evidence is brought before the Secretary of War that such State has had within its limits, during two months of the year, organized camps of instruction in which were assembled a number of troops not less than one-twentieth of the enrolled militia force of the State. 4. Requiring the President to furnish to State governors, upon their requisitions, such army officers as may be desired to aid commanders of State camps of instruction, so that the information collected in the federal army may be extended to all State organizations. 5. Giving authority to the President to muster into the service of the United States, State troops, in all cases in which he is now authorized by law to call forth the militia. (*See* CALLING FORTH.)

DEFILADING—consists in raising the parapets of a fortress or field-work, or in depressing the terre-pleins so much as to conceal the interior of the work from the view of an enemy on an elevated position. It also consists in directing the magistral lines of its parapets toward points, where local impediments, as rivers, marshes, lakes, &c., would prevent a besieger from constructing batteries. The former is defilading by relief, the latter is termed defilading by the trace or plan. When a field-work has been necessarily constructed in such a situation that it may be commanded by some height within range of artillery, the defilading is made by raising the parapet, or constructing traverses in the interior of the work. The necessary trace for a field-work to accomplish these objects is more expeditiously effected by the eye and a few poles and profiles, than by resorting to theoretical and scientific proceedings, which constitute a part of the art of the engineer, and which are indispensable considerations in permanent fortification.

DEFILE. Any narrow passage—as a ford, a bridge, a road

through a village, mountain passes, &c., are defiles. To pass a defile safely, it is necessary first to drive away, as far as possible, the enemy. Under cover of this engagement, other troops pass the defile as soon as they reach it. The aim should be to pass the defile as quickly as possible; whether advancing or retreating. The passage in double columns will facilitate the formation in order of battle on the right and on the left after having passed the defile, and this order has the advantage of occupying both sides of the road. But it cannot be too strongly urged that quickness in the passage is the great consideration, and theoretical movement must give way to this primary object If the defile is a ford or bridge, and the passage in retreat, formations on the bank of the river, after the passage, ought not to take place. Combats separated by a river end in nothing, and the worst possible way of defending a bridge or ford is taking positions too near it. The enemy would certainly unite his artillery upon the opposite bank, and not attempt the passage until he had greatly worsted the defenders of the ford or bridge by his projectiles. The defenders would lose many men, and would probably have been demoralized before coming to close quarters. It is necessary then to wait until a portion of the enemy passes the bridge or ford. If the enemy be then vigorously attacked the defenders will, by a hand-to-hand conflict, render nugatory his artillery on the opposite bank, as well as all of his troops that have not yet crossed. To accomplish this intended purpose, it will only be necessary to place troops at some point, at full cannon range from the bridge, or if the accidents of ground admit of cover, nearer still to the bridge. If a bridge is passed in advancing, the troops which pass first are pushed forward to gain as much ground as possible, and thus favor the passage of other troops, by relieving them of the dangers of the combat. In this case the simplest and most rapid method of crossing is the best. (Consult *Aperçus sur quelques Details de la Guerre*, par MARSHAL BUGEAUD.)

DELINQUENT, (DISBURSING OFFICERS.) Such officers may be dismissed by the President of the United States on failure to render their accounts of disbursements quarterly in the United States, and every six months if resident in a foreign country; (*Act* January 31, 1823.) (*See* DEFAULTER.)

DEMILUNE—is a work constructed to cover the curtain and shoulders of the bastion. It is composed of two faces forming a salient angle towards the country, has two demi-gorges formed by the counterscarp, and is surrounded by a ditch. The demilune is sometimes termed a ravelin.

DEPARTMENT. Any general officer commanding an army, or colonel commanding a separate department, may appoint general court-martial, whenever necessary; (Art. 65.)

Besides the territorial divisions, called Departments, in the Rules and Articles of War, the term is also applied to the following branches of the service: Adjutant-general's, Inspector-general's, Medical, Pay, Ordnance, Quartermaster's, and Subsistence Departments.

DEPARTMENT OF WAR. There shall be an Executive Department, to be denominated the Department of War; and there shall be a principal officer therein, to be called the Secretary for the Department of War; (*Act* Aug. 7, 1789.) "He is to perform and execute such duties as shall, from time to time, be enjoined on, or intrusted to him, by the President of the United States, agreeably to the constitution, relative to military commissions, or to the land forces or warlike stores of the United States, or such other matters respecting military affairs, as the President of the United States shall assign to said *department*. And furthermore, that the said principal officer shall conduct the business of the said department in such manner as the President of the United States shall, from time to time, order or instruct. That there shall be in said department an inferior officer, to be appointed by the said principal officer, to be employed therein as he shall deem proper, and to be called the chief clerk in the Department of War, and who, whenever the said principal officer shall be removed from office by the President of the United States, or in any other case of vacancy, shall, during such vacancy, have the charge and custody of all records, books, and papers, appertaining to said Department. The said principal officer, and every other person to be appointed or employed in said Department, shall, before he enters on the execution of his office or employment, take an oath or affirmation, well and faithfully to execute the trust committed to him;" (*Act* Aug. 7, 1789.) It seems impossible to read this act of Congress, and contend that officers of the army are a portion of the War Department. And the statute book will be searched in vain to find authority given to the Secretary over any officers other than officers of Staff Departments, or over subjects disconnected with the custody of public records, the support and supply of troops, the manufacture and care of warlike stores, the keeping of exact and regular returns of all the forces of the United States, or other kindred administrative matters; such as receiving the proceedings of courts-martial, and laying them before the President of the United States for his approval or disapproval, and orders in the case. There is no act of Congress which authorizes the Secretary of War to *command* the troops, and he being no part of the army, the President, of course, cannot authorize him to do so. But "the Secretary of War is (Peters' Digest of Decisions of Federal Courts, vol. 1, p. 179) the regular constitutional organ of the President for the administration of the military establishment of the nation; and rules and orders *publicly promulgated* through him, must be received as the acts of the Executive, and as such are binding upon all within the sphere of his legal and constitutional authority."

By an act of Congress, approved March 3, 1813, it is provided: "That it shall be the duty of the Secretary of War, and he is hereby authorized, to prepare general regulations, better defining and prescribing the respective duties and powers of the several officers in the adjutant-general, inspector-general, quartermaster-general, and commissary of ordnance departments, of the topographical engineers, of the aids of generals, and generally of the general and regimental staff; which regulation, when approved by the President of the United States, shall be respected and obeyed, until altered or revoked by the same authority." Here was a partial delegation of legislative power; and under this power of legislation so confined to the several staff departments, the Secretary of War, with the approval of the President, established bureaus of the War Department, making the head of each staff department chief of a bureau, in all fiscal and administrative matters connected with his particular department under the general direction of the Secretary of War. The War Department thus centralized all army administration, and efforts have since been made to centralize in the same way the command and government and regulation of the army. But as the 62d article of war declares that when different corps come together, the officer highest in rank shall command the whole, and *give orders for what is needful to the service,* unless otherwise specially directed by the President of the United States, according to the nature of the case," while the 61st article gives the command to the senior regimental officer within his regiment, when other troops are not present, such centralization, if not a violation of law, would be a violation of all military principles, destructive alike to discipline and military spirit. For (says Odier): "Commands given immediately by the highest authority cause agitation rather than action. The superior authority becomes weakened in proportion as the eye becomes accustomed to it. Fear of it ceases, and when the highest authority habituates itself to doing every thing, as soon as it ceases to be sufficient to do all, there is nothing done. All degrees of rank and command have their degree of importance. Authority must regularly ascend and descend. Every inferior grade is the *lieutenant* of its superior grade, even to the oldest soldier, who replaces the corporal. Obedience is reciprocal to authority." Rules established by Congress, defining the rights, powers, and duties of all officers and soldiers, are much needed. (*See* SECRETARY OF WAR.)

DEPLOYMENT. All tactical manœuvres intended to pass from close column to the order of battle are deployments. Deployments, however convenient or brilliant, which cause the soldier to turn his back to the enemy, are not suited to war. (Consult Infantry and Light Infantry and Rifle tactics for the prescribed deployments.)

DEPOSITION OF WITNESSES—when not of the line or staff of the army, may be taken in cases not capital, provided the prosecutor and accused are present at the taking of the same, or duly notified; (ART. 74. *See* WITNESS.)

DEPOT. The colonel of ordnance, under the direction of the Secretary of War, is authorized to establish depots of arms, ammunition, and ordnance stores, in such parts of the United States, and in such numbers, as may be deemed necessary; (*Act* Feb. 8, 1815.)

Three recruiting depots have also been established under the direction of the Secretary of War, but a system of regimental depots is much needed. In England and in France, regimental depots have been found indispensable. In France, upon taking the field, a regiment leaves in depot the quartermaster and the accounting officer of the corps, the clothing officer, workmen, and stores; infirm men, those too old for war, and uninstructed recruits; these make the depot; the wounded and sick are sent there to be re-established; new levies are received there, and detachments of able-bodied and instructed men are successfully directed from the depot towards the army. The depot, like the stomach, receives, elaborates, and gives life to its members. It is at the depot that the clothing, and shoes, and all the wants of the regiment are provided; it is there that the accountability is centralized, that the papers are kept; it is at the depot that all regimental administration goes on; and for that purpose the major of the regiment remains there, and likewise commands. In England, the depot company is one left at home by regiments embarking for India, for the purpose of recruiting. There are four *reserve companies* for all foreign stations except India, which remain at home under the command of the senior major. A roster is regularly kept of the officers at the depot; and to insure that each individual embarks in his proper turn to join the service companies, a figure marking his place on the roster, is annexed to every officer's name in the monthly returns transmitted to the adjutant-general. Regimental records, with the attestations and service records of the men doing duty with the regiment abroad, are left at the depot, and filled up at stated periods.

DERRICK—consists of a spar which is always kept in an oblique position; one end of it on the deck of a ship, the other supported by guys, and generally used to hoist heavy weights. (*See* GIN.)

DESERTER. Punishable by stripes, by sentence of general court-martial. Not punishable by death in time of peace. May be tried and punished, although the term of enlistment may have elapsed previous to apprehension. (ART. 20, and *Acts* March 16, 1802, May 29, 1830, May 16, 1812, and March 2, 1833.)

Of a deserter from the enemy, we demand his name, his country; the motive of his desertion; the number of his regiment; the name of his colonel; his immediate general; that of the commander-in-chief; the strength of his particular corps; that of the whole army; whether distributions are regular; how many cartridges each man has; how many guns there are; whether there are many sick or wounded in the camp of the enemy; whether the soldiers have confidence in their chief, and whether he is well treated by them.

DETACHED BASTION—is one which is separated from the enceinte by a ditch.

DETACHED WORKS—are those which are constructed beyond the range of the musketry of the main works; and as a constant and steady communication with them cannot be kept up during a siege, they are frequently left to their own resources; nevertheless, they ought to exercise a general influence on the defence of the place.

DETACHMENT. (*French Origin.*) BARDIN, *Dictionnaire de l'Armée de Terre* thus defines it: A word which has the same origin as *attach*. It implies any fraction of a body, or an entire corps charged particularly with functions which are dependent for their duration upon circumstances in war or actual service. The Romans expressed by the word *Globus** nearly the meaning of detachment. The movable columns of the French army were *detachments* formed sometimes of whole corps, sometimes of fractions of corps. We call also detachments, the escorts of convoys of prisoners, those for evacuations, certain extra duties, some maritime expeditions, a patrol, &c. Agreeably to the definition given in the instructions of the year six, the separation of many men from a single or from different corps, and the subsequent reunion of those men under a military chief, constitutes a detachment, and it is so considered, whether upon a voyage, or stationed in a depot of a corps

---

* A troop; a squadron, or party of soldiers; a knot of men who jointly carry on any design. AINSWORTH's *Latin Dictionary.*

or in garrison; whether in cantonment, or whether in reference to the means of transportation that may be necessary for it. In some cases, picket and small detachments have the same signification. The following illustrations of the meaning of detachment are drawn from various sources:—

Rules and Articles of War passed Sept. 20, 1776.

ART. XII. Every officer commanding in any of the forts, barracks, or elsewhere, where the corps under his command consists of *detachments* from different regiments, or of independent companies, may assemble courts-martial, &c.; [such courts were called detachment courts-martial.] ART. II. SEC. 17. For the future, all general officers and colonels, serving by commission from the authority of any particular State, shall, on all *detachments*, courts-martial, or other duty, wherein they may be employed in conjunction with the regular forces of the United States, take rank, &c.—When regiments or *detachments* are united, either in camp, garrison, or quarters, the eldest officer, whether by brevet or otherwise, is to command the whole; (*Regulations British Army*.) The *detachments* which are, from time to time, sent from the depots at home to regiments abroad, &c.—The periods of the year at which detachments are required to embark for foreign stations, &c.; (*Regulations British Army*.)—Whenever recruits are to be sent from a depot or rendezvous to a regiment or post, a separate muster and description roll, and a separate account of clothing of each *detachment*, will be placed in the hands of the officer assigned to the command of such *detachment*; (*U. S. Army Regulations*.)—Any detachment so far separated from the main body to which it belongs as to render it impracticable for the commander of the latter to make muster and inspection enjoined by the general regulations, is considered as a separate command within the meaning and for the purpose of this regulation.—Where a field-officer is serving with *detached* companies of his regiment, the captains thereof will make their company monthly returns through him, which returns he will transmit with his own personal report to regimental head-quarters; (*Regulations of the War Department, dated* Feb. 10, 1855.)

SEC. * * And the said corps may be formed into as many companies or *detachments* as the President of the United States may direct. (*Act of Congress*.)

"Corps, formed by *detachments*, are the usual method in which brevet officers are employed, as they cannot be introduced into regiments without displacing other officers, or violating the right of succession, both of which are justly deemed injurious in every service. But the reasoning is new by which the employing such officers in detached corps is made an infringement of the rights of regimental officers; (*Letter of General Washington, dated* August 11, 1780.)

DETAIL FOR DUTY—is a roster, or table, for the regular performance of duty either in camp or garrison. The general detail is regulated by the adjt.-general, according to the strength of the several corps. The adjutant of each regiment superintends the detail of the officers and non-commissioned officers for duty, and orderly sergeants detail the privates.

DEVIATION OF FIRING. (*See* FIRING.)

DIMINISHED ANGLE—is that formed by the exterior side and the line of defence in fortification.

DISBURSING OFFICERS. Exclusively of the paymasters of the army, and other officers already authorized by law, no other permanent agents shall be appointed, either for the purpose of making contracts, or for the purchase of supplies, or for the disbursement in any other manner of moneys for the use of the military establishment, but such as shall be appointed by the President of the United States, with the advice and consent of the Senate. But the President may appoint such necessary agents in the recess of the Senate to be submitted for their advice and consent at their next session, provided that the compensation allowed to either shall not exceed one per centum per annum, nor be more than $2,000 per annum; (*Act* March 3, 1809.) All purchases and contracts are made under the direction of the Secretary of War; (*Act* March 3, 1809.) Shall give bonds to be regulated by the President, and may be dismissed by the President on failure to render their account. (*See* DEFAULTER; DELINQUENT.)

DISCHARGE. After a non-commissioned officer or soldier shall have been duly enlisted and sworn, he shall not be dismissed the service without a discharge in writing; and no discharge granted to him shall be sufficient, which is not signed by a field-officer of the regiment to which he belongs, or commanding officer, where no field-officer of the regiment is present; and no discharge shall be given to a non-commissioned officer or soldier, before his term of service has expired, but by order of the President, the Secretary of War, the commanding officer of a department, or the sentence of a general court-martial; nor shall a commissioned officer be discharged the service but by order of the President of the United States, or by sentence of a court-martial; (ART. 11.) Under this article it has been contended that the President may arbitrarily *discharge* any commissioned officer from the service; but as the Rules and Articles of War provide for the punishment of all military crimes, disorders, or neglects, by courts-martial, all arbitrary and capricious action over such matters is thereby necessarily excluded. Besides, *dismission* and *discharge* are essentially different. The latter, in its primitive sense, means relieved of a burden or obligation. Thus, as every individual who enters the army by enlistment or commission must remain in it until regularly *discharged*, under penalty of being considered a deserter, the article declares that no discharge of a commissioned officer is regular but by the order of the President of the United States, or the sentence of a court-martial. Voluntary separations from the service, therefore, or resignations, are only legal when accepted by the President of the United States. No other military authority is competent to release an officer from the obligations he assumes on entering the army, even on his own application. Hence the use of the word *discharge* in the article, so as to embrace voluntary separations authorized by the President, and involuntary separations by sentence of court-martial. But the article gives no power to the President to *dismiss* summarily. Had such been the intention, the authority would have been clearly given, as it has been by the act of Jan. 31, 1823, in the case of delinquent disbursing officers—a power not needed, if it before existed under Article 11. This rule of making the acceptance of an officer's resignation dependent upon the President or highest military authority, is necessary; because an officer who was amenable to punishment for infractions of military law, might otherwise, by the resignation of his commission, escape punishment. The Court of King's Bench in England have decided, therefore, that an officer of the East India Company's service has not the right to resign his commission under *any* circumstances, and whenever he pleased; (case of Capt. Parker; *Prendergast*, p. 248.) In the case of Capt. Vertue, however, (*Prendergast*, p. 250,) while the court held that Capt. Vertue's resignation was invalid, as having been made in pursuance of an improper combination of a large number of officers, yet Mr. Justice Yates intimated that there may be a state of circumstances, under which an officer may have a legal right to resign, and so to obtain a release of exemption from military law.

Such would undoubtedly be the decision of a civil court in the United States. The power given to the President of accepting or withholding his acceptance of a resignation was intended for the maintenance of justice, and not the oppression of individuals; and if that power should be perverted, a court of justice might, and no doubt would, interpose its writ of *habeas corpus*.

DISCIPLINE. It ought to result from a perfect uniformity of rules; for stability, method, exactness, and even routine, are necessary to insure its maintenance; under a perfect discipline, troops in peace and in war, in garrison or in campaign, would be fitted for all the duties of war. To attain this perfection, it is necessary that discipline should rest entirely upon law; it ought to have its roots in patriotism; to be adapted to the character of the people; to the spirit of the age, and the nature of the government. It is essential to make rights and duties inseparable. This absolute necessity, and the importance of regularity of pay, are truths dwelt upon by French writers. Discipline may be distinguished as active and passive. The first derives its power from a military hierarchy or range of subordination, skilfully established and regulated; it is secured by calmness, impartiality, promptness, firmness, and the prestige of character in officers. These qualities are manifested by preventing wrongs rather than by punishing faults, and by abstaining from arbitrary corrections when obliged to chastise. Discipline, intrusted to such authorities enlightened by military experience, will partake of the character of paternal government, and will not be enforced with an unsparing harshness suited only to governments essentially despotic.

The dogma, that military discipline can only be sustained by the aid of severe and unpitying punishment, is far removed from the idea here suggested. That unpitying military discipline seems to have prompted Peter the Great, when he sacrificed a young officer, who triumphantly fought the Swedes without orders. Thus also thought Frederic the Great, when he executed the unfortunate Zietten, who violated an order by keeping a light a little too long in his tent. But such harsh principles are no longer inculcated in the best governed armies of Europe. Passive discipline is the fusion of individual interest in national interest. The first military virtue is *esprit de corps*, with fidelity to the oath taken upon assuming the military character. These duties exact obedience to the laws, and to the lawful orders of the President of the United States, and officers set over us according to law. These laws should command obedience from all inferiors, and distinctly define the extent of all authority. They ought to bind the President or commander-in-chief as well as the simple soldier. RIGHTS and DUTIES must be reciprocal, and be alike established by law, which should, to maintain discipline, " precisely determine the functions, duties, and rights of all military men—soldiers, officers, chiefs of corps, generals." Discipline that has attained this perfection supplies the deficiency of numbers, and gives new solidity to valor; since, although surrounded by dangers, the brave man feels that his leaders and comrades are not less devoted, less vigorous, or less experienced than himself.

Discipline is sometimes used as meaning "system of instruction," but its signification is much broader. Its technical military sense includes not only the means provided for exercise and instruction, but subjection to all laws framed for the government and regulation of the army. The good or bad discipline of an army depends primarily upon the laws established for its creation, as well as its government and regulation.

DISEASE. (*See* SANITARY PRECAUTIONS.)

DISEMBARKATION. In disembarkations, the first essential matter is to determine by *reconnoissance* the proper point for landing—how near the landing can be approached with vessels of light draught, to scour the beach and thus cover the operation; and secondly, the manner in which the men, horses, and some field-artillery are to be disembarked. The landing of heavy ordnance and all supplies is a subsequent matter. Having chosen the point of debarkation, the troops are put into flat-bottomed boats, previously provided, as expeditiously as possible, but without hurry or disorder—*they are to sit down in the boats, and positively ordered not to load until formed on the beach.* Each man should carry three days' provisions cooked in his haversack, at least forty rounds of ammunition, and his canteen filled with water. The men should also carry their intrenching tools. The covering vessels must be liberal with round shot, grape, and canister; and under cover of their fire, the *first line of boats* should pull boldly in, recollecting that the men are to be landed, and that the sooner it is done the better. When a boat grounds, the officer jumps out over the bow, and the men follow also over the bow. If the boat is large, or there are rocks, so as to render it unsafe for an accoutred man to jump, the gang-boards must be used. The men follow the officer to the sheltered spot selected by him for their formation. Without waiting for other boats, the officer will consider his men part of a line of skirmishers, the supports of which are behind. As soon as each boat is clear, she must shove off, and pull to the shipping for a fresh load.

The second division of boats will land as the first, but these will not commence firing until the whole of each company has joined, when they will act as supports, under the command of their proper officers. As soon as a sufficient number of well-united companies are on shore, the irregularly formed skirmishers first landed will be relieved, formed by companies, and sent to their respective battalions. Boats employed landing troops should have neither guns, masts, nor sails; their equipments should be gang-boards, oars, grapnels and painters, boat hooks, bailers, hammers and nails, sheet lead, grease, and canvas; the latter articles to enable them to stop a small shot hole, in case of accident.

The launches of men-of-war are used for disembarking field-artillery, when opposed by the enemy. Two planks are laid from the bow to the stern of the launch, parallel to each other, at the distance of the space of the wheels; a bead is nailed to the inside edge, to prevent the wheels from slipping off. Two gang-boards, which can be laid out or taken on board, are fitted to the bow ends of the planks, so as to reach from them to the shore, as a ramp. These launches are towed by smaller boats. It is very desirable that this portion of artillery, with their officers and men, should be on board men-of-war. Each two-decker can take a couple; the guns are stowed away on the upper deck, the carriages and wheels in the chains, so that the guns can be mounted and ready to be lowered into the boats in a very few minutes. The muzzle of the gun must point forward in the launch, and as soon as the boat touches ground, the gang-boards are put out and the guns run ashore. The artillery should endeavor to gain the shore and land with the troops. It is dragged by the sailors or troops. A sufficient supply of ammunition must be at hand in a boat or two, close to the shore. In an emergency the harness may be at once sent ashore, and if the vessels are near, horses may be made to leap out and swim ashore. Under other circumstances, boats of proper capacity must be provided for the disembarkation of horses, heavy ordnance, &c.; or it may be necessary to establish temporary wharves on trestles, or by means of boats, and to erect shears, cranes, or derricks.

On a smooth, sandy beach, heavy pieces may be landed by rolling them overboard as soon as the boats ground, and hauling them up with sling carts. (*See* EMBARKATION. Consult *Aide Memoire of the Military Sciences;* SCOTT's *Orders and Correspondences during the Campaign in Mexico.*)

DISINFECTANTS. (*See* SANITARY PRECAUTIONS.)

DISMISSION. No sentence of a court-martial in time of peace dismissing a commissioned officer, or which, in war or peace, affects a general officer, shall be carried into execution without the approval of the President of the United States; (ART. 65.) Disbursing officers may be dismissed by the President alone, without the intervention of a court-martial, on failure to account properly for moneys placed in their hands; (*Act.* Jan., 1823.) A general court-martial in time of peace may dismiss, with the approval of the President, in all cases in which they are authorized to sentence to " death or such other punishment as may be inflicted by a general court-martial." (*See* DEATH.) Such court may also sentence a commissioned officer to be cashiered or dismissed the service in the following cases :—1. Drunkenness on duty;

(Art. 45.) 2. Breach of arrest; (Art. 77.) 3. Conduct unbecoming an officer and a gentleman; (Art. 83.) 4. Using contemptuous or disrespectful words against the President of the United States, against the Vice-president thereof, against the Congress of the United States, or against the chief magistrate or legislature of any of the United States in which he may be quartered; (Art. 5.) 5. Signing a false certificate relating to the absence of either officer or soldier, or relative to his or their pay; (Art. 14.) 6. Making a false muster of man or horse; (Art. 15.) 7. Taking money or other thing by way of gratification, on mustering any regiment, troop, or company, or on signing muster rolls. 8. Making a false return to the Department of War, or to any of his superior officers authorized to call for such returns of the state of the regiment, troop, or company, or garrison under his command: or of the army ammunition, clothing, or other stores thereunto belonging; (Art. 18.) 8. Sending and accepting a challenge to another officer or soldier to fight a duel; (Art. 25.) 9. An officer who commands a guard, knowingly and wilfully suffering any person to go forth to fight a duel, and all seconds, promoters, and carriers of challenges shall be punished as challengers; (Art. 26.) 10. Selling, embezzling, misapplying, or wilfully, or through neglect, suffering provisions, arms, &c., to be spoiled or damaged; (Art. 36.) 11. Any commanding officer who exacts exorbitant prices for houses let out to sutlers, or connives at like exactions from others, or who by his own authority and for his private advantage lays any duty or imposition upon, or is interested in, the sale of any victuals, liquors, or other necessaries of life brought for the use of the soldiers, may be discharged the service; (Art. 31.) 12. Failure, by a commanding officer, to see justice done to offenders, and reparation made to the party injured, by officers or soldiers ill-treating any person, or disturbing fairs or markets, or committing any kinds of riots to the disquieting of citizens of the United States; (Art. 32.)

DISMOUNT. To dismount the cavalry, is to use them as infantry. Guards, when relieved, are said to dismount. They are to be marched with the utmost regularity to the parade-ground where they were formed, and from thence to their regimental parades, previously to being dismissed to their quarters. To dismount a piece of ordnance, is to take it from the carriage.

DISOBEDIENCE OF ORDERS—punishable by a court-martial with death or otherwise, according to the nature of the offence; (Art. 9.)

DISORDERS. (*See* ABUSES; CRIMES.)

DISPART—is the difference of the semi-diameter of the base-ring and the swell of the muzzle, or the muzzle-band of a piece of ordnance. (*See* ORDNANCE.)

DISRESPECT TO A COMMANDING OFFICER—punished by court-martial.

DISRESPECTFUL WORDS—used by any officer or soldier against the President, Vice-president, the Congress or the governor of any State where he may be quartered, punishable with cashiering or otherwise, as a court-martial may direct; (ART. 5.)

DISTANCES. *Pacing Distances.*—"If you count the strokes of either of your horse's fore-feet, either walking or trotting, you will find them to be upon an average about 950 to a mile. In a field-book, as you note each change of bearing, you have only to note down also the number of paces (which soon becomes a habit); and to keep count of these, it is only necessary to carry about thirty-five or forty small pieces of wood, like dice (beans or peas will do), in one waistcoat-pocket, and at the end of every 100 paces remove one to the empty pocket on the opposite side. At each change of bearing you count these, adding the odd numbers to the number of hundreds, ascertained by the dice, to be counted and returned at each change of bearing to the other pocket. You should have a higher pocket for your watch, and keep the two lower waistcoat-pockets for this purpose. Now, to plot such a survey, you have only to take the half-inch scale of equal parts, (on the six-inch scale, in every case of instruments,) and allowing *ten* for a hundred, the half-inch will represent a thousand paces. You may thus lay down any broken number of paces to a true scale, and so obtain a tolerably accurate map of each day's journey. The latitude will, after all, determine finally the scale of paces; and you can at leisure adjust each day's journey by its general bearing between different latitudes, and subsequently introduce the details." (Sir THOMAS MITCHELL.)

A traveller, when the last of his watches breaks down, has no need to be disheartened from going on with his longitude observations, especially if he observes occultations and eclipses. The object of a watch is to tell the number of seconds that elapse between the instant of occultation, eclipse, &c., and that, a minute or two later, when the sextant observation for time is made; and all that it actually *does*, is to beat seconds, and to record the number of beats. Now, a string and stone swung as a pendulum will beat time; and a native who is taught to throw a pebble into a bag at each beat will record it; and, for operations that are not tedious, he will be as good as a watch. The rate of the pendulum is, of course, determined by taking two sets of observations, with three or four minutes' interval between them; and, if the distance from the point of suspension to the centre of the stone be thirty-nine inches, and if the string be thin, and the stone very heavy, it will beat seconds very nearly indeed. The observations upon which the longitude of the East African lakes now depends (1859) are lunars timed with a string and a stone, in default of a watch.

Units of length.—A man should ascertain his height; height of his eye above ground; ditto, when kneeling; his fathom; his cubit; the span, from ball of thumb to tip of one of his fingers; the length of the foot, and the width of two, three, or four fingers. In all probability, some one of these is an even and a useful number of feet or inches, which he will always be able to recollect, and refer to as a unit of measurement. A stone's throw is a good standard of reference for greater distances. Cricketers estimate by the length between wickets. Pacing should be practised. It is well to dot a scale of inches on a pocket-knife.

Angles to measure.—A capital substitute for a very rude sextant is afforded by the outstretched hand and arm. The span between the middle finger and the thumb subtends an angle of 15°, and that between the forefinger and the thumb an angle of $11\frac{1}{4}°$, or one point of the compass. Just as a person may learn to walk *yards* accurately, so may he learn to span out these angular distances accurately; and the horizon, however broken it may be, is always before his eyes to check him. Thus, if he begins from a tree, or even from a book on his shelves, and spans all round until he comes to the tree or book again, he should make twenty-four of the larger spans and thirty-two of the lesser ones. These two angles of 15° and $11\frac{1}{4}°$ are particularly important. The sun travels through 15° in each hour; and therefore, by "spanning" along its course, as imagined, from the place where it would stand at noon, (aided in this by the compass,) the hour before or after noon, and, similarly, after sunrise, or before sunset, can be instantly reckoned. Again, the angles 30°, 45°, 60°, and 90°, all of them simple multiples of 15°, are by far the most useful ones in taking rough measurements of heights and distances, because of the simple relations between the sides of right-angled triangles, whose other angles are 30°, 45°, &c. As regards $11\frac{1}{4}°$, or one point of the compass, it is perfectly out of the question to trust to bearings taken by the unaided eye, or to steer a steady course by simply watching a star or landmark, when this happens to be much to the right or the left of it. Now, nothing is easier than to span out the bearing from time to time.

Squaring.—As a triangle whose sides are as 3, 4, and 5, must be a right-angled one (since $5^2 = 3^2 + 4^2$), we can always find a right angle very simply by means of a measuring tape. We take a length of twelve feet, yards, fathoms, or whatever it may be, and peg the two ends of it, close together, to the ground. Next a peg is driven in at the third division, and then the third peg is held at the seventh division of the cord, which is stretched out till it becomes taut, and the peg is driven in. These three pegs will form the corners of a right-angled triangle.

Measurements, &c.—The breadth of a river may be measured without instruments and without crossing it, by means of the following useful problem from the French "Manuel du Génie," which requires pacing only:

To measure A B (Fig. 106), produce it any distance to D; from D, in any direction, take any equal distances, D C, C d, and produce B C to b, making C b = C B; join d b and produce it to a, where A C produced intersects it; then a b is equal to A B. In practice, the points D C, &c., are marked by bushes planted in the ground, or by men standing.

Colonel Everest, the late surveyor-general of India, has pointed out the following simple way of measuring an angle, and therefore a triangle:

A B is the base, R R the river, C an object on the other side; (Fig. 107.) He paces any length A a'; and an equal length A a''; also a' a'', which is the chord of a' A a''. In other words—

$$\sin. \frac{A}{2} = \frac{a' \, a''}{2 \, A \, a'}:$$

in the same way B is found. A B being known,

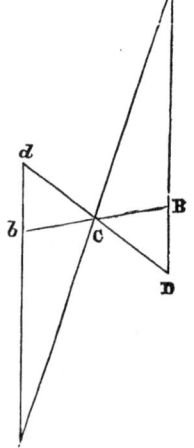

FIG. 106.

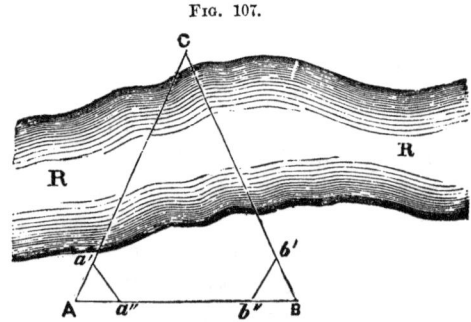

FIG. 107.

the triangle A B C is known, and the breadth of the river can be found. The problem can be worked out, either by calculation or by protraction. (GALTON's *Art of Travel.* See STADIA; SURVEYS; TARGET; VELOCITIES.)

DISTRIBUTION—means, generally, any division or allotment made for the purposes of war, and minor arrangements made for the supply of corps.

DISTRICT. One of those portions into which a country is divided, for the convenience of command, and to insure a co-operation beween distant bodies of troops.

DITCH—sometimes called the Fosse—is the excavation made round the works, from which the earth required for the construction of the rampart, parapet, and banquette is obtained. In besieging a fortification, when the ditch is dry, and a descending gallery has been constructed, the passage of the ditch consists of an ordinary sap pushed from the opening in the counterscarp wall to the slope of the breach, and, when necessary, it is carried on to crown the summit of the breach. If the ditch be full of water, and the locality favors its being drained, every means must be used to break the batardeaux, to cause the water to flow away entirely or in part. If none of the batteries can see the batardeaux, the sluices must be sought and destroyed by shells, or by mining. Should the assailants be unable to breach the batardeaux or to destroy the sluices, a bridge or causeway must be thrown across. This is one of the most difficult operations in a siege. The bridge or causeway, with its epaulement, is constructed with pontoons or casks, or, if without them, with fascines, hurdles, gabions, and sand-bags, openings being left in the causeway to allow the free flowing of the water, if it be a running stream, or can be made so by the defenders. A wet ditch may sometimes be crossed by a raft of sufficient length, which should be constructed along the counterscarp, and attached by one end to the bottom of the descent. The raft is then allowed to swing round with the current, if there be one, or is rowed or pulled round, if there is not one, so as to form a connection across the ditch with the breach.

The following experiment for crossing a wet ditch was successfully tried at Chatham by Sir Charles Pasley:—Two hundred large casks were prepared, with their heads taken out; they were lashed by fours, end to end, so as to form hollow piers, about 18 feet in length, of unequal diameters, in consequence of the unequal size of the casks. Each pier was launched in succession from a great gallery, representing that of the counterscarp in a regular siege. These piers had guys at each end, by which they were hauled round into their intended position, and there sunk by means of sand-bags. After this, the intervals between the upper tiers of casks were filled in with long fascines, and others were laid over these at right angles, till a general level was obtained, when strong skids were laid over all, and a 24-pounder, on a travelling carriage, was dragged through the gallery, and passed along these skids to the other side. In this manner, a piece of water, representing a wet ditch, was bridged over with ease and comparative expedition. This experiment was afterwards tried with full success in the Mast Pond of Chatham Dockyard, where a very strong current was produced, much stronger than could occur in the ditches of any fortified place. It is stated, that there was no perceptible depression in the bridge as the 24-pounder passed over. The same experiment was tried with common gabions, lashed together, end to end, in like manner, forming hollow piers or cylinders, which were similarly sunk one over another until the upper layer rose above the water, and were covered with fascines and skids. These, also, bore a 24-pounder, which caused a depression of more than 6 inches in the part over which it was passing. The gabions were very weak and old. The piers of casks were fastened as follows: on being placed end to end, staples were driven into each cask, about 10 inches from their ends, in three equi-distant parts of their circumference; strong spun-yarn, connecting the staples, lashed the four casks together. Six or eight bushel sand-bags were necessary to sink each pier with ease, yet without making it sink too rapidly. To get them into the water, they were launched on ways made of planks. In making the gabion bridge, each pier consisted of four gabions lashed end to end like the casks, by spun-yarn, at three equi-distant points of the circumference. These were not loaded to make them sink. It was found, from the irregularity of their surface, that the second pier merely forced the first out from the bank to make room for itself; the third the second, and so on, until the tier of gabions connected the two scarps. On rolling other piers on the top of them, the lower ones sunk to the bottom, and brushwood and fascines were laid in the intervals of the gabions to form a level surface; (HYDE's *Fortifications*.)

DIVISION. In the ordinary arrangement of the army, two regiments of infantry or cavalry shall constitute a brigade, and shall be commanded by a brigadier-general; two brigades, a division, and shall be commanded by a major-general. Provided always that it shall be in the discretion of the commanding general to vary this disposition whenever he shall judge proper; (*Act* March 3, 1799; *Sec.* 8.)

DOMICILE. By law every man's domicile is in the country where he has his permanent residence, or to which he ordinarily returns for the purpose of residence after occasional absence; and in case of his death, the right of succession to his goods and chattels and personal property of all sorts is regulated by the law of the country of his domicile, although he may happen to die beyond its limits. As regards military men, their employment on duty involving only temporary absence in intention would not, on common principles, cause a change of domicile; and as the laws of different States of the Union vary on the subject of the right of succession to property, the subject is of great interest to military men. Recently, an officer who was a native of South Carolina died intestate in the city of New York, and no heirs being forthcoming, his estate was taken possession of by the public administrator, although the Rules and Articles of War enacted by Congress provide that, in such cases, an officer of the army at the station shall take possession of the effects for purposes of administration.

"Personal property, in point of law, has no locality, and in case of the decease of the owner, must go wherever in point of fact situate, according to the law of the country where he had his domicile." (ROBERTSON's *Law of Personal Succession.*)

The 14th Lord Somerville entered the army in 1745, and continued in the service till the peace of 1763, during which period he accompanied his regiment to England, Scotland, and Germany, both in quarters and on active duty. At his death in 1796, a question arose, whether, under the circumstances, his domicile was English or Scotch; and the Master of the Rolls, (Sir R. P. Arden,) in giving judgment, said : "I am clearly of opinion Lord Somerville was a Scotchman upon his birth, and continued so to the end of his days. He *never ceased* to be so, never having abandoned his Scotch domicile, or established another. The decree, therefore, must be, that the succession to his personal estate ought to be regulated according to the law of Scotland." His honor must consequently have been of opinion, that a Scotchman entering the British army does not thereby lose his original Scotch domicile; and since the union of England and Scotland, the army is certainly as much that of Scotland as of England.

Sir Charles Douglas, a Scotchman by birth and original domicile, left his native country at the age of twelve, to enter the navy. From that time to his death, he was in Scotland only four times : 1st, as captain of a frigate; 2dly, to introduce his wife to his friends, on which occasion he staid about a year; 3dly, upon a visit; and 4thly, when, upon his appointment to a command upon the Halifax station, he went in the mail coach to Scotland, and died there in 1789. He was not for a day resident there in any house of his own; nor was he ever there except for temporary occasions. He also commanded the Russian navy for about a year, and was afterwards in the Dutch service.

He had no fixed residence in England till 1776, in which year he took a house at Gosport, where he lived as his home when on shore. This was his only residence in the British dominions; and when he went on service he left his wife and family at Gosport. At his death it became necessary to decide whether his domicile was Scotch or English, because he had made a will, bequeathing a legacy to his daughter, with certain conditions, which were void by the law of Scotland, but valid by the law of England. The House of Lords decided that his original domicile was Scotch, and that though he did not lose it in this first instance, by becoming an officer in the British navy, he abandoned it by entering a foreign service, and acquired a Russian domicile; that on returning to England, and resuming his position as a British officer, he acquired an English domicile, but did not recover his Scotch domicile, that his subsequent visits to Scotland, not being made *animo manendi*, did not revive his Scotch domicile, and that the succession to his property, as that of an Englishman, was therefore to be governed by the law of England, in which country he last acquired a domicile.

In connection with this subject, it may be proper to notice an opinion expressed by the Master of the Rolls, during the argument of Lord Somerville's case—that an officer entering the military or naval service of a foreign power, with consent of the British government, and taking a qualified oath of allegiance to the *foreign* state, does not thereby abandon or lose his native domicile.

In Forrest *v.* Funston, the defendant was a lieutenant in the king's army, and held a situation of master gunner at Blackness Castle in Scotland, where he had the charge of considerable military stores, with an apartment for his residence. He was a native of Strabane in Ireland; and it was held by the Court of Session, that though it was his duty to reside at Blackness, he did not by the possession of his office acquire a Scotch domicile. With respect to the East India Company's Service, the question of domicile does not turn upon the simple fact of the party being under an obligation, by his commission, to serve in India; but when an officer accepts a commission or employment, the duties of which necessarily require residence in India, and there is no stipulated period of service, and he proceeds to India accordingly, the law from such circumstances presumes an intention consistent with his duty, and holds his residence to be *animo et facto* in India.

In the recent case of General Forbes, in the Court of Chancery, the subject of domicile in its relation to military men was extensively discussed before the Vice-chancellor Wood. Nathaniel Forbes, afterwards General Forbes, was born in Scotland of Scotch parents; his father being possessed of an ancestral estate called Auchernach, on which there was then no house. In 1786, Nathaniel Forbes, being then a minor, and a lieutenant on half-pay in the 102d foot, a disbanded regiment, contracted a marriage with a Scotch lady. He shortly afterwards obtained an appointment in the service of the East India Company; and in December, 1787, he sailed for India, where he continued until 1808. He then obtained a furlough, and returned with his wife to Scotland. On the death of his father in 1794 he had succeeded to the family estate in Scotland; and during his furlough he built a house there, and furnished it, and made some improvements in the grounds. In 1812 he returned with his wife to India, and remained there for several years. The wife left India in 1818: and in 1822 her husband, who had then attained the rank of a general officer, and was colonel of a regiment, also quitted India, according to the rules of the service, with the intention of never returning to that country; and he never did return thither. During the whole of his service under the East India Company General Forbes retained his commission and rank of a lieutenant in the king's army. His domicile was without doubt originally Scottish. After his final return from India he had an establishment at a hired house in Sloane-street, London. He also kept his house at Auchernach furnished: and had some servants there also. He likewise became a justice of the peace and a commissioner of taxes in Scotland: and kept his pedigree and papers (including his will) at Auchernach, where he was in the habit of residing half the year, and where he had constructed a mausoleum in which he wished to be buried. But his health did not permit him to reside constantly at Auchernach, where his establishment was also not suitable for his wife; and his house in Sloane-street was manifestly his chief establishment, and his wife resided there. He died in 1851. His wife thereupon laid claim to a share of his property according to the Scotch law of succession, and contended that, in the events which had happened, he must be considered to have died possessed of his original Scottish domicile. The substantial question in the case was, whether his domicile was in England or in Scotland. If he had been a single man his final domicile would probably have been considered Scottish. But the court held that Sloane-street, having been his chief establishment, and the abode of his wife, must be taken to have been the seat of his domicile. In pronouncing judgment upon the case, the learned Vice-chancellor ruled the following points: 1. That the Scottish domicile of General Forbes, notwithstanding his having gone to India during his minority, in the service of the East India Company, continued until he attained the age of twenty-one:

on the principle that a minor cannot change his domicile by his own act. 2. That, on attaining twenty-one, he acquired an Anglo-Indian domicile; and thereupon his Scottish domicile ceased: on the principle that a service in India, under a commission in the Indian army, of a person having no other residence, creates an Indian domicile. 3. That the circumstance of his being a lieutenant on half-pay in a disbanded king's regiment, did not affect the question. 4. That the Anglo-Indian domicile of General Forbes continued unchanged until his departure from India in 1822: the furlough, or limited leave of absence, implying by its nature that it was his duty to return to India on its expiration. 5. That in 1822 the Anglo-Indian domicile of General Forbes was abandoned and lost: the possibility of his being called upon, as colonel of a regiment, to return at some indefinite time to active service in India, being too remote to have any material bearing upon the question. 6. That he had acquired by choice a new domicile in England on his return from India.

DRAGOONS. There are two regiments of dragoons in our army. (*See* ARMY; CAVALRY.)

DRAG-ROPE. This is a 4″ hemp rope, with a thimble worked into each end, one of the thimbles carrying a hook. Six handles, made of oak or ash, are put in between the strands of the rope, and lashed with marline. It is used to assist in extricating carriages from different positions; by the men, for dragging pieces, &c. Length 28 feet.

DRAWING. (*See* RECONNOISSANCE.)

DRILL. The manœuvres and tactical exercises of troops.

DRUNKENNESS ON DUTY. Any commissioned officer who shall be found drunk on his guard, post, or other duty, shall be cashiered. Any non-commissioned officer or soldier so offending, shall suffer such corporal punishment as shall be inflicted by a court-martial; (ART. 45.)

DUEL. Sending and accepting a challenge, or, if a commanding officer, permitting knowingly a duel, or seconding, promoting, or carrying challenges in order to duels, punishable with cashiering, if commissioned officers, and with corporal punishment in the case of non-commissioned officers and soldiers; (ARTS. 25, 26.) (*See* CHALLENGES.)

DUTY. In all military duties, the tour of duty is invariably from the eldest downwards. Brigade duties are those performed by one regiment in common with another. Regimental duties are those performed by the officers and companies of a regiment among themselves. A court-martial, the members of which have been assembled and sworn, is reckoned a duty, although they may have been dismissed without trying any person. If an officer's turn for picket, general court-martial, or fatigue, happens when he is upon any other duty, he is not obliged to make good that picket, &c., when he comes off, but his tour passes him; however, if an officer is on the inlying picket, he is liable to be relieved, and placed on other duties. Officers cannot exchange their duties without permission of the commanding officer. A guard, detachment, or picket, having once marched off the place of parade, is reckoned to have performed a duty, though it may have been dismissed immediately afterwards. Officers, on all duties under arms, are to have their swords drawn, without waiting for any word of command for that purpose.

DYSENTERY. (*See* SANITARY PRECAUTIONS.)

E.

ECHELON. An arrangement of battalions, so that each has a line of battle in advance or in rear of its neighboring battalion. (Consult *Infantry Tactics*, vol. 3. See also MANŒUVRES IN COMBAT.)

ELEVATION. The elevation of a work is the projection of its face on a vertical plane by horizontal rays. It shows the height or depth of a work, and also its length, when the plane of projection is parallel to the face. Applied to a piece of ordnance, the elevation is the inclination of the axis of the piece above the plane on which the carriage stands.

EMBARKATION. Field-batteries should always be embarked by the officers and men belonging to them, who will then know where each article is stowed. Articles required to be disembarked first, should be put in last. When there are several vessels laden with ordnance and ordnance stores for an expedition, each vessel should have on each quarter, and on a signal at mast head, a number that can be easily distinguished at a distance. The same numbers should be entered on the list of supplies shipped in each vessel. The commander will then know exactly what resources he has with him. Articles shipped must be divided among vessels according to circumstances; but, as a general rule, place in each vessel every thing required for the service at the moment of disembarkation, so that there will be no inconvenience, should other vessels be delayed.

If boats are to be employed in the embarkation, and the boats are much lower than the top of the wharf, the guns and ammunition boxes will be lowered into the boat by means of cranes; but when the gunwales are nearly level with the wharf, the ammunition boxes may be more expeditiously put on board by hand, and if there are no cranes, the guns may be parbuckled into the boats. Men told off to the carriages, will prepare them for embarkation. Each carriage, when called for, is to be run forward to the boat or crane; the gun unlimbered and dismounted; the ammunition boxes, shafts, wheels, &c., &c., to be taken off; the washers and linch-pins carefully put away. If they are left in the axle-tree they are liable to be lost. When a battery is embarked in different vessels, every part should be complete, and a proportion of general stores on each. Should two batteries be on the same vessel, they should be stowed on different sides of the vessel.

The embarkation of horses is more difficult than that of guns, particularly if it be necessary first to take them alongside the vessel in boats. In bad weather the guns and carriages are easily hoisted, but not the horses. If the embarkation of both cannot go on, therefore, at the same time, the horses should be embarked first. Horse ships are always provided with slings for hoisting in the horses; they are made of stout canvas, and are about $6\frac{1}{2}$ or 7 feet long, and from $2\frac{1}{4}$ to $2\frac{1}{2}$ feet wide. It may be necessary to embark horses: 1st, when the transports can come alongside the wharf, and the horses are taken on board at one operation; or, 2d, when the transports cannot come alongside the wharf, and the horses are embarked first in boats; or, 3d, when the horses are embarked in boats, from an open beach.

The first case is the best, easiest, and most expeditious—resembling in all respects the hoisting a cask in and out of the hold of a vessel. Horses should generally be blindfolded for this purpose, as this prevents their being frightened or troublesome. In the second case there are two operations: first, lowering the horse into the boat, and, after the passage of the boat to the vessel, hoisting the horse into the transport. Sheers or derricks are absolutely necessary for this purpose, because the tackle must be of such a description as to raise the horse off the ground instantaneously, which a crane cannot do. The head of the derrick must incline inwards while the horse is rising; but when he is high enough, the head of the derrick or sheers must be forced out, to bring the horse directly over the boat. Horses may, in this way, be embarked in boats from a beach. Sand or straw must be put into the boats to preserve their bottoms, and to prevent the horses from slipping. The horses should stand athwart, the head of one horse being on the starboard side, and the head of the next to him on the larboard side. The conductors must sit on the gunwale or stand between the horses. Decked gun-boats or coasting vessels are very convenient for this purpose when there are time and materials for the necessary preparation, as they not only hold a greater number of horses, but can come alongside of a wharf, and the horses, by means of a ramp, may be walked aboard. The disembarkation of horses is carried on by the same means as their embarkation. (*See* DISEMBARKATION. Consult *Army Regulations* for the rules governing troops embarked on transports.)

EMBEZZLEMENT—either of public property or money, punishable in the case of an officer with cashiering, and making good the loss; if a non-commissioned officer, by reduction to the ranks, corporal punishment, and making good the loss; (ART. 36 and ART. 39.)

By SEC. 16 of Act approved Aug. 6, 1846, using in any manner for private purposes, loaning or depositing in bank any public money, and any failure to pay over or to produce public money intrusted to persons charged with its safe keeping, transfer, and disbursement, is made prima facie evidence of embezzlement, and declared to be felony. The taking of receipts and vouchers without paying the amount which they call for, and all persons advising or participating in said act, are also declared guilty of embezzlement by the same section.

EMBRASURE. An embrasure is an opening cut through the parapet to enable the artillery to command a certain extent of the surrounding country. The space between every two of these openings, called the *merlon*, is from 15 to 18 feet in length. The opening of the embrasure at the interior is two feet, while that towards the country is usually made equal to half the thickness of the parapet. The interior elevation of the parapet, which remains after cutting the embrasure, is called the *genouillere*, and covers the lower part of the gun carriage. The *plongee*, or slope given to the *sole*, is generally less than the inclination given to the superior slope of the parapet, in order that the fire from the embrasure may meet that of the musketry from the parapet at a point within a few feet from the top of the counterscarp.

Fig. 108 represents the rear elevation of a two-gun portion of an elevated battery revetted with gabions. In this figure the two gabions at the necks of the embrasures are made to assume a small degree of slope which may usually be done, because the gabions, one with another, occupy rather less than the regular average space of 2 feet each, when placed very close together, so that those of the upper tier will generally admit of being closed at top, and eased at bottom, to favor this arrangement. If not, the neck of the embrasure may be made of equal width throughout, without attempting the kind of slope alluded

FIG. 108.

to; but the gabions which form the cheeks of the embrasures should have a slope gradually increasing from the neck towards the front, until the fifth gabion (more than five will seldom be used) has a slope of at least one-third of its height.

Fig. 109 is the plan of a portion of parapet and embrasure, showing the arrangement of gabions above adverted to.

Fig. 110 shows in elevation the arrangement of the gabions and of the sand-bags above them, as well as the genouillere or solid part of the

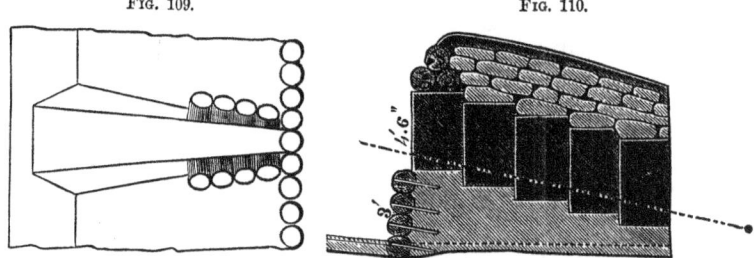

FIG. 109.　　　　FIG. 110.

embrasure, below the sole of it, in a construction that frequently arises in sieges, especially in the offensive crowning batteries on the crest of the glacis, where the depression of the sole of the embrasure is considerable, to allow of the guns being pointed to spots of the wall some distance below them.

EMOLUMENTS. (*See* PAY.)

ENCAMPMENT. (*See* CAMP.)

ENCEINTE—is the body of the place, or the first belt of ramparts and parapets that inclose the place.

ENFILADE. To sweep the whole length of the face of any work or line of troops, by a battery on the prolongation of that face or line.

ENGINEER CORPS. (*See* ARMY for its organization.) The functions of the engineers being generally confined to the most elevated branch of military science, they are not to assume, nor are they subject to be ordered on, any duty beyond the line of their immediate profession, except by the special order of the President of the United States; but they are to receive every mark of respect to which their rank in the army may entitle them respectively, and are liable to be transferred, at the discretion of the President, from one corps to another, regard being paid to rank; (ART. 63.)

The engineers are charged with planning, constructing, and repairing all fortifications and other defensive works; with disbursements of money connected with these operations. In time of war, they present plans for the attack and defence of military works; lay out and construct field defences, redoubts, intrenchments, roads, &c.; form a part of the vanguard to remove obstructions; and in retreat, form a part of the rear guard, to erect obstacles, destroy roads, bridges, &c., so as to retard an enemy's pursuit. (*See* SAPPERS AND MINERS.) (Consult LAISNÉ, *Aide Memoire à l'Usage des Officiers du Génie.*)

ENGINEERS, TOPOGRAPHICAL. (See ARMY for their organization.) The duties of the corps consist in surveys for the defence of the frontiers, and of positions for fortifications, in reconnoissances of the country through which an army has to pass, or in which it has to operate; in the examination of all routes of communication by land or by water, both for supplies and military movements; in the construction of military roads and permanent bridges connected with them; and the charge of the construction of all civil works, authorized by acts of Congress, not specially assigned by law to some other branch of the service. (Consult SALNEUVE, *Cours de Topographie à l'Usage des Eléves de l'Ecole d'Etat Major.* R. S. SMITH's *Topographical Drawing.*)

ENLISTMENTS—are voluntary, and made for five years; (*Act* June 17, 1850.) Any non-commissioned officer or soldier who shall enlist himself in any other regiment, troop, or company, without a regular discharge from the regiment, troop, or company in which he last served, to be considered a deserter; (ART. 22.) Whenever enlistments are made at or in the vicinity of military posts on the western frontier, and at remote and distant stations, a bounty equal in amount to the cost of transporting and subsisting a soldier from the principal recruiting depot in the harbor of New York, to the place of such enlistment be, and the same is hereby allowed to each recruit so enlisted, to be paid in unequal instalments at the end of each year's service, so that the several amounts shall annually increase, and the largest be paid at the expiration of each enlistment; (*Act* June 17, 1850.) The amounts and instalments have been fixed in the regulations for the Pay Department. (*See* RE-ENLISTMENT.)

ENSIGN. Lowest grade of commisssioned officers of infantry.

ENTANGLEMENT. Abattis, so called, when made by cutting only partly through the trunks, and pulling the upper parts to the ground, where they are picketed.

ENTICING. Any person whatever who shall procure or entice a soldier to desert the service of the United States, may be fined not exceeding $300, or imprisoned any term not exceeding one year, at the discretion of any court having cognizance of the same; (*Act* March 16, 1802.)

EPAULEMENT. An elevation thrown up to cover troops from the fire of an enemy. It is usually composed of gabions filled with earth, or made of sand-bags, &c.

EPAULETTE. Badge of rank, of bullion, worn by officers on the shoulders. The Army Regulations prescribe these badges under authority given by law to the President to establish the uniform of the army.

EPROUVETTE, (PENDULUM.) The best method of testing the projectile force of gunpowder, is to ascertain by experiment its effects when used in the same quantities in which it is to be employed in service. This method has been adopted by establishing, at the Washington Arsenal, a cannon pendulum and a musket pendulum, which are used for proving samples of powder sent from the manufactories. The apparatus shows the initial velocity of a ball fired from a cannon or musket.

In the ordinary eprouvette, gunpowder of *small grain* and low specific gravity gives the highest range, whilst the *ballistic pendulum* shows that the greatest initial velocity in a shot from a heavy cannon is produced by powder of great *specific gravity* and *coarse grain*. (*Ordnance Manual.*)

EQUIPAGE, CAMP AND GARRISON—are tents, kitchen utensils, axes, spades, &c. (*See* CLOTHING.)

EQUIPMENT. The complete dress of a soldier, including arms, accoutrements, &c.

ESCALADE, AND SURPRISE OF A FORTIFIED PLACE. A place is taken by *surprise*, whenever a sufficient number of men are secretly introduced into it to cause the defenders to abandon or surrender it. It is taken by *escalade*, when ladders are used to cross the walls. (Fig. 111.)

The surest way of succeeding in a surprise, is to have a perfect knowledge of the interior of the place, or to be accompanied by reliable guides, who know those parts of the place which may be penetrated with least difficulty. Such parts are ordinarily dilapidated portions of the body of the place; houses contiguous to the walls, the windows of which are not barred, &c., &c. Aqueducts and sewers have also sometimes been used for the introduction of armed men, unknown to the garrison. But when a place is badly guarded, all parts are accessible with ladders, and it is sometimes best to choose the highest walls for the escalade, as the enemy will probably, from a feeling of security, be less vigilant at such parts of the body of the place. Thus, at the siege of Badajoz in 1812, the English escaladed the highest walls in the city, and penetrated into the interior, while the attack directed upon breaches in the lower walls, although vigorously made, was repulsed. When

Fig. 111.

it is considered how slow a process it is to bring up ladders to the counterscarp, in order to descend by them into the ditch, then to cross the ditch, and to rear the ladders against the escarp, and to mount them, it is evident that success will, in a great measure, depend upon the number of men that can mount at the same moment; in other words, upon the number of ladders. A ladder beyond a certain length becomes unwieldy, and the rearing of it difficult. The distance from the foot of the ladders to the wall should be at least equal to one-fourth of their height. If the distance be greater, the ladder will be easily broken under the weight of the men mounting them; if much less, they will be so erect that the soldiers, as they ascend, must be continually in danger of falling headlong down. The scaling ladders introduced by Sir Charles Pasley, are in pieces of 12' 8" and 7' 6" in length, fitting into each other with strong double iron sockets, and tied by stout ropes. These can be arranged for any length, and quickly adjusted. Ladders made of long spars are awkward to carry; especially if there be narrow sharp turnings in approaching the point of escalade: nor can long sound spars be always procured. It is desirable that ladders should be made of light, tough wood: teak wood is too heavy. If a guy-rope be attached to each side of the ladder, they greatly assist in adjusting and fixing it against the wall: the men told off for the guy-ropes should stand close to the wall, within the slope of the ladder; these guy-ropes should be fixed at 5 or 6 feet below the top of the ladder, to prevent their being cut by the enemy on the wall. The total lengths of the ladders should exceed the height to be escaladed by 3 or 4 feet, in order that the men may step easily off the ladders on to the parapet or wall. Many failures have occurred from ladders being too short. It is desirable to have a pair of stout lifting bars, 3 or 4 feet long, with hooks, for each ladder. When an escalade is to take place, be sure to practise the men intended for the service thoroughly in carrying, in fixing, in ascending, and descending the ladders; descending, for going down a counterscarp; ascending, for getting up an escarp. Always use as many ladders as possible. If there be a counterscarp to descend, leave half the ladders there, while the other half are used against the escarp, that no time may be lost. Ascend the ladders together, on as large a front as possible. When an escalade is opposed by an enemy, take care that a good firing party covers the escalade, with especial directions to fire upon any work that may flank the ladders. Avoid night attacks, except under peculiar circumstances: the example of gallant men is lost at night, whilst timidity is infectious. Make all arrangements under cover of darkness, but assault at day-break.

At the moment of the escalade, the ladders should be filled with soldiers, and it is necessary, therefore, that they should be underpropped about the middle. Soldiers exercised in *gymnastics* are capable of mounting high walls with arms and accoutrements, by means of a hook, helved to a pole sufficiently long to reach the top of the wall. This exercise is practised by some French troops, and the walls of the citadel of Montpellier are thus escaladed with the greatest facility.

Precipitous rocks may be escaladed by grasping bushes and roots, or by planting the bayonet in the crevices of the rocks, in order to reach the top. Such escalades are very dangerous when an enemy defends the height, as heavy stones may be rolled down upon the assailants; but activity and ingenuity accomplish much, as was shown by the French in the attack upon Fort Scharnitz near Innspruck. They tied their haversacks round their heads, and, protected by this buckler, they scrambled up the rocks, despite the stones precipitated upon them. And still later the difficult ascent at Alma was scaled by French troops, in the face of Russian artillery and infantry.

The most favorable time for a *surprise* is that of a winter night, when there is no moon. A long march may then be made without discovery, and the troops may arrive an *hour before day*. This is the propitious moment for the execution of the design. It is then that men sleep most profoundly; and it is at that hour the attacking force may begin in the dark, and end the work by daylight; such favorable circumstances are much increased by heavy wind and rain during the night, as the clanking of arms and other inevitable noises made by the troops cannot be heard by the garrison, and the latter, besides, are more disposed to negligence. It is extremely important for the men to be able to recognize each other in the darkness, and the simplest means of doing so is to put the shirt outside the dress, or to tie a white band around the arm.

The party must be furnished with petards, axes, and levers, to force open doors; with beams and ladders, to overthrow and scale walls. Hurdles and fascines are necessary to cross muddy ditches, or broad planks may be used as a substitute for hurdles. With fascines small ditches and pools are filled up. All these articles should be carried by the men from the last halting-place. Wagons and animals would lead to discovery, and are therefore left at a safe distance, while every precaution is taken to maintain silence in the assailing party. The soldiers should also not light their pipes, as the fire can be seen from a long distance in the dark. Barking dogs must be quieted without the use of fire-arms, and every one must be on the alert.

The dispositions made for the attack will vary with circumstances, but in general it is well to divide the force into three parts: the first to penetrate into the city; the second to remain without and protect, if necessary, the retreat of the first; and the third to take such position as is most likely to prevent aid from reaching the enemy.

When the first division has penetrated the city by escalade or otherwise, it surrounds at once some of the adjacent quarters, and holds the outlets of the principal streets, whilst detachments quickly open the gates to the troops outside, after having taken or killed the guards. As soon as the gates are opened, and sufficient numbers are at hand, the troops spread themselves in the city, after leaving good reserves, upon which to retreat in case of check. The house of the commandant, barracks, arsenal, and the guards of the interior are at once sought, to prevent, if possible, any re-union of the defenders, and to paralyze all their efforts by the seizure of the commanding officer. If time and means of recovering from his stupor and concentrating his force in the interior of the city be left to the enemy, great risk will be run of being driven out, as the attacking force is necessarily everywhere weak, from the great number of points occupied.

The famous example of Cremona, where Prince Eugene, after having made himself master of a great part of the city, and after having seized Marshal Villeroi, who commanded there, was nevertheless then driven out by the defenders, shows that all is not lost to the defenders when the enemy has seized the exterior posts. Another example may be cited in the surprise of Bergen-op-Zoom in 1814, by Gen. Graham, where, although the surprise was successful, yet the assailants, in the end, were obliged by the garrison to surrender after considerable loss.

Much may then be done by defenders even under such circumstances, but much more may be accomplished by the most unceasing vigilance, and this quality, instead of being relaxed in stormy nights, should be then redoubled. (Consult *Cours de Tactique, par le General* DUFOUR.)

ESCARP, (or SCARP)—is the side of the ditch next to the place, which, in permanent fortifications, is usually faced with masonry.

ESCORT. (*See* CONVOY.) There are also funeral escorts; escorts of honor; color escorts; &c., &c.

ESPLANADE. Empty space for exercising troops in fortified places.

ESPRIT DE CORPS. The brotherhood of a corps; military and regimental pride. Nothing is so prejudicial to it, as the failure to unite the companies of a regiment. It might also be promoted by recording the distinguished services of a regiment on its colors. (*See* SOLDIER.)

EVACUATE. To withdraw from a town or fortress, in consequence either of a treaty or a capitulation, or of superior orders.

EVIDENCE—is that which makes clear, demonstrates, or ascertains the truth of the very fact or point in issue; (3. *Bl. Comm.*, 367.) Evidence may be considered with reference to, 1, the *nature* of the evidence; 2, the *object* of the evidence; 3, the *instruments* of evidence; and, 4, the effect of evidence.

As to its *nature*, evidence may be considered with reference to its being, 1, the primary evidence; 2, secondary evidence; 3, positive; 4, presumptive; 5, hearsay; and, 6, admissions.

1. *Primary evidence.* The law generally requires that the best evidence the case admits of shall be given; (1 *Stark. Ev.*, 102, 390.)

2. *Secondary evidence* is that species of proof which is admissible on the loss of primary evidence. Before it is admitted, proof must be made of the loss or impossibility of obtaining the primary evidence.

3. *Positive* evidence is that which, if believed, establishes the truth of a fact in issue, and does not arise from any presumption. Evidence is positive when the very facts in dispute are communicated by those who have actual knowledge of them by means of their senses; (1 *Stark.* 19.)

4. *Presumptive* evidence is that which is not direct, but where, on the contrary, a fact which is not positively known, is presumed from one or more other facts or circumstances which are known; (1 *Stark.* 18.)

5. *Hearsay* is the evidence of those who relate not what they know themselves, but what they have heard from others. As a general rule, hearsay evidence of a fact is not admissible. But evidence given on a former trial by a person since dead is admissible, as is also the dying declarations of a person who has received a mortal injury. A few more exceptions may be found in Phillips' *Ev.*, chap. 7; 1 Stark. *Ev.*, 40.

6. *Admissions*, which are the declarations made by a party for himself or those acting under his authority. These admissions are generally evidence of facts declared, but the admissions themselves must be proved.

The *object* of evidence is to ascertain the truth between the parties. Experience shows that this is best done by the following rules, which are now binding in law: 1. The evidence must be confined to the point in issue; 2. The substance of the issue must be proved, but only the substance is required to be proved; 3. The affirmative of the issue must be proved. A witness, on being admitted in court, is first subjected to the examination of the party in whose behalf he is called. This is termed the *examination in chief*. The principal rule to be observed by the party examining is, that leading questions are not to be asked. The witness is then cross-examined by the other party. The object of cross-examination is twofold: to weaken the evidence given by the witness as to the fact in question, either by eliciting contradictions or new explanatory facts; or, secondly, to invalidate the general credit of the witness. In the latter case it is a general rule, that a witness may refuse to answer any question, if his answer will expose him to criminal liability. The general practice of English courts also seems to authorize his refusal to answer any question which will disgrace him. The credit of a witness may likewise be impeached by the general evidence of others as to his character; but in this case no evidence can be given of particular facts which militate against his general credit. Witnesses are excluded from giving evidence by: 1. Want of reason or understanding; 2. Want of belief in God and a future state; 3. Infancy; and, 4. Interest. Besides witnesses, records and private writings are also *instruments* of evidence.

1. Records, in all cases where the issue is *nul tiel reord*, are to be proved by an exemplification duly authenticated; that is, an attestation made by a proper officer, by which he certifies that a record is in due form of law, and that the person who certifies it is the officer appointed by law to do so. In other cases an examined copy, duly proved, will in general be evidence.

2. Private writings are proved by producing the attesting witness, or, in case of his absence, death, or other legal inability to testify, as if, after attesting the paper, he becomes infamous, his handwriting may be proved. When there is no witness to the instrument, it may be proved by evidence of the handwriting of the party, by a person who has seen him write, or in a course of correspondence has become acquainted with his hand. Parol evidence is admissible to defeat a written instrument on the ground of fraud, mistake, &c.; or to apply it to its proper subject matter, or, in some instances, as ancillary to such application, to explain the meaning of doubtful terms, or rebut presumptions arising extrinsically. But in all cases the parol evidence does not usurp the place or arrogate the authority of the written instrument. (Consult generally *Treatises on Evidence* by PHILLIPS and STARKIE; BOUVIER's *Law Dictionary;* BRANDE's *Encyclopædia*.)

EVOLUTIONS. (*See* MANŒUVRES.)

EXECUTION OF LAWS. On all occasions when the troops are

employed in restoring or maintaining public order among their fellow-citizens, the use of arms, and particularly fire-arms, is obviously attended with loss of life or limb to private individuals; and for these consequences, a military man may be called to stand at the bar of a criminal court. A private soldier also may occasionally be detached on special duty, with the necessity of exercising discretion as to the use of his arms; and in such cases he is responsible, like an officer, for the right use or exercise of such discretion. One of the earliest reported cases on this subject occurred in 1735, when Thomas Macadam, a private sentinel, and James Long, a corporal, were tried before the Admiralty Court of Scotland, upon a charge of murder under the following circumstances: They were ordered to attend some custom-house officers, for their protection in making a legal seizure; and being in a boat with the officers in quest of the contraband goods, one Frazer and his companions came up with them, leaped into the boat, and endeavored to disarm the soldiers. In the scuffle, the prisoners stabbed Frazer with their bayonets, and threw him into the sea. For this homicide the prisoners were tried and convicted of murder by a jury; and the Judge-admiral sentenced them to death. But the High Court of Justiciary reversed this judgment, on the ground that the homicide in question was necessary for securing the execution of the trust committed to the prisoners. The report of this case contains the following remarks upon it by Mr. Forbes, afterwards Lord President of the Court of Session of Scotland; and they appear to be of great importance to military men: "Where a man has *by law* weapons put into his hands, to be employed not only in defence of his life when attacked, but in support of the execution of the laws, and in defence of the property of the Crown, and the liberty of any subject, he doubtless may use those weapons, not only when his own life is put so far in danger that he cannot probably escape without making use of them, but also when there is imminent danger that he may by violence be disabled to execute his trust, without resorting to the use of those weapons; but when the life of the officer is exposed to no danger, when his duty does not necessarily call upon him for the execution of his trust, or for the preservation of the property of the Crown, or the preservation of the property or liberty of the subject, to make use of mortal weapons, which may destroy His Majesty's subjects, especially numbers of them who may be innocent, it it is impossible from the resolution of the Court of Justiciary to expect any countenance to, or shelter for, the inhuman act." This quotation, in the latter part of it, has a direct bearing on the case of the unfortunate Captain Porteus, whose trial took place in the following year, and whose melancholy fate is the groundwork of Sir Walter Scott's "Heart of Mid Lothian." In the year 1736, the collector of customs on the coast of Fife made a seizure of contraband goods of considerable value, which were condemned and sold. Two of the proprietors of these goods took an opportunity of robbing the collector of just so much money as these goods had sold for. They regarded this as merely a fair reprisal, and no robbery; but they were nevertheless taken up, tried, and condemned to death for the fact. With the exception of some smuggling transactions, in which they had been concerned, the prisoners were men of fair character; and the mob expressed much dissatisfaction with their sentence, and the prospect of their execution. On the Sunday preceding the day appointed for the execution, the prisoners were taken to a church near the gaol, attended by only three or four of the city guards, to hear divine service. None of the congregation had assembled, and the guards being feeble old men, one of the prisoners made a spring over the pew where they sat, while the other, whose name was Wilson, in order to facilitate his companion's escape, caught hold of two of the guards with his hands, and seized another with his teeth, and thus enabled his companion to join the mob outside, who bore him off to a place of safety. Wilson then composedly resumed his own seat, without making any attempt to recover his own liberty. This generous conduct of Wilson created a strong public feeling in his favor; and the magistrates of Edinburgh soon learned that an attempt would be made by the mob to rescue him at the place of execution. They therefore procured some of the regular forces on duty in the suburbs to be posted at a convenient distance from the spot, so as to support the city guard, in case they should be vigorously attacked. The officer, whose turn it was to do duty as captain of the city guard, being deemed unfit for the critical duties of the day, Captain Porteus, unfortunately for himself, was appointed to the command on the occasion. His men were served with ball-cartridge; and, by order of the magistrates, they loaded their pieces when they went upon duty. The execution took place without any disturbance until the time arrived for cutting down the body, when the mob severely pelted the executioner with stones, which hit the guards as they surrounded the scaffold, and provoked them to fire upon the crowd. Some persons at a distance from the place of execution were thus killed. As soon as the body was removed, Captain Porteus withdrew his men, and marched up the West Bow, which is a narrow winding passage. The mob, having recovered from the fright occasioned by the previous firing, followed the guard up this passage, and pelted the rear with stones, which the guards returned with some dropping shot, whereby some where killed, and others wounded. On reaching the guard-house they deposited their arms in the usual form, and Captain Porteus went with his piece in his hand to the Spread Eagle Tavern, where the magistrates were assembled. On his arrival there, he was charged with the murder of the persons who had been slain by the city guards, on the allegation that he had commanded the guards to fire. The mob was very riotous, and called for justice upon him; and the magistrates, after adjourning to the council chamber, committed him to the Tolbooth for trial. The strongest feeling existed against him on the part of the mob, until the hour of his trial before the High Court of Justiciary arrived, when, to their great satisfaction, he was found guilty, and condemned to be hanged. The higher classes of society, however, unaffected by the popular prejudice against the unfortunate prisoner, exerted themselves strenuously in his behalf, and succeeded in obtaining a reprieve. This created the greatest discontent among the lower orders, who, on the night before the day originally appointed for the execution, broke open the gaol, dragged the unhappy Captain Porteus down stairs by the heels, carried him to the common place of execution, and there, throwing a rope over a dyer's pole, hanged him with many marks of barbarity. The perpetrators of this outrage were never discovered, and the subject gave rise to very warm debates in Parliament, particularly in the House of Lords, with respect to the conduct of the city magistrates and officers.

It was quite clear, however, with reference to the criminality of Captain Porteus, that he had ordered his men to fire without sufficient cause or justification; and, under such circumstances, he was in point of law justly found guilty of murder.

Ensign Hugh Maxwell, of the Lanarkshire Militia, was tried in 1807, before the High Court of Justiciary of Scotland, for the murder of Charles Cottier, a French prisoner of war at Greenlaw, by improperly ordering John Gow, a private sentinel, to fire into the room where Cottier and other prisoners were confined, and so causing him to be mortally wounded. It appeared that Ensign Maxwell had been appointed to the military guard over 300 prisoners of war, chiefly taken from French privateers. The building in which they were confined was of no great strength, and afforded some possibilities of escape. The prisoners were of a very turbulent character, and to prevent their escape during the long winter nights, an order was given that all lights in the prison should be put out by nine o'clock, and that if this was not done at the second call, the guard were to fire upon the prisoners, who were often warned that this would be the consequence of disobedience with regard to the lights. On the night in question, there was a tumult in the prison, but of no great importance; and Ensign Maxwell's attention having been on that account drawn to the prisoners, he observed a light burning beyond the appointed hour, and twice ordered it to be put out. This order not being obeyed he ordered the sentry to fire, but the musket merely snapped. He repeated the order; the sentinel fired again, and Cottier received his mortal wound. At this time there was no symptom of disorder in the prison, and the prisoners were all in bed. The general instructions issued from the adjutant-general's office in Edinburgh, for the conduct of the troops guarding the prison, contained no such order as that which Ensign Maxwell had acted upon; and it appeared that the order in question was a mere verbal one, which had from time to time, in the hearing of the officers, been repeated by the corporal to the sentries, on mounting guard, and had never been countermanded by those officers, who were also senior to Ensign Maxwell. The Lord Justice Clerk described the case to the jury as altogether the most distressing that any court had ever been called upon to consider, and laid it down most distinctly, that Ensign Maxwell could only defend himself by proving specific orders, which he was bound to obey without discretion; or by showing that in the general discharge of his duty he was placed in circumstances, which gave him discretion, and called upon him to do what he did. His lordship was of opinion that both these grounds of defence failed in the present case; and the jury having found Ensign Maxwell guilty of the minor offence of *culpable homicide*, with a recommendation to mercy, the court sentenced him to nine months' imprisonment. Ensign Maxwell's conduct certainly exhibited none of those gross features which characterize murder; but at the same time he was guilty of a rash and inconsiderate act, which, if he had not been engaged at the time in military duty, though he was mistaken in the exercise of it, would probably have been held to amount to murder. In Maxwell's case, the soldier who fired the shot was not prosecuted for the act, nor was he liable to such prosecution.

It is laid down in a book of authority, that if a ship's sentinel shoot a man, because he persists in approaching the ship when he has been ordered not to do so, it will be murder, unless such an act was necessary for the ship's safety. And it will be murder, though the sentinel had orders to prevent the approach of any boats, had ammunition given to him when he was put on guard, and acted on the mistaken impression that it was his duty. In Rex *v.* Thomas, the prisoner was sentinel on board H.M.S. *Achille*, when she was paying off. The orders to him from the preceding sentinel were to keep off all boats, unless they had officers in uniform in them, or unless the officer on deck allowed them to approach: and he received a musket, three blank-cartridges, and three balls. The boats pressed, upon which he repeatedly called to them to keep off; but one of them persisted, and came close under the ship, and he then fired at a man in the boat and killed him. It was put to the jury to find whether the sentinel did not fire under the mistaken impression that it was his duty; and they found that he did. But the case being reserved for the opinion of the judges, their lordships were unanimous that it was murder. They thought it, however, a proper case for a pardon: and further, they were of opinion that if the act had been necessary for the preservation of the ship, as if the deceased had been stirring up a mutiny, the sentinel would have been justified.

The cases already cited turned upon the improper exercise of discretion by the officers concerned. But in the following case, though not attended with actual consequences involving a criminal charge, the discretion in the use of arms was wisely exercised, and indicated great presence of mind, and correctness of judgment.

Some years ago, the public journals of London recorded the meritorious behavior of a private sentry, upon the occasion of a riotous mob assembled at the entrance of Downing-street, with the intention of attacking the government offices in that quarter of the town. This man standing alone presented his musket, and threatened to fire upon the crowd, if the slightest attempt were made to approach the particular office for the defence of which he was placed on duty, and succeeded by the terror thus created, though at a great risk of consequences to himself, in keeping the rioters at bay until a larger force arrived to assist him. The soldier's conduct was publicly much approved. It was also clearly legal according to Macadam's case; and if after the announcement of his intentions the mob had pressed forward to execute their purpose, he would have been held justified at law in firing at the rioters upon his own responsibility. The Duke of Wellington, as Constable of the Tower, testified his marked approbation of this man's conduct, by promoting him at once to a Wardership at that fortress.

During the Irish insurrection of 1848, Smith O'Brien was arrested at the railway station of Thurles, on a charge of high treason. A public passenger train was on the point of starting for Dublin, and the engineeer was mounted on the engine, with the steam up, and every thing in readiness for the immediate prosecution of the journey. The scene of the arrest lay in the disturbed distret, which was in the occupation of the troops employed to suppress the insurrection and prevent its extension. General Macdonald's aide-de-camp, having been apprised of the arrest, proceeded instantly to the station, and there commanded the engineer to dismount from the engine, and to stop the train; it being of the utmost importance to the public safety and service that the news of the arrest should not be carried along the line of railway, as the country people might assemble in great numbers and destroy the rails, and rescue the prisoner, or otherwise impede the conveyance of the prisoner to Dublin. Such interference would obviously have occasioned great loss of life, besides the danger to the public service at such a season. The engineer at first refused to obey the aide-de-camp's orders, whereupon the officer presented his pistol at the engineer, and threatened him with instant death if he persisted in his refusal. The man then dismounted; but it is conceived that the officer pursued a correct line of conduct, and exercised upon the occasion a sound discretion, which would have been a good legal defence to him, if he had ultimately proceeded to execute his threat upon the engineer. "Power in law (says Sir Edward Coke) means power with force."

The right of officers or soldiers to interfere in quelling *a felonious riot*, whether with or without superior military orders, or the direction of a civil magistrate, is quite clear, and beyond the possibility of mistake. This subject, however, was formerly little understood; and military men failed in their public duty through excess of caution.

George III. and his Attorney-general (Wedderburn) both deservedly acquired high credit for their energy in the crisis of the riots of 1780. When the king heard that the troops which had been marched in from all quarters were of no avail in restoring order, on account of a scruple that they could not be ordered to fire till an hour after the Riot Act had been read, he called a cabinet council, at which he himself presided, and propounded for their consideration the legality of this opinion. There was much hesitation among the councillors, as they remembered the outcry that had been made by reason of some deaths from the interference of the military in Wilkes's riots, and the eagerness with which grand juries had found indictments for murder against those who had acted under the command of their superiors. At last the question was put to the Attorney-general, who attended as assessor, and he gave a clear, unhesitating, and unqualified answer to the effect, that if the mob were committing a felony, as by burning down dwelling-houses, and could not be prevented from doing so by other means, the military, according to the law of England, might and ought to be ordered to fire upon them: the reading of the Riot Act being wholly unnecessary and nugatory under such circumstances. The exact words used by him on this occasion are not known; but they must have been nearly the same which he employed when he shortly afterwards expounded from the judgment seat the true doctrine upon the subject. The requisite orders were issued to the troops, the conflagrations were stopped, and tranquillity was speedily restored.

This eminent lawyer having become Chief Justice of the Court of Common Pleas, with the title of Lord Loughborough, delivered a charge to the grand jury on the special commission for the trial of the rioters of 1780, in the following terms: "I take this public opportunity of mentioning a fatal mistake into which many persons have fallen. It has been imagined, because the law allows an hour for the dispersion of a mob to whom the Riot Act has been read by the magistrate, the better to support the civil authority, that during that time the civil power and the magistracy are disarmed, and the king's subjects, whose duty it is at all times to suppress riots, are to remain quiet and passive. No such meaning was within view of the legislature, nor does the operation of the act warrant such effect. The civil magistrates are left in possession of all those powers which the law had given them before. If the mob collectively, or a part of it, or any individual within or before the expiration of that hour, attempts, or begins to perpetrate an outrage amounting to felony, to pull down a house, or by any other act to violate the law, it is the duty of *all present*, of whatever description they may be, to endeavor to stop the mischief, and to apprehend the offender."

"A riot (says Mr. Justice Gaselee) is not the less a riot, nor an illegal meeting, because the proclamation of the Riot Act has not been read; the effect of that proclamation being to make the parties guilty of a capital offence if they do not disperse within an hour; but if that proclamation be not read, the common law offence remains, and it is a misdemeanor; and all magistrates, constables, and even private individuals are justified in dispersing the offenders; and if they cannot otherwise succeed in doing so, they may use force."

After the suppression of the great riots of London in 1780, by the aid of the troops, as already mentioned, the government was acrimoniously attacked both in and out of parliament, on the ground that the employment of a military force, to quell riots by firing on the people, could only be justified, if at all, by martial law proclaimed under a special exercise of the royal prerogative; and it was thence argued that the nation was living under martial law. But Lord Mansfield, the Chief Justice of the King's Bench, addressed the House of Lords on this subject, and placed it in its true light. "I hold (said his lordship) that His Majesty, in the orders he issued by the advice of his ministers, acted perfectly and strictly according to the common law of the land, and the principles of the Constitution. . . . . Every individual in his private capacity may lawfully interfere to suppress a riot, much more to prevent acts of felony, treason, and rebellion. Not only is he authorized to interfere for such a purpose, but it is his duty to do so : and if called upon by a magistrate, he is punishable in case of refusal. What any single individual may lawfully do for the prevention of crime and preservation of the public peace, may be done by any number assembled to perform their duty as good citizens. It is the peculiar business of all constables to apprehend rioters, to endeavor to disperse all unlawful assemblies, and in case of resistance, to attack, wound, nay kill those who continue to resist;—taking care not to commit unnecessary violence, or to abuse the power legally vested in them. Every one is justified in doing what is necessary for the faithful discharge of the duties annexed to his office, although he is doubly culpable if he wantonly commits an illegal act under the color or pretext of law. The persons who assisted in the suppression of those tumults are to be considered mere private individuals acting as duty required. My lords, we have not been living under martial law, but under that law which it has long been my sacred function to administer. For any violation of that law the offenders are amenable to our ordinary courts of justice, and may be tried before a jury of their countrymen. Supposing a soldier or any other military person who acted in the course of the late riots, had exceeded the power with which he was invested, I have not a single doubt that he may be punished, not by a court-martial, but upon an indictment to be found by the Grand Inquest of the City of London or the County of Middlesex, and disposed of before the ermined judges sitting in Justice Hall at the Old Bailey. Consequently the idea is false, that we are living under a military government, or that, since the commencement of the riots, any part of the laws or of the Constitution has been suspended or dispensed with. I believe that much mischief has arisen from a misconception of the Riot Act, which enacts that after proclamation made persons present at a riotous assembly shall depart to their homes ; those who remain there above an hour afterwards shall be guilty of felony and liable to suffer death. From this it has been imagined that the military cannot act, whatever crimes may be committed in their sight, till an hour after such proclamation has been made, or, as it is termed, 'the Riot Act is read.' But the Riot Act only introduces a new offence—remaining an hour after the proclamation—without qualifying any pre-existing law, or abridging the means which before existed for preventing or punishing crimes."

In the case of Handcock v. Baker, which was an action brought against the defendants, who were not constables, for forcibly detaining and confining the plaintiff, in order to prevent him from murdering his wife, Mr. Justice Heath made the following observations: "It is a matter of the last consequence that it should be known upon what occasions bystanders may interfere so as to prevent felony. In the riots which took place in 1780, this matter was much misunderstood, and a general persuasion prevailed that no indifferent person could interpose without the authority of a magistrate; in consequence of which much mischief was done which might otherwise have been prevented." And in the same case Mr. Justice Chambre said: "There is a great difference between the right of a private person in cases of intended felony and breach of the peace. It is lawful for a private person to do any thing for the prevention of a felony." And in so doing it becomes quite immaterial whether the persons wounded or slain are taking any active part in the riot. In the case of Clifford v. Brandon, which was an action by a barrister of great eminence against the box-keeper of Covent Garden Theatre, who had arrested him in the theatre for wearing in his hat a ticket with O.P. on it—this being a badge of the party by whom the celebrated O.P. riots relative to the prices of admission were carried on—and nothing else having been proved against him—the Lord Chief Justice, Sir James Mansfield, said: "If any person encourages, or promotes, or takes part in riots, whether by words, signs, or gestures, or by wearing the badge or ensign of the rioters, he is himself to be considered a rioter, he is liable to be arrested for a breach of the peace. In this case all are principals."

But notwithstanding the existence of a clear right and duty on the part of military men voluntarily to aid in the suppression of a riot, it would be the height of imprudence to intrude with military force, except upon the requisition of a magistrate, unless in those cases where the civil power is obviously overcome, or on the point of being overcome, by the rioters.

With regard to the requisition of military aid by the civil magistrate, the rule seems to be, that when once the magistrate has charged the military officer with the duty of suppressing a riot, the execution of that duty is wholly confided to the judgment and skill of the military officer, who thenceforward acts independently of the magistrate until the service required is fully performed. The magistrate cannot dictate to the officer the mode of executing the duty; and an officer would desert his duty if he submitted to receive any such orders from the magistrate. Neither is it necessary for the magistrate to accompany the officer in the execution of his duty.

The learning on these points may be gathered from the charge of Mr. Justice Littledale to the jury, in the trial of the mayor of Bristol, for breach of duty in not suppressing the riots at that city in 1831. "Another charge (said His Lordship) against the defendant is, that upon being required to ride with Major Beckwith, he did not do so. In my opinion he was not bound to do so in point of law. I do not apprehend it to be the duty of a justice of the peace to ride along and charge with the military. A military officer may act without the authority of the magistrate, if he chooses to take the responsibility; but although that is the strict law, there are few military men who will take upon themselves so to do, except on the most pressing occasions. Where it is likely to be attended with a great destruction of life, a man, generally speaking, is unwilling to act without a magistrate's authority; but that authority need not be given by his presence. In this case the mayor did give his authority to act; the order has been read in evidence; and he was not bound in law to ride with the soldiers, more particularly on such an occasion as this, when his presence elsewhere might be required to give general directions. If he was bound to make one charge, he was bound to have made as many other charges as the soldiers made. It is not in evidence that the mayor was able to ride, or at least in the habit of doing so; and to charge with soldiers it is not only necessary to ride, but to ride in the same manner as they do; otherwise it is probable the person would soon be unhorsed, and would do more harm than good: besides that, if the mob were disposed to resist, a man who appeared in plain clothes leading the military would be soon selected and destroyed. I do not apprehend that it is any part of the duty of a person who has to give general directions, to expose himself to all kinds of personal danger. The general commanding an army does not ordinarily do so, and I can see no reason why a magistrate should. A case may be conceived where it might be prudent, but here no necessity for it has been shown."

This subject was also luminously expounded by the late Lord Chief Justice Tindal, in his charge to the grand jury on the special commission held at Bristol, on the 2d of January, 1832, for the trial of the parties implicated in the formidable riots and devastations committed in that city during the autumn of the previous year: "It has been well said that the use of the law consists, first, in preserving men's persons from death and violence; next, in securing to them the free enjoyment of their property; and although every single act of violence, and each individual breach of the law, tends to counteract and destroy this its primary use and object, yet do general risings and tumultuous meetings of the people in a more especial and particular manner produce this effect, not only removing all security, both from the persons and property of men, but for the time putting down the law itself, and daring to usurp its place. . . . . . In the first place, by the common law, every private person may lawfully endeavor, of his own authority, and without any warrant or sanction of the magistrate, to suppress a riot by every means in his power. He may disperse, or assist in dispersing, those who are assembled; he may stay those who are engaged in it from executing their purpose; he may stop and prevent others whom he shall see coming up, from joining the rest; and not only has he the authority, but it is his bounden duty, as a good subject of the king, to perform this to the utmost of his ability. If the riot be general and dangerous, he may arm himself against the evil-doers to keep the peace. Such was the opinion of all the judges of England in the time of Queen Elizabeth, in a case called 'The Case of Arms,' (Popham's *Reports*, p. 121,) although the judges add, that 'it would be more discreet for every one in such a case to attend and be assistant to the justices, sheriffs, or other ministers of the king in doing this.' It would, undoubtedly, be more advisable so to do; for the presence and authority of the magistrate would restrain the proceeding to such extremities, until the danger was sufficiently immediate, or until some felony was either committed or could not be prevented without recourse to arms; and at all events the assistance given by men who act in subordination to, and in concert with, the civil magistrate, will be more effectual to attain the object proposed, than any efforts, however well intended, of separate and disunited individuals. But if the occasion demands immediate action, and no opportunity is given for procuring the advice or sanction of the magistrate, it *is the duty of every subject to act for himself, and upon his own responsibility* in suppressing a riotous and tumultuous assembly; and he may be assured that whatever is honestly done by him in the execution of that object, will be supported and justified by the common law. And whilst I am stating the obligation imposed by the law on every subject of the realm, I wish to observe that the law acknowledges no distinction in this respect between the soldier and the private individual. The soldier is still a citizen, lying under the same obligation, and invested with the same authority to preserve the peace of the king as any other subject. If the one is bound to attend the call of the civil magistrate, so also is the other; if the one may interfere for that purpose when the occasion demands it, without the requisition of the magistrate, so may the other too; if the one may employ arms for that purpose, when arms are necessary, the soldier may do the same. Undoubtedly the same exercise of discretion which requires the private subject to act in subordination to, and in aid of, the magistrate, rather than upon his own authority, before recourse is had to arms, ought to operate in a still stronger degree with a military force. But where the danger is pressing and immediate, where a felony has actually been committed, or cannot otherwise be prevented, and from the circumstances of the case no opportunity is offered of obtaining a requisition from the proper authorities, the military subjects of the king, like his civil subjects, not only may, but are bound to do their utmost, of their own authority, to prevent the perpetration of outrage, to put down riot and tumult, and to preserve the lives and property of the people."

It is one result of the law, as laid down by the foregoing authorities, that a military officer refusing or failing, on a proper occasion, to bring into action against a riotous or an insurrectionary mob, the force under his command, would be guilty of an indictable offence at common law, and might be prosecuted accordingly for breach of duty, independently of his liability to military censure.

The most recent case on this subject arose out of the conduct of the military at Six-mile Bridge, in the County of Clare, during the parliamentary election for that county in the year 1852. At the ensuing Spring Assizes held at Ennis in February, 1853, an indictment for murder was preferred against the magistrate and the officers and men whose conduct was impeached; but the grand jury threw out the bill: and the case is here noticed only for the sake of the charge delivered to them by Mr. Justice Perrin, who thus commented upon the law in its application to the offence of which the military were accused:

"It appears that there was an escort of soldiers, consisting of forty men, with two sergeants, as a safe-guard for some persons going to the hustings at Six-mile Bridge, under the command of a captain and a lieutenant, and the conduct of a magistrate—a very difficult and a very nice service. With respect to the requisition, its terms, grounds, or sufficiency, the soldiers could have no knowledge. The orders of the general, which they are bound to obey, and not permitted to canvass, were obligatory on them; and for its sufficiency they are not responsible, and you are happily relieved from any inquiry into that matter. Under that order, and the command of Captain Eager, and the conduct of Mr. Delmege, they assembled. They proceeded to Six-mile

Bridge, and were there, with their arms in their hands, in obedience to orders. Those orders will not justify any unlawful conduct or violence in them, but it accounts for their presence there in arms : for ordinary persons going on such an occasion as that to the hustings would act very indiscreetly and very dangerously, if, perhaps, not very illegally, to arm themselves with deadly weapons, in order to meet obstruction or opposition, if it were expected. But the soldiers were bound, and were there under orders ; and that which in other persons might denote a previous evil or deadly intention, you will see, plainly suggests none in them, for they must obey their orders as soldiers. *There was nothing illegal in their proceeding through the crowd with the freeholders, possibly like any other body of freeholders and their companions, but doing or offering no unnecessary violence, nor were they to be subject to any violence beyond others.* They had no right to force a way through the crowd by violence, nor to remove any obstruction by arms, still less by discharging deadly fire-arms. They had no right to repel a trespass on themselves, or on the escort, by firing or inflicting mortal wounds. You will observe the distinction I take between removing an obstruction and repelling a trespass in another part of the case. *They had a right to lay hold of, as every subject of Her Majesty has, and to arrest persons guilty of any assault or trespass, or other act tending to a riot, either to restrain or make them amenable.* There is no distinction between soldiers and others in that respect, Lord Mansfield says, and his attention was very much called to this subject, touching the military engaged, not as soldiers, but, he says, as citizens, and I say, as subjects of Her Majesty. No matter whether their coats be red or brown, they are employed not to subvert, but to preserve the laws which they ought to prize so highly, taking care not to commit any unnecessary violence, or to abuse the power vested in them. Every one is justified in doing what is necessary for the faithful discharge of his duty, although he is deeply culpable if he wantonly commits any illegal act under the color or pretext of law. Those persons who assist in the suppression of tumults are to be considered as mere private individuals, acting as duty requires. It is a mistake to suppose that having resort to soldiers, is introducing martial law or military government. Suppose a soldier, or any other military person, who acted in the course of the late occurrence, had exceeded the powers with which he was invested, there is no doubt that he may be punished, not by a court-martial, but by an indictment, to be found by the Grand Inquest of the County of Clare, and to be disposed of before the criminal judge, acting with the assistance of the jury, in the court of the county. If assaulted, or struck with stones, they had a right to repel force by force, but not with deadly or mortal weapons; though if provoked by blows, so as to lose the command of their tempers—though more forbearance, perhaps, would be expected from soldiers than from others—if they did, when so provoked, use the mortal weapons in their hands, not with any previous premeditation on their parts so to use them—and I have marked the distinction between soldiers and others under such circumstances—in such repulsion or affray, the law, in consideration of the provocation and the frailty of human nature, reduces the crime, which would otherwise be murder, to manslaughter. And if it should still further appear that, having been so assailed and attacked, they had been guilty of no aggression, and repelling force by force, the violence proceeded so far that, without any misconduct on their part, their lives were threatened, and in actual danger; and if it appears that, in order to save themselves and their lives, they were obliged to fire, and did fire, in the defence of their lives, and slay, the homicide is excusable and justifiable. *But in order to warrant that finding by the jury, or that proceeding by the soldiers, you must be convinced by actual proof that their conduct had been all through correct, and by actual proof—not the saying nor the opinions of any individual—that their lives were in danger, and were saved by the firing, and only by the firing.* In order to warrant such a finding as that, you must entertain that conviction founded upon the evidence given before you. The facts evincing danger imminent to their lives, and which could be prevented only by the firing, must be established by clear evidence, demonstrating that such danger existed, and could be preserved only by resorting to that deplorable remedy. *In considering that matter, you will recollect that there were of the party forty soldiers fully armed, with fixed bayonets, under the command of two officers and two sergeants;* and further, that it is at least doubtful whether there was any legal command upon them to fire. No command was given by their officers—I think that is admitted on all hands. *And further, you must recollect that the firing cannot be justified upon the ground merely that otherwise the freeholders might either have escaped or been withdrawn. That would afford no justification for slaying the assailants.* You will also consider where the matter occurred—in this respect favorable to the accused—a narrow lane. In another point of view, (but that is a matter for inquiry,) it is said to have been near the courthouse, and near an open road, where there was a large body of police, and a strong detachment of soldiers stationed, and where several magistrates were in attendance. You will also consider the matter I have before taken into consideration, whether the soldiers fired without orders, and whether they showed the steadiness and forbearance that they ought. I need not again repeat to gentlemen of your intelligence, that when I state any thing, I merely state what I have been informed; and I will not state a word as to that, but you will look to the evidence before you. If it shall appear to you that shots were fired, and some persons were killed, at a considerable distance from the lane, and out of that lane, and by some of the soldiers who had occupied and immediately come from it, and gained the open ground without any continued resistance—where there was no pretence of danger to their lives, and the persons were, some at a great distance, and some of them with their backs turned—if that state of facts appeared, without previous excitement and previous provocation, it would amount to a case of murder; but it will be for you to say whether such a state of facts as to some individual soldiers should appear—whether there was any previous excitement and provocation (which, as I before told you, would reduce the killing, though it would not justify it, to manslaughter) continuing for a sufficient time, and preventing the blood from cooling. You will consider how far that consideration in your mind operates, and leads you to the conclusion that they acted, not from a deliberate intention to take away life, but from the excitement and warmth produced by previous provocation. That would reduce the crime to manslaughter. Therefore, gentlemen, as to those persons who were slain on what is called the Lodge Road, or near Miss Wilson's, your inquiry will be: first, as to whether any persons were slain; next, by whom they were slain: because, unless it appears that the whole body of soldiers were forward, and if it should appear there were only a few there, it will be your duty to inquire with respect to them if it make any distinction in the finding—to identify and particularize those individuals. If you should find that the homicide was of the worst description, and that they had unnecessarily, and without provocation and excitement to excuse, and also a warmth of blood, for which there is allowance made, you could not visit their act upon the whole body; and, therefore, it will be material for you to ascertain who those individual persons were. That is as much and as important a part of the bill as any other. Then, gentlemen, if they be distinguishable, it is your duty to do so. If you find them guilty of a higher degree of offence than any of the others, you must be able to distinguish them: for you cannot find a general verdict against all upon that. With respect to those slain in the lane, if you are convinced that the soldiers were not the aggressors, but that when they fired they were unlawfully assailed, so as to be in real danger of their own lives, and could not otherwise save them—as

I before mentioned, it would amount to justifiable homicide, and ought to be so found. But if you think that, though they were not the aggressors, and that they were assailed and struck, and, being thereby provoked, repelled force by force, with the affray thickening, and they receiving blows, either from weapons in the hands, or from stones cast upon them—that they were provoked so, and repelled force by force, so as to get their blood so heated that they fired and slew them—I think then you ought to find a bill of manslaughter against all, that is, against every man who is proved to you to have discharged his musket on that occasion; but you must have such proof, of course. And whatever you find in respect to those slain in the lane—manslaughter or homicide in self-defence—you ought to find a bill of manslaughter, at the very least, against every soldier who is proved to have fired in the broad street, or what is called the Lodge Road. These are the observations that I think it right to suggest for your assistance. I cannot, of course, in my imperfect view of the facts, give you such advice and assistance as I would give a jury upon a case which I had heard; but I will be ready and happy, if you find any difficulty in applying any thing I have said upon the evidence, to give you such further assistance as I can, and answer any questions which you shall put to me on the subject."

It may, perhaps, be useful to subjoin a general order issued to the commander-in-chief at Madras, in April, 1825, during the government of Sir Thomas Munro, shortly after a melancholy affair at Kittoor, in which one or two civil servants of the East India Company lost their lives under circumstances which, in the opinion of the public authorities, indicated, both in the civil and military functionaries, a want of general knowledge respecting the subject of the order.

"The Honorable, the Governor in Council, deems it necessary to lay down the following rules relative to the exercise of the authority with which civil magistrates, and other officers acting in a similar capacity, are vested, for calling out military force to preserve the peace of the country:

"1. The first and most important rule is, that no civil officer shall call out troops until he is convinced, by mature consideration of all the circumstances, that such a measure is necessary.

"2. When the civil officer is satisfied of the necessity of the measure, he should, before carrying it into execution, receive the sanction of government, unless the delay requisite for that purpose is likely to prove detrimental to the public interests. In that case, also, he should fully report the circumstances to government.

"3. When the civil officer may not deem it safe to wait for the orders of government, he should address his requisition for troops, not to any subordinate military officer, but to the officer commanding the division, to whom he should communicate his object in making it, and all the information he may possess regarding the stength and designs of those by whom the public peace is menaced or disturbed. His duty is confined to these points. *He has no authority in directing military operations.*

"4. The officer commanding the troops has alone authority to determine the number and nature of those to be employed; the time and manner of making the attack, and every other operation for the reduction of the enemy.

"5. Whenever the officer commanding the division may think the troops at his disposal inadequate to the enterprise, he should call upon the officer commanding the neighboring division for aid, and report to government and to the commander-in-chief.

"6. No assistant or subordinate magistrate is authorized to call out troops. When any such officer thinks military aid necessary, he must refer to his superior, the principal magistrate of the district.

"The foregoing rules are to be observed, when it can be done without danger to the public safety. Should any extraordinary case occur, which admits of no delay, civil and military officers must then act according to the emergency and the best of their judgment. Such cases, however, can rarely occur, unless when an enemy becomes the assailant; and therefore occasion can hardly ever arise for departing from the regular course of calling out troops, only by the requisition of the principal civil magistrates of the province, to the officer commanding the division.

"Ordered, that the foregoing resolutions be published in general orders to the army, and be communicated for the information and guidance of such civil officers as they concern." (Consult PRENDERGAST. *See* CALLING FORTH MILITIA; OBSTRUCTION OF LAWS; INSURRECTION; MARSHALS; POSSE COMITATUS.)

EXEMPTS FROM MILITIA DUTY. The Vice-president of the United States; the officers, judicial and executive, of the government of the United States; the members of both houses of Congress, and their respective officers; all custom-house officers, with their clerks; all post-officers and stage-drivers, who are employed in the care and conveyance of the mail of the post-office of the United States; all ferrymen employed at any ferry on the post road; all inspectors of exports; all pilots and mariners actually employed in the service of any citizen or merchant within the United States; and all persons who are or may be exempted by the laws of the different States; (*Act* May 8, 1792.)

EXPEDITION—is an enterprise undertaken either by sea or by land against an enemy, the fortunate termination of which principally depends on the rapidity and unexpected nature of its movements. To be successful, the design and preparations for an expedition should, as far as may be practicable, be carefully concealed; the means employed be proportioned to the object in view; the plan carefully arranged, and its execution intrusted to a general whose talents are known to fit him for such a command, and who possesses a perfect knowledge of the scene of action.

EXPENSE MAGAZINES—are small powder magazines containing ammunition, &c., made up for present use. There is usually one in each bastion.

EXTERIOR SIDE—is the side of the polygon, upon which a front of fortification is formed.

EXTERIOR SLOPE—is a slope given to the outside of the parapet. It is found by experience that earth of common quality will naturally acquire a slope of 45°, even when battered by cannon. This inclination is therefore given to the slope.

EXTRA ALLOWANCES. Officers shall not receive any additional pay, extra allowance, or compensation in any form whatever, for disbursements of public money, or any other service or duty whatsoever, unless the same shall be authorized by law, and the appropriation therefor explicitly set forth; that is, for such additional pay, extra allowance, or compensation; (*Act* Aug. 23, 1842.)

EXTRA EXPENSES. Where any commissioned officer shall be obliged to incur any extra expense in travelling, and sitting on general courts-martial, he shall be allowed one dollar and twenty-five cents per day, if not entitled to forage, and one dollar if so entitled; (*Act* Jan. 29, 1813.)

F

FACE OF A GUN. The superficies of the metal at the extremity of the muzzle.

FACES OF A BASTION—are the two sides extending from the salient to the angle of the shoulder.

FACES OF A SQUARE. The sides of a battalion when formed in square.

FACINGS. The movement of soldiers to the right, left, right about, left about, &c.

FALSE ALARMS. Punishable. (*See* ALARM.)

FALSE CERTIFICATES. Punishable with cashiering; (ART. 14.) (*See* CERTIFICATE.)

FALSEHOOD. The *onus probandi* in all accusations lies with the accuser. If A accuses B of having told a falsehood, A must prove it by legal evidence.

FARRIER AND BLACKSMITH. Allowed to cavalry regiments. (*See* ARMY; VETERINARY.)

FASCINES—are long cylindrical fagots of brushwood, and when designed for supporting the earth of extensive epaulements, are called *saucissons*, and are about 18 feet long, and ten inches thick; those for the revetment of the parapets of batteries are eight or ten feet long; those for covering wet or marshy ground from 6 to 9 feet long. (*See* REVETMENT for construction of fascines.)

FATIGUE DUTY. Soldiers on fatigue duty allowed an extra gill of whiskey; (*Act* March 2, 1829.)

That the allowance of soldiers employed at work on fortifications, in surveys, in cutting roads, and other constant labor, of not less than ten days, authorized by an act approved March second, eighteen hundred and nineteen, entitled "An act to regulate the pay of the army when employed on fatigue duty," be increased to twenty-five cents per day for men employed as laborers and teamsters, and forty cents per day when employed as mechanics, at all stations east of the Rocky Mountains, and to thirty-five cents and fifty cents per day, respectively, when the men are employed at the stations west of those mountains.—Approved August 4, 1854.

FAUSSE BRAIE—is a second enceinte, exterior to, and parallel to the main rampart, and considerably below its level.

FEVER. (*See* SANITARY PRECAUTIONS; MEDICINE.)

FIELD. In a military sense, the scene of a campaign or battle.

FIELD DAY. A term used when a regiment is taken out to the field, for the purpose of being instructed in the field exercise and evolutions.

FIELD MARSHAL. The highest military rank excepting that of captain-general.

FIELD OFFICERS. Colonels, lieutenant-colonels, and majors, are called field officers. They should always be mounted, in order to give ground for movements, circulate orders, and correct pivots.

FIELD WORKS. Their object is to provide a body of troops, or a town, with a secure protection against a sudden assault of superior numbers by the interposition of a parapet of some material capable of resisting the effects of projectiles. This parapet may be made of very miscellaneous materials, but is usually of earth, excavated from a ditch, which will itself be an obstacle to attack. The usual figure of a parapet with its ditch is shown in Fig. 112.

The exterior slope $e f$, which is always exposed to the action of the

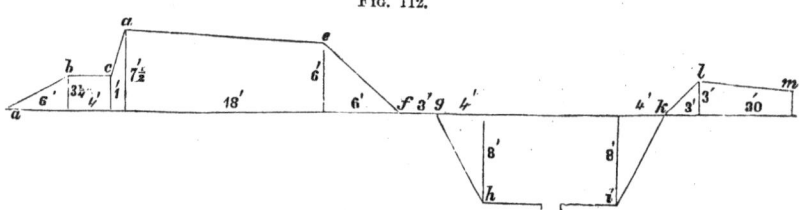

FIG. 112.

weather, and during an engagement to enemy's shot, must have that inclination or slope which the materials composing it would assume when poured loosely from a height, and at which they would therefore stand without any additional support. This inclination for earth of ordinary tenacity, is about 45°; *i. e.*, the base on which the slope stands is equal to its height, or it has a depression of 1 in 1. The parapet would afford the best cover if its superior slope, $d\ e$, were horizontal, or rather parallel to the plane of site; but in this case a musket-shot, fired along its surface, could not reach the ground within a very considerable distance in front of it; a gentle inclination is therefore given to it, and experience has fixed this slope at a depression of 1 in 6. The interior slope, $d\ c$, of this parapet must be nearly vertical, that soldiers may lean against it and fire easily over it. It must, therefore, be supported by a wall of some material, called a *revetment*. The base of this slope is usually one-fourth the height. It has a depression, therefore, of 4 in 1. A step, $b\ c$, called the *banquette*, is added, of a height sufficient to enable a man of ordinary stature to fire conveniently over the crest, and sloping away gently towards the rear to facilitate the alternate advance and retirement of each soldier to discharge and load his firelock. The base of this slope is usually 1½ to 2 times the height. The depression is, therefore, 1 in 1½ or 2. The thickness of a parapet, that is, of its superior slope, must be sufficient to withstand the effects of the projectiles likely to be discharged against it. To afford security against

Musketry	its thickness must be	5 feet.
6-pounders	" "	6 "
9-pounders	" "	9 "
12-pounders	" "	12 "
18-pounders	" "	18 "
24-pounders and heavier guns	" "	20 to 24 feet.

In field-works, which are seldom made to resist heavy artillery, a thickness of parapet of 11 feet will generally be sufficient.

The height of a parapet will greatly depend upon its position. It will readily be seen from Fig. 112, that a bullet striking the parapet near the upper part will have to traverse a small portion only of the thickness of the parapet in order to pass through.

It becomes necessary, therefore, to give to a parapet a height rather greater than that to which cover is required. Hence on a plain where the attacking and defending parties are on the same level, the height of a parapet, to furnish cover to men 6 feet high, is usually 7½ feet. Should the parapet be situated upon the brow of a hill, the defenders could obtain cover to any desired extent by merely retiring from it. In this case a height sufficient to protect the soldiers while firing is all that will be necessary; this will usually be from 4 to 6 feet. (Fig. 113.)

Should these conditions be reversed, that is, should the attacking party be in possession of the higher ground, a height of parapet up to 10 or 12 feet may be indispensable, and when the slope of the ground is considerable, even this will afford cover to a small distance only behind it; (Fig. 114.) It may be said generally then that the height of parapets varies from 4 to 12 feet, and the thickness from 4 to 25 feet.

FIGS. 113, 114.

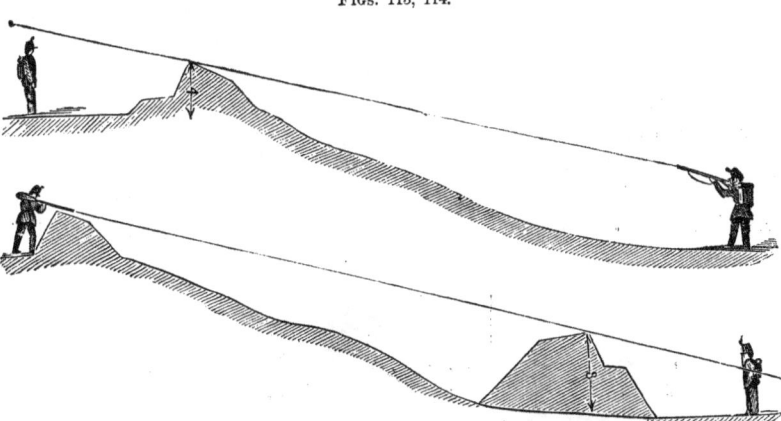

In the defence of field positions the following considerations require special notice:—

1st. The period likely to elapse before the position is attacked.

2d. The number of troops by whom the position is to be held.

3d. The number of men available for the construction of the work, and the nature of the materials at hand.

On the first of these considerations will depend the height and thickness of the parapet, depth and width of the ditch, and the nature of the obstacles which may be added, as only a certain amount of work can be executed in a given time, and a work of even feeble profile thoroughly complete will be capable of a better defence than a stronger work only partially executed. The extent which it may be desirable to give to the work will be limited by the number of men available for its defence. There must, at least, be sufficient to man the whole of the parapet, and a reserve, in addition, is almost essential. The length of crest line measured in yards, must not exceed half the number of men allotted for its defence. When either labor or materials are scarce, it may be necessary to reduce the profile, and to contract the extent of the work below that which would be desirable under other circumstances; but in this case the details should be so arranged as to admit of subsequent additions, should circumstances allow it, so as to bring the whole work to that condition which might have been desirable, though unattainable in the first instance. When time, labor, and materials are abundant, a good parapet and ditch should always be made to secure the defenders. The dimensions and construction of such a parapet have already been given. But cover can be obtained for a limited number of men in a more expeditious way. Thus a man will be equally protected from an enemy's fire, by standing behind a parapet 6 feet high, or in a trench 3 feet deep, with a bank of earth 3 feet high in front of him. Now to dig a trench 3 feet deep, and throw the earth to the front so as to form a bank 3 feet high, may be performed by the same number of men in at most $\frac{1}{3}$ of the time required for the construction of a complete parapet 6 feet high. A trench and breastwork then will be generally used when the time is limited, and when cover and not the creation of an obstacle is the principal object of the work. Fig. 115

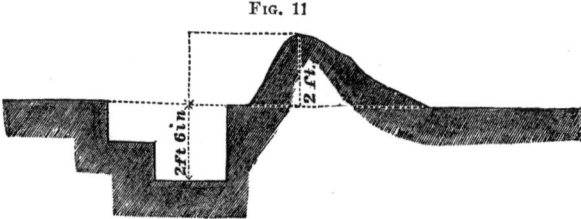

Fig. 11

represents a section of the slightest work of this nature which can be of any service. Here a trench $2\frac{1}{2}$ feet deep is dug, and the earth thrown to the front forms a rough parapet 2 feet high. The trench can contain one rank only, and the total cover being $4\frac{1}{2}$ feet high, the men will not be safe except when sitting or stooping. A trench and breastwork of these dimensions can be completed in about 1½ hours. The next section (Fig. 116) is more serviceable; the total height of cover in this

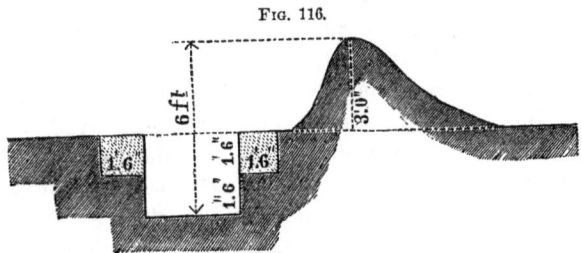

Fig. 116.

case is 6 feet. The men will be safe therefore so long as they remain in the trench, which provides room for one rank only at a time. The completion of this work would require about 3 hours.

Fig. 117 is a section of a breastwork and trench of a capacity suffi-

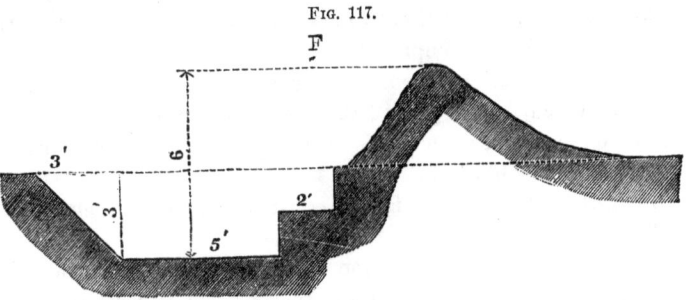

Fig. 117.

cient for most of the purposes for which works of this nature are usually required. The trench is wide enough to contain two ranks of men at the same time, and affords cover 6 feet in height. Such a work can be executed in about 5 hours.

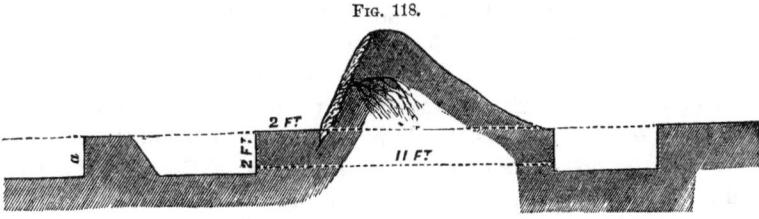

Fig. 118.

Fig. 118 is a profile adapted to marshy or rocky situations where shallow trenches only are practicable.

This work can be constructed very rapidly when labor is abundant, as two working parties, one in front and the other in rear, can be employed at the same time. The work to be performed then will generally be the excavation of a trench or ditch, and the formation of a parapet or breastwork, with the earth thrown out of it. It will in most cases be executed by the troops themselves, though sometimes laborers may be obtained. In constructing a simple trench and breastwork one row of workmen only can be advantageously employed at the same time, and it will be found desirable to place them 6 feet apart; as at this distance each man can use his arms freely, without interfering with or injuring his neighbor. When the saving of time is of more consequence than economy of labor, the diggers may be placed 4 feet apart, and the completion of the work will be accelerated, though not in proportion to the increase in the number of workmen. An ordinary laborer or common soldier can excavate one cubic yard, $i.\ e.$ 27 cubic feet, in any but the hardest soils per hour; and can continue working at this rate for 8 hours. Should the soil be loose or sandy, so that the pickaxe is seldom required, this estimate may be nearly doubled. The trench or breastwork will be completed in the time in which each man will finish his portion, that is, a portion equal in length to the interval between any two adjacent diggers: therefore the number of hours will be equal to the number of cubic yards in such portion. Whence the following rule is at once obtained:

To find the time required for the construction of a trench or parapet, in ordinary soil—

Multiply the area of the section of the trench in square feet by the interval between the diggers (not less than 6 feet), and divide this product by 27, the quotient is the number of hours required for the construction of the work. Conversely, to find the area of the section of the trench or breastwork which can be executed in a given time—

Multiply the number of hours by 27, and divide the product by the interval (in feet) between the diggers, the result will be the area, in square feet, of the section of the trench or breastwork.

It will frequently happen that cover can be speedily obtained, and positions rendered defensible in a very short time, by taking advantage of the hedges, ditches, or walls, which may be met with, or of the obstacles which may be presented by the natural features of the ground. General rules for proceeding under all the various circumstances which may occur cannot be given, but the following examples will show what may be effected in certain cases, and indicate the character of the operations usually required. Fig. 119 represents a common hedge and ditch turned into a breastwork to be defended from the hedge side. If the hedge be thick and planted on a bank, as is generally the case, and especially if the ditch be tolerably deep and contain water, the breastwork will be rendered strong at the expense of little labor. A shallow trench should be excavated behind the hedge, and the earth thrown up to raise the bank sufficiently to form a rough breastwork some 18 inches thick at the top. Should the hedge be more than 6 feet high, it should be cut to that height, and the branches interwoven with the lower part to strengthen it. A hedge to be defended from the ditch side (Fig. 120) is a ready-made trench and breastwork, and will become a convenient work by a little scarping of the sides and widening and levelling of the bottom of the ditch, and by the addition, if necessary, of a banquette.

Fig. 119.

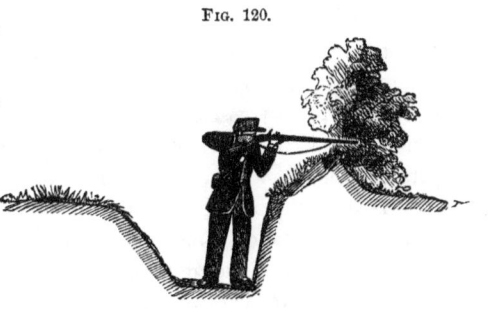

Fig. 120.

A good nine-inch brick wall is musket-shot proof. Such a wall 4 feet high will require no alteration, but may be used as a parapet by forming loopholes with sand-bags laid on the top, Fig. 121. Should there be time, a ditch should be dug in front, and the earth thrown up against the front of the wall to prevent the enemy from using the loopholes against the defenders. A wall 15 feet high can be pierced with two tiers of loopholes, one at 8 feet above the ground, the other at the top of

Fig. 121.

the wall. In rear a scaffolding must be erected of two stages to serve as banquettes. Such an arrangement is shown in the diagram, (Fig. 122.)

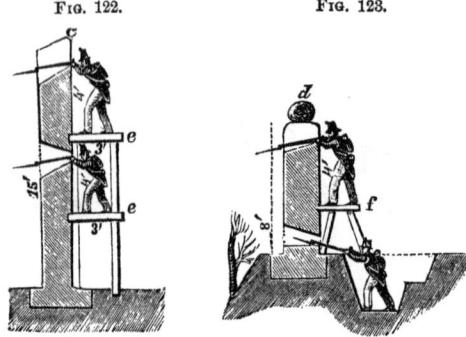

FIG. 122.   FIG. 123.

A wall 8 feet high may also be pierced with two tiers of loopholes as shown in Fig. 123. A trench must be dug in this case, to enable the defenders to make use of the lower tier of loopholes, and a scaffolding erected to serve as a banquette for the upper. On an emergency, materials of almost any conceivable description, as sacks or casks of earth, of sand, of coal, or even of corn or flour, bales of cotton, of cloth, packs of wool, mattresses, trusses of hay, fagots, carts or wagons of stable litter, brick rubbish or paving stones, may be formed into parapets of defence, while the approach of an enemy may be rendered exceedingly difficult, by a judicious combination of obstacles which, under urgent circumstances, may be extemporized of trees, bushes, posts, wagons, wheels, strong palings, chairs, tables, and miscellaneous articles of furniture, with iron rails, pitchforks, and agricultural implements, carefully arranged in the front, and secured by chains or ropes strongly picketed to the ground. Every soldier should be able to form for himself a rifle pit. This can be accomplished by digging a hole in the ground about 3 feet deep and 3 feet square at the top, with a little step to enable him to get in or out with ease. The excavated earth should be thrown up to the front to form a protection. A loophole should be made by three sand-bags; two placed longitudinally, and one across.

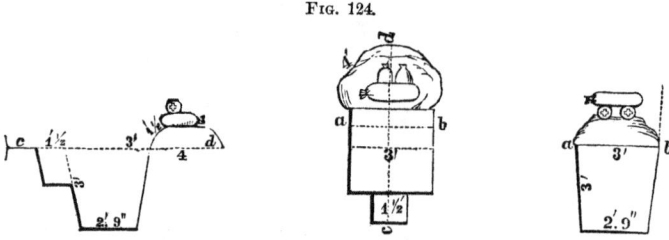

FIG. 124.

A rifle pit of this construction is shown in plan, section, and elevation in Fig. 124. Approach to field-works should be rendered difficult by the formation of obstacles of various kinds, so that troops when coming to the assault may be detained under heavy fire as long as possible while they are endeavoring to force or surmount the obstacle. Contrivances of this nature are very numerous. (*See* ABATIS, TROUS-DE-LOUP, CROWS'-FEET, CHEVAUX-DE-FRIZE, INUNDATIONS.) In defensive warfare it is frequently necessary to intrench towns and villages, to secure them from the incursions of small parties, or to serve as points of support for the movements of troops. If a town or village be commanded on all sides, or even by great elevations on one side, if the houses be of wood and the roofs thatched, so as to be easily set on fire, such a position should be avoided. Neither should a detachment of troops occupy a town or village too extensive for their number, unless a part of the village can be easily and effectually separated from the rest. The number of the detachment should at least equal the number of yards in the exterior line of works by which the village is surrounded. To place a village in a state of defence, the first object will be to complete a continuous line of defensive works, by which it may be entirely surrounded. To this end advantage is taken of all buildings, fences and walls, near the exterior edge. The buildings, when substantial, may serve as bastions to flank the connecting lines of works, and when due preparations have been made will become strong positions. The walls and hedges must be strengthened by banks of earth, and will form curtains connecting the stronger portions. All openings remaining must be closed by parapets, strengthened by ditches, abatis, palisading, and such obstacles as the locality may present, and the streets must be barricaded at intervals. Barricades may be constructed of materials of almost any kind of earth, of timber, of paving stones, of wagons of stable litter; (the wheels should be taken off.) In buildings occupied for defence the doors and windows should be blocked up with sand-bags, supported by frames of wood, and the glass must be removed from the windows. Should there be no projecting wings or porches, it will be necessary to obtain a flanking fire by the construction of balconies projecting from the windows, and furnished with loopholes in the sides and bottom, so that a flanking fire can be brought to bear on the ground at the foot of the wall. This arrangement is shown in the diagram, (Fig. 125.) The beams supporting the gallery or balcony are bolted to the flooring within; the balcony is surrounded with good oak boarding of 4" or 5" thick. That the communications of the defenders may be free,

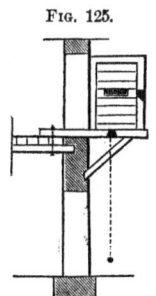

FIG. 125.

all interior hedges and walls which can in any way impede their movements must be levelled, so that they may be able to bring support rapidly to any point pressed by an enemy. Those hedges which it may be desirable to retain must be strengthened in the manner already pointed out. The strength of the position may (when circumstances admit) be greatly increased by the formation of an interior keep, whither the defenders may retire and obtain favorable terms of capitulation should they be unable to withstand the assaults of their assailants. A substantial building within the town, as a gaol, may be converted into a keep by blocking up unnecessary openings; by covering entrances or any unflanked portions of the walls with tambours; by loopholing the walls and surrounding them if possible with a ditch, palisade, and abatis. In the absence of a building of this nature, it will be desirable to construct a redoubt, of as strong a character as time will allow. If the village be of considerable extent, and a position can be found which cannot be commanded from the neighboring buildings, the redoubt may be of earth, as in an ordinary field-work. While the actual defences of the village are thus being prepared, parties will be occupied on the ground without, in creating obstacles and entanglements in the immediate vicinity of the place, and in removing and levelling all obstructions between such obstacles and the limits of rifle range. The greatest obstacle which can be presented to an attacking force, will, in future, be a long level tract, fully commanded by a sweeping fire. It is, in fact, difficult to see how an assaulting body could pass over such a tract of 1,000 or 800 yards in extent, to attack a work in daylight without being annihilated. To remove every object, whether tree or bush, rising ground, dry ditch, or hedge, which could afford cover or concealment to a rifleman, will be an object of primary importance in executing the arrangements for defence. Ditches full of water, or which can be filled, may generally be left, as they impede, and cannot assist the assaulting party. Fig. 126 gives an illustration of the means, already described, usually applicable for placing a village in a state of defence.

A very little time devoted to the study of the subject, would enable an officer in command of a picket or charged with the defence of an outpost to determine the construction of all the works that are requisite for protection and defence. THE SELECTION OF THE POST is what will first engage attention, and the following considerations must have their weight in determining the point:

The inequalities of the ground, and the objects upon it, such as buildings or fences, &c., should be of such a nature, and in that relative situation to each other, as to be convertible into a fortified post with

FIE.] MILITARY DICTIONARY. 293

FIG. 126.

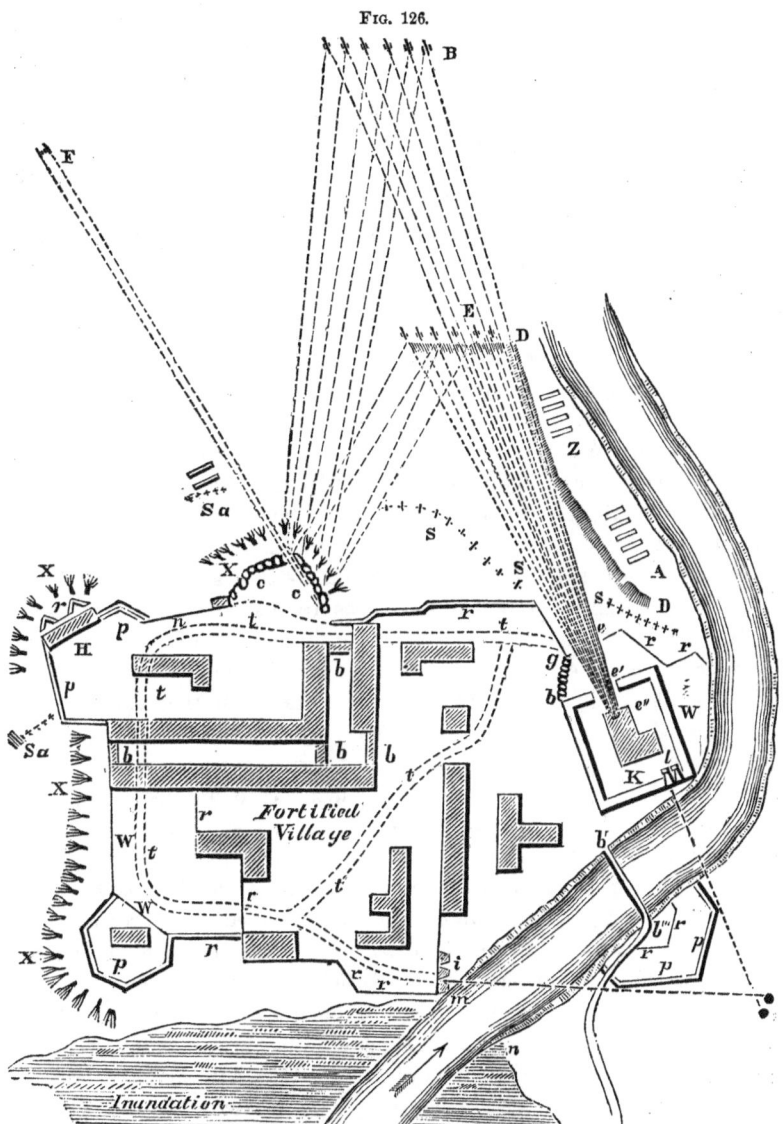

DEFENCE OF A FORTIFIED VILLAGE.

w, loopholed walls; P, parapets and ditches; *c*, ditto of casks; *x*, abatis; *r*, stockades; *b*, barriers; *t t*, free communication, road or passage; H, fortified house; K, keep.

ATTACK OF THIS FORTIFIED VILLAGE.

D D, flying sap-parallel or trench of cover; B, open field battery, first opened at about 350 yards' distance; E, ditto, advanced to breach; F, one 9-pounder and one 24-pounder howitzer, to enfilade flanking defences $e\ e'\ e''$, breaches; A, storming party; Z, supporting ditto; *s s s*, firing party and skirmishers; S *a*, false attacks, to divert the attention of the garrison at the moment of the real assault.

THE LEAST POSSIBLE LABOR, AND IN THE SHORTEST TIME. The position should not be commanded, especially on the flanks or in the rear, within the ordinary range of a field-piece. There should be plenty of materials on the spot for the construction of temporary works, and for forming obstructions in front of them. The soil should be of a nature that is easily worked, if it is foreseen that any trenches or ditches will have to be executed. It should generally be DIFFICULT OF ACCESS, and yet offer the MEANS OF RETREATING in security. And should be in a situation for fulfilling the object for which the detachment is to be posted.

In arranging the general plan of defensive works, the following points will require more particular attention :—It must be ascertained from a minute examination of the position, what figure will give the greatest quantity of fire over the most accessible points of attack, and the general contour of the intrenchment should make available buildings or fences on the ground. THE OBJECT THE WORK IS EXPECTED TO FULFIL in reference to the supporting force; the distance from that force; or whether it is to be left to itself to hold an enemy in check as long as possible; or whether it is to be defended to the last extremity. Its SITUATION WITH RESPECT TO THE ENEMY as to distance, &c.; whether it is likely to be attacked by overwhelming forces, or only subject to the brusque attack of cavalry or infantry in smaller bodies; whether artillery is likely to be brought up against it, for in that case earthen works, when merely for the purposes of cover, are in some respects better than buildings or stockades; the parapets, too, must be thicker; —whether it can be surrounded, for in such a case it must be inclosed all round, &c. THE NUMBER OF MEN THERE WILL BE FOR ITS DEFENCE, taking it as an established rule, that it is better to have a force concentrated, than too much distributed, and therefore injudicious to make works of a greater extent than can be well manned and vigorously defended. For instance, in small works there might be a file of men for every pace or yard in the length of their breastwork, and in larger ones the same, with a reserve of from one-fourth to one-sixth of the whole in addition. On some such general basis, a calculation of the proportionate extent of a work might be made. All this of course depends very much upon circumstances. THE NUMBER OF MEN, whether soldiers or inhabitants, that can be collected together for working, and whether there are tools enough for them, so as not to undertake more work than can be well done. And, which is a very important point, THE TIME THERE IS TO DO IT IN. Whether an immediate attack is to be apprehended, or otherwise, for this will decide not only the nature of the works, but the parts of them that require the first attention; as will be more apparent when the details of execution are brought under consideration. THE NATURE OF THE MATERIALS that can be had on the spot, or procured in the neighborhood. This will have a great influence on the details of the plan to be pursued, and will afford opportunity for the display of considerable tact and intelligence, in appropriating and adapting the means at hand for carrying the general plan into effect, and securing its objects with the LEAST POSSIBLE LABOR. No one who is not conversant with work of this description, can have any idea of the great saving of time and labor that may be effected, by taking advantage of what might appear at a casual glance to be very unimportant and local features; such, for instance, as gentle undulations in the ground.

Details of Execution.—The following description of tools and stores would be found more or less necessary, where temporary works were to be thrown up. They are classed in three divisions, that their separate uses may be apparent.

Class 1. Field Exercise Tools.

Shovels,
Pickaxes,
Felling-axes,
Bill-hooks,

For sinking trenches, forming breastworks, felling timber, making abatis and obstructions, &c.

Class 2. For Houses, Walls, &c.

Sledge-hammers,
Hand-borers,
Crowbars,
Saws.
Augers,
Spike-nails,

For forming loopholes, breaking through walls; preparing timber for barricades, stockade work, &c.

Class 3. General service and purposes of defence.

Sand-bags,
Rockets,
Small shells,
Hand-grenades,

The sand-bags for blocking up windows, forming loopholes, &c.; the rockets and shells for defence of houses and intrenchments.

The proportions of these necessary to be demanded would of course vary with the description of work which might be anticipated. For example, in throwing up earthen works in an open country, a pickaxe and shovel for every man that could be employed on the breastworks would be wanted. If an abatis could be formed, and there were fences to be cut up and levelled, one-third of the men would be advantageously employed with felling-axes and bill-hooks. In a case where houses were to be placed in a state of defence, walls would have to be broken through for making loopholes, and windows, doors, and passages to be barricaded; here crowbars, hand-borers, sledge-hammers, spike-nails, and saws would be required in greater proportion than spades and pickaxes. Sand-bags are included as being very useful for many purposes, such as protecting men when firing over a parapet or breastwork, quickly blocking up the lower parts of windows, &c.

A man will carry one hundred empty sand-bags, weighing about 60 lbs., each of which will contain a bushel of earth, and when *full* they are *musket-proof*. Rockets, small shells, and grenades, are mentioned as being very powerful and attainable auxiliaries in the defence of posts and houses; and one great advantage of them is, that any body who has common sense may use them, or at least be instructed in the requisite precautions in a few minutes. A CERTAIN DIVISION OF LABOR must also be attended to, and a man should always have a tool put into his hand that he has been accustomed to use; carpenters should therefore be employed where saws and axes are wanted; miners and blacksmiths where walls are to be broken through; laborers where the spade and pickaxe come into play. Those who never handled tools of these descriptions, would be most usefully employed in collecting materials. It would be well also to select such men for the first tour of duty, as patrols, and sentries, and to employ the best workmen in overcoming the greatest difficulties, which are usually found in the commencement. A little foresight will not be misapplied in considering these points. It is essential to obtain the assistance of the inhabitants in executing works of this description, and an officer should always have authority to enforce their attendance, and to pay them in proportion to their exertions. They should also be required to bring with them whatever tools they can best use, or that are most wanted.

A stick may be cut to measure lines, and stakes will be driven to show the slope and general form of the profile necessary in each particular case. Whatever form is to be given to a work, it is traced upon the ground by laying off its angles according to the number of their degrees, and its sides are designated by little furrows dug with the mattock or spade along cords stretched in the proper direction. To profile a work is to figure upon the ground its elevation by means of poles and laths nailed together; (Fig. 127.) The officer who directs the work ought to take with him four or five soldiers who carry mattocks, 100 pickets, twenty poles ten or twelve feet long, twenty laths, some camp colors, and a cord 65 feet in length. There ought also to be a carpenter, who carries hammer, nails, and a saw.

FIG. 127.

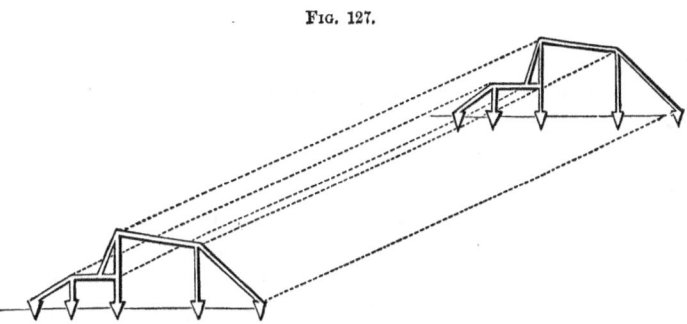

Field-works necessary or desirable in the operations of an army in the field to strengthen lines of battle, keep open lines of communication, protect bridges from destruction, &c., will generally be constructed under the supervision of engineers. They may have any extent, from a simple redan, or a battery, to a line or several lines of works, some of considerable magnitude, extending over a position of ten or twenty miles. It will only be possible here to give a brief description of the works usually adopted for these purposes.

Field-works, then, are usually arranged in three classes:—

First-Class, consisting of works open at the gorge—

Redan	Double Redan
Redan with flanks	Tenaille Head
Lunette	Bastion Head

Second Class, consisting of works inclosed all round—

Redoubt
Bastion Fort

Third Class, consisting of lines both continuous or at intervals—

Lines of Redans	Lines of Bastions
Lines of Tenailles	Lines at intervals
Indented Lines à la Cremaillère	

A redan is a work of the simplest kind. It consists of two faces of parapet and ditch, forming a salient angle. Redans serve to cover bridges, causeways, avenues, &c., and being quite open at the gorge, are only suited for positions in which their extremities rest on rivers or other obstacles, so that they cannot be turned, or else when protected by the full sweeping fire of works in their rear. Redans in front of other works are generally mere covers for an advanced post; for example, if a strong redoubt occupies the commanding summit of a hill, its elevation and position usually prevent the deep hollows and approaches by the valleys being fully seen from its faces. Redans may then be advantageously constructed on the lower knolls, or under features of the hill, to command all the hollows, which cannot always be reached by the fire of the main redoubt.

Lines.—Continuous lines of rampart, parapet, and ditch, are sometimes used to connect important redoubts, or to cover the front of a position, and they may have, according to circumstances, a variety of tracings. To cover any considerable extent of country with continuous lines is generally considered injudicious, but must not be altogether condemned; as in particular cases, especially on ground unfavorable for manœuvring, it may be an advantageous constructon. Continuous lines require a great expenditure of labor in their construction, and a large force is necessary for their defence; if forced at one point, the whole is lost, and they interfere greatly with the offensive movements of the troops they cover. When circumstances oblige any considerable extent of country to be defended, lines at intervals are more generally adopted. Lines at intervals are a series of detached works arranged in two or more rows, mutually supporting each other, and each capable of enduring an independent attack. In lines at intervals the most advanced positions are usually occupied by simple works open at the gorge as Redans and Lunettes, within range of each other, that is, not more than 600 or 700 yards apart. These works, being open at the gorge, can be fully commanded by the works in rear, which can bring a fire upon every point within them; if taken by an enemy, they cannot, therefore, be held by him until the latter works are also subdued. The second line of works are generally a series of redoubts, adapted in shape to the features of the ground, 400 or 500 yards behind the salient works, covering their intervals, and protecting their faces and ditches by a powerful flanking fire. If necessary, a third line of works on similar principles may be added. The works in the second line, *i. e.* the redoubts, must be made as strong in rear as in front, or an enemy would not fail to attempt to carry them by an attack on the rear, and the faces of all the works should, as far as possible, be directed on ground which the enemy cannot occupy, so as to be protected from his enfilade fire. The annexed diagram (Fig. 128) exhibits a tract of ground defended by lines at intervals, and will convey an idea of the general arrangement of works of this nature.

In the construction of these and all other field-works, the following maxims must be strictly observed: 1st. That the works to be flanked, are never to be beyond the range of the weapons of the works flanking them, that is, never out of the effective range of musketry. 2d. That the angles of defence should be about right angles. 3d. That the salient

FIG. 128.

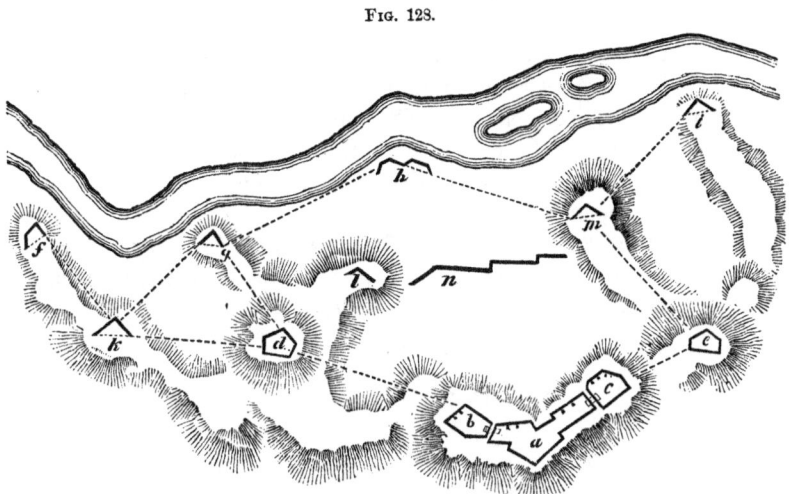

angles of works should be as obtuse as circumstances will permit 4th. That, although ditches cannot always be as fully flanked, as in permanent fortification, yet that partial flanking must be carried as far as possible. 5th. That in the construction of field-works, reference should not only be had to the direct and immediate obstacles that the work itself presents to the enemy, and the positive effects of fire on the approaches to it; but likewise the relative value of the work must be considered, as to the support it can give to, or receive from, other works. 6th. That the outline of a field-work should be proportioned to the number of men intended to defend it. 7th. The ground over which an enemy must pass to the attack should, if possible, be seen both in front and flank. (Consult HYDE's *Fortifications;* JEBB's *Attack and Defence; Traité Théorique et Pratique de Fortification Passagère,* &c., par M. ERNEST DE NEUCHEZE, *Capitaine,* &c.; MAHAN's *Field Fortifications; Aid Memoir to the Military Sciences,* Edited by a Committee of the Corps of Royal Engineers.)

FILE—generally means two soldiers, a front and rear rank man. Each man occupies in line about 21 inches; 10 files require a space of 7 paces; 100 files, 70 paces. The French designate men ranged in four ranks, as follows: the front rank men as *chefs de file;* the second rank, *serres demi files;* the third *chefs demi file;* and the rear rank *serres files.*

FINDING. Before a court-martial deliberates upon the judgment, the judge-advocate reads over the whole proceedings of the court; he then collects the votes of each member, beginning with the youngest. The best mode of doing so is by slips of paper. The Articles of War require a majority in all cases, and in case of sentence of death, two-thirds. It is not necessary to find a *general* verdict of guilt or acquittal upon the whole of every charge. The court may find a prisoner guilty of part of a charge, and acquit him of the remainder, and render sentence according to their finding. This is a *special* verdict; (HOUGH's *Military Law Authorities*.)

FIRE, (VARIETIES OF.) Direct fire is when the battery of guns is ranged parallel to the face of the work, or the line of troops to be fired at, so that the shot strike it perpendicularly.

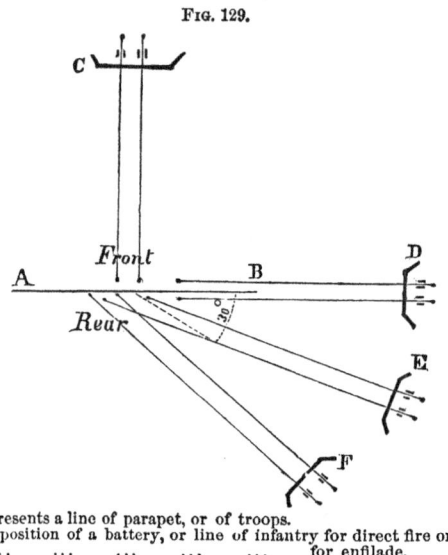

FIG. 129.

A B represents a line of parapet, or of troops.
C is the position of a battery, or line of infantry for direct fire on A B.
D ... ... for enfilade.
E ... ... for slant.
F ... ... for reverse.

ENFILADE.—Enfilade fire is when the battery is ranged perpendicularly to the prolongation of the crest of a parapet, or to a line of troops, so that the shot flies in the same direction, or parallel to the line or parapet, sweeping along from one end to the other.

OBLIQUE.—Oblique fire is when the battery of guns is ranged so as to form an angle with the front of the object to be struck.

PLUNGING.—Plunging fire is when the shot is fired from a position considerably higher than the object fired at.

RICOCHET.—Ricochet fire is firing with a slight elevation, and with small charges, in a direction enfilading the face of the work, so that the shot are pitched over the parapet, and bound along the rampart from end to end, with destructive effect on the guns and gunners.

REVERSE.—Reverse fire is when the shot strikes the interior slope of the parapet at an angle greater than 30°.

SLANT.—Slant fire is when the shot strikes the interior slope of the parapet, forming with it a horizontal angle, not greater than 30°.

VERTICAL.—Vertical fire is that in which the shot or shell describes a lofty curve through the air before it falls; such is the fire from mortars.

FIRE BALL. Made like a light-ball, except that, being intended to light the works of an enemy, it is also loaded with a shell.

FIRING. In the discharge of fire-arms, it is necessary to know the position and relations existing between the three following lines (Fig. 130): 1st, the *line of sight*, which is the prolongation of the visual

FIG. 130.

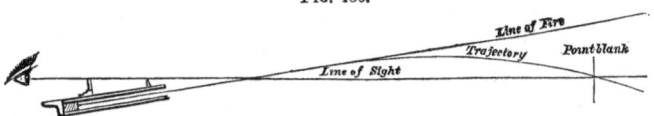

ray passing through the highest points of the breech and the muzzle; 2d, the *line of fire*, which is the prolonged axis of the piece; and 3d, the trajectory described by the projectile.

The *point-blank range* is the second intersection of the trajectory with the line of sight.

The causes of deviation in firing are:

(1.) From the construction of the arm.	Causes which can be corrected.	Wrong position of the sight. Calibre not exact. Barrel imperfect. Too hard on the trigger. Windage.
	Which cannot be corrected.	The recoil. Vibrations of the barrel, (spring of barrel.)
(2.) From the charge powder.	Not exact measure. Form of grain and variable quality of powder. Its deterioration from dampness in transportation, &c. More or less ramming. Sticking along the bore, from becoming foul and damp. Getting foul or dirty.	

(3.) From the ball.	Not being of the exact weight and calibre. More or less deformed in loading, or on leaving the barrel. Not having the centre of gravity in the centre of the figure, (spherical ball.)
(4.) From the atmosphere.	The effect of wind. The temperature; moisture in, and density of the air. The position of the sun. Difference of level between the target and gun.

For the same kind of arm, the dimensions, charges, weights, projectile, &c., being constant, the point-blank may be considered as constant, and serves as a point of reference in firing at different distances.

With a piece having a point-blank, that is, any piece having an angle in front, made by the line of sight and the line of fire, it is necessary, in firing at a point-blank object, to aim directly at the object. If the object be situated within the point-blank range, it will be necessary to aim below. If the object be situated beyond the point-blank, we must aim above the object.

As the end of the gun obstructs the view of the object, in aiming above the point to be reached, and, moreover, as it is difficult to determine at a certain distance the elevation that ought to be given to the line of sight, a *hausse* or *tangent scale* is placed upon the breech of the cannon, which, by enlarging its diameter, increases the angle of sight and consequently the point-blank range. The tangent scale is now generally used with guns and howitzers, and the hausse, or rear sight, has also been attached to small arms of 1855. In addition to the tangent or hausse some simple instrument may be used for determining distances. (*See* STADIA.)

Fired under angles of $4° 15'$, $4° 30'$, and $4° 50'$, the new rifle musket, altered rifle, and altered musket have, respectively, a range of 1,000 yards. (*See* HAUSSE.) The elongated musket balls do not cease to ricochet on level ground at a distance of 1,000 yards. A strong wind, blowing perpendicularly to the direction of the rifle-musket ball, will deflect it from its course 12 feet in 1,000 yards; about 3 feet in 500 yards, and $1\frac{1}{2}$ feet in 200 yards. The effect of wind on the pistol-carbine balls is somewhat greater for the same distance.

When two oblong balls are fired from the new rifle musket or altered rifle, with the ordinary service charge of 60 grains, they separate from each other and from the plane of fire about 4 feet in a distance of 200 yards. If the piece be held firmly against the shoulder, no serious inconvenience will be felt; but for the two balls it is necessary, in aiming, to give the barrel greater elevation in the proportion of 6 feet for 200 yards. In cases of emergency, two balls might be employed against masses of infantry or cavalry, at distances not exceeding 300 yards. The angle of maximum range for the mortar is nearly 42°. The *angle of fall* is the angle made by the last element of the trajectory with the ground, and when this angle is small, the projectile rebounds upon the earth and performs a series of *ricochets*, increasing in number as the angle of incidence diminishes, or as the ground is firm and elastic.

The point-blank ranges of siege and garrison guns, with ordinary charges, are respectively eight hundred yards for the 24-pounder, seven hundred and seventy-five yards for the 18-pounder, and seven hundred yards for the 12-pounder. For field-artillery, the point-blank ranges are seven hundred and fifty yards for the 12-pounder, and six hundred and seventy-five yards for the 6-pounder.

The point-blank is increased or diminished by the hausse or tangent scale, and is then called the artificial point-blank. The practical rule in aiming field-guns by means of the tangent is: give one-twelfth of an inch on the instrument for each twenty-five yards beyond point-blank.

The direct fire is employed in breaching parapets or walls, against troops in column, and in most cases where the object of attack is possessed of considerable depth or thickness.

The enfilade fire, with heavy ordnance, full charges and solid shot, is especially effective in those circumstances which admit of its adoption; a single shot having been known to disable several guns, or to strike down a whole rank of men.

Enfilade fire à ricochet is generally employed to dismount guns on parapets, protected by traverses, at ranges varying from 400 to 600 yards.

The ricochet and vertical fires, being intended to act upon a surface, and not an isolated point, may be executed during the night, as well as by daylight. (*See* TARGET. Consult THIROUX; KINGSBURY'S *Artillery and Infantry; Reports of Experiments by Ordnance Department*, U. S. A., 1856; HYDE'S *Fortification*.)

FLAG. The flag of the United States shall be thirteen horizontal stripes, alternate red and white. The Union shall be a number of white stars in a blue field, corresponding with the number of States in the Union. Upon the admission of a State to the Union, another star is added to the flag on the 4th of July next succeeding her admission; (*Act* April 4, 1818.)

All flags captured from an enemy to be displayed in such public place as the President may deem proper; (*Act* April 8, 1814.)

FLAGS OF TRUCE—are frequently sent by an enemy with the design of gaining information. To prevent this, it is usual for outposts to halt the flag of truce, and if he is merely the bearer of a letter, receipt for it, and order the party to depart, preventing all conversation with sentries. It may sometimes, however, be necessary to send the bearer of the flag to head-quarters. in this case, his eyes are bandaged, and he is forwarded with an escort.

Flags of truce are used when an enemy is in position, on a march or in action. The flag ought always to be preceded by a trumpeter 25 paces in advance, and when within range of the guns of the sentinels or videttes, he halts, returns his sword to its scabbard, and at the same moment raises and flourishes a white flag or handkerchief. If he is not signalled to retire, he continues to advance step by step until ordered to halt. If he remarks that it is sought to draw him into a snare, he retires at a gallop with his trumpet as soon as he is *certain* of the bad intention. When consent is given to receive him, he submits to all measures that may be exacted of him for the fulfilment of his mission.

If it is during an action that a flag proceeds from the ranks of the enemy, the ranks that he leaves halt and cease their fire. He proceeds towards the chief of the adverse force, and at a suitable distance returns his sabre to its scabbard, and raises his flag. If he is not signalled to retire, and if the fire ceases in his front, he continues to advance and executes his orders. Some serious motive is indispensable for sending a flag during an action, for the enemy is apt to believe that it is a stratagem, and therefore fires upon the flag, and follows up his aim more vigorously, while the opposite party have lost time.

FLANK. The right or left side of a body of men, or place. Flank presupposes a formation more or less deep. A flank march is upon the prolongation of the line to which a body faces. Thus, when we say the enemy, by a flank march, outflanked our right wing, it is understood that the enemy, by marching parallel to our line of battle, put himself in position upon our extreme right.

To disturb the flanks of a column or army is to throw an opposing force upon either side of the route that it follows. By this manœuvre the march of the column is retarded, or it is forced to halt; its baggage is sometimes seized, and terror and disorder fall upon the masses.

Flank (To)—is to cover and defend the flanks. We flank a camp by posts placed on the right and left; a corps d'armée is flanked by detachments which take roads parallel to the routes followed by the larger body; smaller columns are flanked by flankers on the right and left, who keep in view the columns, warn them of the approach of an enemy, discover ambuscades, skirmish with them, and fall back when needed upon the mass of the troops.

FLANK OF A BASTION—is that side which connects the face and curtain. It is one of the principal defences of the place, as it protects the curtain, the face, and flank of the opposite bastion, and the passage of the ditch.

FLÈCHE—is a simple species of field-work. It consists of two faces forming a salient angle. One simple rule for their construction is to select a spot for the salient and throw up a breastwork on either side, forming an angle of not less than 60°, and allowing one yard for each file.

FOOT—in a military sense, implies infantry soldiers.

FORAGE. The hay, corn, fodder, and oats required for the subsistence of the horses of an army. Generals, field-officers, cavalry-officers, and staff-officers receive a commutation in lieu of forage for each horse allowed by law, owned, and kept in service. (*See* PAY.) The maximum ration of forage is fourteen pounds of hay or fodder and twelve pounds of oats, corn, or barley. The established forage ration is furnished by the quartermaster's department. The food of horses however, like that of men, must be modified according to circumstances, by changing established proportions or by substituting one article of food for another. A knowledge of the different descriptions of food capable of maintaining a horse in working condition is essential. Forage in garrison or established quarters is ordinarily obtained under contract; but in the field the resources of the country occupied must be made immediately available. War deranges the proportions commonly maintained between demand and supply, and cripples agricultural industry. It is for the military administrator to counteract as far as possible this tendency, and not alone to seize upon all the resources of supply, but to render them continuously productive. Under the very best arrangements, however, few countries when they become the theatre of contending armies can long support the drain upon them, and afford sufficient sustenance for the immense number of animals which accompany an army, and a partial supply must under the most favorable circumstances be drawn from without. While the army is acting in the immediate vicinity of the sea-board there is little difficulty in maintaining this supply, but when it advances inland, and the means of water transport fail, it becomes a matter of extreme difficulty to provide the requisite transport for so bulky an article as forage. The artillery can render some assistance in this respect, and should be required to carry in their wagons at least three days' supply, but the cavalry soldier cannot always encumber himself with his forage ration, and at best can only be expected to carry three days' allowance of oats or barley, relying upon the supply department for his hay. Although hay has been packed by hydraulic pressure, the necessity of a further reduction of bulk, both as a question of economy and of convenience, has always been apparent. This consideration, and representations of the waste incurred at the seat of war in the unloading of grain, and its transport to the front, led Mr. Julyan, asst.-com.-gen., B.A., to apply his inventive mind to the manufacture of what is now known as the "Amalgamated Field-forage." This consisted of a preparation of chopped hay, bruised oats, bran, &c., in the proportions usually issued to cavalry horses, thoroughly mixed together, subjected to a chemical process for the expulsion of fixed air, and compressed by hydraulic power into thick cakes of great solidity. It was cut up into rations of 22 lbs. each, and four of such pieces were packed in one canvas cover, which was convertible into a nose-bag. From these bags the horses were to have been fed, the forage being restored to its original bulk and condition by moderate friction and a few minutes' exposure to the air. This preparation thus combined the advantages of extreme portability, full nutritious property, cheapness, and (from its being almost impervious to air and fire, as well as from its peculiar form) exemption from the accidents, deterioration, and losses to which forage in its ordinary state is subject.

FORAGE MASTER. (*See* WAGON MASTER.)

FORAGING—is properly the collection of forage or other supplies systematically in towns or villages, or going with an escort to cut nourishment for horses in the fields. Such operations frequently lead to engagements with the enemy. Foraging parties are furnished with reaping hooks and cords. The men promptly dismount, make bundles with which they load their horses, and are prepared for any thing that may follow. The word *foraging* is sometimes inaccurately used for marauding. When foraging is effected in villages, it is best not to take the party into the village, but to send for the chief persons and stipulate with them that the inhabitants shall bring the required forage and other stores out to the troops. If the inhabitants do not promptly comply with this moderate command, it is necessary to take the troops into the village. In this event, all possible means must be taken to prevent disorder, as for instance:

1. A certain number of houses are assigned to each company, so that the commander of the detachment may hold each company responsible for the disorders committed within its limits.

2. Guards are posted and patrols sent out, who arrest any foragers guilty of disorder.

3. If the form of the village permits, a part of the detachment remains at the centre to pack the horses and load the wagons as fast as the other men bring the forage from the houses.

In places where an attack may be expected, the foraging is conducted as follows: Either fatigue parties are sent with wagons, or parties of cavalry with their own horses; in both cases a special escort is added for the protection of the foragers. In all cases, the strength of the escort depends upon the degree of danger, the space over which the foraging is to extend, and the distance from the enemy. During the march of foragers to and from the foraging ground, if they consist of a fatigue party with wagons, an escort is added, which acts in conformity with the rules for escorting convoys. If the foragers consist only of cavalry with their own horses, then on the outward march they move in one body, observing the precautions prescribed for movements near the enemy; on the return march, if the horses of the foragers are packed and led, the detachment acting as escort should not pack more than 40 pounds on their horses, so that the load may not prevent them from acting against the enemy. One hundred and twelve pounds may be packed on a horse, and the horse must be led; 56 pounds are packed in two trusses. Sometimes the escort, or a part of it, may be sent out early to the foraging ground, to take measures for the security of the foragers before they arrive. For the safety of the foragers when at their work, the escort is divided into two or three parts, according to circumstances; one part places a chain of outposts and sends out patrols, to guard the whole ground; another furnishes the supports of the outposts, and if there are infantry or mounted rifles with it they occupy the points which cover the approaches; the third part is placed in reserve near the centre of the ground, that it may easily reach any point attacked. If the enemy attacks while the foraging is going on, the escort should go to meet him or defend itself in position, endeavoring to stop him until the foragers have finished their work, and are drawn out on the road for their return march; then the escort commences its retreat, acting as a rear guard, and endeavoring to keep the enemy as far from the foragers as possible. If it is impossible to hold the enemy in check long enough to finish the work, they should at least send forward and protect all the foragers who have packed their horses or loaded their wagons; the rest join the escort. If there is a probability of driving off the enemy by uniting all the foragers to the escort, it is best to abandon the forage already packed, and to begin foraging anew after having repulsed the enemy. It is permitted to abandon the forage entirely only in extreme urgency, when there is absolutely no other way of saving the foragers. If the enemy is repulsed, we must not be induced to pursue him except far enough to prevent a renewal of the attack, but must endeavor to complete the foraging. The foraging must not be extended over any ground not guarded by the escort. If the escort is too weak to cover the whole space designated for foraging, the ground is divided into parts, and the foraging effected in the different portions successively. If the foraging ground is at a considerable distance from the camp, it will be a proper precaution to post a special detachment in support half way. Foraging in places occupied by the enemy is undertaken only upon the entire exhaustion of the ground occupied by our own troops. Such foraging is covered by offensive operations, so that, having driven in the enemy's advanced troops or other parties, we may rapidly seize all the supplies to be found in the vicinity. This is called *forced foraging*. The strength and composition of a detachment for forced foraging must be such that it can overwhelm the enemy's troops, and remain long enough in position to enable the accompanying detachment of foragers to complete their work and retreat out of danger. The main conditions of success in such an enterprise are suddenness, rapidity, and determination in the attack, promptness in the work of the foragers, and tenacity in holding the position taken from the enemy as long as necessary. Success will be greatly facilitated by partial attacks made upon different points of the enemy's position while the foraging is going on. Attacks upon foragers should be sudden and rapid, in order, by not giving the escort time to defend the points attacked, to produce confusion among the foragers and thus prevent them from working. The approach of the attacking party should be concealed, rapid, and compact; that is, it should not send out parties to any great distance in front or on the flanks, and, as a general rule, should not divide its force prematurely, but only the moment before the attack. The force of a detachment sent to attack foragers depends chiefly upon the object of the attack—that is, whether it is designed to capture the foragers, or only to prevent them from foraging by alarming them, or to prevent them from carrying off forage already packed. It is in all cases advantageous to begin with several simultaneous false attacks by small parties, to perplex the enemy and oblige him to divide the escort; then to direct the main party of the detachment upon the principal point of the enemy's arrangements, overthrow his weakened escort, and penetrate to the road of retreat, so as either to cut off and destroy a part of the escort and foragers, or to force them to abandon their work and fly, by threatening to cut them off. If from the disproportion of force it is impossible to prevent the foraging entirely, the attacking party confines itself to delaying the work; its operations, therefore, should consist in partial attacks upon several points, in order to alarm and disperse the foragers by breaking through the outposts at several points. Upon meeting a considerable force of the enemy these attacking parties should at once retreat, and renew the attack in a different place. In such operations a portion of the attacking detachment should be kept together and held in reserve, as a support and rallying point for the small parties. If they do not succeed in preventing the foraging, they may try to attack the foragers on the return march; observing in this case the rules laid down for attacks upon convoys; (McCLELLAN's *Military Commission to Europe.*)

FORCE. Any body of troops.

FORDS. In examining and reporting upon a ford, the main points to be considered are: the firmness and regularity of the bottom, its length, width, and direction; the depth, (and its increase by tides or floods,) the rapidity of the current, the facilities of access, security from attack, and the means of rendering it impassable: a ford should always be tried personally before making a report on its capabilities. The *depth* of fords for cavalry should not be more than 4 feet 4 inches, and for infantry 3 feet 3 inches; but if the stream is not very rapid, and the direction of the crossing is down-stream, the latter may pass by holding on to the horses, even if the depth is four feet. Should the stream be very rapid, however, depths much less than these could not be considered fordable, particularly if the bottom is uneven. Carriages with wheels 5 feet in diameter may cross a ford 4 feet deep; but if it is necessary to keep their contents dry, the depth should not be more than 2, or at most $2\frac{1}{2}$ feet. Fords are generally to be found above or below a bend, and often lie in lines diagonally across the river; small gravel forms the best bottom; and rock, on the contrary, the most dangerous, unless perfectly regular and not slippery. They may be sounded by means of a boat having a pole attached. But cavalry or good swimmers may effect it with lances or poles, carefully feeling their way before advancing. Parts which may be too deep, or even the whole width, if the river is narrow, may be rendered fordable by throwing in fascines parallel to the direction of the current, and loading them with stones, which must afterwards be covered with smaller material to render the surface level. The approaches should also be levelled, and where the soil is soft, rendered firm by covering them with fascines, &c.,

so that the troops may advance with a broad front, and rapidly mount the further bank. The extent and direction of the ford should be clearly marked out by means of poles firmly fixed, and these may be notched, so that a dangerous rise in the river may be observed. If the current is rapid, a number of these placed along the upper edge of the ford, and connected by ropes, will also be useful to prevent men on foot being swept away; and boats and horsemen should also be in readiness to rescue them. The force of the current may be broken by the cavalry crossing a little above them; but if the bottom is sandy, the cavalry should cross after the infantry and artillery, as the passage of the former deepens a ford sometimes very materially. The opening and shutting of the mill-sluices will sometimes alter the depth of fords, and floods may even entirely destroy them; they can be rendered impracticable by means of large stones, harrows, planks with spikes, sharp stakes driven in so as to be concealed by the water, abatis, &c., or by cutting trenches across; (*Aide Memoire.*)

FORGE. One travelling forge and one battery wagon accompany each field-battery. They are furnished with the tools and materials required for shoeing horses and for the ordinary repair and preservation of carriages and harness. The total weight of the forge when loaded is 3,383 lbs., that of the battery wagon loaded is 3,574 lbs.

FORLORN HOPE. Officers and soldiers who generally volunteer for enterprises of great danger, such as leading the attack when storming a fortress.

FORT—is an inclosed work of the higher class of field-works. The word, however, is loosely applied to other military works.

FORTIFICATION. A fortification in its most simple form consists of a mound of earth, termed the *rampart*, which encloses the space fortified; a *parapet*, surmounting the rampart and covering the men and guns from the enemy's projectiles; a *scarp wall*, which sustains the pressure of the earth of the rampart and parapet, and presents an obstacle to an assault by storm; a wide and deep *ditch*, which prevents the enemy from approaching near the body of the place; a *counterscarp wall*, which sustains the earth on the exterior of the ditch; a *covered way*, which occupies the space between the counterscarp and a mound of earth, called a *glacis*, thrown up a few yards in front of the ditch for the purpose of covering the scarp of the main work. The work by which the space fortified is immediately enveloped is called the *enceinte*, or *body of the place*. Other works are usually added to the *enceinte* to strengthen the weak points of the fortification, or to lengthen the siege by forcing the enemy to gain possession of them before he can breach the body of the place. These are termed *outworks*, when enveloped by the covered way, and *advanced works*, when placed exterior to the covered way, but in some manner connected with the main work; but if entirely beyond the glacis and not within supporting distance of the fortress, they are called *detached works*. In a bastioned front the principal outwork is the *demi-lune*, which is placed in front of the curtain; it serves to cover the main entrance to the work, and to place the adjacent bastions in strong re-enterings. The *tenaille* is a small low work placed in the ditch, to cover the scarp wall of the curtain and flanks from the fire of the besiegers' batteries erected along the crest of the glacis.

The *places of arms* are points where troops are assembled in order to act on the exterior of the work. The *re-entering places of arms*, are small redans arranged at the points of juncture of the covered ways of the bastion and demi-lune. The *salient places of arms*, are the parts of the covered way in front of the salients of the bastion and demi-lune. Small permanent works, termed *redoubts*, are placed within the demi-lune and re-entering places of arms for strengthening those works. Works of this character constructed within the bastion, are termed *interior retrenchments;* when sufficiently elevated to command the exterior ground, they are called *cavaliers*.

Caponnieres are works constructed to cover the passage of the ditch from the tenaille to the gorge of the demi-lune, and also from the demi-lune to the covered way, by which communication may be maintained between the enceinte and outworks. *Posterns* are underground communications made through the body of the place or some of the outworks. *Sortie passages* are narrow openings made through the crest of the glacis, which usually rise in the form of a ramp from the covered way, by means of which communication may be kept up with the exterior. These passages are so arranged that they cannot be swept by the fire of the enemy. The other communications above ground are called ramps, stairs, &c. *Traverses* are small works erected on the covered way to intercept the fire of the besiegers' batteries. *Scarp and counterscarp* galleries are sometimes constructed for the defence of the ditch. They are arranged with loopholes, through which the troops of the garrison fire on the besiegers when they have entered the ditch, without being themselves exposed to the batteries of the enemy.

In seacoast defences, and sometimes in a land front for the defence of the ditch, embrasures are made in the scarp wall for the fire of artillery; the whole being protected from shells by a bomb-proof covering overhead; this arrangement is termed a *casemate*. Sometimes double ramparts and parapets are formed, so that the interior one shall fire over the more advanced: the latter in this case is called a *fausse braie*. If the inner work be separated from the other, it is called a *retrenchment*; and if it has a commanding fire, a *cavalier*. The *capital* of a bastion is a line bisecting its salient angle. All works comprehended between the capitals of two adjacent bastions, are called a *front*.

In the Prussian system of fortification, the defence of the ditch being provided for by casemated caponnieres, the necessity for breaking up the outline of the enceinte into a succession of salient and re-entering angles, as in the bastion tracings, is altogether removed. The enceinte may, therefore, have that outline which in the particular case is most advantageous for defence, and best adapted to the natural features of the position. This will generally be a polygon, more or less regular, according to the regularity or irregularity of the site. The caponnieres for the defence of the main ditch may either be on the centre of the front, or at the alternate salient angles; the latter, as being more secure from an enemy's distant fire, appears the better position. The length of the exterior side may be of almost any magnitude, though 600 yards are, perhaps, as great as under any ordinary circumstances would be requisite. The enceinte is a massive rampart and parapet, fronted by a revetment, from 24 to 30 feet in height, which is sometimes wholly or partially loopholed for musketry. The centre of the ditch is occupied by the casemated caponniere, a massive work of masonry, capable of containing two stages of five guns each, one on either face; so that the ditch on either side of the caponniere is swept by the fire of ten guns.

The advocates for the Prussian system claim for it the following advantages: 1st. When the range of musketry is given up as the standard length of a line of defence, and that of artillery substituted for it, the exterior sides of the polygons of fortification may evidently be much extended. 2d. The Prussian engineers prefer the construction of casemated flanks for the defence of ditches, as being more secure than the ordinary flanks of the bastion system; that is, the guns are protected from enfilade and vertical fire from a distance, and cannot be counter-battered by direct fire, until the assailant crowns the glacis. They use caponnieres for the defence of the main ditch, and for the ditches of the ravelin. 3d. The ravelins can be made as salient as the detached ravelins of Chasseloup and Bousmard; while the caponnieres or casemated projections by which their ditches are defended, protect the body of the place from the breaching batteries of the enemy on the counterscarp, at the salient angles of the ravelins. These ravelins are more under the fire of the enceinte, than detached ravelins; they contain a greater interior space; there is a saving of masonry at the gorge: and fewer troops secure the work from assault. 4th. In the attack of these fronts, the approaches are opposed on the capital of the ravelin, by three mortars in casemates under the parapet, cutting off the salient of the ravelin, and by guns on the terre-plein above. The glacis is protected on each side, by the fire of 90 yards of the enceinte, and from 80 yards of the faces of the ravelin, which (being covered by the advanced portions of greater elevation) is very difficult to enfilade. 5th. The establishment of batteries on the counterscarp of the salient angle of the ravelin, is rendered very difficult by countermines, and by a double tier of fire along the whole width of the ditch, viz., from the caponniere and from the enceinte behind it; even supposing this caponniere to be silenced, its massive ruins would prevent a serious breach being made in the enceinte. 6th. The attempts of an enemy to lodge himself on the advanced part of the ravelin are opposed by countermines, prepared in the work during its construction, and by the retrenchment behind: moreover, any endeavor to establish a battery in the narrow part of the angle, would be opposed by the fire of the whole enceinte behind the ravelin;—by that of the casemated keep;—and by sorties having their flanks fully protected. 7th. The permanent possession of the ravelin can only be obtained after the destruction of the keep, (which commands every part of the interior, and is not seen from the exterior;) and until this is accomplished the enemy cannot make his approaches on the glacis, for the purpose of constructing his breaching batteries against the enceinte; or he would be taken both in flank and in reverse. 8th. The great caponniere flanking the ditch of the enceinte is independent of the keep of the ravelin, (which, after being taken, would be open to the fire of the enceinte and its detached escarp;) while its double tier of guns, sweeping the whole width of the ditch, can only be opposed by batteries directly in front. The establishment of these batteries, and of others for breaching the escarp at the salient, would, of course, require the capture of two ravelins, between which the approaches would be sheltered from the collateral works; but the ground would be diminished in extent on advancing near the place, and consequently expose the troops (concentrated in larger numbers) to a more destructive fire. 9th. From the great projection of the ravelin, and the obtuseness of the angles of the polygon, the effects of ricochet on the enceinte are prevented in an octagon, as the prolongations of the sides of the polygon, or the enceinte, are intercepted by the ravelins; which ravelins might (in cases where the ground is favorable) be made to project still further, so as to cover the ditch from enfilade by distant batteries, and thus secure the great caponnieres from annoyance. 10th. The salient angles of the enceinte may also be retrenched by a detached loopholed wall, which would bring a great extent of fire on the breach. 11th. The Prussians consider that, by these arrangements, they obtain much superiority over the ordinary bastion systems, including those of Bousmard and Chasseloup de Laubat. That greater means of resistance are obtained at a comparatively small expense, which means might be increased when required, by cavaliers, by interior retrenchments, and by a covered way, with redoubts. 12th. The armament required would be comparatively small, as in the flanks or caponnieres, which completely enfilade the main ditches at a short range, a few pieces only would be necessary to prevent a coup-de-main, while a full supply to resist a serious attack might be brought by easy and secure communications. A few guns placed on the salients of the ravelins would be sufficient to keep off an enemy until he had broken ground; while the whole disposable guns of the place might easily be brought upon the enceinte on that side, and the second part of the collateral ravelins. 13th. The fatigue attending the usual arrangements would also be greatly diminished by the easiness and security of the communications. The garrison need not be numerous, as they are not required to expose themselves in outworks beyond the main ditch; they are protected by casemates in the flank defences, which are sufficiently strong to allow of their concentrating nearly the whole force on the points of importance, and which, being concealed from the enemy, do not give known points to his vertical fire.

Fort Alexander, which crowns a height commanding the town of Coblentz, (Fig. 131,) is a beautiful specimen of the German system. The position around Coblentz occupies the four opposite angles, made by the Moselle and the Lahn, which rivers empty themselves into the Rhine, nearly opposite to each other; for the Lahn runs into the Rhine about a league above Coblentz. The general form of the ground is very favorable for the offensive or defensive operations of an army in possession of it, and its fortresses; and many of the high roads from the most important towns in Germany pass in this direction; whilst the country is so difficult of access, that it is next to impossible to avoid the main road. Coblentz is situated in the angle formed by the junction of the Moselle with the Rhine. It extends about three-fourths of a mile in each direction. The enceinte of the town is secure against a coup-de-main. Its rampart forms a succession of salient and re-entering angles, which being obtuse are little liable to enfilade; while the ditches are flanked by good casemated batteries, having three guns in each flank. The gateways are strong casemated barracks, containing

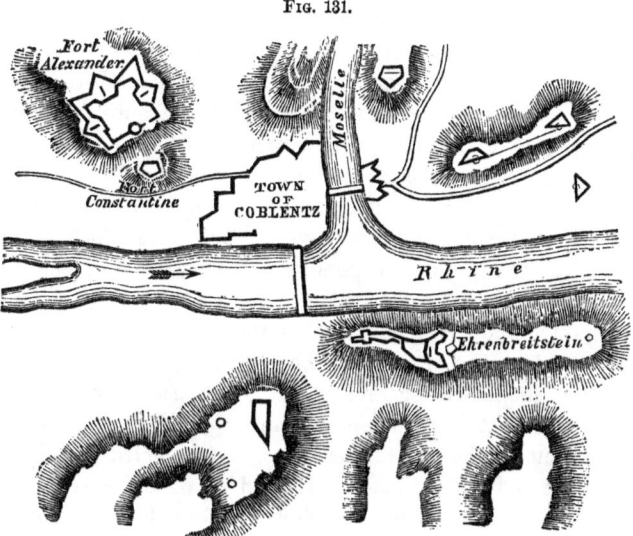

Fig. 131.

batteries to flank the ditches and approaches. These casemates are separated from the ramparts on each side, and form a kind of citadel: the profile of the rampart is nearly similar to Carnot's: the wall is well covered. Should the neighboring works on the heights be reduced, the town would be commanded and exposed to an enemy's fire. It is, however, no easy matter for an enemy to get possession of these commanding sites. The two most important of these are, Ehrenbreitstein on the right bank, and Fort Alexander on the left bank, of the Rhine.

Ehrenbreitstein occupies a commanding rocky site, 400 feet above the river, inaccessible on three sides, and on the approachable side from the north, it is defended by strong double works; having abundant casemates for its garrison, stores, and artillery. It is the key of the whole position, commanding all the surrounding works within its range, and having smaller works detached from it, for looking into hollows, that cannot be seen from the main works. It has a fine well, 300 feet deep. The faces of the works defending the only approachable side, can mount forty-three pieces of ordnance in casemates; the ditches are well defended by casemated batteries; and the escarps are about 35 feet in height. It is altogether a most formidable work. The piers that separate the casemates and support the arches are made to project right through to the front of the revetment, which is 10 feet thick: and the courses, instead of being horizontal, are laid in successive arches, the joints forming rays from a centre. The whole is built of rough stone, and grouted in, so as to settle in time into a solid mass.

Fort Alexander with its dependencies, commands all the approaches to Coblentz between the rivers. The principal front of this work has its exterior side about 650 yards, and its interior side about 500 yards in length. The ravelins and the counterguards have their faces directed so, that their prolongations do not fall upon the plateau in front, but upon the hollows and ravines, &c., from which they cannot be enfiladed. The flanking caponniere is very strong, being a casemated work for two tiers of guns; each flank has five guns in the lower tier for flanking the ditch, and five in the upper tier for flanking the terre-pleins of the counterguards. The casemates in the faces or angular parts are loopholed for musketry. Each caponniere serves as a good barrack for 160 men, besides stores. This work is completely covered in front by the counterguard or ravelin, which is only two feet lower than the body of the place. Each flank of the enceinte contains six casemates for guns to flank the ditches before them. The faces and ditches of the ravelins are flanked by solid casemated caponnieres, which cover the body of the place from any batteries that might be established at the rounding of the counterscarp of the ravelin. The ditches of the counterguards are flanked by casemated batteries, placed in the faces of the ravelins. The body of the work is an oblique parallelogram, about 5° from a right angle: the side fronts are about 420 yards, and the rear front 500 yards in length, in order to suit the ground. There is a strong casemated tower at the gorge connected with a communication from Fort Constantine. There is no covered way; the counterguards answer the purpose. Good ramps and other arrangements are made in the countersloping glacis and its salients, favorable for sorties. It is calculated that 5,000 men would be sufficient to man all these works on both sides of the river; while it is evident that a vast army might be securely cantoned within the circuit of the works. A great number of trees have been planted all around Fort Alexander; the roots of which, left in the ground, would defy the ordinary work of sappers and miners; and would therefore prove formidable obstacles in the process of a regular attack, while the timber would be invaluable in a siege; (HYDE's *Fortification*.)

FORTIFICATION (FRONT OF)—consists of all the works constructed upon any one side of a regular polygon, whether placed within or without the exterior side; or, according to St. Paul, all the works contained between any two of the oblique radii. Some authors give a more limited sense to the term "front of fortification," by confining it to two half bastions joined by a curtain. If the polygon be regular, that is, if all the sides be of equal length, and the fronts of the same description, it is called a regular work; but if they differ, it is called an irregular work.

FORTIFICATION (IRREGULAR)—is that, in which, from the nature of the ground or other causes, the several works have not their due proportion according to rule; irregularity, however, does not necessarily imply weakness.

FORTIFICATION (NATURAL)—consists of such objects formed by nature, as are capable of impeding the advance of an enemy; and a station is said to be naturally fortified, when it is situated on the top of a steep hill, or surrounded by impassable rivers, marshes, &c.

FORTIFICATION (REGULAR)—is that in which the works are constructed on a regular polygon, and which has its corresponding parts equal to each other.

FORTRESS. A fortress is a fortified city or town, or any piece of ground so strongly fortified as to be capable of resisting an attack carried on against it, according to rule.

FOUGASS. Charges of gunpowder are frequently placed at the bottom of a pit or shaft dug in the ground over which an enemy must pass to the attack. In these cases they take the name of fougasses. The chief difficulty attending the use of fougasses is to explode them at the instant when the enemy is passing over, as any variation in the time of explosion from this instant renders them altogether useless. It is, therefore, recommended to place an obstacle over them, as an abatis or chevaux-de-frize, so that the fougasses may be exploded while the enemy is occupied in forcing his way over. Sometimes a fougass is made of several loaded shells placed in a box, with a charge of powder under. The box should be pitched, to keep the charge dry. (Fig. 132.)

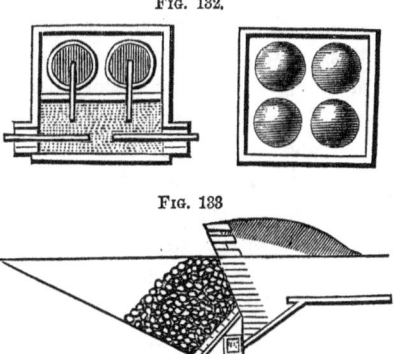

FIG. 132.

FIG. 133.

A stone fougass (Fig. 133) is made by excavating a shaft 6 feet deep, inclined to the horizon at an angle of about 45°. At the bottom place a charge of 55 lbs. (a cubic foot) of powder, then a strong shield of wood at least 6 inches thick, in front of the charge, and over the shield throw in three or four cubic yards of pebbles, of not less than half a pound weight each. A sufficient body of earth must be placed vertically, above the charge, and retained over the upper part of the shaft, near the edge, by a revetment of sods, to insure the effect taking place in the right direction. Fougasses are usually fired by means of an augot, or casing tube, containing a hose or saucisson, &c., led up the side of the pit or shaft, and then parallel to the surface of the ground, at a depth of two or three feet; or they may be fired, at the proper moment, by means of a loaded musket with its muzzle in the powder, and a wire or string fastened to the trigger.

Analogous to fougasses were the Russian powder-boxes used at Sebastopol, Fig. 134. Each consisted of a double deal box, of a capacity sufficient to contain 35 lbs. of powder, water-tight, and effectually secured from the penetration of damp; into the top of each box was inserted a vertical tin tube, connected with a horizontal tin tube at the surface of the ground.

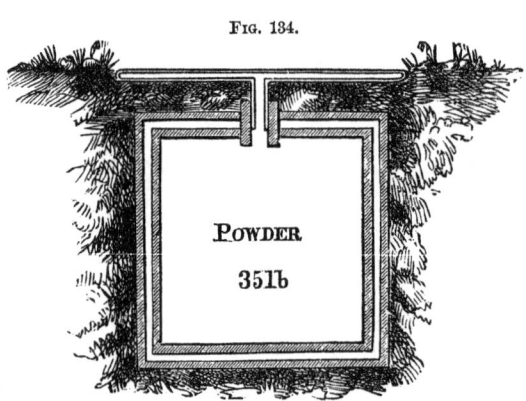

Fig. 134.

Within the latter was a glass tube, filled with sulphuric acid, and coated with a composition of chlorate of potassa, sugar, sulphur, and gum water, which immediately takes fire on coming in contact with the acid. The space between the interior of the tin tube and the exterior of the glass tube, as well as the vertical tin tube, is filled with gunpowder. A little earth spread lightly over the whole completes the arrangement. A person walking over the ground, and treading on the tin tube, crushes it, and the glass tube contained in it, causing the escape of the sulphuric acid, and the explosion of the gunpowder.

FRAISES—are palisades placed horizontally or obliquely, at the edge of a ditch on either side, or projecting from the exterior slope of a parapet. If the slope be very long, there are sometimes two rows of fraises used.

FRAUD. Association of any officer with another officer convicted by a court-martial of fraud or cowardice shall be deemed scandalous; (ART. 85.) (*See* COWARDICE.)

Fraud consists in unlawfully, designedly, and knowingly appropriating the property of another with a criminal intent. It is any trick or artifice employed by one person to induce another to fall into an error or detain him in it, so that he make an agreement in contracts contrary to his interest. The fraud may consist in the misrepresentation or in the concealment of a material fact; (BOUVIER's *Law Dictionary*.)

FRAYS. (*See* QUARRELS.)

FRICTION PRIMER FOR CANNON—consists of a tube charged with gunpowder, to the top of which is fastened a cup containing friction powder, composed of two parts of chlorate of potassa, and one of sul. of antimony, which is exploded by means of a slider pulled out with a lanyard. The *tube, cup,* and *slider* are made of sheet brass. The lanyard, for pulling off the primer, is a piece of strong cod line (about .2 in. thick) 12 feet long; to one end is attached a small *iron hook*, with an eye for the line, and to the other end a wooden toggel, .75 in. diameter, and 4 in. long. If injured by moisture, the primers become serviceable again when dried, and they have the great advantage of portability and certainty of fire.

FRONTIER. (*See* DEFENCE, NATIONAL.)

FUMIGATION. To correct and purify an infectious or confined atmosphere, such as is often found in transports, fumigations are necessary. The materials recommended for the purpose are brimstone with sawdust; or nitre with vitriolic acid; or common salt with the same acid. One fluid ounce of sulphuric acid mixed with two fluid ounces of water, and then poured over four ounces of common salt, and one ounce of oxide of manganese in powder, these latter ingredients being previously placed in hot sand, are also recommended. Burning charcoal is also a good disinfectant. (*See* SANITARY PRECAUTIONS.)

FUNERALS. Army Regulations prescribe the honors to be paid at funerals.

FURLOUGHS. The term is usually applied to the absence with leave of non-commissioned officers and soldiers. (*See* ABSENCE WITH LEAVE.)

FUZE—is the means used to ignite the bursting charge of shells. They are classified as *Time, Concussion,* and *Percussion* Fuzes. The time fuze is composed of a case of paper, wood, or metal, inclosing a burning composition. It is cut or bored to a length proportioned to the intended range of the shell, so that it shall burn down and explode the bursting charge, just as the shell strikes the ground, or earlier if desirable. Instead of driving the fuze composition into a wooden tube as formerly, and requiring a saw to give the fuze its proper length according to range, the shell is now supplied with a plug of hard wood or metal, having a hole reaped out exactly the size of a paper case containing the composition. By varying this composition, the same length suffices for all the ranges or times of burning required. And these having the different compositions in paper cases of as many different colors, the cannoneer at a field-piece may, in an instant, insert into the plug the colored fuze required for the desired range. Similar fuzes have been adopted for the columbiads, the plugs being of bronze instead of wood. Three kinds of time fuzes are employed in the United States Service, viz., the Mortar Fuze, the Borman Fuze, and the sea-coast fuze. The best and simplest form of the percussion fuze is the ordinary percussion cap placed on a cone affixed to the point of the projectile. The arrangement should be protected by a safety cap to prevent the percussion cap taking fire by the discharge of the piece.

"Bickford's fuze" is a small tube of gunpowder, sewed round with tarred twine, and then pitched over. It is not injured by damp, and when well made, will burn under water, and is used for firing the charges of mines, &c. The Gomez Patent Electric Safety train or fuze is made in the form of a tape, inclosing a chemical compound that burns at the rate of one mile in four seconds; it may be used like the Bickford fuze. (*See* RIFLED ORDNANCE.)

G

GABIONNADE. A work constructed with gabions.

GABIONS—are cylindrical baskets of various dimensions, open at both ends, used to revet the interior slopes of batteries, the cheeks of embrasures, and to form the parapet of trenches. (*See* REVETMENT for the construction of gabions.)

GALLERY. In permanent fortification, a passage or communication to that part of a mine where the powder is lodged. The principal gallery, from which others originate, is constructed under the banquette of the covered way, and follows that portion of the works throughout its whole extent. Another gallery is formed in a direction parallel to the first at 50 or 60 yards' distance, and communicates with the first by means of other galleries perpendicular to it. Galleries are lined with masonry. When finished they are about six feet high and four and a half feet wide.

GARRISON—designates the troops employed in a strong place for its security, and it is also applied to the place itself when occupied by troops. The President may employ such troops of the United States as he may judge necessary as garrisons of fortifications; (*Act* March 20, 1794.)

GENERAL. Rank above lieutenant-general. There is no such grade in the United States army.

GENERAL OFFICERS. All officers above the rank of colonel. Any sentence of a court-martial affecting a general officer must be approved by the President. (*See* COURT-MARTIAL.)

GENOUILLÈRE. From the French *genou*, knee. It is that part of the parapet of a battery which remains above the platform and under the gun, after the opening of the embrasure.

GEOMETRY. The science which teaches the dimensions of lines, surfaces, and solids. It is a necessary introduction to fortification and mechanics. It enables us to ascertain the distances of inaccessible objects, the dimensions of a given surface, the contents of a given solid; to compute the distances and motions of the planets; to predict celestial phenomena; and to navigate a ship from any given point to another on the surface of the globe.

Geometry, besides other divisions, is divided into *ancient* and *modern*: ancient geometry being that form of demonstration and investigation which was employed by the Greeks, and of which Euclid's Elements form a well-known example; modern geometry is that in which algebra, or the differential and integral calculus, is employed. We also speak of pure geometry, practical geometry, and applied geometry. Descriptive geometry was first employed by Monge, and subsequently by other French geometers, to express that part of science which consists in the application of geometrical rules to the representation of the figures, and the various relations of the forms of bodies, according to certain conventional methods. It differs from ordinary perspective, inasmuch as the design or representation is made in such a manner that the exact distance between the different points of the body represented can always be found, and consequently all the mathematical relations resulting from the form and position of the body may be deduced from the representation.

In descriptive geometry, the situation of points in space is represented by their projections on two planes, at right angles to each other, called the *planes of projection*. It is usual to suppose one of the planes of projection to be horizontal, in which case the other is vertical; and the projections are called horizontal or vertical, according as they are on the one or the other of these planes. According to this system, any point whatever in space is represented by drawing a perpendicular from it to each of the planes of projection: the point on which the perpendicular falls is the projection of the proposed point. As contiguous points in space form a line, so the projections of those points, which are also contiguous, form a line in the same manner, which is the projection of the given line. Hence as two projections only are required for the determination of a point in space, they are also sufficient for the determination of any curve whatever, whether of single or double curvature.

The same mode of representation cannot be employed with regard to surfaces; for, as the projections of the contiguous points of a surface cover a continuous area on both planes of projection, there is nothing to indicate that any particular point on one of the planes of projection corresponds to one point more than another on the second plane, and consequently that it belongs to one point more than another in space. But if we conceive the surface which is to be represented to be covered with a system of lines succeeding one another according to a determinate law, then, by projecting these lines on each of the two planes, and marking the correspondence of the one projection with the other, the projections of all the different points of the surface will have an evident dependence on each other, and the surface will be rigorously and completely determined.

Some elementary surfaces may, however, be represented in a much more simple way. The plane, for example, is completely defined by the straight lines in which it intersects the two planes of projection. These lines are denominated the *traces* of the plane. A sphere is also completely defined by the two projections of its centre, and the great circle which limits the projections of its points. A cylinder is defined by its intersection (or trace) with one of the planes of projection, and by the two projections of one of its ends. A cone is represented by its intersection with one of the planes of projection and the two projections of its summit.

The most immediate application of descriptive geometry is the representation of bodies, of which the forms are susceptible of rigorous geometrical definition. Sculpture, architecture, painting, and all the mechanical arts, the object of which is to give to matter certain determinate forms, borrow from descriptive geometry their graphical procedures, by the aid of which all the parts of an object are faithfully represented in relief before the object itself is executed. But it was chiefly in consequence of its application to civil and military engineering, and to fortification, that this branch of geometry received a distinctive appellation, and is considered of much importance in the Polytechnic school of France, and our own Military Academy. (Consult DAVIES' *Descriptive Geometry*.)

GIN. The derrick, sheers, and gin have one common object, viz.: to find a fulcrum in space, to which the pulley, in the shape of block and tackle, is to be applied. In the derrick and sheers this is effected on one and two legs, and stability is given by guys. The gin usually consists of three long legs, two of which are joined together by cross bars, and the third, called the pry pole, elevates the gin. A pulley is supported at the top, round which a rope is passed for elevating the weight. Fig. 135 shows the manner of working the gin. There are three kinds of gins used in service: the field and siege, the garrison, and the casemate. The last two differ from each other only in height; the first differs from the others in construction and size. Either of them may be used as *derrick* or *sheers*. The garrison and casemate gins differ from the siege gin in having two braces of iron instead of three wooden cross-bars or braces, and in having the *pry* pole inserted between the legs, which are kept together by the clevis bolt. The upper pulley (generally treble) is hooked to the clevis. (For description, setting up, and mechanical manœuvres with gins, consult *Instruction in Heavy Artillery*.)

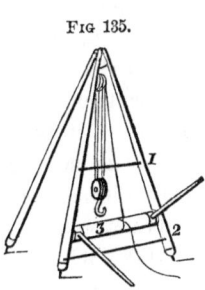

FIG 135.

GIRDER. In building, the principal beam of a floor for supporting the binding or other joists, to lessen their bearing or length.

GLACIS. The superior slope of the parapet of the covered way, extended in a gentle declivity to the surrounding country. It is seldom used in field-works. (*See* FORTIFICATION.)

GLANDERS. A virulent and dangerous disease among horses, principally shown in a mucous discharge from the nostrils. To prevent this infectious disorder from spreading, it is necessary at once to remove the horse from his stall, and thoroughly wash with soap and water the rack, manger, and every part of the stall from which the horse has been removed. When the parts are thus made clean, they must also be covered with a quick-lime wash immediately after it is mixed, and afterwards three coats of oil colors given to it. The same precautions are taken in FARCY. (*See* VETERINARY.)

GORGE. The gorge of a fortification or gorge of a work is the opening on that side of the work corresponding to the body of the place, or the side whence comes the defence. In isolated works, the gorge is sometimes intrenched. The gorges of works not attached to a fortress, but which are its dependencies, are in general open, or without parapets, in order that the enemy may not cover himself from the fire of the place if he should seize such detached works. If the works are liable to surprise, and their gorges cannot be shut, a row of palisades are planted there, and mines are prepared so as to overthrow the enemy if he should seize the work, and attempt to construct a lodgement there. The gorge of a bastion is usually an open space between the extremities of the flanks of the bastion. The larger this gorge is, the better is the defence; for when the ruined bastion is about to fall by siege into the hands of the enemy, the defenders can construct defensive works or dig small ditches in the gorge of the abandoned bastion. Such resistance sometimes drives the besiegers to the necessity of battering in breach the curtain.

GORGE OF MOUNTAINS—is the passage, more or less compressed, between two mountains which are used as a passage-way into valleys. Gorges are important military points. If they lead to an intrenched camp, it is necessary to fortify them, and post there grand guards; these positions are the principal theatres for affairs of posts. A gorge should never be entered without previous examination.

GOVERNMENT. The Constitution of the United States provides that Congress shall make rules for the government and regulation of armies. By government is understood not only the body of fundamental laws of a State, but also the body of persons charged with the management of the executive power of a country, direction, power or authority which rules a community, administration, rule, management; (WORCESTER's *Dictionary*.)

Government of the military (says BARDIN, *Dictionnaire de l'Armée de Terre*) is that branch of the code which embraces the creation and regulation of the military *hierarchy*, or the gradual distribution of inferior authority. The power of making rules of government is that of SUPREME COMMAND, and from this living principle proceeds the localization of troops, their organization and distribution; rules for rewards and punishments; and generally all rules of government and *regulation* whatsoever, which the legislature may judge necessary, to maintain an efficient and well-disciplined army.

All authority over the land forces of the United States must therefore be derived from Congress. For, although the President is the commander-in-chief, yet his functions, as such, must be regulated by Congress, under the 17th clause of Sec. 8 of the Constitution, as well as under the general authority of Congress to make rules for the government and regulation of the land forces. The President cannot be divested of power which Congress may assign to any inferior military commander, because the authority of the greater includes that of the less. But all authority over the land and naval forces save the *appointment* of the commander-in-chief rests with Congress, and no authority can be exercised not delegated by Congress, except such as may be fairly deduced from powers given for the effective discharge of the duties annexed to his office. (*See* ADMINISTRATION, and references under that head; ADJUTANT; ADJUTANT-GENERAL; AID-DE-CAMP; APPOINTING POWER; ARMY; ARMY, (*Regular;*) ARMY REGULATIONS; ARTICLES OF WAR, and references under that head; ARTILLERY; ASSIGNMENT; BOOTY; BOUNTY; BREVET; BRIGADE; BRIGADIER-GENERAL; BRIGADE-INSPECTOR; CADET; CAPTAIN; CAVALRY; COLONEL; COMMAND; COMMANDER OF THE ARMY; COMMANDER-IN-CHIEF; COMMISSARY OF SUBSISTENCE; COMMISSION; CONGRESS; CONSTITUTION; CORPORAL; CORPS; COURT-MARTIAL and references under that head; COURT OF INQUIRY; DEFENCE, (*National;*) DEPARTMENT; DETACHMENT; DISCIPLINE; DIVISION; ENGINEERS CORPS; ENGINEERS, (*Topographical;*) ESPRIT DU CORPS; FIELD OFFICERS; FLAGS; FORAGE MASTER; GARRISON; GENERAL OFFICERS; GRATUITY; GRENADIERS; HIERARCHY; INDEMNIFICATION; INDIAN; INFANTRY; JUDGE-ADVOCATE; LAW, and references under that head; LAW, (*Martial;*) LIEUTENANT; LIEUTENANT-COLONEL; LIEUTENANT-GENERAL; LINE; LOSSES; MAJOR; MAJOR-GENERAL; MARINE CORPS; MEDICAL DEPARTMENT; MILITIA; NON-COMMISSIONED OFFICERS; OATH; OBEDIENCE; OFFICERS; ORDERS; ORDNANCE DEPARTMENT; ORDNANCE SERGEANTS; ORGANIZING; PARDON; PAY; PAY DEPARTMENT; PAYMASTER-GENERAL; PENSION; PLATOON; POST; PRESIDENT; PROMOTION; PUNISHMENT; QUARTERMASTER'S DEPARTMENT; QUARTERMASTER-GENERAL; RAISE, and its references; RANK; REGIMENT; REGULATION, and its references; REMEDY; REPRIEVE; RETAINERS; RIFLEMEN; SAPPERS; SECRETARY OF WAR; SENIOR; SERGEANT; SERVICE, and its references; SOLDIER; STAFF; STATE TROOPS; STANDARDS; STORE-KEEPERS; SUBSISTENCE DEPARTMENT; SUPERIOR; SUPERINTENDENT; SUPERNUMERARY; SURGEON; SURGERY, (*Military;*) SUTLERS; TRADE; TRAIN; TRANSFERS; TRAVELLING ALLOWANCES; TREATY; UNIFORM; VETERAN; VICE-PRESIDENT; VOLUNTEERS; WAR; WARRANT.)

GRAND DIVISION. A division composed of two companies in battalion manœuvres.

GRAPE-SHOT. A certain number of cast-iron balls put together by means of two cast-iron plates, two rings, and one pin and nut. Canister has superseded the use of grape in field-guns. Grape-shot are used with the 8-in. howitzers and the columbiad of that calibre, by adopting the sabot of the sea-coast howitzer, which serves for both pieces. The grape for these 8-in. pieces is made of 6-pd. shot.

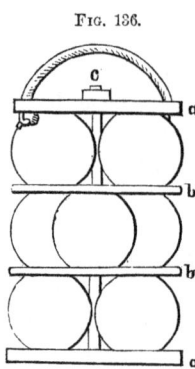

Fig. 136.

GRAPPLING-IRONS—consist of from four to six branches bent and pointed, with a ring at the root. A rope being fastened through this ring, any object at which the grappling-irons are thrown, may be dragged nearer.

GRATUITY. In the French service whenever a non-commissioned officer is promoted, he is given a gratuity, called *Gratification de Première Mise d' Officier*, in order to provide his equipment as officer. In the same manner, at the beginning of a campaign, a sum of money is given to all officers of the French army, according to grade, as an equipment fund; it is called *Gratification d'entrée en Campagne, ou Indemnité d'entrée en Campagne*.

GRAVITY, GRAVITATION. These terms are used to express the mutual tendency which all bodies have to approach each other if not opposed by other resistance.

Force of Gravity—Motion of falling bodies: Let t be the time of descent in seconds, of a body falling freely, in vacuo; h, the space described in the time t; v, the velocity acquired at the end of that time, and g the velocity acquired at the end of one second of time; then:

$$h = \tfrac{1}{2} g t^2; \qquad v = g t = \sqrt{2 g h}$$

The velocity g, which is the measure of the force of gravity, varies with the latitude of the place, and with its altitude above the level of the sea. The force of gravity at the latitude of 45° = 32.1803 feet; at any other latitude L; g = 32.1803 feet — 0.0821 cos. $2 L$. If g' represents the force of gravity at the height h above the sea, and r the radius of the earth, the force of gravity at the level of the sea will be $g = g' (1+ \dfrac{5\,h}{4\,r})$.

In the latitude of London, at the level of the sea, g = 32.191 feet.
    do.   Washington,   do.   do. g = 32.155 feet.

GRENADE. A shell thrown by hand or in baskets from stone mortars. A hand-grenade is a small shell about $2\tfrac{1}{2}$ inches in diameter, which, being set on fire by means of a short fuze and cast among the enemy's troops, causes great damage by its explosion. They may be thrown 26 yards. Rampart-grenades are larger, and are used to roll down ramparts, &c.

GRENADIERS. The right flank company of a regiment.

GRIEVANCES. (*See* WRONGS.)

GROOVES. Spiral grooves or "rifles" cut into the surface of the bore of fire-arms, have the effect of communicating a rotary motion to a projectile around an axis coincident with its flight. This motion increases the range of the projectile, and also corrects one of the causes of deviation by distributing it uniformly around the line of flight. For expanding projectiles, experiment shows that broad and shallow grooves with a moderate twist give range, endurance, accuracy of fire, and facility in loading and cleaning the bores. The United States have therefore adopted for arms three grooves, each in *width* equal to the lands, or $\frac{1}{6}$ of the circumference of the bore; and uniformly decreasing in *depth* from the breech where it is .015 in., to the muzzle, where it is .005 inch; with a uniform *twist*, one turn in six feet for long barrels or the musket, and one turn in four feet for short barrels or the carbine. The proper twist to be given to the grooves, depends on the length, diameter and initial velocity of the projectile used; but the most suitable twist can only be determined by experiment.

GUARDS—are used for security and police by troops in the field, in camps, garrisons, and quarters. Guards are designated as advance or van, and rear guards; outposts and picket guards; quarter, camp, and garrison guards; and general officers' guards. The tour of service of guards is usually twenty-four hours. Sometimes a guard is detached from a single corps, and sometimes from several corps. In either case during the tour of service, the guard receives orders from the commanding officer and officers of the guard. It is for the time detached from its corps. (The description and duties of guards are given in Army Regulations.)

GUERILLA. (*See* PARTISAN.)

GUIDES. Men employed to give intelligence respecting a country and the various roads intersecting it. All armies employed in an enemy's country find it to their advantage to use guides.

GUIDES, (TACTICAL.) The duties of guides are given in the Tactics.

GUIDONS. Each company of cavalry has a silken guidon prescribed in Army Regulations.

GUN-COTTON—is common cotton, steeped in a mixture of sulphuric acid and nitric acid, and when properly soaked, is well washed in running water, and then dried. The explosive force of three parts of gun-cotton equals that of eight parts of gunpowder. Major Mordecai's experiments at Washington in the years 1845, 1847, and 1848, to determine the fitness of gun-cotton as a substitute for gunpowder in the military service, show: 1. Explosive cotton burns at 380° Fahr., therefore it will not set fire to gunpowder when burnt in a loose state over it. 2. The projectile force of explosive cotton, with moderate charges, in a musket or cannon, is equal to that of about twice its weight of the best gunpowder. 3. When compressed by hard ramming, as in filling a fuze, it burns slowly. 4. By the absorption of moisture its force is rapidly diminished, but the force is restored by drying. 5. Its bursting effect is much greater than that of gunpowder, on which account it is well adapted for mining operations. 6. The principal residua of its combustion are water and nitrous acid; therefore the barrel of a gun would be soon corroded if not cleaned after firing. 7. In consequence of the quickness and intensity of its action when ignited, it cannot be used with safety in the present fire-arms. 8. An accident on service, such as the insertion of two charges before firing, would cause the bursting of the barrel; and it is probable that the like effect would take place with the regular service-charges if several times repeated.

GUNNERS. For the service of field and heavy ordnance, there is with each piece one man called a gunner, who gives all the executive commands in action. He is answerable that the men at the piece perform their duties correctly. (Consult *Instruction for Field and Heavy Artillery*.)

GUNNER'S CALIPERS. Made of sheet brass, with steel points. The graduations show diameters of guns, shot, &c.

GUNNER'S PERPENDICULAR. This is made of sheet brass; the lower part is cut in the form of a crescent, the points of which are made of steel; a small spirit level is fastened to one side of the plate, parallel to the line joining the points of the crescent, and a slide is fastened to the same side of the plate, perpendicular to the axis of the level. The instrument is useful in marking the points of sight on siege guns and mortars, when the platform is not level.

GUNNER'S PINCERS. Iron with steel jaws, which have on the end of one a claw for drawing nails, &c.

GUNNER'S QUADRANT, (wood.) A graduated quadrant of six inches radius, attached to a rule 23.5 inches long, (Fig. 137.) It has a *plumb-line* and *bob*, which are carried, when not in use, in a hole in the end of the rule covered by a brass plate. The quadrant is applied either by its longer branch to the face of the piece, or this branch is run into the bore parallel with the axis, and the elevating screw turned or the quoin adjusted until the required degree is indicated.

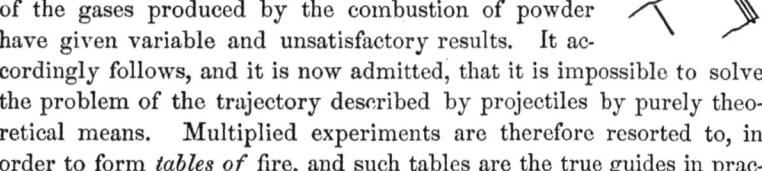

FIG. 137.

GUNNERY. Laws regulating the resistance of the air are complicated and undetermined. The attempts also made to determine the volume and tension of the gases produced by the combustion of powder have given variable and unsatisfactory results. It accordingly follows, and it is now admitted, that it is impossible to solve the problem of the trajectory described by projectiles by purely theoretical means. Multiplied experiments are therefore resorted to, in order to form *tables of* fire, and such tables are the true guides in practical gunnery.

The maximum range of the largest cannon fired under an angle of 45° does not exceed 8,000 yards: siege guns fired under smaller angles give ranges varying from 3,000 to 4,500 yards. The range of field-pieces in their ordinary fire is from 1,790 to 2,200 yards. Tables of ranges are given in Ordnance and Artillery Manuals, for the mountain howitzers, field-guns and howitzers, heavy ordnance, and Hale's war rockets. These tables give ranges at different elevations, the charges of powder, the weight of the shot, spherical case shot or shell in each case. They show the *time of flight* of the shell, and consequently the length of fuze required; and also at what angles of elevation, in the 8 or 10-in. columbiads, shot cease to ricochet upon the water. (*See*, for such tables, articles: ARTILLERY; BALLISTICS; FIRING; INITIAL VELOCITY; ORDNANCE; RIFLED ORDNANCE; ROCKETS.)

GUNPOWDER. In the United States, the proportion of ingredients for the military service are: 76 or 75 of saltpetre, 14 or 15 charcoal, and 10 of sulphur; for sporting, 78 or 77 saltpetre, 12 or 13 charcoal, and 10 sulphur. The powder is coarse or fine grained. In the United States, to every 10 grains troy weight of powder, there are 150 grains of cannon powder, 1,100 musket powder, 6,000 rifle, and 73,000 sporting. The size of the grain is tested by sieves. Musket power is now recommended for all small arms.

A new powder, invented by Capt. Rodman, Ordnance Dept., shows great ingenuity, and has given most important results. An ordinary grain of powder burns from the surface to the centre, and the largest portion of the gas is evolved in the $\frac{2}{100}$ part of a second. The force of the charge is therefore expended upon the projectile before it is sensibly moved, and there is a corresponding strain upon the gun. Capt. Rodman thought, if powder could be made to burn on an *increas-*

ing instead of a decreasing surface, so that the gas should be evolved completely but not so rapidly before the projectile left the piece, the same velocity would be communicated, and the strain would be distributed uniformly over the whole piece. To accomplish this, he formed the "dust" into a cake, and inserted into it numerous small wires, which, being pulled out, left corresponding avenues for the passage of flame and ignition of the mass; thus making the interior surface of combustion *increasing* instead of *decreasing*. The enormous pressures from large charges of powder have thus been entirely obviated by the introduction into service of Rodman's hollow caked powder, or its substitute, the large-grained powder, each grain being six-tenths of an inch. This discovery, with the idea of Capt. Rodman of cooling cast-iron cannon from the interior by means of a current of cold water flowing through a hollow core, has enabled him to cast a 15-in. columbiad which, after three hundred rounds, with a charge of 40 lbs. of powder, showed no appreciable enlargement of either bore or vent, and causes Capt. Rodman to believe that the piece will bear 1,000 rounds without material injury; (BENTON; *Experiments on Gunpowder by* MAJ. MORDECAI, *Ordnance Dept.*)

GUNS—are long cannon without chambers, having their calibres determined by the weight of their balls. (*See* CALIBRE; ORDNANCE.)

GUNTER'S CHAIN—is the chain commonly used for measuring land. It is 66 feet or 4 poles in length, and is divided into 100 links, each of which is joined to the adjacent one by three rings; and the length of each link, including the connecting rings, is 7.92 inches. The advantage of this measure consists in the facility which it affords for numerical calculations. The English acre contains 4,840 square yards; and Gunter's chain being 22 yards in length, the square of which is 484, it follows that a *square chain* is exactly the tenth part of an acre. A square chain, again, contains 10,000 square links, so that 100,000 square links are equal to an acre; consequently, the contents of a field being cast up in square links, it is only necessary to divide by 100,000, or to cut off the last five figures, to obtain the contents expressed, in acres; (BRANDE's *Encyclopedia*.)

GUY. A rope used to swing any weight, or to keep steady any heavy body, and prevent it from swinging while being hoisted or lowered.

H

HAIL. A sentinel hails any one approaching his post, with "*Who goes there?*"

HALT. A rest during a march, and a word of command in tactical manœuvres.

HAND. A measure four inches in length. The height of a horse is computed by so many hands and inches.

HANDSPIKES. The *trail* handspike for field carriages is 53 inches in length; the *manœuvring* handspike for garrison and sea-coast carriages and for gins is 66 inches; for siege and other heavy work it is made 84 inches long and 12 lbs. weight; the *shod* handspike is particularly useful in the service of mortars and of casemate and barbette carriages; the *truck* handspike for casemate carriages, (wrought iron;) the *roller* handspike, for casemate carriages. It is made of iron, 1 inch round, the point conical, whole length 34 inches.

HARBORING AN ENEMY. Punishable with death or otherwise, according to sentence of a court-martial; (Art. 56.)

HAUSSE or BREECH SIGHT—is a graduated piece attached to the barrel near the breech, which has a sliding piece retained in its place by a thumb screw, or by the spring of the slider itself. This slider should have an opening through which the gun can be conveniently aimed; and is raised to such a height as we think will give the necessary elevation for the distance. The term coarse sight means a large portion of the front sight, as seen above the bottom of the rear-sight notch; and a fine sight is when but a small portion is seen. The effect of a coarse sight is to increase the range of the projectile.

Graduation of rear-sights.—If the form of the trajectory be known, the rear-sight of a fire-arm can be graduated by calculation; the more accurate and reliable method, however, is by trial. Suppose it be required to mark the graduation for 100 yards: the slider is placed as near the position of the required mark as the judgment of the experimenter may indicate; and, with this elevation, the piece is carefully aimed, and fired, say ten times, at a target placed on level ground at a distance of 100 yards. If the assumed position of the slider be correct, the centre of impact of the ten shot-holes will coincide with the point aimed at; if it be incorrect, or the centre of impact be found below the

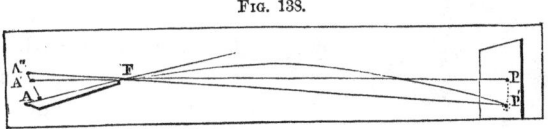

Fig. 138.

point aimed at, then the position of the slider is too low on the scale. Let P be the point aimed at, and P' the centre of impact of the cluster of shot-holes, we have, from close similarity of the triangles, $A'F : FP :: A'A'' : PP'$; from which we can determine $A'A''$ the quantity that must be added to AA', to give the correct position of the graduation mark for 100 yards. If the centre of impact had been above P, the trial mark would have been too high. Lay off the distance AA'' above A'', on the scale, and we obtain an approximate graduation for 200 yards, which should be corrected in the same way as the preceding, and so on. The distance PP' is found by taking the algebraic sum of the distances of all the shots from the point P, and dividing it by the number of shots. It will be readily seen that an approximate form of the trajectory may be obtained by drawing a series of lines through the different graduation marks of the rear-sight, and the top of the front-sight, and laying off from the front-sight, on each line, the corresponding range; (BENTON.)

HAVERSACK. Bag issued to soldiers for carrying rations.

HAY. The forage ration is fourteen pounds of hay, and twelve pounds of oats, corn, or barley. Cattle will eat many sorts of herbage when cut small, but refuse it if uncut. They will eat reeds, seaweed, leaves, &c.

FIG. 139.

To cut Chaff, (Fig. 139.)—Tie a sickle against a tree, with its blade projecting; then, standing in front of the blade, hold a handful of reeds across it with both hands, one hand on either side of the blade; pull it towards you, and the reeds will be cut through; drop the cut end, seize the bundle afresh, and repeat the process. In this way, after a little practice, chaff is cut with great ease and quickness. A broken sickle does as well as a whole one, and a knife may be used, but the curve of its edge is ill adapted for the work. (*See* FORAGE.)

HEIGHT. Elevation, as to occupy or to crown a height; the height of a soldier, &c. (*See* DISTANCES; SURVEYING.)

HELMET. Defensive armor or covering for the head used by heavy cavalry.

HIERARCHY, (MILITARY.) The essential element for the government and service of an army is a military hierarchy, or the creation of different grades of rank, to which different functions and powers are assigned, the lower in regular subordination to the next higher in the ascending scale. It should be founded on the principle that every one acts in an army under the orders of a superior, who exercises his authority only within limits established by law. This authority of the superior should be greater or less according to rank and position, and be proportioned to his responsibilities. Orders should be executed without hesitation; but responsibilities should be confined to him who gives orders in virtue of the superior authority with which he is invested; to him who takes the initiative in an order; to him who does not execute an order that he has received; and to him who usurps a command or continues illegally to exercise its functions.

The grades of the military hierarchy are: 1. The President of the United States; 2. The Lieut.-general; 3. Major-generals; 4. Brigadier-generals; 5. Colonels; 6. Lieutenant-colonels; 7. Majors; 8 Captains; 9. Lieutenants; 10. Cadets; 11. Sergeants; 12. Corporals; 13. Privates. The military hierarchy is determined and consecrated within its sphere of action by: 1. Grades of rank created by military laws; 2. By other laws regulating the exercise of rank; 3. By military insignia; 4. By military honors; and 5. By the military oath. (*See* PRESIDENT OF THE UNITED STATES, AND OTHER GRADES OF THE HIERARCHY; BREVET; COMMISSION; COMMAND; GOVERNMENT; LINE; OATH; OBEDIENCE; OFFICER; ORDERS; RANK; REGULATION.)

HIRING OF DUTY. Punishable at the discretion of a regimental court-martial; (ART. 47.)

HOLSTERS. Cases attached to the pommel of the saddle, to hold a horseman's pistols.

HONORS, (MILITARY)—have been prescribed by the orders of the President, and are paid by troops to the President and other public functionaries, to military officers according to grade, to the colors of a regiment and when two regiments meet. (Consult *Army Regulations*.)

HONORS OF WAR. This expression is used in capitulations; and the chief of a post, when compelled to surrender, always demands the honors of war in testimony of the vigor of his defence. As these terms depend on the disposition of the victorious general, their limits vary; but in some instances garrisons have been allowed to march out, with colors flying, drums beating, some field-pieces, caissons loaded, and baggage. In other cases the garrison marches out to a certain distance, and piles its arms, and is either released as prisoners upon parole, or then becomes prisoners in fact.

HOOF. (*See* HORSE.)

HORN WORK—is a work composed of two half bastions and a curtain or a front of fortification, with two long sides called branches or wings, directed upon the faces of the bastions or ravelins, so as to be defended by them. This work is placed before a bastion or ravelin, and serves to inclose any space of ground or building, which could not be brought within the enceinte.

HORSE. In selecting a horse choose one from 5 to 7 years old, (the latter age preferable,) and from 15 to 16 hands high.

The saddle horse should be free in his movements; have good sight; a full, firm chest; be surefooted; have a good disposition, with boldness and courage; more bottom than spirit, and not be too showy.

The draft horse should stand erect on his legs, be strongly built, but free in his movements; his shoulders should be large enough to give support to the collar, but not too heavy; his body full, but not too long; the sides well rounded; the limbs solid, with rather strong shanks, and feet in good condition.

To these qualities he should unite, as much as possible, the qualities of the saddle horse; should trot and gallop easily; have even gaits, and not be skittish. The most suitable horse for the pack-saddle is the one most nearly approaching the mule in his formation. He should be very strong-backed, and from 14 to 15 hands high.

Horses with very long legs, or long pasterns, should be rejected, as well as those which are poor, lank, stubborn, or vicious.

The mule is preferable to the horse in a very rough country, where its surefootedness is an important quality. There are two kinds: the mule proper, or product of the jackass and mare, which is preferable to the product of the horse and ass. The former brays, the latter neighs.

The mule may be usefully employed from its fourth year to beyond its twenty-fifth. It is usually from $13\frac{1}{2}$ to 15 hands high; is hardy, seldom sick, fears heat but little; is easy to keep; is very surefooted, and especially adapted for draught or packing.

Before choosing horses, their attitudes and habits should be examined in the stable. Leaving the stable, they should be stopped at the door in order to examine their eyes, the pupils of which should contract when struck by the light. Out of the stable, they should neither be allowed to remain quiet, nor to be worried. Care should be taken against being deceived by the effects of the whip, cries, &c. The positions of a horse, his limbs, age, and height, should be examined at different times. He should be walked about with a long rein, observing the action of his rear extremities when he moves off, of his fore ones when approaching, and of both when moving with his flank towards you. The examination should be repeated at a trot, observing in what manner the horse gathers himself; whether he interferes, rocks in his motions, or traverses his shoulders or haunches. Rein him backwards, make one of the men get on him, and see if he is difficult to mount, and whether or not he bears too hard on the bit. Make him gallop a little, to judge of his wind, and see whether his flanks heave. Have his feet washed and examined carefully. Strike upon the shoe to determine whether he is easily shod or not.

AGE.—The age of a horse is determined by the appearance of his teeth. When he is 5 years old, his mouth is nearly perfect with a full set (40) of teeth, 20 in each jaw; six of these are in front, and called *nippers*, or cutting teeth; a tush on each side of these, and on each side of the back part of the jaws six molars, or grinding teeth.

At the birth of the colt, the 1st and 2d grinders have appeared, and in the course of seven or eight days after, the two central nippers force their way through the gums. In the course of the first month, the 3d grinder appears above and below, and shortly after another of the incisors on each side of the first two.

At the end of two months, the central nippers reach their full height, and before another month the second pair will overtake them. They then begin to wear away a little, and the outer edge, which was at first somewhat raised and sharp, is brought to a level with the inner one. So the mouth continues until some time between the 6th and 9th month, when two other nippers begin to appear, making 12 in all, and completing the colt's mouth. After this, the only observable difference, until between the 2d and 3d year, is the wear of these teeth.

These teeth are covered with a polished and very hard enamel, which spreads over that portion above the gum. From the constant habit of nipping grass, and gathering up the animal's food, a portion of the enamel is worn away, while in the centre of the upper surface of the teeth, it sinks into the body of the tooth, forming a little pit. The inside and bottom of this pit, being blackened by the food, constitute the *mark* of the teeth, by the gradual disappearance of which, from the wearing down of the edge, we are enabled, for several years, to judge of the age of the animal.

The teeth, at first presenting a cutting surface, with the outer edge rising in a slanting direction above the inner, soon begin to wear down, until both surfaces are level; and the *mark*, originally long and narrow, becomes shorter, wider, and fainter. Fig. 140 represents the appearance of the animal's mouth at 12 months. The four middle teeth are almost level, and the corner ones becoming so. The mark in the two middle teeth is wide and faint; in the two next, darker, longer, and narrower; and in the extreme ones it is darkest, longest, and narrowest. This appearance of the nippers, together with the coming of four new grinders, enables the age of the colt to be pretty nearly calculated.

Six months after, the mark in the central nippers will be much shorter and fainter; that in the two other pairs will have undergone an evident change, and all the nippers will be flat.

At two years old, this change will be still more manifest, and the lower jaw of the colt will present the appearance represented in Fig. 141. About this period, too, a new grinder appears, making 20 in all,

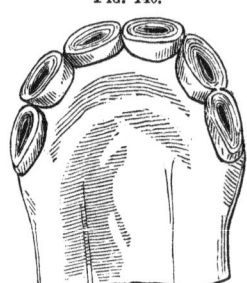

Fig. 140.

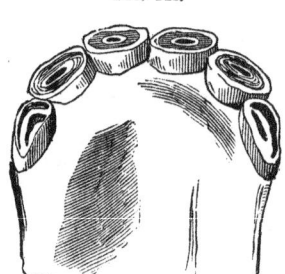

Fig. 141.

and a still more important change takes place. This consists in the formation of the permanent teeth which gradually come up from beneath, *absorb*, and take the place of the temporary, or milk teeth, as they are called, and finally push the top parts of these latter out of their places. These permanent teeth are much larger and stronger than the first ones.

The teeth are replaced in the same order that they originally appeared, and consequently, at the end of the second year, the first grinders are replaced by permanent and larger ones; then the central nippers, and so on. At the end of the third year, the colt's mouth will present the appearance shown in Fig. 142. The central teeth are larger than the others, with two grooves in the outer convex surface, and the mark is long, narrow, deep, and black. Not having yet attained their full growth, they are rather lower than the others. The mark in the two next nippers is nearly worn out, and it is wearing away in the extreme ones.

A horse at three years old ought to have the central permanent nippers growing; the other two pairs wasting; six grinders in each jaw, above and below—the first and fifth level with the other, and the sixth protruding. The sharp edge of the new incisors will be very evident when compared with the neighboring teeth.

As the permanent nippers wear, and continue to grow, a narrower portion of the cone-shaped tooth is exposed to attrition, and they look as if they had been compressed. The mark, of course, gradually disappears as the pit is worn away.

At three years and a half, or between that and four, the next pair of nippers will be changed. The central nippers will have attained nearly their full growth. A vacuity will be left where the second stood, or they will begin to peep above the gum, and the corner ones will be diminished in breadth, worn down, and the mark becoming small and faint. At this period, too, the second pair of grinders will be shed.

At four years, the central nippers will be fully developed; the

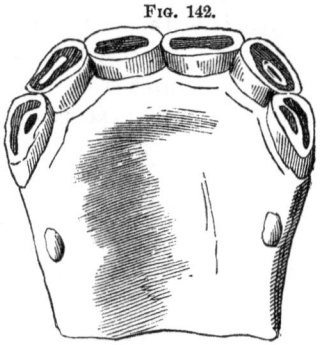
Fig. 142.

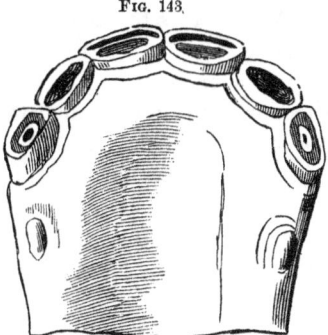
Fig. 143.

sharp edge somewhat worn off, and the mark shorter, wider, and fainter. The next pair will be up, but they will be small, with the mark deep, and extending quite across them. The corner nippers will be larger than the inside ones, yet smaller than they were, flat, and the mark nearly effaced. The sixth grinder will have risen to a level with the others, and the tushes will begin to appear. See Fig. 143. The small size of the corner nippers, the want of wear in the others, the little growth of the tush, the smallness of the second grinder, the low forehand, the legginess of the colt, and the thickness and little depth of the mouth, will prevent the horse from being passed off as over four years old.

The tushes are much nearer the nippers than the grinders, but this distance increases with the age of the animal. The time of their appearance is uncertain, and it may vary from the fourth year to four years and six months.

At four years and a half the last important change takes place in the mouth. The corner nippers are shed, and the permanent ones begin to appear. The central nippers are considerably worn, and the next pair are commencing to show signs of usage. The tush has now protruded, and is generally a full half-inch in height. After the rising of the corner nippers the animal changes its *name*—the colt becomes a horse, and the filly a mare.

At five years the corner nippers are quite up, with the long deep mark irregular on the inside, and the other nippers bearing evidence of increased wear. The tush is much grown, the grooves have nearly disappeared, and the outer surface is regularly convex, though the inner is still concave, with the edge nearly as sharp as it was six months before. The sixth molar is quite up, and the third wanting, which last circumstance will be of great assistance in preventing deception. The three last grinders and the tushes are never shed. Fig. 144 represents the mouth of a 5-year old horse.

At six years the mark on the central nippers is worn out, though a difference of color still remains in the centre of the tooth, and although a slight depression may exist, the deep hole with the blackened surface and elevated edge of enamel will have disappeared. In the next incisors the mark is shorter, broader, and fainter; and in the corner teeth the edges of the enamel are more regular, and the surface is evidently worn. The tush has attained its full growth of nearly an inch in length; convex outwards, concave within, tending to a point, and the extremity somewhat curved. The third grinder is fairly up, and all the grinders are level.

At seven years, the mark is worn out in the four central nippers,

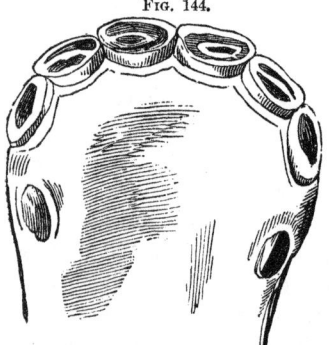
Fig. 144.

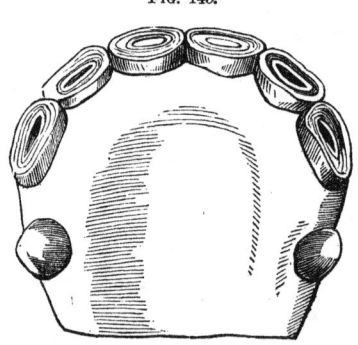
Fig. 145.

and fast wearing away in the corner ones. The tush is becoming rounded at the point and edges; still round outside, and beginning to get so inside. (Fig. 145.)

At eight years old, the tush is rounded in every way; the mark is gone from all the bottom nippers, and nothing remains in them that can afterwards clearly show the age of the horse.

An operation is sometimes performed on the teeth of horses, to deceive purchasers in regard to age. This, called *bishoping*, after the inventor, consists in throwing a horse, 8 or 9 years old, and with an engraver's tool digging a hole in the almost plane surface of the corner teeth, of the same shape and depth of those seen in a 7-year old horse. The holes are then burned with a heated iron, leaving a permanent black stain. The next pair of nippers are also sometimes lightly touched. An inexperienced person might be deceived by the process; but a careful examination will disclose the irregular appearance of the cavity—the diffusion of the black stain around the tushes, the sharpened edges and concave inner surface of which can never be given again—and the marks on the upper nippers. After the horse is 8 years old, horsemen are accustomed to judge of his age from the nippers in the upper jaw, where the mark remains longer than in the lower jaw teeth; so that at 9 years of age it disappears from the central nippers; at 10 from the next pair, and from all the upper nippers at 11. During this time, too, the tushes are changing, becoming blunter, shorter, and rounder; but the means for determining accurately the age of a horse, after he has passed 8 years, are very uncertain.

The general indications of old age, independent of the teeth, are deepening of the hollows over the eyes, and about the muzzle; thinness and hanging down of the lips; sharpness of the withers; sinking of the back; lengthening of the quarters; and the disappearance of windgalls, spavins, and tumors of every kind.

The perpendicularity with which a horse habitually stands, determines his good qualities and endurance. Viewed in profile, his front legs should be comprised between two verticals: the one, A, (Fig. 146,) let fall from the point of his shoulder, and terminating at his toe; the other, B, from the top of the withers, and passing through the elbow. A line, C, passing through the fetlock-joint, should divide the limb into two equal parts. The hind legs should be comprised between two verticals, A′ falling from the hip, and B′ falling from the point of the buttock; the foot at very nearly equal distances from these two lines. A line, C′, let fall from the hip-joint, should be equally distant from these two lines A′, B′.

Viewed in front, a vertical let fall from the point of the shoulder, should divide the leg along its central line. In rear, a vertical from the point of the buttock, should divide the leg equally throughout its entire length.

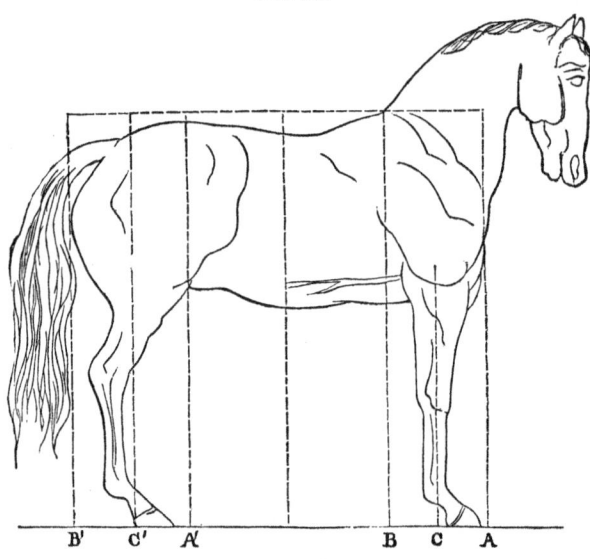

FIG. 146.

The height of the horse, measured from the top of the withers to the ground, should be equal to his length from the point of the shoulder to the point of the buttock. His chest, looking at him from the front, should be broad; and viewed from the rear, he should be broad, with good muscle, and strongly built.

"The thoroughbred horse enters into every other breed, and *adds or often gives to it its only value*. For a superior charger, hunter, or saddle horse, three parts or one-half should be of pure blood; but for the horse of all work, less will answer. The road horse, according to the work required of him should, like the hunter, possess different degrees of blood. The best kind of coach horse is foaled by mares of some blood, if the sire is a three-fourth or thoroughbred stallion of sufficient size and substance. Even the dray horse, and every other class of horse, is improved by a partial mixture of the thoroughbred.

"The first point of a good hunter is that he should be light in hand. For this purpose, his head must be small; his neck thin, especially beneath; his crest firm and arched, and his jaws wide. The head will then be well set on. It will form a pleasant angle with the neck, which gives a light and pleasant mouth."

"The road horse or hackney should be a hunter in miniature, with these exceptions: his height should rarely exceed fifteen hands and an inch. He will be sufficiently strong and more pleasant for general work below that standard. He should be of more compact form than the hunter, of more bulk according to his height. It is of essential consequence that the bones beneath the knee should be deep and flat, and the tendon not *tied in*. The pastern should be short, and less oblique or slanting than that of the hunter or race-horse. The foot should be of a size corresponding with the bulk of the animal, neither too hollow nor too flat, and open at the heels. The forelegs should be perfectly straight; for a horse with his knees bent will, from a slight cause and especially if overweighted, come down. The back should be straight and short, yet sufficiently long to leave comfortable room for the saddle between the shoulders and the *huck*, without pressing on either. Some persons prefer a hollow-backed horse. It is generally an easy one to go. It will canter well with a lady; but it will not carry a heavy weight, or stand much hard work. The road horse should be high in the forehead, round in the barrel, and deep in the chest."

A horse travels the distance of 400 yards at a walk, in $4\frac{1}{2}$ minutes; at a trot, in 2 minutes; at a gallop, in 1 minute. He occupies in the ranks a front of 40 inches, a depth of 10 feet; in a stall from $3\frac{1}{2}$ to $4\frac{1}{2}$ feet front; at a picket, 3 feet by 9. Average weight of horses 1,000 lbs. each. A horse carrying a soldier and his equipments, (say 225 lbs.,) travels 25 miles in a day, (8 hours.) A *pack* horse can carry 250 to 300 lbs. 20 miles a day. A *draught* horse can draw 1,600 lbs. 23 miles a day, weight of carriage included. Artillery horses should not be made to draw more than 700 lbs. each, the weight of the carriage included. The ordinary work of a horse for 8 hours a day may be stated at 22,500 lbs. raised one foot in a minute. In a *horse* mill, the horse moves at the rate of 31 feet in a second. The diameter of the path should not be less than 25 or 30 feet. Daily allowance of water for a horse is four gallons. A horse-power in steam engines is estimated at 33,000 lbs. raised 1 foot in a minute; but as a horse can exert that power but 6 hours a day, one steam horse-power is equivalent to that of four horses.

The actual mode of taking wild horses is by throwing the lasso, whilst pursuing them at full speed, and dropping a noose over their necks; by which their speed is soon checked, and they are choked down. Mr. Rarey's sixpenny book tells all that can be told on the subject of horse-breaking; but far more lies in the skill and horse-knowledge of the operator, than in the mere theory. His way of mastering a vicious horse, is by taking up one fore-foot, and bending his knee, and slipping a loop over the knee until it comes to the pastern-joint, and then fixing it tight. The loop must be caused to embrace the part between the hoof and the pastern-joint firmly, by the help of a strap of some kind, lest it should slip. The horse is now on three legs, and he feels conquered. If he gets very mad, wait leisurely till he becomes quiet; then caress him, and let the leg down, and allow him to rest. Then repeat the process. If the horse kicks in harness, drive him slowly on three legs. In breaking-in a stubborn beast, it is convenient to physic him until he is sick and out of spirits, or to starve him into submission. Salt keeps horses from straying, if they are accustomed to come up to the camp and get it. But it is a bad plan, as they are apt to hang about, instead of going off to feed. They are so fond of it, that they have been known to stray back to a place where they had been licking it, in front of the doors. (Consult GIBBON; SKINNER's *Youatt;* BRANDE's *Encyclopedia; Memorial des Officiers d'Infanterie et de Cavalerie.* See PAY; VETERINARY.)

HORSEMANSHIP—consists in perfect mastery of the horse. The principles laid down by *Baucher* in his method of horsemanship, published in Philadelphia in 1851, profess to give *any* horse in less than three months:

1. General suppling; 2. Perfect lightness; 3. Graceful position; 4. A steady walk; 5. Trot, steady, measured, extended; 6. Backing as easily and as freely as going forward; 7. Gallop easy with either foot, and change of foot by the touch; 8. Easy and regular movement of the haunches, comprising ordinary and reversed *pirouettes;* 9. Leaping the ditch and the bar; 10. Making the horse raise his legs diagonally as in a trot, but without advancing or receding; 11. Halt from the gallop by the aid of, first, the pressure of the legs, and then a light support of the hand. "The education of the men's horses, being less complicated than that of those intended for the officers, would be more rapid. The principal things will be the supplings and the backing, followed by the walk, the trot, and the gallop, while keeping the horse perfectly in hand."

Horsemanship in war consists in address in the exercise of arms while skilfully using the proper paces of the horse in different accidents of ground, with ability in the rider to obtain immediate obedience in all movements that may be rationally demanded. To accomplish this, constant exercise is required of both horse and cavalier, and the individual instruction now prescribed in the French army gives this skilfulness, and habituates horses to separations from each other, and to instant yielding to the will of the rider. (Consult BAUCHER; *Cavalry Tactics; Travail Individuel.*)

HORSE EQUIPMENTS.

STATEMENT OF THE COST OF HORSE EQUIPMENTS, PATTERN 1859, FURNISHED BY THE ORDNANCE DEPARTMENT.

The regulations require that requisitions for Horse Equipments shall follow the form prescribed for ordnance requisitions. Stirrups, saddle-bags, girths, and surcingles, to be entered separately instead of under the head SADDLE in the following list. CURB BRIDLES to embrace the various kinds of curb bits, scutcheons, curb chains, and leather fittings complete. WATERING BRIDLES to include every thing else instead of using separate heads for halters, blankets, &c., &c.

	Price per piece.	Price per set.	Amount.
SADDLE.	$ cts.	$ cts.	
Saddle tree covered with raw hide with metal mountings attached.	4 13	4 13	
Saddle flaps with brass screws, each	1 10	2 20	
Back straps, with screws, rivets, and D's, each	58	1 16	
Girth strap, long	50	50	
" " short	30	30	
Cloak straps, each	25	1 50	
Stirrup leathers, each	70	1 40	
Sweat leathers, each	70	1 40	
Stirrups with hoods, each	60	1 20	
Carbine socket and strap	72	72	
Saddle-bags	3 75	3 75	
Crupper	1 75	1 75	
Girth	80	80	
Surcingle	1 17	1 17	
Total cost	....		$21 98
BRIDLE.			
*Bit, No. 1, $5 } average per 100 sets	4 20	4 20	
" Nos. 2, 3, and 4, $4 }			
Brass scutcheon with company letter, each	5	10	
Reins	80	80	
Headpiece	85	85	
Front	10	10	
Curb chain with hooks	20	20	
Curb chain safe	8	8	
Total cost	....		6 33
HALTER.			
Headstall, complete	2 00	2 00	
Hitching strap	50	50	
Total cost	....		2 50
WATERING BRIDLE.			
Snaffle bit, chains, and toggles	90	90	
Watering rein	80	80	
Total cost	....		1 70
Spurs	45	90	
Spur straps	10	20	
Total cost	....		1 10
Curry comb	20	20	
Horse brush, wooden back	94	94	
Picket pin	20	20	
Lariat rope	1 15	1 15	
Total cost	....		2 49
Total cost of equipment	....	....	36 10
Blanket for cavalry service, dark, with orange border, 3 lbs., at 70 cents per lb	2 10	2 10	
Blanket for artillery, scarlet, with dark blue border, 3 lbs., 70 cents per lb	2 10	2 10	
Nose-bag	75	75	
Hitching strap	20	20	

* No. 1 is Spanish; Nos. 2, 3, and 4, are American.

HOSPITALS—are under the immediate direction of their respective surgeons. The general regulations of the army prescribe the allowance of attendants; the issues to hospitals, &c., &c. (*See* AMBULANCE; SURGEON; SURGERY.)

HOT SHOT. The charges for hot shot are from $\frac{1}{4}$ to $\frac{1}{6}$ the weight of the shot. With small velocities, the shot splits and splinters the wood, so as to render it favorable for burning. With great velocity, the ball sinks deep into the wood, is deprived of air by the closing of the hole, and chars instead of burning the surrounding wood. It should not penetrate deeper than 10 or 12 inches. Red-hot balls do not set fire to the wood until some time after their penetration. They retain sufficient heat to ignite wood after having made several ricochets upon water. The wads are made of clay or hay. Clay wads should consist of pure clay, or fuller's earth free from sand or gravel well kneaded with just enough moisture to work well. They are cylindrical and one calibre in length. Hay wads should remain in the tub to soak, at least ten or fifteen minutes. Before being used, the water is pressed out of them. When hay wads are used, vapor may be seen escaping from the vent on the insertion of the ball; but as this is only the effect of the heat of the ball on the water contained in the wad, no danger need be apprehended from it. With proper precautions in loading, the ball may be permitted to cool in the gun without igniting the charge. The piece, however, should be fired with as little delay as possible, as the vapor would diminish the strength of the powder. FURNACES FOR HEATING SHOT are erected at the forts on the sea-coast. These furnaces hold sixty or more shot. The shot being placed, and the furnace cold, it requires one hour and fifteen minutes to heat them to a red heat; but after the furnace is once heated, a 24-pdr. shot is brought to a red heat in twenty-five minutes; the 32-pdr. and 42-pdr. shot require a few minutes longer. Three men are required to attend the furnace: one takes out the hot shot, and places them on the stand to be scraped; another scrapes them and puts them in the ladle; and the third supplies cold shot and fuel; (GIBBON.)

HOURS OF SITTING. (*See* COURTS-MARTIAL.)

HOUSINGS. The cloth covering for saddles prescribed as part of the uniform of the army in regulations.

HOWITZER. A chambered cannon. (*See* CALIBRE.)

HURDLES. Pickets three feet high united by pliable twigs, so as to make a breadth of two feet. They are used to render batteries firm; to pass over boggy ground or muddy ditches. (*See* REVETMENT.)

HURTER. The hurter is a piece of timber, from six to ten inches square, placed along the head of a gun platform, at the foot of the interior slope of the parapet, to prevent the latter from being injured by the wheels of the gun-carriage.

HUSSARS. Light cavalry.

HUTS—are frequently constructed by troops on retiring to winter-quarters. The quarters occupied by United States troops on our frontiers are generally huts made by the troops. There have recently been built portable houses, the parts of which correspond, and which are readily put up. The experiment is not yet a success. (*See* ADOBE; CAMP; CARPENTRY; SAW-MILL.)

I

ICE. Ice two inches thick will bear infantry; four inches thick, cavalry or light guns; six inches heavy field-guns; 8 inches 24-pdr. guns on sledges; weight not more than 1,000 lbs. to a square foot. Water that is slightly frozen is made to bear a heavy wagon by cutting reeds, strewing them thickly on the ice, and pouring water upon them. When the whole is frozen into a firm mass, the process must be repeated.

IMPRISONMENT. Officers may be sentenced to imprisonment by a general court-martial in any case where the court may have discretionary authority. General, garrison, and regimental courts-martial may sentence soldiers to imprisonment, solitary or otherwise, with or without hard labor for various offences enumerated in the Articles of War. A garrison or regimental court-martial, in awarding imprisonment, is limited to a period not exceeding thirty days. When a court awards solitary imprisonment as a punishment, it is necessary that the words "Solitary Confinement" should be expressed in the sentence.

INDEMNIFICATION. In the French and English armies, there is an indemnification established for losses in the military service, and other allowances are also made in the nature of indemnifications; as for *furniture; fuel* and *light; forage; expenses of divine worship; command money* to general and field officers; *quarters; expenses upon routes; provisions; gratuity at the beginning of a campaign; field allowances; mess; carriage of baggage; blood money; permanent pensions; temporary pensions,* or *gratuities* in lieu thereof; *rewards* for meritorious conduct; and pensions to widows and children of officers.

In the United States service, the law provides that if a horse be lost in battle, an officer may receive not exceeding two hundred dollars for his horse, and allowances are made for quarters, fuel, forage, provision, and transportation of baggage, and command money in certain cases.

INDIANS. The red man of America is so called, and as the troops of the United States have always been the pioneers of civilization, their contact with the Indians is always more or less immediate. The problem of the disappearance of the race is fast being solved; and every humane mind must contemplate with sorrow the destitution to which the Indians have been driven. Something, it is believed, may be done for them by the system of policy proposed in the article on national defence, and that policy would be greatly promoted if the United States maintained on our frontier a few Indian regiments, officered by details from the army. The successful adoption of this policy in India by the English, and in Algiers by the French, proves its practicability, and no men would make better light cavalry and light infantry than the Indians on our western frontier.

The President is authorized to cause army rations to be issued to Indians; (*Act* June 30, 1834.)

All purchases on account of Indians, and all payments to them of money or goods, shall be made by such person as the President shall designate for that purpose. And the superintendent, agent, or sub-agent, together with such military officer as the President may direct, shall be present, and certify to the delivery of all goods and money required to be paid or delivered to said Indians. And the duties required by any section of this act of military officers, shall be performed without any other compensation than their actual travelling expenses; (*Act* June 30, 1834.)

Army surgeons may be employed by the Secretary of War to vaccinate Indians; (*Act* May 5, 1832.)

A foreigner going into Indian territory without a passport from the War Department, superintendent, agent, sub-agent, or from the officer commanding the nearest military post, or remaining intentionally therein after the expiration of his passport, is subject to forfeit and pay the sum of one thousand dollars; (*Act* June 30, 1834.)

It shall be lawful for the military force of the United States to be employed, in such manner and under such regulations as the President may direct, in the apprehension of every person found in the Indian territory in violation of any of the provisions of this act, and cause him to be conveyed for trial to the nearest civil authority; and the military force may also be employed in the examination and seizure of stores, packages, and boats, with spirituous liquor or wine, and in preventing the introduction of persons and property into the Indian country contrary to law. Provided that no person apprehended by the military force as aforesaid shall be detained longer than five days after arrest, and before removal for surrender to the civil authority; (*Act* June 30, 1834.)

When goods or other property are seized under this act, the process of prosecutions shall be the same as in the case of goods, &c., brought into the United States in violation of the revenue laws; (*Act* June 30, 1834.)

Persons attempting to settle in Indian territory may be removed by military force; (*Act* 1832. *See* TREATY.)

INFANTRY. Its depth of formation has progressively diminished since the centre and wings have been armed alike, and the use of pikes discontinued. The formation in lines has fitted infantry for action on all kinds of ground, and the invention of *massing*, the condensation of ranks, and formations by size, have given it a perfect *ensemble*. Its march has gained in rapidity by the simplification of evolutions, the resort to guides, and turning upon PIVOTS; it acts more skilfully in affairs of plains and outposts, by the rapidity of its changes of direction, formations in order of battle, and alternate ployments and deployments. The general adoption of tactical inversions, it is thought, would add still more to this skilfulness.

The improved rifle-musket, with thorough target practice, gives to infantry immense advantages over cavalry and artillery. The effective range of the new musket permitting skirmishers to open fire at 1,000 yards, fields of battle will cover more ground than formerly, and the use of smaller columns than battalions of eight and ten companies will probably be resorted to. An organization of battalions of six companies of 100 men each, in two ranks, in lieu of the former, would be an improvement; and in the United States service this might be accomplished by adding two companies with two battalion-adjutants and sergeant-majors to each regiment. The front of each battalion would not be too great. Columns would be formed by division in mass. There would be three such divisions, and the square formed would have $\frac{1}{3}$ its rifles in the first and fourth fronts, and $\frac{1}{6}$ each in the other two fronts. Such well-instructed men, in firing, would be perhaps able to show, as in the experiment at Hythe, that a piece of artillery with its men and horses might, at 810 yards, be completely disabled by 30 riflemen in three minutes, and also be an overmatch for cavalry.

Infantry has always guarded the frontier in war; it supports cavalry in great reconnoissances; furnishes swimmers when the cork jacket is resorted to; is employed both in the attack and defence of fortresses;

slings the musket and throws grenades; mounts heights by escalade; escorts and attacks convoys; supports foraging parties; defends abatis; is at home in all accidents of ground; finishes operations begun by artillery; crowns heights which horses and pieces of artillery cannot reach; decides the fate of battles, sometimes with the aid of cavalry, and sometimes alone. Costing little, active, occupying relatively little ground; readily lodged, maintained, and renewed, it is easily subsisted, and often finds in its knapsacks, haversacks, and utensils carried by the men, all its wants supplied, when separated from baggage trains.

It has been made a question whether excellent cavalry may not beat mediocre infantry, and whether excellent infantry would not be overthrown by mediocre cavalry?

There is this great difference between infantry and cavalry: infantry has always changed its tactics at the same time with its arms, whereas cavalry cannot change its manner of fighting, although it has more than once attempted the forms of infantry tactics.

Cavalry cannot operate as a whole, except upon unbroken ground; it is unsuited to firing; the *order of battle* is its great means of action; the sabre or lance is its only reliance; the invention of powder has not improved the art it exercises. Squares of cavalry are useless; the circular formation which has been conceived is a chimera; defence is not its strength; movement is its life, an unbroken field its element, and the charge its principal means of offence. But within range of the rifle, at 1,000 yards, it must be destroyed before reaching its object.

The elementary tactics of infantry consists in securing its rear and its flanks; in never being entirely disfurnished of its fire; in attacking with the bayonet; in defending itself by firing within proper range, and progressively, rather than simultaneously; using the aid of the grenade and rocket, and in resorting to the bayonet, as prescribed in the bayonet exercise. In the offensive movements of a field of battle, infantry ought never to be disfurnished of its fire, except when the enemy falls back, and it is known that his retreat is not a stratagem to draw the fire of the assailants, in order to push down upon them masked cavalry.

Infantry being suited for close or distant combat, the aim of its tactics is to prescribe the best order for the shock, and the best orders for firing. The chef-d'œuvre of art consists in the most rapid and successful transformations of these orders; in the mechanism of changes of front; and in the ployments and deployments of columns of attack and the formation of squares against cavalry.

In campaign, infantry preferably occupies broken ground, woods, &c. A trench, abatis, or chevaux-de-frise is sufficient to secure its safety. In crossing plains, its head and flanks should be covered by cavalry; in retreat, the infantry forms the rear guard, to protect the column of cavalry. For this purpose it occupies hills or ravines, or, standing firm in heavy masses, the cavalry defiles until it has gained ground suited to cavalry operations. When the cavalry has reached such a position, it deploys, faces to the rear to cover in its turn the retreat of the infantry.

Didactic authors, as well as historians, recognize the superiority of infantry. VOLTAIRE calls it the *soul* of armies; MACHIAVEL, the *sinew*; it is the principal force and lever of power in time of war; it can act alone; other arms move to second it: thus good infantry is the true strength of nations; every one in an army feels its importance; its posts guard the army; its duties are, of all others, the most constant, the most simple, the most easily regulated, and the most certain and most important.

The duties of engineers and artillery require more learning; those of cavalry, in war, are sometimes more dashing and brilliant; but the services of infantry are always in demand. In attack and defence of all kinds; the descent into the ditch; or the defence of the breach, the trench, and the rampart; the insult of palisades, or the fire from the parapet; in ambuscades; or on any field of battle whatever, infantry must exercise its skilfulness and attest its valor. Valleys, fords, defiles, water-courses, ravines, abatis, forests, heights, plains, parallels, camps, outworks, covered ways, advance guards, and rear guards, are all in turn its theatre of action. All kinds of troops mutually aid each other, and it is the skilful combination of their efforts which constitutes, in part, the science of the general-in-chief. To make good infantry, it is essential that it should pass some months in a *camp of instruction*. The soldier must be taught to take care of his arms and accoutrements, to march, to fire well, to build huts, to handle the axe, spade, and shovel, to make cartridges, fascines, hurdles, and gabions, suited to field-works, to cook, and to consider his knapsack, haversack, &c., as part of himself. (*See* DISCIPLINE; ARTICLES OF WAR; TACTICS; MANŒUVRES IN COMBAT. Consult BARDIN.)

INFORMANT. In case a civil person is the complainant, he becomes the principal witness before a court-martial, and after giving his evidence may remain in court, in order that the judge-advocate may refer to him; (HOUGH.)

INITIAL VELOCITY. The velocity with which a projectile leaves the piece, that is, the space in feet then passed in a second, is called its *initial* velocity; the space passed over in a second at any succeeding point of the trajectory its *remaining* velocity, and the *terminal* velocity is the velocity with which it strikes the object. The greatest initial velocities do not exceed four or five hundred yards, and are given by charges not exceeding one-third the weight of the ball; the feeblest are produced by charges of about one-twenty-fourth the weight of the ball. The musket pendulum used at Washington Arsenal has shown the initial velocity of the elongated ball for the rifle-musket to be 963 feet per second, and that of the pistol-carbine 603. For ordinary practice, where the weight of the powder and the projectile alone vary, initial velocities may be considered *directly proportional to the square root of the weight of powder divided by the square root of the weight of the projectile.*

In the experiments made at Washington by Major Mordecai with the gun and ballistic pendulums combined for the purpose of ascertaining the initial velocities produced by equal charges of powder in the same piece of ordnance on balls of different weights, it was found that, with a 24-pounder gun and a charge of 4 lbs. of powder, the windage being .175 inch, the initial velocity of a shell filled with lead and weighing 27.68 lbs., was 1,325 feet; of a marble ball weighing 9.29 lbs., was 2,154 feet; and of a lignum vitæ ball weighing 4.48 lbs., was 2,759 feet. The two first of these velocities are nearly in the inverse ratio of the square roots of the weights of the shot; but the two last are nearly as the cube roots of the weights inversely. (Consult BENTON. *See* BALLISTICS.)

TABLE OF INITIAL VELOCITIES WITH SERVICE CHARGES.

KIND OF CANNON.	Charge of Powder.	KIND OF PROJECTILE.			REMARKS.
		Shot.	Shells.	Spher'l Case.	
	lbs.	feet.	feet.	feet.	
6-pdr. Field................	1.25	1,439		1,857	When the initial velocities of shot, shell, and spherical case shot are given, the weight of the charge refers to shot.
12-pdr. Field................	2.50	1,486		1,486	
12-pdr. Field Howitzer.....	1.00			953	
24-pdr. Siege Gun.........	6.00	1,680	1,054		
	8.00	1,870	1,670		
8-inch Siege Howitzer......	4.00		907		
32-pdr. Sea-coast Gun.......	8.00	1,640	1,450		
15-inch Columbiad	40.00		1,328		

INJURIES, LIABILITY FOR PRIVATE INJURIES. In the exercise of professional duty by military officers, injuries may frequently be occasioned to other officers, or to private individuals, whose legal remedies are here considered. As between officers themselves, the language of the Articles of War is sufficiently comprehensive to bring most of such cases within the cognizance of a court-martial; but a court-martial has no power to award pecuniary damages for injurious conduct. Its jurisdiction is criminal, and its judgments are penal. It may happen, too, that the common feeling of the service, to which the offending or the complaining party belongs, would in many cases render an application to such a tribunal utterly fruitless; as the general sentiment of the members of a particular profession or class of society, respecting a matter of professional or corporate right or conduct, is often found to be at variance with the public law of the land. Civil actions are therefore maintainable against commissioned officers, for exceeding their powers, or for exercising them in an oppressive, injurious, and improper manner, whether towards military persons or others. Extreme difficulties, however, lie in the way of plaintiffs in actions of this nature; for no such action is maintainable for an injury, unless it be accompanied by malice or injustice: and the knowledge of this, (says Mr. Baron Eyre,) while it can never check the conduct of good men, may form a check on the bad. Where an officer (says the same learned judge) makes a slip in form, great latitude ought to be allowed; but for a corrupt abuse of authority none can be made.

It will be convenient to consider the law upon this subject: 1st, as it applies to wrongs committed by officers towards persons under military authority; and, 2dly, as it applies to persons not subject to such authority. Some of the decisions that will be quoted were pronounced in cases where naval officers were concerned; but the principle of the decisions applies equally to both services. I. *Wrongs towards Persons under Military Authority.*—A notion appears to have at one time extensively prevailed that an officer could have no remedy against ill treatment received from his superiors in the course of professional duty, except by bringing the offending party to a court-martial, and subjecting him to the penalties of the Articles of War. This opinion, however, was quite unfounded in point of law; and such a state of things might often be productive of the worst consequences. The question was distinctly raised in Grant *v.* Shand, where an action was brought by an officer in the army against his superior officer for oppressive, insulting, and violent conduct. The plaintiff was directed to give a military order: and it appeared that he sent two persons, who failed. The defendant thereupon said to the plaintiff, " What a stupid person you are," and twice struck him; and although the circumstances occurred at Gibraltar, and in the actual execution of military service, it was held by the learned judge at the trial that the action was maintainable; and a verdict was found for the plaintiff. An application was afterwards made to the Court of King's

Bench to set aside the verdict; and Lord Mansfield, the chief-justice, was very desirous to grant a new trial; but the court, after argument, refused to disturb the verdict. So also an action will lie for unjust treatment under the form of discipline, as in Swinton *v.* Molloy, where the defendant, who was captain of the *Trident* man-of-war, put the purser into confinement, kept him imprisoned for three days without inquiring into the case, and then released him on hearing his defence. The purser brought his action against Captain Molloy, for this unlawful detention in custody; and, upon the evidence, Lord Mansfield said, that such conduct on the part of the captain did not appear to have been a proper discharge of his duty, and therefore that his justification under the discipline of the navy had failed him. The jury gave £1,000 damages. In the foregoing case no want of uprightness was attributed to Captain Molloy; and the decision rested wholly on the circumstance of his having committed an injustice, although without a corrupt intention. Cruelty or unnecessary severity, when wilfully committed in the exercise of superior authority, are also good causes of action. Thus in Wall *v.* Macnamara, the action was brought by the plaintiff, as captain in the African corps, against the defendant, Lieutenant-governor and Military Commandant of Senegambia, for imprisoning the plaintiff for the space of nine months at Gambia, in Africa. The defence was a justification of the imprisonment under the Mutiny Act, for the disobedience of orders. At the trial it appeared that the imprisonment of Captain Wall, which was at first legal, namely, for leaving his post without leave from his superior officer, though in a bad state of health, was aggravated with many circumstances of cruelty, which were adverted to by Lord Mansfield, in the following extract from his charge to the jury: "It is admitted that the plaintiff was to blame in leaving his post. But there was no enemy, no mutiny, no danger. His health was declining, and he trusted to the benevolence of the defendant to consider the circumstances under which he acted. But supposing it to have been the defendant's duty to call the plaintiff to a military account for his misconduct, what apology is there for denying him the use of the common air in a sultry climate, and shutting him up in a gloomy prison, when there was no possibility of bringing him to a trial for several months, there not being a sufficient number of officers to form a court-martial? These circumstances, independent of the direct evidence of malice, as sworn to by one of the witnesses, are sufficient for you to presume a bad, malignant motive in the defendant, which would destroy his justification, had it even been within the powers delegated to the defendant by his commission." The jury thereupon found a verdict, for Captain

Wall, with £1,000 damages. An undue assumption of authority in matters not within the range of military discipline, is also a good ground of action against a superior officer. This appears from the case of Warden v. Bailey, where the plaintiff was a permanent sergeant in the Bedford regiment of local militia, of which the defendant was the adjutant. In November, 1809, the lieutenant-colonel issued a regimental order for establishing an evening school at Bedford. He appointed the sergeant-major the master, and ordered all sergeants and corporals, including the plaintiff, to attend and pay eight-pence a week towards the expenses of the school. The plaintiff and some other of the scholars having afterwards omitted to attend, several were tried by court-martial and punished. The plaintiff, however, was only reprimanded, and he promised regular attendance in future. Shortly afterwards he was ordered to attend a drill on parade, when the defendant, who appears to have been a shopkeeper, shook his fist at the plaintiff, called him a rascal, and told him he deserved to be shot. The defendant then directed a sergeant to draw his sword and hold it over the plaintiff's head, and if he should stir to run him through; and, by the defendant's direction, a corporal took off the plaintiff's sash and sword. The plaintiff was then conducted, by the defendant's order, to Bedford gaol, with directions that he should be locked up in solitary confinement, and kept on bread and water. He was thus imprisoned for three days. He was then brought up before the colonel and the defendant, and other officers of the regiment, and again remanded to the gaol. The plaintiff's health having been impaired by the continuance of this treatment for several weeks, he was afterwards conducted to his own house, and there kept a close prisoner until January, 1810, when he was escorted by a file of corporals from Bedford to Stilton, to be tried by court-martial for mutinous words spoken on parade at the time of his arrest, and for thereby exciting others to disobedience. He was tried accordingly, but liberated in March, 1810. Upon this he brought his action against the adjutant for the wrongful imprisonment, when an objection was taken that the question of the propriety of the arrest was not within the jurisdiction of the civil courts The Court of Common Pleas, however, overruled this objection. Sir James Mansfield, C. J.: "It might be very convenient that a military officer might be enabled to make the men under his command learn to read and write,—it might be very useful, but is not a part of military discipline. Then, further, there is a tax of 8d. a week for learning to read and write. . . . . . The subject cannot be taxed, even in the most indirect way, unless it originates in the Lower House of Parliament." Mr. Justice Lawrence:

"It is no part of military duty to attend a school, and learn to write and read. If writing is necessary to corporals and sergeants, the superior officers must select men who *can* write and read; and if they do not continue to do it well, they may be reduced to the ranks. Nor is it any part of military duty to pay for keeping a school light and warm: this very far exceeds the power of any colonel to order." In a subsequent stage of the same case, when it was attempted to justify or defend the mutinous expressions used by Warden on parade as above stated, on the ground of the illegality of the order which gave rise to them, the court held, that although Warden had been unlawfully arrested for disobedience to that order, such a circumstance afforded no warrant for insubordinate language on Warden's part, and therefore no exemption from military arrest and punishment for the same. " Nor will he (said Lord Ellenborough, C. J.) be less an object of military punishment, because the order of the lieut.-colonel, to which this language referred, might not be a valid one, and such as he was strictly competent to make. . . . . There may be disorderly conduct to the prejudice of good order and military discipline, in the manner and terms used and adopted by one soldier in dissuading another soldier not to obey an order not strictly legal. If every erroneous order on the part of a commanding officer would not only justify the individual disobedience of it by the soldier, but would even justify him in making inflammatory and reproachful public comments upon it to his fellow-soldiers, equally the objects of such order with himself, is it possible that military order and discipline could be maintained?" The common defence of officers, against whom actions of this nature are brought, is a justification of their conduct as agreeable to the discipline of the service, and contributory to the maintenance of that discipline. And there can be no doubt, that where the conduct brought into question is not an oppressive, malicious, or unreasonable exercise of power, and does not amount to an excess or abuse of authority, an action is wholly unsustainable. The principles upon which the Courts of Law proceed in actions arising out of the abuse of military power, will receive further illustration from the language of Lord Mansfield, in summing up the evidence to the jury in Wall *v.* Macnamara. His lordship thus expressed himself: "In trying the legality of acts done by military officers in the exercise of their duty, particularly beyond the seas, where cases may occur without the possibility of application for proper advice, great latitude ought to be allowed; and they ought not to suffer for a slip of form, if their intention appears by the evidence to have been upright. It is the same as when complaints are brought against inferior civil magistrates, such as justices of the peace, for acts done by them in the exercise of their civil duty. There the principal inquiry to be made by a court of justice is, *how the heart stood?* and if there appear to be nothing wrong there, great latitude will be allowed for misapprehension or mistake. But, on the other hand, if the heart is wrong,—if cruelty, malice, and oppression appear to have occasioned or aggravated the imprisonment, or other injury complained of, they shall not cover themselves with the thin veil of legal forms, nor escape under the cover of a justification the most technically regular, from that punishment, which it is your province and your duty to inflict on so scandalous an abuse of public trust." It is no legal objection to an action for the abuse of military authority, that the defendant has not been tried and convicted by a court-martial, for that argument holds in no case short of felony. The infliction of an unjust or illegal sentence, pronounced by a court-martial, is a good cause of action by the prisoner, against all or any of the members of the court, and all persons concerned in the execution of the sentence; such a sentence, if it exceeds the authorized measure of punishment, being not merely invalid for the excess, but absolutely void altogether. The most remarkable case on record of this kind is that of Lieutenant Frye, of the Marines, who, after an unnecessary previous imprisonment for fourteen months, was brought to trial before a naval court-martial at Port Royal in the West Indies, and sentenced to be imprisoned for fifteen years, for disobedience of orders, in refusing to assist in the imprisonment of another officer, without an order in writing from the captain of Her Majesty's ship *Oxford*, on board of which Lieutenant Frye was serving. At the trial the written depositions of several illiterate Blacks were improperly received in evidence against him, in lieu of their oral testimony, which might have been obtained and sifted by cross-examination; and the sentence pronounced was itself illegal for its excessiveness, the Act 22 George II., which contains the naval Articles of War, not allowing any imprisonment beyond the term of two years. On the return to England of Admiral Sir Chaloner Ogle, the president of the court-martial, Lieutenant Frye brought an action against him in the Court of Common Pleas for his illegal conduct at the trial, when the jury, under the direction of the Lord Chief-Justice Willes, gave a verdict for the plaintiff, with £1,000 damages. The Chief-Justice at the same time informed Lieutenant Frye that he might have an action against all or any of the other members of his court-martial; and Lieutenant Frye accordingly issued writs against Rear Admiral Mayne and Captain Renton, upon whom the same were served as they were coming ashore at the conclusion of the proceedings of the day at another court-martial, of which they were acting members, for the trial of Vice-admiral Lestock, for his conduct in a naval engagement with the French fleet off Toulon, in the early part of the same year. This was deemed a great insult by the members of the sitting court-martial, who accordingly passed some resolutions or remonstrances in strong language, highly derogatory to the chief-justice, which they forwarded to the Lords of the Admiralty, by whom the affair was reported to the king. His Majesty, through the Duke of Newcastle, signified to the Admiralty " his great displeasure at the insult offered to the court-martial, by which the military discipline of the navy is so much affected; and the king highly disapproved of the behavior of Lieutenant Frye on the occasion." The Lord Chief-Justice, as soon as he heard of the resolutions of the court-martial, ordered every member of it to be taken into custody, and was proceeding to uphold the dignity of his court, in a very decided manner, when the whole affair was terminated in Nov., 1746, by the members of the court-martial signing and sending to his lordship a very ample written apology for their conduct. On the reception of this paper in the Court of Common Pleas, it was read aloud, and ordered to be registered among the records as a " memorial," said the Lord Chief-Justice, " to the present and future ages, that whoever set themselves up in opposition to the laws, or think themselves above the law, will in the end find themselves mistaken." The proceedings and the apology were also published in the *London Gazette* of 15th Nov., 1746. At a naval court-martial for the trial of Mr. Crawford, a midshipman of Her Majesty's ship *Emerald*, for contempt and disobedience to the orders of his superior officer, Captain Knell, the court inadvertently found Mr. Crawford guilty only of having been *disorderly when a prisoner at large*, which formed *no part of the offence of which he was accused;* and he was reprimanded accordingly. Mr. Crawford thereupon brought an action against the captain for damages; and the learned judge who presided at the trial, having made some severe animadversions on the illegality of the proceedings, the jury awarded heavy damages. A similar action was brought against Colonel Bailey, colonel of the Middlesex militia, for improperly flogging a private in the militia, and the jury gave £600 damages. In Moore *v.* Bastard also, an action was brought against the president of a court-martial for imprisoning the plaintiff upon an alleged charge of subornation of perjury. The jury gave £300 damages. An action was tried in 1793 before Mr. Barron Perrot, at the spring assizes for the county of Devon, against the officers of the Devon militia, for inflicting 1,000 lashes on the plaintiff, in pursuance of their sentence pronounced against him at a court-martial, held to try him upon a charge of mutiny; the only act proved being that the plaintiff had written a letter to the colonel of the regiment, which was not communicated to any one else, telling him that the men of the regiment were discontented. The jury gave £500 damages; and the case is quoted with approbation by Mr. Justice Heath, who also intimated, that if the plaintiff had died under the punishment, all the members of the court-martial would have been liable to be hanged for murder. There was also another case of an action against Captain Touyn, a naval officer, in which the plaintiff recovered damages for the infliction of several dozen lashes without a court-martial, for a single offence, thereby exceeding the custom which had prevailed in the navy, that commanding officers might inflict one dozen lashes (called a starting) without a court-martial. No action, however, will lie for merely bringing a man to a court-martial, nor for the previous arrest or suspension; such acts being clearly within the limits of military authority, and exercisable, like all other such powers, in a discretionary manner, under the safeguards and at the risks provided by the Articles of War. A commanding officer has, of necessity, a discretionary power to arrest, suspend, and bring to trial by court-martial, any person under his orders. But though this power is indispensable, and its limits cannot, like those of the power of punishment, be exceeded in point of extent, it may, nevertheless, be oppressively, or improperly used; and therefore, by the Articles of War, such conduct is of itself a distinct military offence, triable by a military jurisdiction. This was the opinion of the Judges of the Exchequer Chamber, in the case of Sutton v. Johnstone, and it seems also to be a just inference from the judgment in the same case, that when an officer is expressly charged and found guilty before a court-martial, of having improperly brought another to trial before a similar tribunal, an action is sustainable for the special damage resulting from the offence; but that, until the officer procuring the first trial has been found guilty of improper conduct by a court-martial, a court of law cannot interfere; no civil tribunal being capable of appreciating, with sufficient delicacy, the circumstances which attend the exercise of military power, or of accurately discriminating the grounds of its application. *Want of probable cause* for the accusation is the only basis on which an action for a malicious prosecution before a court-martial can rest; and when that is shown, malice will be inferred by the law. An acquittal, however, by the court-martial, of the party who brings the action, is not conclusive as to the want of probable cause. At the same time, such an acquittal is an essential preliminary to the action, for though the accuser may have been actuated by the most clear and undisguised malice, yet if he substantiates his original charge to the satisfaction of a court-martial, the accused has no *locus standi* in a civil court, even upon the fullest evidence of his prosecutor's malice, it being impossible to say that there was a want of probable cause, after a court-martial has adjudged that there was a positive cause. Innocence and uprightness of intention will therefore, on the one hand, be no defence to an action of this nature, when there appears to have been a want of probable cause for the prosecution before the court-martial; while, on the other hand, the most malicious, or even corrupt intention, will not subject the accuser to a civil action, where he succeeds in establishing the criminal charge before the military tribunal. A wrongful imprisonment being, in the language of the law, a *tort*, savoring of crime, it is held that if two commit a *tort*, and the plaintiff recovers against one, he cannot recover against the other for the same *tort*. This rule was applied in the above-mentioned case of Warden *v.* Baily, where another action was brought against the colonel of the Bedford militia for the same transaction, and the court held that the imprisonment inflicted by the defendant, the adjutant, terminated on the plaintiff being brought up before the colonel on the third day, and being then remanded by him, so that the adjutant was held not liable for more than the first three days' imprisonment, and the colonel not liable, except from the time of the commencement of the remand ordered by himself. It should be observed, however, that no civil action will lie, in the first instance, against a commissioned officer for a discretionary exercise of military authority while in the performance of actual duty in the field in time of war. Where a discretionary power is clearly vested by military usage in the officer whose conduct is impeached, questions as to the exercise of such authority are so essentially military, that the civil tribunals decline to consider them without the previous judgment of a court-martial. This was settled in the case of Barwis *v.* Keppel, in which the plaintiff was a sergeant in the second battalion of the first regiment of foot guards. The defendant, Colonel Keppel, was the second major of that battalion; and in the absence of his superior officers he had the command of it. In 1760, the battalion was ordered to Germany, under the command of the defendant, to form part of the king's forces serving under Prince Ferdinand. In September, 1761, the prince, being in hourly expectation of a battle, issued an order that all deserters from the enemy should be immediately sent to head-quarters without a moment's delay. The plaintiff had full notice of this order; and three French deserters having surrendered to him, he detained them six hours without bringing them to head-quarters or reporting their arrival. For this neglect of orders the plaintiff was tried by court-martial, and sentenced to be suspended from his rank of sergeant for a month, and to do the duty and receive the pay of a private soldier during the same time. On the sentence being reported to Colonel Keppel, he did not confirm it, but made an order at the foot of the sentence in the following terms:— "But, as Sergeant Barwis could not be ignorant of the duke's order concerning deserters, and Colonel Keppel thinking his neglect might have been attended with the utmost bad consequences, orders that he be broke, and that Corporal Billow be appointed sergeant in his room." This order was carried into execution, and the plaintiff served accordingly as a private until his battalion returned to England. Colonel Keppel was appointed, in 1762, to command an expedition against the Havannah; and, on his return to England, Barwis brought an action against him for maliciously and improperly reducing him (Barwis) to the ranks. A verdict was found for the plaintiff, with £70 damages, subject to the opinion of the Court of Common Pleas, upon the question, whether the action was maintainable. The court held, that as the whole matter took place abroad, and in the field, in open war, the conduct of the defendant, Colonel Keppel, could not be tried in a civil court. *Per curiam*: "By the Act of Parliament to punish mutiny and desertion, the king's power to make articles of war is confined to his own dominions. When his army is out of his dominions, he acts by virtue of his prerogative, and without the Statute or Articles of War, and, therefore, you cannot argue upon either of them, for they are both to be laid out of this case; and, *flagrante bello*, the common law has never interfered with the army; *silent leges inter arma*. We think (as at present advised) that we have no jurisdiction at all in this case; but if the plaintiff's counsel think proper to speak more fully to this matter, we are willing to hear him." The report contains the following memorandum:—"But plaintiff, seeing the opinion of the court against him, acquiesced, and the judgment was for the defendant, *ut audivi*."

It was intimated, however, by the two Chief-Justices, Lord Mansfield and Lord Loughborough, on a subsequent occasion, that if the conduct of Colonel Keppel had been previously condemned by a court-martial, an action at law would have been maintainable against him, although the transaction in question took place in the field, and in open war.

Again, with respect to the exercise of military power by commanding officers in the execution of actual service, and the right of action against them on such grounds, the following observations fell from the court in Sutton *v.* Johnstone: "Commanders, in a day of battle, must act upon delicate suspicions; upon the evidence of their own eye; they must give desperate commands; they must require instantaneous obedience. In case of a general misbehavior, they may be forced to suspend several officers, and put others in their places. A military tribunal is capable of feeling all these circumstances, and understanding that the first, second, and third part of a soldier's duty is obedience. But what condition will a commander be in, if upon the exercising of his authority he is liable to be tried by a common-law judicature? . . . . Not knowing the law, or the rules of evidence, no commanding or superior officer will dare to act; their inferiors will insult and threaten them. . . . . Upon an unsuccessful battle, there are mutual recriminations, mutual charges, and mutual trials. . . . . . Party prejudices mix. If every trial is to be followed by an action, it is easy to see how endless the confusion, how infinite the mischief must be. The person unjustly accused is not without his remedy. He has the properest among military men. Reparation is done to him by an acquittal; and he who accused him unjustly is blasted forever, and dismissed the service. These considerations induce us to turn against introducing this action."

It may be gathered, also, from the case of Sutton *v.* Johnstone, which was an action between naval officers, that, unless a court-martial shall first expressly decide that it was physically impossible for an officer to execute the orders delivered to him in the field or on actual duty, he has no right of action against his commanding officer for bringing him to a court-martial on a charge of disobedience to those orders, even though the court-martial may have acquitted him of misconduct.

Delay in bringing an officer to a court-martial, after he has been put under arrest, is also no ground of action against the officer ordering the arrest; this being a point of purely military conduct and authority, of which a court-martial alone can properly judge. But if a court-martial should condemn the commanding officer's conduct on such an occasion, an action against him would probably lie. Captain Sutton, of H. M. S. *Isis*, brought an action against Commodore Johnstone, for maliciously bringing him to a court-martial on charges of disobedience to orders during an engagement with a French force in 1781. It appeared that Captain Sutton, after his arrest at the close of the engagement, was carried with the squadron to India, where he was detained in arrest for two years, during a lengthened cruise and various naval operations, before he was eventually sent to England by Admiral Sir Richard Hughes, to be tried. His trial was thus delayed for two years and a half; and great stress was laid on these circumstances, as an unnecessary aggravation of his arrest. But the court said: "The delay is charged to be contrary to the defendant's duty as commander-in-chief. There is no rule of the common or statute law applicable to this case. It is a mere military offence. It is the abuse of a military discretionary power; and the defendant has not been tried for it by court-martial. A court of common law cannot in such a case assume an original jurisdiction. It is like the case of Barwis v. Keppel; this objection we think fatal."

But, although questions regarding the use or abuse of military discipline can thus in some instances be discussed in the civil courts, the learned judges of those tribunals have deprecated the resort to such proceedings in ordinary circumstances; and in Warden v. Bailey, where the court entertained the case, and ordered a new trial, the Chief-Justice, Sir James Mansfield, said, "I must express the strongest wish that the cause will not be again tried, for all disputes respecting the extent of military discipline are greatly to be deprecated, especially in time of war; they are of the worst consequence, and such as no good subject will wish to see discussed in a civil action; they ought only to be the subject of arrangement among military men." In the case which gave rise to the foregoing observations, the learned judges allowed that a considerable amount of unnecessary violence and indignity had taken place.

A recent case of Walton v. Major Gavin of the 16th Lancers, for alleged false imprisonment, gave rise to a very important question with reference to the Article of War which directs that no officer commanding a guard, or provost-marshal, shall refuse to receive or keep any prisoner committed to his charge by any officer or non-commissioned officer belonging to the queen's forces, which officer or non-commissioned officer shall, at the same time, deliver an account in writing signed by himself, of the crime with which the prisoner is charged. And, after very elaborate argument, it was held by Lord Campbell, C. J., and Mr. Justice Coleridge and Mr. Justice Wightman, (Erle, J. dissenting,) that a commanding officer, receiving into his custody a person subject to military law and accused of desertion by a non-commissioned officer who signed the charge, was justified in detaining the prisoner, notwithstanding any irregularity in the proceedings antecedent to his own reception of the prisoner, and was not bound to inquire into the legality of such proceedings. Judgment was therefore given for the defendant. The principle appears to be the same which is applied to the governor or keeper of any ordinary prison, who on receiving a prisoner with a warrant, regular in point of form, for his detention, is justified in receiving him without inquiring whether the magistrate who signs the warrant is duly qualified to act as a justice, or whether in a poaching case the bird mentioned in the warrant, as the *corpus delicti*, was properly designated a partridge.

Negligence in the use of military arms or weapons is also a good cause of action. In Weaver *v.* Ward, the case was, that the plaintiff and defendant were both soldiers of the trained bands of London. While Ward's band was skirmishing, by way of military exercise, with their muskets charged with powder, against another train-band to which Weaver belonged, Ward's musket was discharged in such a manner as to wound the plaintiff, who thereupon brought an action of trespass against Ward. The defence made by Ward was, that he was in training by order of the Lords of the Council, and skirmishing in obedience to military command, and that the injury happened casually, by misfortune, and against his will. But this was decided not to be enough. *Per curiam:* "No man shall be excused of a trespass except it may be judged utterly without his fault. As if a man by force take my hand and strike you, or if here the defendant had said that the plaintiff ran across his piece when it was discharging, or had set forth the case with the circumstances, so as that it had appeared to the court that it had been inevitable, and that the defendant had committed no negligence to give occasion to the hurt."

As a general rule, all language traducing or defaming a man in the way of his profession or calling is actionable, as it tends to his pecuniary damage or loss.

The communication to the Judge-advocate General, by the president of a court-martial, of their opinion, in the form of a censure, respecting the prosecutor's charges, and his conduct in preferring them, is not a libel, and cannot be made the subject of an action at law. This point was decided in 1806, in the case of Jekyll *v.* Moore. Captain Jekyll, of the 43d regiment, had preferred certain charges against Colonel Stewart of the same regiment, who was accordingly tried by a general court-martial, of which Sir John Moore was president. The judgment of the court was, that "the court do most fully and most honorably acquit him:" but to this sentence the following remarks were subjoined: "The court cannot pass without observation the malicious and groundless accusations that have been produced by Captain Jekyll against an officer whose character has, during a long period of service, been so irreproachable as Colonel Stewart's; and the court do unanimously declare that the conduct of Captain Jekyll, in endeavoring falsely to calumniate the character of his commanding officer, is most highly injurious to the good of the service." Captain Jekyll contended that the foregoing passage formed no part of the matter submitted to the judgment of the court, and was, therefore, a libel on him. He accordingly brought his action for it in the Court of Common Pleas, against Sir John Moore, but the whole court was of opinion that no such action could be maintained. Sir James Mansfield, chief-justice: " In order to enable the court-martial to decide upon the charges submitted by the king, they must hear all the evidence, as well on the part of the prosecution as of the defence; and after hearing both sides, are to declare their opinion whether there be any ground for the charges. If it appear that the charges are absolutely without foundation, is the president of the court-martial to remain perfectly silent on the conduct of the prosecutor, or can it be any offence for him to state that the charge is groundless and malicious? It seems to me that the words complained of in this case form part of the judgment of acquittal, and consequently no action can be maintained upon it."

It may perhaps be fairly inferred from the foregoing decision, that if a court-martial pass a censure upon the prosecutor, with reference to a matter which is not expressly connected with the charge under trial before such court-martial, or with the proceedings of the court, the case would stand upon a different footing, and would probably be held actionable on the principle of Mr. Crawford's case already noticed.

Confidential communications from the members of a military court of inquiry to the superior military authorities are likewise privileged, and furnish no ground of action to the officer whose conduct is implicated in the documents.

Neither is the promulgation of a sentence in the gazette by a competent official person to be deemed a libel on the officer named in the paper. In 1807 Lord Wm. Bentinck, governor of Madras, issued the following public order: " The Honorable the Court of Directors having resolved to dismiss Colonel Oliver of this establishment from the service of the Honorable Company, for gross violation of the trust reposed in him as Commanding Officer of the Molucca Islands, the Right Honorable the Governor in Council directs that the name of Colonel Oliver be erased from the Army List of this Presidency, from the 20th June last." In 1811, Colonel Oliver brought an action at Westminster against Lord William Bentinck for the publication of this order, on the ground of its containing libellous matter injurious to the plaintiff. But the Court of Common Pleas decided it to be no libel. Sir James Mansfield, chief-justice.: " How should an officer in India know why he was dismissed, if the reason assigned is not to be made known? If the Court of Directors were peremptorily to dismiss him, without assigning a reason, that would be a greater hardship on the defendant. . . . One should be very sorry to have any thing like a judgment in favor of a plaintiff in such an action as this, than which a more foolish or a more mischievous one cannot easily be imagined; it is much better for the Company, for the country, and for the plaintiff himself, that the cause of his dismissal should be stated, than that it should be supposed that the East India Company did it *suo arbitrio*."

" On the same principle, (says Mr. Justice Heath, in the same case,) when a delinquent, guilty of some enormity, has been brought to a court-martial, the commander-in-chief is not chargeable with libel for directing the sentence to be read at the head of every regiment."

It is decided also, that any communications made by private individuals to superior officers, for the *bona fide* purpose of obtaining redress of grievances, or otherwise invoking the exercise of authority over other officers, will be deemed privileged communications, and no libels.

The principle of the law on this subject, was declared by the court, in Cutler *v.* Dixon, to be this, that, " if actions should be permitted in such cases, those who have just cause of complaint, would not dare to complain for fear of infinite vexation."

But where the author of a written communication traducing another person in his professional character has himself no interest in the matter, the *bona fides* of the proceeding will be no defence against an action. In Harwood *v.* Green, the plaintiff was master of the *Jupiter* transport; and the defendant, a lieutenant in the navy acting as government agent on board, wrote a letter to the secretary at Lloyd's, imputing to Harwood misconduct and incapacity in the management of the vessel. In consequence of this letter, Harwood brought an action against Lieutenant Green for a libel. Lieutenant Green defended himself on the ground that his letter was a privileged communication. But the Lord Chief-Justice Best declared his opinion to the jury, that an officer in the navy had not, as such, the right to make any communication to Lloyd's, but only to the government, by whom, if the matter were important, it might be again communicated to Lloyd's; and the jury gave Harwood a verdict with £50 damages.

It may be useful to mention here, as a legal point giving rights of redress between military men, that a superior officer cannot safely deal for his own advantage, in money matters, with a junior officer under his command. The influence which a senior officer can exercise over his junior is such as to destroy, or at least to control, in the purview of a Court of Equity, that entire freedom which is essential to the perfection of a bargain or contract; and if a regimental officer places himself in a position where such influence may operate to the prejudice of the junior, the transactions between them are liable to be set aside for want of fairness or conscientiousness. This is the rule applied to dealings between a guardian and his ward, a physician and his patient, a landlord and his steward, a clergyman and a penitent, and all other cases where the existence of a just and unavoidable influence may lead to abuse.

II. *Wrongs towards Persons not under Military Authority.*—Injuries may be occasioned to persons not subject to military authority, by officers mistaking or exceeding their powers, or exercising them with malice, negligence, or unskilfulness; but for acts of this kind a remedy lies only in the civil courts; the military tribunals, as already observed, having no power to grant pecuniary compensation by way of damages, and non-military persons having no *locus standi* as prosecutors before such courts, which are instituted solely for the maintenance of order and discipline among the armed forces.

In cases of the kind now under consideration, it is quite immaterial whether the cause of action has arisen within the realm, or beyond the seas; though this proposition was not finally established until the year 1774, when the great case of Fabrigas *v.* Mostyn was determined in the Court of King's Bench, and put an end to all further question or doubt upon the subject. The plaintiff was a native of Minorca, of which island the defendant, General Mostyn, was governor. The general had by his own absolute authority imprisoned the plaintiff and banished him from the island without a trial. The defence was, that in the peculiar district of Minorca, where the offence occurred, no ordinary court or magistrate had jurisdiction. But the proof of this defence failed, and the jury gave the plaintiff £3,000 damages. The objection, however, was taken that the action did not lie, by reason of the foreign locality of the cause of it, and the point was twice argued at great length; but judgment was eventually pronounced against General Mostyn, in accordance with the verdict of the jury. It should be noticed also that, as General Mostyn happened to be a governor, his appointment gave him the character of a viceroy, so that *locally and during his government* no civil or criminal action lay against him. On principles of public justice, therefore, it was necessary that a remedy should be had in England.

The *undue assumption* or *mistaken* exercise of authority by officers towards non-military persons, is a clear ground of action against them in the civil courts, even though there be no malice accompanying the transaction.

Captain Gambier, of the navy, under the orders of Admiral Boscawen, pulled down the houses of some sutlers on the coast of Nova Scotia, who supplied the seamen of the fleet with spirituous liquors. The act was done with a good intention on the part of the admiral; for the health of the sailors had been affected by frequenting these houses. Captain Gambier, on his return to England, incautiously brought home in his ship one of the sutlers whose houses had been thus demolished. The man would never otherwise have got to England; but on his arrival he was advised to bring an action against Captain Gambier. He did so, and recovered £1,000 damages. But as the captain had acted by the orders of Admiral Boscawen, the representatives of the admiral defended the action, and paid the damages and costs. This was a favorable case, unaccompanied by any malicious feeling; but the parties concerned did not attempt to disturb the verdict.

Admiral Sir Hugh Palliser was defendant in a similar action for destroying fishing huts on the Labrador coast. After the treaty of Paris, the Canadians, early in the season, erected huts for fishing, and by such means obtained an advantage over the fishermen who came from England. It was a nice question upon the rights of the Canadians. But the admiral, on grounds of public policy, ordered the huts to be destroyed. An action was brought against him in England by one of the injured parties, and the case ended in arbitration. But on the part of the admiral it was never contended that the action did not lie by reason of the subject-matter of it having occurred beyond the seas.

"I remember," said Lord Mansfield, "early in my time being counsel in an action brought by a carpenter in the train of artillery against Governor Sabine, who was governor of Gibraltar, and who had barely confirmed the sentence of a court-martial, by which the plaintiff had been tried and sentenced to be whipped. The governor was very ably defended, but nobody ever thought that the action would not lie; and it being proved that the tradesmen who followed the train were not liable to martial law, the court were of that opinion, and the jury found the defendant guilty of the trespass, as having had a share in the sentence, and gave £700 damages."

The following case, involving the same principle, occurred in India, and was there tried before the Supreme Court of Madras. Mr. H. Smith was agent, at Secunderabad, of a mercantile house at Madras, from whom he received a very handsome salary. He became indebted to a soldier of H. M.'s 33d regiment for some work intrusted to him, and a dispute having arisen between them as to the amount, this led to a violent altercation between Mr. Smith and the superintendent of the bazaar acting under local military regulations. Lieutenant-colonel Gore thereupon sent a file of men to arrest the plaintiff, who was accordingly seized about six o'clock in the evening, and marched from his house through the streets of the cantonment to the main guard at Secunderabad, where he was kept till twelve o'clock the next day. In consequence of these proceedings, he brought an action against Colonel Gore for false imprisonment. Secunderabad was an open cantonment for a part of the subsidiary force serving in the territories of the Nizam; the force consisting partly of British and partly of native troops. It had barracks, and the men were hutted. It was also upon a field establishment, constantly ready for immediate service. The Article of War then in force, being the 22d in the 11th section of the Statute 27 Geo. II., was thus intituled, "Of duties in quarters, in garrison, and in the field;" and it enacted, "that all sutlers and retainers to the camp, and all persons whatsoever serving with forces *in the field*, though not enlisted soldiers, are to be subject to orders, according to the rules and discipline of war." Sir Thomas Strange, C. J.: "The question was, whether the troops, *being cantoned*, were in the state to which the cited Articles of War applied. The court thought they were not. It might have been a field force, being upon a field establishment, so as to be ready to move at the shortest notice. There might be great similarity in the arrangements adopted for an army, whether in the field or cantoned. A respectable witness, Brigade-major Lyne, intimated as much. Still, so far as the court could form a judgment upon a question of this nature, there seemed to be a difference between a camp and a cantonment, which appeared material. . . . . . When in the field, not only the army, but its appendages, must be under the immediate control of the officer commanding it, according to the rules and discipline of war. So situated, the sutler, who chose to follow the camp, identified himself in a manner with the soldier for every purpose almost but that of fighting. . . . . . The plaintiff called upon the court to say, whether the force in question, under the command of the defendant, was at the time in the field. It seemed impossible to say that it was, without confounding ideas apparently very distinct. . . . . . The defendant appeared to have acted under a mistake of his authority, for which he was liable to answer, as it had been productive of serious injury to the plaintiff." Judgment was therefore given against Colonel Gore, with fifty pagodas damages.

In the foregoing case reference was made to an action brought by Mr. Robert Bailie, an up-country trader in the province of Bengal, against Major-general Robert Stewart, for an assault and false imprisonment. Mr. Bailie had resided within the cantonments of Cawnpore for many years, and dealt in European articles, which he principally disposed of to the military stationed there. In October, 1797, upon a complaint made to him by one of the people of his Zenanah, he tied up and very severely flogged one of his *chowkydars*. For this act Major-general Stewart ordered Mr. Bailie to be tried by court-martial; and as he acknowledged to have used no less than six switch whips in the flogging, alleging as his reason, that as they were new whips, he was afraid of breaking them and spoiling their sale, the court-martial sentenced him to five days' imprisonment, and to make an apology to the commanding officer. This sentence General Stewart, though he did not approve of it, confirmed; and issued orders for Mr. Bailie to depart the camp as soon after his enlargement as possible. The Supreme Court of Calcutta held Mr. Bailie to be a sutler within the meaning of the Articles of War, so as to render him amenable to military law. But in the above-mentioned action of Smith *v.* Lieut.-col. Gore, the chief-justice, Sir T. Strange, declined to be governed by the decision in General Stewart's case, as the note furnished to the court did not clearly show whether or not the army was in the field when the transaction occurred.

An *unreasonable* or *malicious* exercise of power will, in like manner, render an officer liable to an action for damages. An instance of this occurred in the year 1783, when an action was brought against General Murray, governor of Minorca, for improperly suspending the judge of the Vice-admiralty Court of that island. The general had professed himself ready to restore the judge on his making a particular apology; and, on reference to the home authorities, the king approved of the suspension, unless the governor's terms were complied with. There was no doubt as to General Murray's power to suspend the judge for proper cause; yet, on the proof of his having unreasonably and improperly exercised the authority, and notwithstanding the king's approbation of his proceedings, damages to the amount of £5,000 were awarded against him by a jury; and, as Mr. Baron Eyre observed, it never occurred to any lawyer that there was any pretence for questioning the verdict.

Negligence or *unskilfulness* in the exercise of an officer's duty may also be a cause of action for damages in respect of private injuries thus occasioned; and in such cases the approval of an officer's conduct by the government, or by the superior military authorities, will neither relieve him from liability to an action, nor have any influence upon the decision of the courts of Westminster Hall. Those tribunals investigate such matters on independent evidence, according to their own rules, and pay no regard to the previous conclusions of official functionaries, however high their rank may be.

It is a rule of English law, in unison with the law of nations, by which all civilized states are governed, that no officer engaged in military operations in his country's cause, by the order or with the sanction of the constituted authorities, shall incur any individual or private responsibility for acts done by virtue of his commission or official instructions. Such transactions being of a public nature, redress or satisfaction for injuries to which they give birth, is to be sought by public means alone, from the sovereign power of the belligerent or offending state, according to the principles of international law, and the general usages of civilization, which never suffer such matters to be litigated before ordinary tribunals.

If, in time of peace, the citizens of a friendly foreign state sustain a private injury at the hands of a naval or military officer serving under the orders of the British government, but unauthorized by his commission or instructions to do the act complained of, the ordinary tribunals of England afford the same redress against him as in the case of a British subject similarly aggrieved; and this rule applies even in those cases where the violated rights of the foreigner are such as the law of England denies or prohibits to its own subjects.

But if the British government have expressly instructed the officer to commit the act which constitutes or gives occasion to the grievance, the matter becomes an affair of state which is not cognizable by the courts of law, and must be adjusted by diplomatic arrangement between the two governments concerned. In such cases also it is quite sufficient, if the officer's proceedings, though not originally directed or authorized by the terms of his instructions, are afterward sanctioned and adopted by the government; for this renders them public acts, over which courts of law have no jurisdiction. (Consult PRENDERGAST's *Law relating to Officers of the Army.*)

INJURING PRIVATE PROPERTY. (*See* WASTE or SPOIL.)

INLYING PICKET. A body of infantry or cavalry in campaign, detailed to march, if called upon, and held ready for that purpose in camp or quarters.

INSPECTORS-GENERAL. There are two inspectors-general of the army with the rank of colonel. Assistant adjutants-general are ex-officio assistant inspectors-general. The duties of inspectors-general are prescribed by Army Regulations. In the French army, a certain number of general officers are annually designated to make inspections, and such inspections embrace every thing relative to organization, recruiting, discharges, administration, accountability for money and property instruction, police, and discipline of the several corps of the army. At these inspections all wrongs are redressed, and each inspection is continued from eight to ten days. The inspector examines and studies the condition of the corps under arms, as well as off parade; he receives all applications for discharge, and for the retired list. He notes those who merit promotion, rewards, or reprimands. He assembles the council of administration, and verifies their accounts; visits the storehouses, quarters, hospitals, prisons; inspects the clothing, arms, &c., &c., and, in fine, scrutinizes every thing which it is desirable should be known. He gives his orders to the regiment for the ensuing year, and makes a detailed report of what he has seen and done.

INSURRECTION. (*See* CALLING FORTH MILITIA.) It will be observed that whenever the President of the United States is authorized by law to *use* the military force in cases of insurrection or obstruction to the laws, he must first, by proclamation, have commanded the insurgents to disperse and retire peaceably to their respective abodes within a limited time; (*Act* Feb. 28, 1795. *See* OBSTRUCTIONS TO THE LAWS.)

INTERIOR FLANKING ANGLE—is formed by the line of defence and the curtain.

INTERIOR SIDE—is the line drawn from the centre of one bastion to that of the next, or the line of the curtain produced, to the two oblique radii of the front.

INTRENCHED CAMP. A position is so called when occupied by troops, and fortified for their protection during the operations of a campaign.

INTRENCHMENT. A ditch or trench with a parapet; field-works. In permanent fortification, intrenchments are made in various parts of the works to prolong the defence, as a breast-work and ditch at the gorge of the bastion, &c.

INUNDATION. An inundation or collection of water is produced by forming across a stream one or more dams.

INVASION. (*See* CONSTITUTION; CALLING FORTH MILITIA; NATIONAL DEFENCE.)

INVERSION. In case a column, marching right in front, shall be under the necessity of forming into line faced to the reverse flank by the promptest means, the command is given: Halt! By inversion right into line wheel, battalion guide right. This movement will give an order of battle with the left company occupying the right of the battalion, and the right the left.

Inversions are very important in the field, and they offer such great advantages, that Bonaparte strongly advised their employment in many circumstances. Our tactics admit the employment of inversions in the formations to the right and left in line of battle, and also in the successive formations, except in that of *faced to the rear into line of battle*. When used, the first command always begins, *By inversion*. (*See* INFANTRY.)

INVEST. To take the initiatory measures to besiege a town, by securing every road and avenue leading to it, to prevent ingress or egress.

IRON PLATES. In the experiments made against the "Undaunted," at Portsmouth, the following results were obtained:—Six wrought-iron 68-lb. shot were fired with a charge of 16 lbs. at 200 yards, the iron plates being $4\frac{1}{2}$ in. thick; four of these shot broke the plates, but did not penetrate the timber; two passed entirely through both plates and timber. Forty-three cast-iron 68-lb. shot were fired against other plates of similar thickness. Of these, four passed through the plates but not the timber. Nine passed through both; but there was only one case of a shot taking good effect after striking an uninjured plate. Thus of the four shots that passed through the plates without penetrating the timber, only one went through a plate that had not been previously weakened.

The shot that penetrated entirely through the plates and the timber had all passed through plates previously weakened. No penetration was effected by red-hot 68-lb. shot, with a charge of 10 lbs. The 3 and $2\frac{1}{2}$-in. plates were all penetrated by 68-lb. shot and shells.

The following conclusions have been drawn from experiments:—

1st. That thin plates of wrought iron are proof against any shells; for, though the shells may pass through the plates, they will be in a broken state.

2d. That being proof against shells will avail little, unless vessels are likewise proof against solid shot; for shells would, of course, not be fired against ships proof against them, whereas the destructive effects produced by fragments of shot and of plates, and the great damage done to the scantling of the ship by solid shot, appear more like the result of a shell than of a shot.

3d. That rifled projectiles produce greater effect than spherical projectiles of the same weight at long than at short ranges, on account of the rifled elongated projectiles—the resistance to which is a minimum—retaining more of their initial velocity than spherical projectiles at the same distance.

4th. That the thickness of plates required to resist shot fired from the heaviest nature of guns, must not be less than $4\frac{1}{2}$ in.

5th. That, to secure the resistance of the plates and the impenetrability of the sides of a ship, it is indispensable that the plates be strongly backed by masses of the strongest and most resisting timber, as, in all the cases to which reference has just been made, it appears that the plates are easily broken when the support is removed from behind them, by the crushing, fracturing, and damaging effects of the impacts of the shot; (Sir HOWARD DOUGLAS.)

With the knowledge of these data, an iron-clad ship, "Le Gloire," has been built in France, carrying 38 rifled 50-pounders, and France, it is said, will soon have 300 rifled guns in such vessels.

In England, the iron-clad "Warrior," 420 feet long and over 6,000 tons' burden, has been built. The new principle introduced in England, of inclining the iron-clad sides inwardly, so as to make an angle with the horizontal of from 35° to 40°, will cause the shot to glance off, with little injury to the sides. In addition to this, it is proposed to suppress the port-holes, and place the guns in rotating iron cupolas, from which, by a rotatory of 180°, they fire over the bulwarks on either broadside —the gunners being perfectly sheltered under these shot-proof covers; (BARNARD's *Sea-coast Defence*.) The great objection to such an arrangement is its unwieldiness, and the opinion of distinguished officers— that iron plates are only practicable for floating batteries, gunboats, and other vessels of small draft of water, for special purposes, may prove the better opinion, notwithstanding the great outlay made by the French and English governments.

J

JOISTS. The timbers of a floor, whereto boards or lathing for ceiling are nailed. They either rest on the wall or on girders, or sometimes on both. (*See* CARPENTRY.)

JOURNAL, OR ITINERARY. *Directions for keeping the journal of a march west of the Mississippi.*—The journal should be kept in a pocket note-book; or, if one cannot be obtained, in a book made of sheets of paper folded to half the letter size. The record is to run from the bottom to the top of each page. The horizontal divisions in the column headed "*Route*," represent portions of a day's march. The distance, in miles, between each of the horizontal divisions, will be noted in the column headed "*Distance*," which will be summed up at the top of each column, and the sum carried to the bottom of the next column. The notes within each horizontal division are to show the general directions of the march, and every object of interest observed in passing over the distance represented thereby; and all remarkable features, such as hills, streams with their names, fords, springs, houses, villages, forests, marshes, &c., and the places of encampment, will be sketched in their relative positions. The "*Remarks*" corresponding to each division will be upon the soil, productions, quantity and quality of timber, grass, water, fords, nature of the roads, &c., and important incidents. They should show where provisions, forage, fuel, and water can be obtained; whether the streams to be crossed are fordable, miry, have quicksands or steep banks, and whether they overflow their banks in wet seasons; also the quality of the water; and, in brief, every thing of practical importance. When a detachment leaves the main column, the point on the "*Route*" will be noted, and the reason given in the *Remarks*. The commander of the detachment will be furnished with a copy of the journal up to that point, and will continue it over his new line of march.

JOURNAL of the march of [*here insert the names of the regiments or companies composing the column,*] commanded by ———— ————, from [*here insert the point of departure*] to [*the stopping place,*] pursuant to [*here give the No. and date of order for the march.*]

Date.	Hour.	Weather.	Distance.	Route.	REMARKS.
1860. July 8.	5. A. M.	Cloudy, with wind.—Cold early in morning.—Cloudy.	Total, 19 — 3	S.S.E.	Road rocky; but little grass; good water. Plenty of timber on summit of hills, extending 3 miles; road to right of hills.
	1 P. M.		8	High timbered Peak △ Camp No. 1. Springs.	Good shelter for camp at foot of peak; fuel plenty. Springs of sweet water, with good grass near. Road to this point rather more sandy.
	10.		3	MSS	Road runs through a cañon ¼ mile long, to right of a small stream; marsh on left of stream; water sweet; grass excellent. Halted to graze two hours. No Indian signs.
	6.30.		1	Mt. P—— × Det.	Companies F, G, and I, 3d ————, detached at Mt. P——, under command of ————, (see par. 3, General Orders, No. ——,) to take road to ————. A small creek, easily forded.
July 7.	6. 4.30.		4	M River S.S.W.	Road turns short to right at top of hill after crossing river; crossing good, but a little boggy on right bank. This bottom shows signs of recent overflow, when it must have been impassable; banks low; water sweet; no wood near crossing; road hard and good up to river.

JOURNAL, *(Continued.)*

Date.	Hour.	Weather.	Distance.	Route.	REMARKS.
1860.			Total, 47		
July 9.	4.30 A. M.	Rain.	5	*Fork in Road.*	At the point where the road forks, turn to the right. The left-hand road leads to a deep ravine, which cannot be crossed.
	4.30 P. M.	Very pleasant; cloudy in the P. M.	3	△ *Camp No. 2.* *Springs Ravine* S.S.E.	After the road strikes the ravine, it runs one mile along its bank before coming to the crossing place. The camping ground is at springs, half a mile beyond the ravine. Old Indian signs at the springs.
	3 P. M.		15	× *Grave.* Mt. T— S.S.E.	Road less rocky; last three miles rather sandy; no water. Passed at the point marked + an Indian grave.
July 8.	9. 6.30 A. M.		5	*Springs.* S.S.W.	Road still rocky; good springs, where casks should be filled. No more water for twenty miles after leaving springs. Occasional hills to left of road; no wood or grass.
			19		

JUDGE-ADVOCATE. There is one judge-advocate selected from the captains of the army with the brevet rank and pay of a major of cavalry. The judge-advocate, or some person deputed by him, or by the general, or officer commanding the army, detachment, or garrison, shall prosecute in the name of the United States, but shall so far consider himself as counsel for the prisoner, after the said prisoner shall have made his plea, as to object to any leading question to any of the witnesses, or any question to the prisoner, the answer to which might tend to criminate himself. The judge-advocate administers the prescribed oaths to the court and witnesses; (ART. 69.)

The appropriate functions of the judge-advocate, as an essential officer in all general courts-martial, are various in their nature; and as the Articles of War do not describe them with much precision, it is proper to resort to the less positive, though equally binding authority, of established usage and practice.

The Articles of War are silent on the subject of the judge-advocate's assisting the court with his counsels and advice as to any matters of form or law; it nevertheless is his duty, by custom, to explain any doubts which may arise in the course of its deliberations, and to prevent any irregularities or deviations from the regular form of proceedings. The duty assigned the judge-advocate by Art. 69, is more especially incumbent on him in cases where the prisoner has not the aid of professional counsel to direct him, which generally happens in the trials of private soldiers, who, having had few advantages of education, or opportunities for mental improvement, stand greatly in need of advice under circumstances often sufficient to overwhelm the acutest intellect, and embarrass or suspend the powers of the most cultivated understanding. It is certainly not to be understood that, in discharging this office, which is prescribed solely by humanity, the judge-advocate should, in the strictest sense, consider himself as bound to the duty of counsel, by exerting his ingenuity to defend the prisoner, at all hazards, against those charges which, in his capacity of prosecutor, he is, on the other hand, bound to urge, and sustain by proof; for, understood to this extent, the one duty is utterly inconsistent with the other. All that is required is, that in the same manner as in civil courts of criminal jurisdiction, the judges are understood to be counsel for the person accused, the judge-advocate, in courts martial, shall do justice to the cause of the prisoner, by giving full weight to every circumstance or argument in his favor; shall bring the same fairly and completely into the view of the court; shall suggest the supplying of all omissions in exculpatory evidence; shall engross in the written proceedings all matters which, either directly or by presumption, tend to the prisoner's defence; and finally, shall not avail himself of any advantage which superior knowledge or ability, or his influence with the court may give him, in enforcing the conviction, rather than the acquittal, of the person accused.

When a court-martial is summoned by the proper authority, for the trial of any military offender, the judge-advocate, being required to attend to his duty, and furnished with articles of charge or accusation, on which he is to prosecute, must, from the information of the accuser, instruct himself in all the circumstances of the case, and by what evidence the whole particulars are to be proved against the prisoner. Of these, it is proper that he should prepare, in writing, a short analysis, or plan, for his own regulation in the conduct of the trial, and examination of the witnesses. He ought then, if it has not been done by some other functionary, to give information to the prisoner of the time and place appointed for his trial, and furnish him, at the same time, with a copy of the charges that are to be exhibited against him, and likewise a correct detail of the members of the court. The judge-advocate ought then to hand in to the adjutant-general, or staff-officer charged with the details, a list of witnesses for the prosecution, in order that they may be summoned to give their attendance at the time and place appointed.

It is proper, likewise, that he should desire the prisoner to make a similar application, to insure the attendance of the witnesses necessary for his defence. These measures ought to be taken as early as possible, that there may be sufficient time for the arrival of witnesses who may be at a distance. When the court is met for trial, and the members are regularly sworn, the judge-advocate, after opening the prosecution by a recital of the charges, together with such detail of circumstances as he may deem necessary, proceeds to examine his witnesses in support of the charges, while at the same time he acts as the recorder or clerk of the court, in taking down the evidence in writing at full length, and as nearly as possible in the words of the witnesses. At the close of the business of each day, and in the interval before the next meeting of the court, it is the duty of the judge-advocate to make a fair copy of the proceedings; which he continues thus regularly to engross till the conclusion of the trial, when the whole is read over by him to the court, before the members proceed to deliberate and form their opinions. The sentence of the court must be fairly engrossed and subjoined to the record copy of the proceedings; and the whole must be authenticated by the signature of the president of the court, and that of the judge-advocate.

It is required by the Articles of War, (ART. 90,) that "every judge-advocate, or person officiating as such, at any general court-martial, shall transmit, with as much expedition as the opportunity of time and distance of place can admit, the original proceedings and sentence of such court-martial, to the Secretary of War; which said original proceedings and sentence shall be carefully kept and preserved in the office of the said secretary, to the end that the persons entitled thereto, may be enabled, upon application to the said office, to obtain copies thereof." The judge-advocate sends the proceedings to the Secretary of War through the adjutant-general.

The judge-advocate cannot be challenged. He may be relieved at any time. He should, in complicated cases, arrange and methodize the evidence, applying it distinctly to the facts of the charge. Besides applying the evidence fairly to each side of the question, he should inform the court as to the *legal* bearing of the evidence, for there may have been admitted evidence which ought to be rejected from their minds as illegal; (Hough's *Military Law Authorities.*)

JURISDICTION. All officers, conductors, gunners, matrosses, drivers, or other persons whatsoever, receiving pay, or hire, in the service of the artillery, or corps of engineers of the United States, shall be governed by the aforesaid rules and articles, and shall be subject to be tried by courts-martial, in like manner with the officers and soldiers of the other troops in the service of the United States; (Art. 96.)

The officers and soldiers of any troops, whether militia or others, being mustered and in pay of the United States, shall at all times and in all places, when joined or acting in conjunction with the regular forces of the United States, be governed by these Rules and Articles of War, and shall be subject to be tried by courts-martial, in like manner with the officers and soldiers in the regular forces, save only that such courts-martial shall be composed entirely of militia officers; (Art. 97.)

No person shall be liable to be tried and punished by a general court-martial for any offence which shall appear to have been committed more than two years before the issuing of the order for such trial, unless the person, by reason of having absented himself, or some other manifest impediment, shall not have been amenable to justice within that period; (Art. 88.)

JURISDICTION, (Concurrent.) Can courts-martial and civil courts have concurrent jurisdiction over offences committed by soldiers? Or, in other words, if a soldier is guilty of an offence which renders him amenable for trial before the civil courts of the land, can he also be tried for that offence (if its specification should establish a violation of the Rules and Articles of War) by a court-martial?

By the Constitution of the United States Congess is authorized " to make rules for the government and regulation of the land and naval forces; " and Congress, pursuant to this authority, has established rules and articles for the government of the armies of the United States. These rules are an additional code, to which every citizen who becomes a soldier subjects himself for the preservation of good order and military discipline. The soldier, however, is still a citizen of the United States. He has not, by assuming the military character, become, as in many European countries, a member of a privileged body who may claim trial for all offences by courts-martial. He is still amenable to the ordinary common law courts for any offences against the persons or property of any citizen of any of the United States, such as is punishable by the known laws of the land; (ART. 33.) An examination of the Rules and Articles of War will show that the offences therein described, and against which punishment is denounced, are purely military. They are crimes which impair the efficiency of the military body, and even in cases, in which they would be recognized as offences by the ordinary common law courts, they could not be considered the *same offences.*

Take, for instance, Article 9, which inflicts the punishment of death or other punishment, according to the nature of his offence, upon any officer or soldier who shall strike his superior officer. Here is an offence punishable under the known laws of the land as an assault and battery, and, as such, it could be tried by the common law courts. But such trial would not prevent a court-martial from afterwards taking cognizance of it under Article 9; for the offence before the common law court would be striking an *equal,* while before the military court it would have essentially changed its character.

Again, suppose an officer had been guilty of stealing, he might be prosecuted before the common law court for the felony, and afterwards charged with conduct unbecoming an officer and a gentleman, and dismissed the service. It can hardly be contended that the offences in either of the cases cited would be the *same* before the different courts; and if not, Article 87, which forbids a trial a second time for the same offence, could not be pleaded in bar of trial. Recognizing, then, the principle that the soldier, as citizen, is subject to the common law courts for offences committed against the well-being of the State, it must also be recollected that he is subject to trial by a court-martial for any violation of the Rules and Articles of War.

In the case of "Eels, plaintiff in error; *v.* the People of the State of Illinois," it was urged that the act of the State of Illinois under which Eels was tried was void, as it would subject the delinquent to a double punishment for the same offence, the crime with which he was charged being actionable under a law of the United States. The Supreme Court decided that, admitting the plaintiff in error to be liable to an action under the act of Congress, it did not follow he would be twice punished for the same *offence,* and gave the following definition of that term:

"An offence in its legal signification means the transgression of a law. A man may be compelled to make reparation in damages to the injured party, and be liable also to punishment for a breach of the public peace in consequence of the same act, and may be said, in common parlance, to be twice punished for the same offence. Every citizen of the United States is also a citizen of a State or Territory. He may be said to owe allegiance to two sovereigns and may be liable to punishment for an infraction of the laws of either. The same act may be an offence or transgression of the laws of both. Thus an assault upon the marshal of the United States and hindering him in the execution of legal process is a high offence against the United States, for which the perpetrator is liable to punishment; and the same act may also be a gross breach of the peace of the State, a riot, assault, or a murder, and subject the same person to a punishment under the State laws for a misdemeanor or felony. That either or both may, if they see fit, punish such an offender cannot be doubted. Yet it cannot be truly averred that the offender has been twice punished for the same offence, but only that by one act he has committed two offences, for each of which he is justly punishable. He could not plead the punishment by one in bar to a conviction by the other; consequently, this court has decided, in the case of Fox v. the State of Ohio, (5 Howard, 432,) that a State may punish the offence of altering or passing false coin as a cheat or fraud practised on its citizens; and, in the case of the United States v. Marigold, (9 Howard, 560,) that Congress, in the proper exercise of its authority, may punish the same act as an offence against the United States.

K

KEEP. To keep troops is to maintain organized forces.

KIT. A cant word among soldiers to express the necessary articles provided for them, and which they are obliged to keep in order.

KITCHEN. For proposed kitchen-cart for field service *see* WAGON.

KNAPSACK. A square frame covered with canvas carried on an infantry soldier's back, containing his clothing and other necessaries, but not his rations.

KNOTS. The three elementary knots, which every one should know, are here represented (Fig. 147)—viz., the Timber-hitch, the Bowline, and the Clove-hitch. The virtues of the timber-hitch are, that, so long as the strain upon it is kept up, it will never give; when the strain is taken off, it is cast loose immediately. The bowline makes a knot difficult to undo; with it the ends of two strings are tied together, or a loop made at the end of a single piece of string, as in the drawing. For slip-nooses, use the bowline to make the draw-loop. The clove-hitch binds with excessive force, and by it, and it alone, can a weight be hung to a smooth pole, as to a tent-pole. A kind of double clove-hitch is generally used, but the simple one suffices, and is more easily recollected.

Fig. 147.

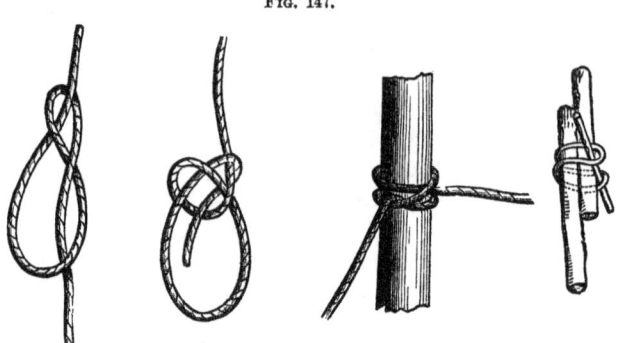

The following additional remarks deserve attention :—A timber-hitch had better have the loose end twisted more than once ; it is liable to slip, if not. To tie a bowline, or any other knot for temporary purposes, insert a stick into the knot before pulling tight. The stick will enable you, at will, to untie the knot—to break its back, as the sailors say—with little difficulty. A bowline is firmer, if doubled ; that is, if the lower loose end in the figure be made to wrap round a second time. A double clove-hitch is firmer than a single one ; that is, the rope should make two turns, instead of one turn, round the pole beneath the lowest loose end in the figure. To make a large knot at the end of a piece of string, to prevent it from pulling through a hole, turn the end of the string back upon itself, so as to make it double, and then tie a common knot. The string may be quadrupled instead of doubled, if required. A *toggle and strap* is a tourniquet. A single or a double band is made to inclose the two pieces of wood it is desired to lash together. Then a stick is pushed into the band and forcibly twisted round. The band should be of soft material, such as the strands of a rope that has been picked to pieces on purpose. The strands must, each of them, be untwisted and well rubbed with a stick to take the kink out of them, and finally twisted in a direction opposite to their original one ; (GALTON's *Art of Travel*.)

L

LADDER BRIDGE—may be formed by running a cart or gun limber into the stream and securing it there, with the shafts in a vertical position, by ropes from both sides of the river ; one end of a ladder from each bank resting upon it, and covering the steps or rungs with planks.

LADDERS. (*See* ESCALADE.)

LANCE. The lance is composed of a sharp steel blade, from 8 to 10 inches long, grooved like a common bayonet with a socket at its base and two iron straps for attaching it to the handle. The handle is of strong light wood, with a tip of iron at its lower end and a leathern loop at its centre of gravity to support and guide the lance. It is usually from $8\frac{1}{2}$ to 11 feet long, and weighs about $4\frac{1}{2}$ lbs. This weapon is not used in the United States service. The Russians have their regular and irregular Cossacks armed with the lance. The Austrians, also, have lancers; but the Polish cavalry use the lance better than any other people. The lance, when not in use, rests in a leather boot attached to the stirrup, the right arm being passed through the leather loop of the lance; or by putting the lower end in the boot and strapping the handle to the pommel of the saddle. Lancers are more formidable than other cavalry because they are able to reach further. Skill in combating a lancer, consists in keeping to his left, in order to shun his lance. Pressed too nearly, the lancer must have recourse to his sabre and let his lance rest upon his arm. The moment in which he attempts to seize his sabre is dangerous to him. The Mexican cavalry are generally lancers.

LANDING. (*See* DISEMBARKATION and EMBARKATION.)

LASHES. A general court-martial may sentence a soldier to receive fifty lashes for desertion. No other crime is punishable with lashes.

LAW—is a rule of action prescribed by a superior power.

Natural law is the rule of human action prescribed by the Creator, discoverable by the light of reason.

Divine law is the law of nature revealed by God himself.

The law of nations is that which regulates the conduct and mutual intercourse of independent nations with each other, according to reason and natural justice. (*See* WAR.)

Municipal or civil law is the rule of civil conduct prescribed by the supreme power in a State, commanding what is right, and prohibiting what is wrong.

The parts of a law are: 1. The declaratory; which defines what is right and wrong. 2. The directory; which consists in commending the observation of right, or prohibiting the commission of wrong. 3. The remedial; or method of recovering private rights, and redressing private wrongs. 4. The vindicatory sanction of punishments for public wrongs; wherein consists the most forcible obligation of human laws.

To interpret a law, we must inquire after the will of the maker; which may be collected either from the words, the context, the subject matter, the effects and consequence, or the spirit and reason of the law.

From the latter method of interpretation arises equity, or the correction of that wherein the law (by reason of its universality) is deficient; (BLACKSTONE's *Commentaries*.)

LAW, (MARTIAL.) By martial law is understood, not laws passed for raising, supporting, governing, and regulating troops, but "it is in truth and reality *no* law, but something *indulged*, rather than allowed as law;" (HALE and BLACKSTONE.) The Constitution of the United States has guarded against the effects of any declaration of martial law within the United States, by providing: " No person shall be held to answer for a capital or otherwise infamous crime, unless on a presentment or indictment of a grand jury, except in cases arising in the land or naval forces, or in the militia, when in actual service in time of war or public danger; nor shall any person be subject for the same offence to be twice put in jeopardy of life or limb; nor shall be compelled, in any criminal case, to be witness against himself, nor be deprived of life, liberty, or property, without due process of law; nor shall private property be taken for public use without just compensation," (ART. 5, *Amendments;*) and further, "In all criminal prosecutions, the accused shall enjoy the right to a speedy and public trial, by an impartial jury of the State and district wherein the crime shall have been committed, which district shall have been previously ascertained by law, and to be informed of the nature and cause of the accusation; to be confronted with the witnesses against him; to have compulsory process for obtaining witnesses in his favor; and to have the assistance of counsel for his defence;" (ART. 6, *Amendments*.)

Within the United States, therefore, the effect of a declaration of martial law would not be to subject citizens to trial by courts-martial, but it would involve simply a suspension of the writ of *habeas corpus*, under the authority given in the 2d clause of Sec. 9 of the Constitution, viz.: "The privilege of the writ of *habeas corpus* shall not be suspended unless when, in cases of rebellion or invasion, the public safety may require it."

The universal practice of all nations has been to give supremacy to the military commander in all sieges. "*Inter arma silent leges*," is then a maxim universally admitted. The public safety in that case imperiously requires that the orders of the commander of the troops should be obeyed, and a commander in the United States is then only justified, *ex necessitate rei*, in suspending the privilege of the writ of *habeas corpus*.

The suspension of this privilege would enable a commander to incarcerate all dangerous citizens; but when brought to trial, the citizen would necessarily come before the ordinary civil courts of the land.

Beyond the United States, troops take with them the Rules and Articles of War, but not the municipal law, to which they are also subject at home. It is necessary, therefore, for a commander, in the absence of laws made by Congress, to declare his own will, commanding what is right, and prohibiting and punishing what is wrong, in the new relation established between his army and the citizens of the foreign country. The following order was the declaration of martial law by Gen. Scott in Mexico:—

HEAD-QUARTERS OF THE ARMY,
National Palace of Mexico, Sept. 17, 1847.

GENERAL ORDERS—No. 287.

The General-in-Chief republishes, with important additions, his General Orders, No. 20, of February 19, 1847, (declaring MARTIAL LAW,) *to govern all who may be concerned.*

1. It is still to be apprehended that many grave offences not provided for in the act of Congress " establishing rules and articles for the government of the armies of the United States," approved April 10, 1806, may be again committed—by, or upon, individuals of those armies, in Mexico, pending the existing war between the two republics. Allusion is here made to offences, any one of which, if committed within the United States or their organized territories, would, of course, be tried and severely punished by the ordinary or civil courts of the land.

2. Assassination, murder, poisoning, rape, or the attempt to commit either; malicious stabbing or maiming; malicious assault and battery; robbery; theft; the wanton desecration of churches, cemeteries, or other religious edifices and fixtures; the interruption of religious ceremonies; and the destruction, except by order of a superior officer, of public or private property, are such offences.

3. The good of the service, the honor of the United States, and the interests of humanity, imperiously demand that every crime enumerated above should be severely punished.

4. But the written code, as above, commonly called the Rules and Articles of War, does not provide for the punishment of *one* of those crimes, even when committed by individuals of the army upon the persons or property of other individuals of the same, except in the very restricted case in the 9th of those articles; nor for like outrages, committed by the same class of individuals, upon the persons or property of a hostile country, except very partially, in the 51st, 52d, and 55th Articles; and the same code is absolutely silent as to all injuries which

may be inflicted upon individuals of the army, or their property, against the laws of war, by individuals of a hostile country.

5. It is evident that the 99th Article, independent of any restriction in the 87th, is wholly nugatory in reaching any one of those high crimes.

6. For all the offences, therefore, enumerated in the second paragraph above, which may be committed abroad—in, by, or upon the army, a supplemental code is absolutely needed.

7. That *unwritten* code is *Martial Law*, as an addition to the *written* military code, prescribed by Congress in the Rules and Articles of War, and which unwritten code all armies, in hostile countries, are forced to adopt, not only for their own safety, but for the protection of the unoffending inhabitants and their property, about the theatres of military operations, against injuries on the part of the army, contrary to the laws of war.

8. From the same supreme necessity martial law is hereby declared as a supplemental code, in and about all cities, towns, camps, posts, hospitals, and other places, which may be occupied by any part of the forces of the United States in Mexico, and in and about all columns, escorts, convoys, guards, and detachments of the said forces, while engaged in prosecuting the existing war in and against the said republic, and while remaining within the same.

9. Accordingly every crime enumerated in paragraph No. 2 above, whether committed:—1. By any inhabitant of Mexico, sojourner or traveller therein, upon the person or property of any individual of the United States' forces, retainer, or follower of the same; 2. By any individual of the said forces, retainer or follower of the same, upon the person or property of any inhabitant of Mexico, sojourner or traveller therein; or 3. By any individual of the said forces, retainer or follower of the same, upon the person or property of any other individual of the said forces, retainer or follower of the same, shall be duly tried and punished under the said supplemental code.

10. For this purpose it is ordered that all offenders in the matters aforesaid shall be promptly seized, confined, and reported for trial, before *Military Commissions*, to be duly appointed, as follows:

11. Every military commission, under this order, will be appointed, governed, and limited, as nearly as practicable, as prescribed by the 65th, 66th, 67th, and 97th of the said Rules and Articles of War, and the proceedings of such commissions will be duly recorded in writing, reviewed, revised, disapproved or approved, and the sentences executed; all, as near as may be, as in the cases of the proceedings and sentences

of courts-martial, *provided*, that no military commission shall try any case clearly cognizable by any courts-martial, and *provided*, also, that no sentence of a military commission shall be put in execution against any individual belonging to this army, which may not be, according to the nature and degree of the offence, as established by evidence, in conformity with known punishments, in like cases, in some one of the States of the United States of America.

12. The sale, waste, or loss of ammunition, horses, arms, clothing, or accoutrements, by soldiers, is punishable under the 37th and 38th Articles of War. Any Mexican, or resident, or traveller in Mexico, who shall purchase of an American soldier either horse, horse-equipments, arms, ammunition, accoutrements, or clothing, shall be tried and severely punished by a military commission, as above.

13. The administration of justice, both in civil and criminal matters, through the ordinary courts of the country, shall nowhere, and in no degree, be interrupted by any officer or soldier of the American forces, except, 1. In cases to which an officer, soldier, agent, servant, or follower of the American army may be a party; and 2. In *political* cases, that is, prosecutions against other individuals on the allegations that they have given friendly information, aid, or assistance, to the American forces.

14. For the ease and safety of both parties in all cities and towns occupied by the American army, a Mexican police shall be established and duly harmonized with the military police of the said forces.

15. This splendid capital—its churches and religious worship; its convents and monasteries; its inhabitants and property, are, moreover, placed under the special safeguard of the faith and honor of the American army.

16. In consideration of the foregoing protection, a contribution of $150,000 is imposed on this capital, to be paid in four weekly instalments of thirty-seven thousand five hundred dollars ($37,500) each, beginning on Monday next, the 20th instant, and terminating on Monday the 11th of October.

17. The Ayuntamiento, or corporate authority of the city, is specially charged with the collection and payment of the several instalments.

18. Of the whole contribution to be paid over to this army, twenty thousand dollars shall be appropriated to the purchase of *extra* comforts for the wounded and sick in hospital; ninety thousand dollars ($90,000) to the purchase of blankets and shoes for gratuitous distribution among the rank and file of the army, and forty thousand dollars ($40,000) reserved for other necessary military purposes.

19. This order will be read at the head of every company of the United States' forces serving in Mexico, and translated into Spanish for the information of Mexicans.

LAW, (MILITARY.) Under the Constitution of the United States, Congress is intrusted with the creation, government, regulation, and support of armies; and all laws passed by Congress for those purposes are military laws. Congress, being also invested with power " to make all laws which shall be necessary and proper for carrying into execution the foregoing powers, and all other powers vested by this constitution in the Government of the United States, or in any department or officer thereof," is supreme in all military matters. The office of commander-in-chief, intrusted by the constitution to the President, must have its functions first defined by Congress. Such military powers only as Congress confers upon him can be exercised. Excepting that, being the commander-in-chief under the constitution, he of course exercises all authority that Congress may delegate to any military commander whatever, by reason of the axiom that the power of the greater includes that of the less.

Many of the functions, thus devolved by the constitution on Congress, in most governments belong to the executive. The king of Great Britain makes rules and articles for the government of armies raised by him with the consent of parliament. Congress, with us, both raises and governs armies. An army raised in Great Britain is the king's army; with us it is the army of the United States. These most essential distinctions should cause Congress to give more of its attention to the army. It should be borne in mind that our rules for the government of the army have been borrowed almost entirely from Great Britain; that the relation of the army to the people is in the two countries entirely distinct; therefore, that rules adapted to an aristocratic government may not be entirely suited to democratic forms. (*See* ACADEMY, (*Military*;) ACCOUNTS; ACCOUNTABILITY, (*System of*;) ADMINISTRATION, and references; ALLOWANCES; APPOINTING POWER; APPROPRIATIONS; ARDENT SPIRITS; ARREARS OF PAY; ARMORIES AND ARSENALS; ARMY; ARMY, (*Regular*;) ARMY REGULATIONS; ARTICLES OF WAR, and references under that head; ASYLUM, (*Military*;) AUDITORS; AUTHORITY, (*Civil*;) BILLET; BOOTY; BONDS; BOUNTY; BREVET; BRIGADE; CADET; CALLING FORTH MILITIA; CAPTAIN; CLERKS; CLOTHING; COLONEL; COMMISSION; CONGRESS; CONSTITUTION; CONSCRIPTION; CONTRACTS; CORPOREAL PUNISHMENT; CORPS; COUNCIL OF ADMINISTRATION; COURT-MARTIAL, and references under that head; COURTS OF INQUIRY; CUSTOM OF WAR; DAMAGE; DEBT; DEFAULTERS;

Defence, (*National;*) Department; Department of War; Depôt; Detachment; Disbursing Officers; Discharge; Discipline; Dismission; Division; Dragoons; Emoluments; Engineer Corps; Engineers, (*Topographical;*) Enlistments; Evidence; Execution of Laws; Exempts from Militia Duty; Extra Expenses; Extra Allowances; Fatigue Duty; Field Officers; Flag; Forage Master; Garrison; General; General Officers; Government, and references under that head; Indemnification; Indian; Insurrection; Jurisdiction; Law; Law, (*Martial;*) Line; Losses; Marine Corps; Marshals; May; Medical Department; Mess; Mileage; Militia; Oath; Obedience; Officer; Orders; Ordnance Department; Ordnance Sergeant; Pay; Pay Department; Paymaster-general; Pension; Pontoon; Post; Posse Comitatus; President; Prize Money; Promotion; Purchasing; Quarters; Quartermaster's Department; Raise, and references under that head; Rank; Ration; Recruiting; Redressing Wrongs; Regiment; Regulations, and references under that head; Reprieve; Retainers; Returns; Revision; Sale; Sappers; Secretary of War; Servants; Service, and references under that head; Staff; Standards; Stores; Storekeepers; Stripes; Subsistence Department; Suit; Superintendent; Supernumeraries; Sutlers; Trade; Transfers; Travelling Allowances; Uniform; Victuals; Vice-president; Volunteers; Wagon-masters; War; Warrant; Waste or Spoil; Whipping; Wills, (*Nuncupative;*) Witness; Widows and Orphans; Women; Worship; Wounds; Wrongs.)

LEAD BALLS—are now generally made by compression, by means of machinery, either at arsenals or at private establishments.

LEAVE. (*See* Absence.)

LEGION. A variable number of men in the Roman army, from four to six thousand, but which always retained its distinctive characteristic of combining all the elements of a separate army. (Consult Bardin, *Dictionnaire de l'Armée de Terre,* and Arnold's *Rome* for a full account of the Roman legion.)

LEVER. The *effective arm* of a lever is the perpendicular distance from the fulcrum to the line of direction of the power or weight. The power is to the weight inversely as the effective arms of the lever:

$$P\ D = w\ d$$

The pressure on the fulcrum is the resultant of the power and weight. The common balance is a simple lever, the arms of which are equal. If the balance is not accurate, the true weight of a body may be found by placing the body in one scale and counterpoising it by any weights in the opposite scale; then remove the body and replace it by known weights until the equilibrium is again restored. The sum of the latter weights will be the weight of the body; (*Ordnance Manual.*)

LIEUTENANT. Rank next below captain.

LIEUTENANT-COLONEL. Rank next below colonel, and above major.

LIEUTENANT-GENERAL. Rank above major-general. Created by *Act* May 28, 1798. Revived by brevet by *Act* Feb. 15, 1855. To expire with present incumbent. Appoints in time of peace not exceeding two aides and one secretary with rank, pay, and emoluments of lieutenant-colonel. In war, entitled to four aides and two secretaries.

LIFTING JACK. A geared screw-jack, for lifting heavy weights, used in mechanical manœuvres of heavy artillery. (Consult *Instruction for Heavy Artillery*.)

LIGHT BALL. A projectile of an oval shape formed of sacks of canvas filled with a combustible composition, which emits a bright flame. Used to light up our own works.

LIGHT INFANTRY. (*See* INFANTRY.)

LIMBER. The forepart of a travelling gun carriage to which the horses are attached. The same limber is used for all field-carriages. It has two wheels and carries the same ammunition chest as the caisson.

LINCHPINS—prevent the wheel from sliding off the axle-tree.

LINE. President Fillmore in general orders, No. 51 of 1851, has given the following satisfactory exposition of the use of the word *line* in our statute book: The 62d Article of War provides that—" If, upon marches, guards, or in quarters, different corps of the army shall happen to join, or do duty together, the officer highest in rank of the line of the army, marine corps, or militia, by commission there, on duty, or in quarters, shall command the whole, and give orders for what is needful to the service, unless otherwise specially directed by the President of the United States, according to the nature of the case." The interpretation of this act has long been a subject of controversy. The difficulty arises from the vague and uncertain meaning of the words " line of the army," which, neither in the English service, (from which most of our military terms are borrowed,) nor in our own, have a well-defined and invariable meaning. By some they are understood to designate the regular army as distinguished from the militia: by others, as meant to discriminate between officers by ordinary commissions and those by brevet; and, finally, by others, to designate all officers not belonging to the staff. The question is certainly not without difficulty, and it is surprising that Congress should not long since have settled, by some explanatory law, a question which has been so fruitful a source of controversy and embarrassment in the service. The President has maturely considered the question, and finds himself compelled to differ from some for whose opinions he entertains a very high respect. His opinion is, that, although these words may sometimes be used in a different sense, (to be determined by the context and subject-matter,) in the 62d Article of War, they are used to designate those officers of the army who do *not* belong to the staff, in contradistinction to those who do, and that the article intended, in the case contemplated by it, to confer the command exclusively on the former. The reasons which have brought him to this conclusion are briefly these: 1st. It is a well-settled rule of interpretation that in the construction of *statutes*, words of doubtful or ambiguous meaning are to be understood in their usual acceptation. Now it must be admitted that, in common parlance, both in and out of the army, the words "line" and "staff" are generally used as correlative terms. 2d. Another rule of construction is, that the same word ought not to be understood, when it can be avoided, in two different senses in different laws, on the same subject, and, especially, in different parts of the same law. Now in another article (74) of this same law, the words "line and staff of the army" are clearly used as correlative and contradistinctive terms. The same remark applies to almost every case in which the words "line" and "staff" occur in acts of Congress. *See*

Act of 1813,	sec.	4,	Cross' Military Laws, p.	165 ;
1813,	"	9,	"	166 ;
1814,	"	19,	"	174 ;
1816,	"	10,	"	190 ;
1838,	"	7,	"	263 ;
1838,	"	8,	"	263 ;
1838,	"	15,	"	264 ;
1838, pars. 7 & 9,			"	268 ;
1846,	sec.	2,	"	283 ;
1846,	"	7,	"	286.

There are many other instances in which the words are so employed, but I have selected these as the most striking. On the other hand, I find but one act of Congress in which the words "line of the army" have been employed to designate the regular army in contradistinction to the militia, and none in which they have been manifestly used as contradistinctive of brevet. 3d. If Congress had meant by these words to discriminate between officers of the regular army and those of the militia, or between officers by brevet and by ordinary commission, it is to be presumed that they would have employed those terms, respectively, which are unequivocal, and are usually employed to express those ideas. 4th. If we look at the policy of the law, we can discover no reasons of expediency which compel us to depart from the plain and ordinary import of the terms: on the contrary, we may suppose strong reasons why it may have been deemed proper, in the case referred to by the article, to exclude officers of the staff from command. In the first place the command of troops might frequently interfere with their appropriate duties, and thereby occasion serious embarrassment to the service. In the next place, the officers of some of the staff corps are not qualified by their habits and education for the command of troops, and alhough others are so qualified, it arises from the fact that, (by laws passed long subsequently to the article in question) the officers of the corps to which they belong, are required to be appointed from the line of the army. Lastly, officers of the staff corps seldom have troops of their own corps serving under their command, and if the words "officers of the line" are understood to apply to them, the effect would often be to give them command over the officers and men of all the other corps, when not a man of their own was present—an anomaly always to be avoided where it is possible to do so. 5th. It is worthy of observation that Article 25, of the first "rules and articles," enacted by Congress for the government of the army, corresponds with Article 62 of the present rules and articles, except that the words " of the line of the army " are not contained in it. It is evident, therefore, that these words were inserted intentionally with a view to a change in the law, and it is probable that some inconvenience had arisen from conferring command indiscriminately on officers of the line or the staff, and had suggested the necessity of this change. It is contended, however, that sec. 10, of the act of 1795, enumerates the major-general and brigadier-general as among the staff officers, and that this construction of the article would exclude them from command, which would be an absurdity. No such consequence would, however, follow. The article in question was obviously designed to meet the case (of not unfrequent occurrence) where officers of different corps of the army meet together with no officer among them who does not belong exclusively to a corps. In such a case, *there being no common superior*, in the absence of some express provision conferring the power, no officer, merely of a corps, would have the right to command any corps but his own: to obviate this difficulty, the article in effect provides that, in such an event, the officer *of the line*, highest in rank, shall command the rest. But if there be a major-general or brigadier-general present, the case contemplated by the article does not exist. No question can arise as to the right of command, because the general officer, not belonging to any particular corps, takes the command by virtue of the general rule which assigns the command to the officer highest in rank. (*See* RANK; COMMAND; BREVET.)

LINE OF DEFENCE—is the line which extends from the angle of the polygon or extremity of the exterior side, through the inner end of the perpendicular, to the flank, of the bastion.

LINE OF LEAST RESISTANCE (THE)—is that which is supposed to extend, from the centre of the charge of a mine, to the nearest surface of the ground.

LINES. A connected series of field-works, whether continuous or at intervals.

LINES AT INTERVALS—are lines composed of separate field-works, so arranged as to flank and defend one another.

LINES CREMAILLÈRE—are composed of alternate short and long faces, at right angles to each other.

LINES OF BASTION—as the name indicates, are formed of a succession of bastion-shaped parapets, each consisting of two faces and two flanks, connected together by a curtain.

LINES OF TENAILLES—consist of parapets, forming a series of salient and re-entering angles.

LINSTOCK. A pointed forked staff used for lighting fort fires; the lower end pointed and shod with iron.

LITTER. If a man be wounded or sick, and has to be carried along upon the shoulders of the others, make a litter for him in the Indian fashion, (Fig. 148;) that is to say, cut two stout poles, each 8 feet long,

FIG. 148.

to make its two sides, and three other cross-bars of 2½ feet each, to be lashed to them. Then, supporting this ladder-shaped framework *over* the sick man as he lies in his blanket, knot the blanket well up to it; and so carry him off. One cross-bar will be just behind his head, another in front of his feet; the middle one will cross his stomach, and keep him from falling out; and there will remain two short handles for the carriers to lay hold on. The American Indians carry their wounded companions by this contrivance after a fight, and in a hurried retreat, for wonderful distances.

LOAD. Command in infantry and artillery instruction. (Consult Tactics of those arms.) In loading small arms the powder should be well shaken out of the paper, to prevent the formation of gas, which, forcing the paper against the sides of the bore, prevents it from leaving with the charge, and endangers the explosion of the next charge when loading, from the lighted paper. There is no danger of heating the piece by rapid firing so as to cause premature explosions, since long before it reaches 600°, the temperature at which gunpowder inflames, it is entirely too hot to handle. In loading cannon the *vent* should always be kept carefully closed, while the loading is going on, especially when sponging, to prevent the current of air from passing out and collecting there pieces of thread, paper, &c., from the cartridge-bag, which would retain fire in the gun, and cause premature explosion the next time the gun was loaded. This precaution is the more necessary, when the sponge fits the bore tight, and acts as a piston. The sponge should be well pressed down against the bottom of the bore, and turned, so as to leave no remnant of the cartridge-bag. In mortars, where a sponge is seldom used, or when it does not fit tightly, the stopping of the vent is not necessary; but it should always be cleared out with the priming wire before the powder is placed in. Mortar-shells should be let down gently so as not to be forced into the chamber, or crush suddenly any powder they may meet. The use of sabots is avoided when firing over the heads of our own men. It may sometimes become necessary to fire a shell from a mortar too large for it; in which case it is wedged in on different sides with pieces of soft wood, and the space between it and the bore filled in with earth.

LOCK. (*See* ARMS.)

LODGEMENT. In a siege *lodgement* signifies the occupation of a position and the hasty formation of an entrenchment thereon to maintain it against recapture. Thus it is said the besiegers, having carried the demi-lune or bastion, effected a lodgement, or the besieged destroyed the lodgements of the enemy. (*See* SIEGE.)

LOGARITHM. The logarithm of a number is the exponent of the power to which another given invariable number must be raised in order to produce the first number. Thus in the common system of logarithms in which the invariable number is 10, the logarithm of 1,000 is 3, because 10 raised to the third power is 1,000. In general, if $a^x = y$ in which equation a is a given invariable number, then x is the logarithm of y. All absolute numbers positive or negative, whole or fractional, may be produced by raising an invariabe number to suitable powers. This invariable number is called the *base* of the system of logarithms: it may be any number whatever greater or less than unity; but having been once chosen, it must remain the same for the formation of all numbers in the same system. Whatever number may be selected for the base, the logarithm of the base is 1, and the logarithm of 1 is 0. In fact if, in the equation $a^x = y$, we make $x = 1$ we shall have $a^1 = a$, whence by definition log. $a = 1$; and if we make $x = 0$ we shall have $a^0 = 1$, whence log. $1 = 0$. The chief properties of logarithms are: that the logarithm of a product is equal to the sum of the logarithms of its factor; the logarithm of a quotient is equal to the difference between the logarithm of the dividend and the logarithm of the divisor; and the logarithm of the power of a number is equal to the product of the logarithm of the number by the exponent of the power; and the logarithm of any root of a number is equal to the logarithm of the number divided by the index of the root. These properties of logarithms greatly facilitate arithmetical operations. For if multiplication is to be effected, it is only necessary to take from the logarithmic tables the logarithms of the factors, and then *add* them into one sum, which gives the logarithm of the required product; and on finding in the table the number corresponding to this new logarithm, the product itself is obtained. Multiplication is thus performed by simple addition. In like manner division is performed by simple subtraction, and by means of a table of logarithms numbers may be raised to any power by simple multiplication, and the roots of numbers extracted by simple division. (Consult BABBAGE, *Logarithms of Numbers;* FARLEY's *Tables of Six-figure Logarithms.*)

LOGISTICS. Bardin considers the application of this word by some writers as more ambitious than accurate. It is derived from Latin LOGISTA, the administrator or intendant of the Roman armies. It is properly that branch of the military art embracing all details for moving and supplying armies. It includes the operations of the ordnance, quartermaster's, subsistence, medical, and pay departments. It also embraces the preparation and regulation of magazines, for opening a campaign, and all orders of march and other orders from the general-in-chief relative to moving and supplying armies. Some writers have, however, extended its signification to embrace STRATEGY.

LOOPHOLED GALLERIES — are vaulted passages or casemates, usually placed behind the counterscarp revetment, and behind the gorges of detached works, having holes pierced through the walls, to enable the defenders to bring a musketry fire from unseen positions, upon the assailants in the ditch. Loopholes, however, are not confined to galleries. In modern fortifications, the revetments, both scarp and counterscarp, are very generally pierced for a musketry fire.

LOOPHOLES—are apertures formed in a wall or stockade, that through them a fire of musketry may be directed on the exterior ground.

LOSSES. In the British army there is a regular provision made for indemnification for losses by fire; by shipwreck; in action with the enemy; by capture at sea; by destruction or capture of a public storehouse; by the destruction of articles or horses, to prevent their falling into the hands of the enemy, or to prevent the spreading of an infectious disorder. In the United States it would seem just that Congress should establish some general rules regulating such matters. The principle of settling all such claims by special legislation cannot but bear hardly on a number of individuals, and also probably in the end imposes greater burdens upon the treasury.

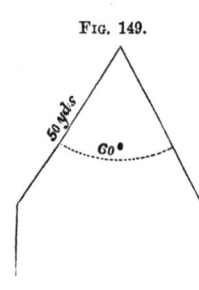

FIG. 149.

LUNETTES—are redans having flanks parallel to their capitals, as in Fig. 149. The faces and flanks may have any moderate extent, according to the purpose for which they are intended; 50 yards for the face, and 25 yards for the flanks, would be a convenient size for many positions.

LYING OUT OF CAMP OR QUARTERS. Punishable, according to the nature of the offence, by a court-martial; (ART. 42.)

M

MACHICOULIS. A projecting wooden gallery from the second story of a house to enable the assailed to fire down on their opponents.

MAGAZINE COVER—of Rifle Musket, 1855. (*See* ARMS, *Small.*)

MAGAZINES. Powder magazines ought to secure an unobstructed circulation of air under the flooring as well as above. The magazine should be opened and aired in clear dry weather; the ventilators should be kept free; and no shrubbery or trees should be allowed to grow so near as to protect the building from the sun.

All batteries of attack require magazines capable of holding ammunition for daily consumption. Fig. 150 is a section of two strong splinter-proof timbers, say 8 to 9 feet long, and 9 to 12 inches in breadth and thickness, resting on sleepers, and giving an interior space of about the dimensions seen in the figure, covered with one or two tiers of fascines, and over them 3 or 4 feet of dung or stiff earth; this simple construction would answer in many cases.

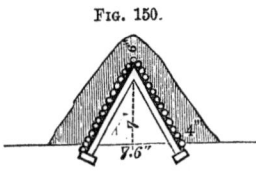

FIG. 150.

By some persons it is considered better to have two small magazines in a battery, made of very stout mining cases, and constructed in the epaulements. Sir John Jones, in his work on "Sieges," says: "Splinter-proof timbers for magazines were cut 12 feet in length, and from 8 to 10 inches in breadth and thickness, and were placed against an epaulement, or parapet, at an angle making the base equal to half the height. They were then covered with a tarpaulin, extending well over the top of the epaulement upon which were laid one or two rows of filled sand-bags, so as to prevent the possibility of the tarpaulin being cut by splinters of shells. A second tarpaulin was usually thrown over the exterior in rainy weather. On this construction, the magazines were found to be perfectly dry, and sufficiently spacious, and of the strength no doubt can remain, as the sand-bag covering was frequently knocked off by large shells, and in no instance were the splinter-proofs broken. The best situations for magazines are on the flanks of the batteries. Nothing can be worse than to place them in rear of the centre of a battery, as then every cartridge has to be carried along the most exposed and dangerous part of the battery, and the number of accidents and casualties which arise therefrom is very great indeed. The artillery always preferred having two magazines formed, rather than to have one exceeding 10 or 12 feet in length; when two were made, they were placed one on either flank, a situation which was found to answer extremely well." (Consult Hyde's *Fortification*; *Ordnance Manual*.)

MAGISTRAL LINE—in a plan, is that which regulates the form of the works. It is that which is first laid down, and from which the other parts of the works are traced. (*See* Cordon.)

MAJOR. Rank between captain and lieutenant-colonel.

MAJOR-GENERAL. Rank between brigadier-general and lieutenant-general.

MALARIA. (*See* Sanitary Precautions.)

MALINGERER. A soldier who feigns illness in order to avoid

his duty. Any soldier, in the English army, convicted of malingering, feigning or producing disease or infirmity, or of being detained in hospital in consequence of materially injuring his health by his own vice or intemperance, and thereby rendering himself unfit for the service; or of absenting himself from an hospital whilst under medical treatment; or of being guilty of a gross violation of the rules of the hospital; or of intentionally protracting his cure; or of wilfully aggravating his disease, is liable to be tried by a court-martial for "disgraceful conduct," and to suffer the punishments attached to that crime.

MANŒUVRE. For prescribed manœuvres consult *Cavalry Tactics; Infantry Tactics; Rifle and Light Infantry Tactics; Instruction for Field Artillery, horse and foot; and Instruction for Heavy Artillery, embracing Mechanical Manœuvres.*

The word manœuvre signifies also movements of entire corps in war executed with general views; and by some writers it is confined to that signification, and the word *evolution* is made to designate the particular means, or the elements of manœuvres; (JABRO.) Manœuvres, according to *Bardin*, are operations in war whether really before an enemy, or simulated on a field of exercise. Their precision and aptness depend upon the skill of the general; the intelligence of his aides-de-camp; upon the chiefs of battalions and their adjutants, and the general guides. Evolutions and manœuvres are, however, often applied in the same sense, and indeed it may well be questioned whether there be any propriety in retaining in books of instruction evolutions which are not used as manœuvres against an enemy.

Manœuvres of Infantry in battle.—The vicious idea that tactical evolutions are not used in war is by no means uncommon, and has frequently caused the loss of battles. It is true that the number of manœuvres used in combats is limited, and that those which are needed can only be judiciously applied by keeping in view moral and physical requirements. The judicious tactician will, therefore, in war eschew: deployments, which cause the soldier to turn his back towards an enemy; countermarches; forming a battalion on the right or left by file into line, and some other movements suited only to parades. One of the most hazardous manœuvres is the formation of columns of great depth and deploying those columns when too near the enemy. Without giving names or places, (says Marshal Bugeaud,) I affirm that I have seen an entire division in column of regiments, which began its deployment within range of the enemy's guns, routed before it finished its manœuvre.

The column is an order of march and manœuvre, rarely an order of battle.—When beyond the range of cannon, and at a distance from the line of battle to be occupied, if the enemy approach and time permits, it is necessary to close in mass, in order to hold the troops in hand for all possible dispositions.

So, in marches near the enemy the columns should march at half distance, when roads permit, in order that they may be less elongated, and all the troops be ready to act promptly. If surprised in this order by the necessity of forming immediately forward into line of battle, or, if without being under this pressing necessity, there is between us and the enemy ground admitting an easy march in line of battle, the column ought to execute forward into line, according to the principles of the tactics. This movement is more prompt and greatly better than closing column in mass, in order to deploy afterwards. In the first case troops only pass over one side of the triangle, whilst by massing the column to deploy afterwards, they must pass over two sides by a complicated manœuvre, which is dangerous from the beginning. In general, it is necessary to shun as much as possible the deployment of great massed columns, for this movement is badly executed even in exercises. It can only be performed far from the enemy, and it is even there inconvenient. It should be renounced in all formations whose object is to take the enemy in flank or reverse, if he be sufficiently near to take measures to prevent success. In that case, the formation of the close columns in mass upon the right or left into line of battle is a necessary manœuvre. This movement, as Marshal Bugeaud suggests, is most important in war; (Fig. 151.) It would have an influence upon battles by the simplicity and rapidity of its execution, and accidents of ground would often be found to conceal the movement from the enemy. It admits of an attack in echelons of battalions against an enemy being commenced as soon as one battalion or the half of a battalion has formed on the right or on the left of the line of the enemy. It also offers the advantage of giving to the line, with the greatest facility, every form that may be wished, and protecting the successive formations by a mass that may be disposed of at pleasure, whether at the extremity of the line to form square against cavalry, or to occupy in advance upon the right or left a commanding position, protecting the flanks of our line. When circumstances, then, compel a march in heavy mass, it is better to present to the enemy a flank of columns, in order to deploy them by formations on the right or on the left into line of battle.

When a line has to pass over a great distance, it is commonly formed into columns of attack. The formation by company in column, in rear of the grenadiers of each battalion, is preferred by Marshal Bugeaud, because it is thus easier to make good dispositions against cavalry. The grenadiers of each battalion make a half wheel, and each battalion, after

Fig. 151.

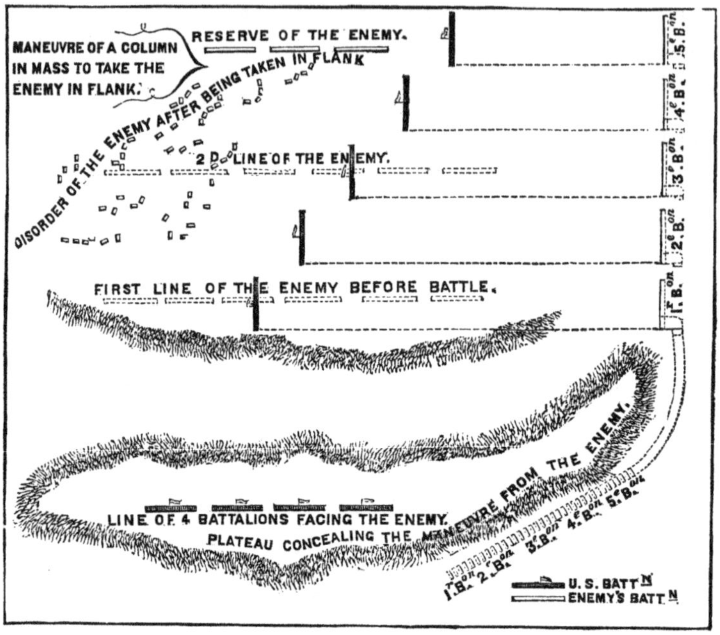

being closed in mass, forms square. But neither the column by companies or divisions ought to be used within range of cannon, whenever there is a possibility of marching in line of battle. It is time that the fact should be admitted, that although the moral effect of the column may be considerable, yet this may be paralyzed by a little manœuvring on the part of the enemy's *line*, which would necessarily obtain great advantage from the superiority of its fire. Small columns, at distances of three battalions from each other marching under cover of the line, may render great services. They would be ready promptly to fill the holes made in the line of battle, and the best means of doing this would be to take the enemy in flank who had pierced them, whenever they could. It is desirable that these columns should each not exceed a half battalion, and be commanded by energetic officers.

The depth of the column adds nothing to the strength of the first battalion composing it, and diminishes that of the mass.—It is, then, vicious to employ more than one battalion, except in the small number of cases where it is necessary to fight in mass, as in carrying a bridge, a defile, an entrenchment, a breach, &c. The other battalions ought to follow at such a distance that they may sustain the attacking battalion without sharing in its disaster or rout, if such should take place. With an interval the chiefs of battalions have time to prepare their troops, and make necessary dispositions; with a single mass the disorder at the head of the column is communicated to the rear almost as readily as an electric spark.

Flank marches, in presence of the enemy, ought always to be made in open column. In this order we are always ready to fight by a simple wheel of each subdivision of the column. Nothing is deranged in the order of battle, whatever may be the strength and number of the lines. Without derangement an excellent disposition may also be made against cavalry. The column will be halted, and each battalion will be closed in mass upon its grenadiers, who make a half wheel. The field-officers, staff, and the officers of grenadiers will be previously warned. Each battalion will form then Marshal Bugeaud's square. The first order will be resumed by taking distances by the head of each battalion; the grenadiers retaking their direction at once.

If deep columns are condemned as an order of attack, those barbarous columns employed in some of the last battles of Napoleon, and particularly at Waterloo, ought to be condemned still more. That column, which appeared to announce the decline of art, consisted in employing all the battalions of a division one behind the other, and thus marching towards the enemy.

Every column has for its object to pass rapidly, and without confusion, into the order of battle, to pass over lightly a given space, and to make prompt dispositions against cavalry. The column against which these remarks are made does nothing of that kind, and if it be attacked upon its flanks, whether by cavalry or infantry, it cannot fail to be destroyed.

Order of battle, march in line of battle, and changes of front.—The line of battle is the true order of battle. It is also the best order of march when in range of cannon, and not exposed to cavalry. It is only in this order that infantry can make use of its fire. If battalions consist of 800 men they will, in a formation of two ranks, be too much extended for most chiefs of battalions. Two companies of each battalion ought then to be formed as columns of reserve. The order in two ranks is beyond question best suited, in oblique attacks, for that part of the line not to be engaged; and with rifle muskets now used the two-rank formation will be found better for that part of the line which is to strike also. Even with old muskets the two-rank formation was used by the British very successfully at Waterloo in squares against cavalry. The fire in two-rank formation is made with more order, more easily, and is better aimed. The march in line of battle ought to be employed whenever the ground permits it, within 1,000 yards of the enemy. We lose then fewer men by cannon, and even if it be desirable to approach the enemy in column, (which is very rare, and should even then be in columns of single battalions,) the march ought still to be in line of battle until within two hundred yards, and then the column of attack ought to be formed while marching. Troops cannot be too much exercised in marching in line of battle. This march is no more difficult than the march of many heads of columns upon the same line, perhaps even less so, for it is difficult to maintain between the columns the intervals necessary for deployments.

Changes of front very near the enemy are rarely perpendicular. The new front nearly always forms with the line of battle an acute angle. In this case, it is necessary to guard against breaking the battalions into column. It is better to use the changes of direction for the line of battle prescribed by the tactics. The two pivot battalions may be thrown upon the new line by companies half faced to the right or left. The other battalions ought to be directed upon the new line by changes of direction which would least expose them to artillery. If, however, we have to guard against cavalry during the execution of the movement, it will be better to break into column the battalions of the leading wing. They will thus form the stem of the battery, and would rapidly make good dispositions against cavalry, as they would only be obliged to close in mass upon the grenadiers and form square.

Changes of front forward are possible under fire, but changes of front to the rear are not so. I believe, (says Marshal Bugeaud,) that the loss of one of our battles in Spain may, in great part, be attributed to a change of front in rear of the left wing, which was attempted at a moment when warmly engaged. The movement rapidly degenerated into a rout; and it could not be otherwise. There are no troops with sufficient *sang-froid* and self-possession to make that movement under the fire of ball and grape. To make the movement, it is necessary first to stop the enemy, and the means of doing that vary with circumstances, and the resources within our command. Charges of cavalry—above all if they threaten the flanks of the enemy's line, would cover the change of front to the rear. If cavalry be not at hand, there is no better means than to advance the second line to the position that it is desired that the front should occupy after its change of front, and withdraw the first line at a run, directing it to form the second line, passing through the intervals of the battalions, now become the first line.

If a line is about coming up with the enemy at the moment of receiving the order to change front, it would be better to finish the charge, by putting the first line of the enemy in rout before executing the movement to the rear. This last principle is applicable to retreats generally: it is often necessary to overthrow an enemy who is too nigh before retiring.

Running movements may, in many cases, save us from destruction. It is necessary, then, to exercise troops in such movements, and make them run in disorder, and re-form at some given point.

Echelons.—The order in echelons is the manœuvre of oblique attacks. By that means we approximate those troops only who are to fight. The remainder are at once threatening and defensive. They hold in check one or many parts of the order of battle of the enemy, and present the best possible protection to the attacking portion. Some echelons to the right and left of that which attacks, are greatly better than any other support. They render, if not impossible, at least very difficult, an attack upon the flank of the attacking portion, as that cannot be assailed without the enemy in turn being taken in flank by echelons. And the latter cannot be turned, except by strong movements, which must weaken the army executing them, and also afford necessary time to guard against them.

Instead of placing flank brigades in advance of the front of the columns or lines that they protect, it is better to place them in rear. Besides the physical advantages of this disposition, there are moral advantages, inasmuch as the latter position enables the echelons to assail, whereas, if they were immediately on the flank of the attack, they might be assailed.

In theory, echelons are placed at regular distances. In practice, the distance is determined by circumstances, and, above all, by the formation of the ground. The regularity of echelons can, therefore, only exist in broad plains. The greater or less distance between echelons depends upon the number of troops, the distances between those of the enemy, and the ulterior views of the general-in-chief; but in general they ought to be within mutual succor, and if cavalry is to be repulsed, they ought to cross fire at about 150 paces after having formed square. The different movements of echelons, the changes of front in each echelon, with the same angle, are very useful in war; it is necessary, therefore, that troops should be exercised in such movements. (*See* BATTLE; CHARGE; CONVOY; DEFILE; INFANTRY; SQUARES. Consult *Aperçus sur quelques Détails de la Guerre, par* MARSHAL BUGEAUD; *Tactique des Trois Armes, par* DECKER.)

MANTLET—is a musket-proof shield, which is sometimes used for the protection of sappers or riflemen during the attack of a fortress. (*See* PENETRATION.)

MANUAL. Exercise of arms; books of reference, as Ordnance Manual, &c.

MARAUDING. (*See* PLUNDER and PILLAGE.)

MARCH. Recruits are taught to march by explaining the principles of the cadenced step in common, quick, and double-quick time. The march in line of battle is the most difficult and most important of the tactical marches. A regiment which can pass over two hundred paces in line of battle without losing its allignment, is well instructed. Marches may be divided into: marches in time of war; marches in route, in time of peace; and tactical marches. Those in time of war are either movements to pass over ground, or else manœuvres to obtain an advantageous position. When an army moves forward to meet an enemy who is still very distant, it will be sufficient to have advanced and rear guards, some flankers, and march in parallel columns over the best routes, each column having its squadrons of cavalry, batteries of artillery, and wagon trains. If the enemy is, however, in the neighborhood, if we march along the front of his camp, or his line of posts, every precaution must be redoubled to gain information of his movements and guard against surprise.

When the march is only a manœuvre, it is often made across fields; through by-roads; then it is necessary to reconnoitre in advance, clear away obstacles, and sometimes even construct little bridges; guides are taken, and information gained from them as well as by *reconnaissances*. Armies are collected together by routes of march, the troops usually marching about 17 miles a day. In general, the marches are made by battalions echeloned at intervals one day's distance from each other. Cavalry ordinarily marches alone and follows the least direct roads, but it is difficult to subsist a numerous cavalry without retarding military operations. Artillery follows the cavalry, or if it has a large convoy, it marches by another route alone. The troops begin to concentrate on the base of operations. Still advancing, the echelons converge, and the troops are cantoned together by *lines* one day's march from each other. The nearer we approach the enemy, the more columns are used; if the country offers parallel debouches, it is always advantageous to march an army corps on many routes, if they are within distance for deployments; but if there is only one means of communication, the different arms are kept 200 yards distant from each other, and the cavalry marches in rear of the column.

On these marches, when a defile is to be passed, the successive passage of each echelon is commanded in advance; and it is a general rule never to crowd troops, so as to paralyze their action, or even render movements difficult; but care must be taken always to keep troops within easy supporting distance of each other.

Sometimes an army is collected very near the enemy. It is necessary then nicely to calculate distances, &c., in order to combine marches for a *simultaneous* convergence of columns on the offensive point.* To bring troops suddenly together, forced marches are made by some of the troops; relays and railways are also used. By forced marches the ordinary day's march is doubled, but under extraordinary circumstances 62 miles have been made in 26 hours. Relays are the use of wagons, &c., obtained by requisition. 250 wagons may carry from 2,000 to 2,300 men. Sometimes the march is made entirely in wagons, and each echelon passes over three days' march in 8 hours. This is done by the troops taking new wagons twice, the old returning empty for other troops.

It is but seldom that any one arm is exclusively employed when near the enemy; it is usual to operate with a combined force of cavalry, infantry, and artillery, so that it may be always possible to employ one or the other arm, according to circumstances and locality. If the main body of the army is composed of the different arms, then the advanced guard is similarly constituted, that it may be able to act in all localities.

The composition of such an advanced guard depends—

1st. Upon the object and nature of its intended operations. During marches in pursuit, it is reinforced by cavalry; but if it is to make an obstinate resistance, it is strengthened with much infantry and artillery. In general, light cavalry are the best for advanced guards, wherever the nature of the ground permits them to operate, but infantry are necessary to support them. Mounted rifles and mounted engineer troops are of great service in advanced guards.

---

* To calculate exactly the time T necessary for the execution of a march:—A column of infantry will generally pass over about five miles in two hours, halts included. A column of cavalry at a walk and trot alternately makes about six miles per hour. Let D then be the distance to be accomplished, d the distance that the men comprising the column pass over in an hour, halts included; l the length of the column; o the delay caused by obstacles; then $t = \dfrac{l}{d}$ will be the time that passes until the left arrives at its destination, and the formula $T = t + o + D$ will give the time sought. One of the elements of o is the lengthening l' of a column in a defile; it is considered by introducing $\dfrac{l'}{d}$ into the formula; o is also the delay caused by marching across fields. These elements may all be estimated and introduced into the formula.

2d. The composition of the advanced guard depends also upon the locality; if the ground is broken, much infantry is required; if it is open, much cavalry; and, in general, light troops.

The order of march of an advanced guard depends principally upon its composition, the order of march of the main body, the locality, &c. The main rule is, that it should never be too much divided, so that there may always be a considerable force in hand to seek the enemy more boldly, and detain him longer. Therefore, even when the main body moves in several columns, the principal part of the advanced guard marches on the main road, sending only small parties on the others to watch the enemy and detach patrols as far as possible in all directions. In an open, level country, the cavalry marches at the head; in a broken country, there is only a small detachment of cavalry at the head, to furnish advanced detachments and patrols. An advanced detachment of cavalry, which sends out patrols in front and on its flanks, moves at the distance of a few miles in front of the advanced guard. Small detachments of cavalry move in a line with it on the other roads; also others on the flanks of the main advanced guard, to secure it against being turned. All the front and flank detachments maintain constant mutual communication by means of patrols, and thus guard the whole space in front of the main body over a great extent. But if the flank columns of the main body march at a great distance from the main road, followed by the advanced guard, then, in addition to this last, each flank column detaches a small advanced guard for its own security.

If the advanced guard is composed of different arms, its distance from the main body depends not only upon its strength, but also on the following circumstances: 1. On its composition. Cavalry may advance much further than infantry. 2. Upon the locality. The more fully the nature of the country secures the advanced guard against being turned, the further may it move from the main body. 3. Upon the object in view. Prior to defensive combats in position, it is advantageous to have the advanced guard as far from the main body as possible, in order to secure time for making the necessary arrangements; but if the main body is already concentrated for a decisive attack upon the enemy, it is sometimes well to be entirely without an advanced guard; during a pursuit, the main body should follow the advanced guard as closely as possible. 4. Upon the order of march of the main body. The longer the time needed by the main body to form in order of battle, on account of the intervals between the columns, the nature of the ground between them, the length of the columns, &c., so much further forward should the advanced guard be pushed. In general, the distance of the advanced guard from the head of the main body should be a little greater than the interval between the outside columns of the main body.

Fig. 152 gives an example of the arrangement of an advanced guard composed of one brigade of light cavalry, 8 battalions of infantry, one battalion of sappers, 6 pieces of horse artillery, and 12 pieces of foot artillery; the main body following in 3 columns.

Whatever slight changes may be made necessary by the nature of the country, can easily be made with the aid of a map and the special information obtained in other ways.

If the country is partially broken and obstructed, it is advantageous to have four or five companies of infantry just behind the leading detachment of cavalry to examine places that are difficult or dangerous for the latter.

Upon the plains, the patrols are of cavalry; in a mountainous region, of infantry. In the latter case, not only the advanced detachments and patrols are of infantry, but also the head and rear of every column; the cavalry and artillery march in the middle, under the protection of the infantry.

In passing through a village, the infantry enter it first, if there are any with the advanced guard; the cavalry either ride rapidly around it, or, according to circumstances, halt a little before reaching the village, and wait until the infantry have passed through.

The passage of important bridges, ravines, and defiles, should be effected in the same manner, the infantry examining them. As soon as the infantry have crossed and formed on the other side, the cavalry send out patrols to a great distance to examine the ground in front before the main body of the advanced guard begins to cross.

The advanced guard having crossed rapidly, forms in front of the passage, to cover the debouche of the main body. The distance of such a position from the passage should be such that, in the event of being attacked, the advanced guard may not be too quickly forced back upon the main body while debouching, and that the latter may have ample time to form without disorder.

Since attacks should be most expected when passing through defiles, or when issuing from them, they should be traversed rapidly, and with the most extended front possible, to prevent the column from stretching out.

An advanced guard possessing a certain degree of independence, without neglecting any of the precautions here laid down, should not be

Fig. 152.

MARCH OF AN ADVANCED GUARD COMPOSED OF 1 BRIGADE OF CAVALRY, (20 COMPANIES,) 2 DIVISIONS OF INFANTRY, (8 BATTALIONS,) 1 BATTALION OF SAPPERS, 6 PIECES OF HORSE AND 12 OF FOOT ARTILLERY.

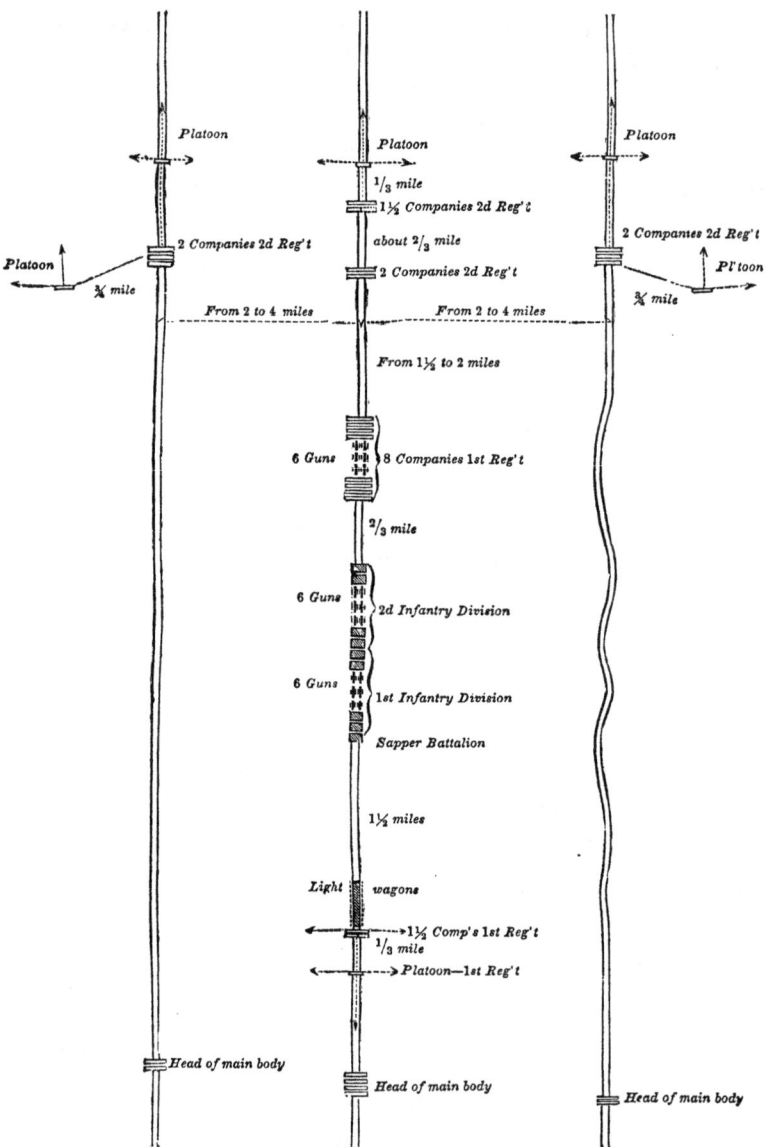

too apprehensive, and, in examining the country, ought not to be detained by objects which cannot conceal the enemy in sufficient force to make him dangerous to the advanced guard.

In very mountainous regions, it is necessary to rely upon the infantry alone; the cavalry and train remaining in rear, and not entering the defiles until they have been occupied. Here the infantry patrols are sent out as far as possible, and occupy the heights from which the direction of the columns may be seen, until relieved by the patrols of the rear guard, which is also of infantry. In this manner the cavalry, which the enemy would attack in such places in preference, is protected. Not a gorge or defile should be left unexamined, for in the mountains an attack may be expected at any moment.

In a wooded country, the commander of the advanced guard takes nearly the same precautions as in the mountains.

If the forest is deep but not broad, detachments of cavalry ride along the skirts, which are occupied by infantry skirmishers as supports; if the forest is dense, but not deep, the infantry lead. The infantry place themselves along the skirts of the wood on both sides of the road; the cavalry then passes through at a fast trot, forms on the plain beyond, and there awaits the rest of the column.

When the road passes through a country but little obstructed by defiles, villages, or other obstacles to the movements of cavalry, and there is no infantry with the advanced guard, mounted rifles are very useful; finally, the enemy, in retreating through such a country, leaves infantry at these obstacles to arrest the pursuit of the cavalry, and delay until the arrival of the infantry; in such cases, mounted rifles or dismounted dragoons will produce sure results by acting against the enemy's infantry.

The main body.—It remains to be said, in reference to this, that the nature of the country must determine its order of march, whether cavalry or infantry are to lead. If the country is broken, particularly if it is wooded, there is great danger in placing the cavalry at the head; for it may not only be unable to act, but, if forced to retreat, may carry disorder into the infantry following.

The artillery should march in the midst of the other troops, but a few pieces may move with the head of the column, to protect it in case of meeting the enemy suddenly.

Infantry, in traversing extensive forests, in which parties of the enemy may easily conceal themselves, replace the flank detachments and patrols of cavalry. (Consult *Aide Memoire d'Etat Major;* McClellan's *Military Companion.*)

MARINE CORPS—when serving with the army, to be supplied by the several officers of the staff of the army; (*Act* Dec. 15, 1814.) The officers of the marine corps may be associated with the officers of the land forces for the purpose of holding courts-martial and trying offenders belonging to either; and in such cases the orders of the senior officer of either corps, who may be present and duly authorized, shall be received and obeyed; (ART. 68.) The marine corps shall at any time be liable to do duty in the forts and garrisons of the United States on the sea-coast, or any other duty on shore, as the President, at his discretion, shall direct; (*Act* July 11, 1798.) The officers, non-commissioned officers, privates, and musicians shall take the same oath and shall be governed by the same rules and articles as are prescribed for the military establishment of the United States and by the rules for the regulation of the navy heretofore, or which shall be established by law, according to the nature of the service in which they shall be employed, and shall be entitled to the same allowance in case of wounds or disabilities, according to their respective ranks, as are granted by the act to fix the military establishment of the United States; (*Act* July 11, 1798.)

MARKER. Soldier who marks the direction of an allignment or pivot points.

MARKSMAN. Good shot; sharp-shooter. (*See* RIFLEMEN; TARGET.)

MARSH POISONS. (*See* SANITARY PRECAUTIONS.)

MARSHALS. The marshals of the several districts and their deputies shall have the same powers in executing the laws of the United States, as sheriffs and their deputies, in the several States, have by law, in executing the laws of the respective States; (*Act* Feb. 28, 1795.) (*See* OBSTRUCTION OF LAWS; POSSE COMITATUS.)

MARTELLO TOWERS—are buildings of masonry, generally circular, and of various dimensions. They are chiefly placed on the sea-coast, having a gun on their summit, mounted on a traversing platform, by which it can fire in any direction.

MARTIAL LAW. (*See* LAW, *Martial*.)

MASKED BATTERY—is when the battery is so concealed or disguised, as not to be seen and recognized by the enemy, until it opens its fire.

MATCH. Slow match is made of hemp, flax, or cotton rope, with three strands slightly twisted. Cotton rope well twisted forms a good match without any preparation, and burns $4\frac{1}{2}$ inches an hour. *Quick* match is made of cotton yarn such as is used in candle-wick, which, after preparation described in the Ordnance Manual, is dredged with meal powder. One yard burns in the open air 13 seconds. Quick match inclosed in tubes burns more rapidly than in the open air, and more so in proportion as the tubes are smaller.

MATTOCK. A pioneer tool, resembling a pick-axe, but having two broad sharp edges instead of points.

MAY. To be permitted; to be at liberty; to have the power. Whenever a statute directs the doing of a thing for the sake of justice or the public good, the word *may* is the same as *shall*. For example, the 23 H. 6 says, the sheriff may take bail—that is construed he *shall*, for he is compellable to do so; (*Carth.*, 293. *Salk.*, 609. *Skin.*, 370.) The words *shall* and *may*, in general acts of the legislature or in private constitutions, are to be construed imperatively, (3 *Alk.*, 166;) but the construction of these words in a deed depends on circumstances; (3 *Alk.*, 282, sec. 1; *Vern.* 152, Case 142; 9 *Porter*, R. 390.)

MEASURES. (*See* WEIGHTS AND MEASURES.)

MEDICAL DEPARTMENT. (*See* ARMY for its organization.) No person can receive the appointment of assistant-surgeon until he has been examined and approved by an army medical board of not less than three surgeons or assistant-surgeons; and no person can receive the appointment of surgeon unless he shall have served five years as asst.-surgeon, and also have been examined by an army medical board constituted as above; (*Act* June 30, 1834.) (*See* AMBULANCE; LITTER; SURGERY.)

MEDICINE, RECIPES, &c., &c. An officer, unless he be a professed physician, need not take a large assortment of drugs. He wants a few powders, ready prepared; which any physician will prescribe for him, such as:—1. Emetic, mild; 2. ditto, very powerful for poison, (sulphate of zinc.) 3. Aperient, mild; 4. ditto, powerful. 5. Cordial for diarrhœa. 6. Quinine for ague. 7. Sudorific, (Dover's powder.) It will save trouble if these be so prepared, that one measureful of each shall be a full average dose for an adult; and if the measure to which they are adapted be cylindrical, and of such a size as just to admit a common lead-pencil, and three-quarters of an inch long, it can at any time be replaced by twisting up a paper-cartridge. In addition to the above powders take cold cream; heart-burn lozenges; lint; a small roll of diachylon; lunar-caustic, in a proper holder, to touch old sores with, and for snake bites; a scalpel and a blunt-pointed bistoury, to open abscesses with, (the blades of these should be waxed, to keep them from rust;) a good pair of forceps, to pull out thorns; a couple of needles, to sew up gashes; waxed thread. A mild effervescing aperient is very convenient.

Seidlitz-powders are perhaps a little too strong for frequent use in a tropical climate. The medicines should be kept in zinc pill-boxes, all of the same diameter, with a few letters punched both on their tops and bottoms, to indicate what they contain, as Emet., Astr., &c.; and the pill-boxes should slip one above another into a long zinc box lined with flannel, and lie there like sovereigns in a *rouleau*. The sulphate of zinc may be invaluable as an eyewash; for ophthalmia is a scourge in many countries. The taste, which should be strongly astringent, is the best guide to the strength of its solution.

For emetics, drink a charge of gunpowder in a tumblerful of warm water or soap-suds, or even tickle the throat.

Vapor-baths are used in many countries, and the Russian plan of making them is often the most convenient. They heat stones in the fire, and put them on the ground in the middle of their cabin or tent; on these they pour a little water and clouds of vapor are given off. Elsewhere, branches are spread on hot wood-embers, and the patient placed on these, wrapped in a large cloth; water is then sprinkled on the embers, which soon covers the patient with a cloud of vapor. The traveller who is chilled or over-worked, and has a quiet day before him, would do well to practise this simple and pleasant remedy.

Ointment.—Simple cerate is equal parts of oil and wax; lard and wax will do.

Seidlitz-powders are made as follows:—

$1\frac{1}{2}$ oz. Carbonate of Soda } For the blue papers.
3 oz. Tartarized Soda

7 drachms Tartaric Acid   For the white papers.

These quantities make 12 sets.

DISEASES.—Fevers of all kinds, diarrhœa, and rheumatism, are the plagues that most afflict soldiers; ophthalmia often threatens them. Change of air, from the flat country up into the hills, as soon as possible after the first violence of the illness is past, works wonders in hastening and perfecting a cure. With a bad diarrhœa, take nothing but broth, and it may be rice, in very small quantities at a meal, until quite restored. The least piece of bread or meat causes an immediate relapse.

REMEDIES.—A great discovery of modern days is the power of quinine to *keep off* fever while travelling across a fever district. It is a widely-corroborated fact, that a residence on the banks of the river, or in low land, is often less affected by malaria than the low hills that overlook it. There are certain precautions which should be borne in mind in unhealthy seasons—as, never to encamp to the leeward of a marsh; to sleep close in between large fires, with a handkerchief gathered round your face, (natural instinct will teach this;) not to start off too early in the morning; to avoid unnecessary hunger, hardship, and exposure.

Drowning.—A half-drowned man must be put to bed in dry, heated clothes; hot stones, &c., to his feet; his head must be raised moderately. Human warmth is excellent, such as that of two strapping men made to lie close up against him, one on each side. All rough treatment is hurtful.

For Snake-bites, tie a string tight above the part, suck the wound, and apply caustic as soon as you can. Or, for want of caustic, cut away with a knife, and afterwards burn out with the end of your iron-ramrod, heated as near a white heat as you can readily get it. The arteries lie deep, and as much flesh may, without much danger, be cut or burnt into, as the fingers can pinch up. The next step is to use the utmost energy, and even cruelty, to prevent the patient's giving way to that lethargy and drowsiness which is the usual effect of snake-poison, and too often ends in death.

Broken Bones.—It is extremely improbable that a man should die, in consequence of a broken leg or arm, *if the skin be uninjured;* but, if the broken end forces its way through the flesh, the injury is a very serious one. Abscesses form, the parts mortify, and the severest consequences often follow. Hence, when a man breaks a bone, do not convert a simple injury into a severe one, by carrying him carelessly. If possible, move the encampment to the injured man, and not *vice versa*. "When a man has broken his leg, lay him on the other side, put the broken limb exactly on the sound one, with a little straw between, and tie the two legs together with handkerchiefs. Thus, the two legs will move as one, and the broken bone will not hurt the flesh so much, nor yet come through the skin;" (DRUITT.)

Excessive Bleeding.—When the blood does not pour or trickle in a steady stream from a deep wound, but in pulses, and is of a bright-red color, all the bandages in the world will not stop it. It is an artery that is wounded; and, unless there be some one accessible who knows how to take it up and tie it, burn deeply into the part, as you would for a snake-bite; or else pour boiling grease into the wound. It is, of course, a barbarous treatment, and far from being sure of success, as the cauterized artery may break out afresh; still, life is in question, and it is the only hope of saving it. After the cautery, the wounded man's limb should be kept perfectly still, and well raised, and cool, until the wound is nearly healed. A *tourniquet*, which will stop the blood for a time, is made by tying a strong thong, string, or handkerchief, firmly above the part, putting a stick through and screwing it tight.

If you know whereabouts the artery lies which it is the object to compress, put a stone over the place and under the handkerchief. The main arteries follow pretty much the direction of the inner seams of the coat sleeves and trousers.

To cure blistered Feet.—" Rub the feet at going to bed with spirits mixed with tallow dropped from a candle into the palm of the hand; on the following morning no blister will exist. The spirits seem to possess the healing power, the tallow serving only to keep the skin soft and pliant. This is Captain Cochrane's advice, and the remedy was used by him in his pedestrian tour;" (MURRAY's *Handbook of Switzerland*.) The receipt is excellent; all pedestrians and all teachers of gymnastics endorse it, and it cannot be too widely known. To prevent the feet from blistering, it is a good plan to soap the inside of the stocking before setting out, making a good lather all over it; and a raw egg broken into a boot, before putting it on, greatly softens the leather. After some hours' walking, when the feet are beginning to be chafed, take off the shoes, and change the stockings; putting what was the right stocking on the left foot, and the left stocking on the right foot. Or, if one foot only hurts, take off the boot, and turn the stocking inside out.

Rarefied Air.—On high plateaux or mountains, travellers must suffer somewhat. The symptoms are described by many South American travellers, where it is called the *puna*. The disorder is sometimes fatal to stout plethoric people; oddly enough, cats are unable to endure it. At villages 13,000 feet above the sea, Dr. Tscudi says that they cannot live. Numerous trials have been made, but the creatures die in frightful convulsions. The symptoms of the puna are giddiness, dimness of sight and hearing, headache, fainting-fits, blood from mouth, eyes, nose, lips, and a feeling like sea-sickness. Nothing but time cures it. It begins to be felt at from 12,000 to 13,000 feet above the sea. M. Hermann Schlagintweit—whose large mountain experience in the Alps and in the Himalaya, up to the height of 20,000 feet or more, is only paralleled by that of his brother—says that he found the headache, &c., to come on when there was a breeze, far more than at any other time. His whole party would awake at the same moment, and begin to complain of the symptoms, immediately on the commencement of a breeze. The symptoms of overwork are not wholly unlike those of the puna, and many young travellers who have felt the first, have ascribed them to the second.

Snow-blindness.—In civilized life blue spectacles are, as is well known, an indispensable accompaniment to snow-mountain expeditions. The Esquimaux adopt the following equivalent: They cut a piece of soft wood to the curvature of the face. It is about two inches thick, and extends horizontally quite across both eyes, and rests on the nose, where a notch is cut to act in the same way as the bridge of a pair of spectacles. This is tied behind the ears. Next a long narrow slit, of the thickness of a thin saw-cut, is made along its middle almost from end to end. Through this slit the wearer can see very fairly. It is narrower than the diameter of the pupil of his eye, and, consequently, the light that reaches his retina is much diminished in quantity.

Scurvy.—Any vegetable diet cures it: lime-juice, treacle, raw potatoes, and acid fruits are especially efficacious. Dr. Kane insists on the value of meat, eaten entirely raw, as a certain anti-scorbutic. It is generally used by the Esquimaux.

Teeth.—Tough diet tries the teeth so severely that a man about to undergo it had much better pay a visit to a dentist before he leaves.

Suffering from Thirst.—Pour water over the clothes of the man, and keep them constantly wet; restrain his drinking, after the first few minutes, as strictly as you can summon heart to do it. In less severe cases, drink water with a tea-spoon; it will satisfy a parched palate as much as if you gulped it down in tumblerfuls, and will disorder the digestion much less.

Suffering from Hunger.—Two or three mouthfuls every quarter of an hour is, to a man in the last extremity, the best thing; and strong broth the best food.

Wasp and Scorpion-stings.—The oil scraped out of a tobacco-pipe is good; should the scorpion be large, his sting must be treated like a snake-bite.

Poisoning.—The first thing is to give a powerful emetic, to throw up whatever poison may still remain unabsorbed in the stomach. Use soap-suds or gunpowder, if proper emetics are not at hand. If there be violent pains and griping, or retchings, give plenty of water to make the vomitings more easy. Nothing now remains to be done but to resist the symptoms that are caused by the poison which was absorbed before the emetic acted. Thus if the man's feet are cold and numbed, put hot stones against them and wrap him up warmly. If he be drowsy, heavy, and stupid, give brandy, and try to rouse him. There is nothing more to be done, save to avoid doing mischief.

Fleas.—"Italian flea-powder," sold in the East, is really efficacious. It is made from the "Piré oti," (or flea-bean,) mentioned in CURZON's *Armenia*, as growing in that country. It is powdered and sold as a specific.

Vermin on the Person.—"We had now been travelling for nearly six weeks, and still wore the same clothing we had assumed on our departure. The incessant pricklings with which we were harassed, sufficiently indicated that our attire was peopled with the filthy vermin to which the Chinese and Tartars are familiarly accustomed, but which, with Europeans, are objects of horror and disgust. Before quitting Tchagan-Kouren, we had bought in a chemist's shop a few sapeks'-worth of mercury. We now made with it a prompt and specific remedy against the lice. We had formerly got the receipt from some Chinese; and, as it may be useful to others, we think it right to describe it here. You take half an ounce of mercury, which you mix with old tea-leaves previously reduced to paste by mastication. To render this softer you generally add saliva; water could not have the same effect. You must afterwards bruise and stir it awhile, so that the mercury may be divided into little balls as fine as dust. (I presume that blue pill is a pretty exact equivalent to this preparation.) You infuse this composition into a string of cotton, loosely twisted, which you hang round the neck; the lice are sure to bite at the bait, and they thereupon as surely swell, become red, and die forthwith. In China and in Tartary you have to renew this salutary necklace once a month;" (Huc's *Travels in Tartary*.)—Galton's *Art of Travel*.

MEMBERS. (*See* Court-Martial.)

MEMBERS, (Supernumerary.) In case supernumerary members are detailed for a court-martial, they are sworn, and it is right that they should sit and be present at all deliberations even when the court is cleared, in order to be prepared to take the place of any absent member. Until then they have no voice; (Hough.)

MENACING WORDS. (*See* Contempt.)

MENSURATION.

MATHEMATICAL FORMULÆ AND DATA.
Lines.

Circle. *Ratio of circumference to diameter,* $\pi = 3.1415926536 = \frac{355}{113}$ *nearly.*

Length of an arc $= \frac{a \pi r}{180}$; r being the radius of the circle, and a the number of degrees in the arc; or, nearly $= \frac{8 c' - c}{3}$; c being the chord of the arc, and c' the chord of half the arc, which is $= \sqrt{\frac{1}{4} c^2 + \text{versine}^2}$.

Length of 1 *degree* $= 0.0174533$; radius being 1.
Length of 1 *minute* $= 0.0002909$.
Length of 1 *second* $= 0.0000048$.

ELLIPSE. *Circumference* = $\frac{100}{200} \pi \sqrt{\frac{1}{2}(a^2 + b^2)}$, nearly; a and b being the axes.

PARABOLA: *Length of an arc*, commencing at the vertex, = $\sqrt{\left(\frac{4 a^2}{3} + \sqrt{b}\right)}$, nearly; a being the abscissa, and b the ordinate.

Surfaces.

Triangle. Half the base × the height; or half the product of two sides × the sine of the included angle, $\left(\frac{1}{2} a b \frac{\sin. C}{R}\right)$; or, $\sqrt{s(s-a)(s-b)(s-c)}$; a, b, c being the sides, and $s = \frac{a+b+c}{2}$

Parallelogram. The base × the height.

Trapezoid. Half the sum of the parallel sides × the height.

Any Quadrilateral. Half the product of the diagonals × the sine of their angle.

Any irregular plane figure bounded by curves. Divide the figure into any *even* number of parts by parallel equidistant ordinates; let a be the sum of the first and last ordinates, b the sum of the *even* ordinates, c that of the *odd* ones, except the first and last; d the common distance between them; then will the area = $\frac{1}{3} d (a + 4b + 2c)$. *Five* ordinates will generally be found sufficient.

Circle. πr^2; or diam.^{2} × .7854; or, circum.^{2} × .07958.

Circular sector. $\frac{r a}{2}$; a being the length of the arc in linear measure.

Circular segment. The difference between the sector, and the triangle formed by the cord and the radii; or $\frac{r a - r^2 \sin. A}{2}$; or nearly = $.4 v (c + \frac{4}{3} \sqrt{\frac{1}{4} c^2 + v^2})$; c being the cord and v the versed sine.

Ellipse. .7854 $a b$; a and b being the axes.

Parabola. $\frac{2}{3} a b$; a being the abscissa, and b the double ordinate.

Right prism or cylinder. Curved surface = height × perimeter of base.

Right pyramid or cone. Half the slant height × perimeter of base.

Frustum of a right prism or cylinder. The perimeter of the base multiplied by the distance from the centre of gravity of the upper section to the base. If the prism or cylinder is oblique, multiply this product by the sine of the angle of inclination.

Frustum of a right pyramid or cone. The slant height × half the sum of the perimeters of the two ends.

Sphere. $4 \pi r^2$; or, diam. × circum.; or, diam.^{2} × 3.1416.

Spherical zone or segment. $2 \pi r h$; or, the height of the zone or segment multiplied by the circumference of the sphere.

Circular spindle. $2 \pi (r c - a \sqrt{r^2 - \frac{1}{4} c^2})$; a being the length of the arc, and c its chord, or the length of the spindle.

Spherical triangle. $\pi r^2 \dfrac{s - 180°}{180°}$; s being the sum of the three angles.

Any surface of revolution. $2 \pi r l$; or, the length of the generating element multiplied by the circumference described by its centre of gravity.

TABLE OF REGULAR POLYGONS.

No. of sides.	Name.	Area.	Radius of circumscribing circle.	Side of inscribed polygon.
3	Triangle.	0.4330127	0.5773503	1.732051
4	Square.	1.0000000	0.7071068	1.414214
5	Pentagon.	1.7204774	0.8506508	1.175570
6	Hexagon.	2.5980762	1.0000000	1.000000
7	Heptagon.	3.6339124	1.1523824	0.867767
8	Octagon.	4.8284271	1.3065628	0.765367
9	Nonagon.	6.1818242	1.4619022	0.684040
10	Decagon.	7.6942088	1.6180340	0.618034
11	Undecagon.	9.3656399	1.7747324	0.563465
12	Dodecagon.	11.1961524	1.9318517	0.517638

The column of *areas*, in the foregoing table, gives the number by which the *square of the side* is to be multiplied, to find the area of the polygon.

The next column gives the multiplier for the *side of a polygon*, to find the radius of the circumscribing circle.

The last column gives the multiplier for the *radius of a circle*, to find the side of the inscribed polygon.

Solids.

Prism or cylinder. Area of base multiplied by the height.

Pyramid or cone. Area of base multiplied by one-third of the height.

Frustum of a pyramid or cone. $\frac{1}{3} h (B + b + \sqrt{B b})$; h being the height; B and b the areas of the two ends. Or, for a conic frustum: $\frac{1}{3} h \times .7854 \times \left(\dfrac{D^3 - d^3}{D - d} \right)$; D and d being the diameters of the two ends.

Frustum of a right triangular prism. The base $\times \frac{1}{3} (H + H' + H'')$.

Frustum of any right prism. The base multiplied by its distance from the centre of gravity of the section.

Cylindrical segment, contained between the base and an oblique plane passing through a diameter of the base: two-thirds of the height multiplied by the great triangular section; or, $\frac{2}{3} r h^2$; r being the radius of the base, and h the area of the height.

Sphere. $\dfrac{4 \pi r^3}{3}$; or, $.5236\ d^3$; r being the radius and d the diameter.

Spherical segment. $\frac{1}{3}\pi h^2 (3r - h) = \dfrac{\pi h}{6}$; $(3 b^2 + h^2)$; b being the radius of the base, h the height of the segment, and r the radius of the sphere: $\dfrac{\pi}{6} = 0.5236$.

Spherical zone. $\dfrac{\pi h}{6}(3 B^2 + 3 b^2 + h^2)$; B, b being the radii of the bases.

Spherical sector. $\frac{1}{3} r \times$ the surface of the segment or zone; or, $\frac{2}{3} \pi r^2 h$.

Ellipsoid. $\dfrac{\pi a^2 b}{6}$; a being the revolving diameter and b the axis of revolution.

Paraboloid. Half the area of the base multiplied by the height.

Circular spindle. $\pi (\frac{1}{6} c^3 - 2 s \sqrt{r^2 - \frac{1}{4} c^2})$; s being the area of the revolving segment and c its chord.

Any solid of revolution. $2 \pi r s$; or, the area of the generating surface multiplied by the circumference described by its centre of gravity.

Any irregular solid, bounded by a curved surface. Use the rule for finding the area of an irregular plane figure, substituting *sections* for *ordinates*.

Cask gauging. 1. — By the preceding rule:

The content of a cask $= \dfrac{\pi}{24} l (d^2 + D^2 + 4 M^2)$; l being the length, d, D, the head and bung diameters, and M, a diameter midway between them, all measured in the clear, inside; $\dfrac{\pi}{24} = 0.1309$.

The same formula may be thus stated: $\frac{1}{6} l (A + B + C)$; l being the length; A and B, the areas of the head and bung sections; and C, that of the section midway between them.

2. Contents of a cask, nearly, $= \dfrac{\pi}{12} l (2 D^2 + d^2)$; or, $l \times$ the area of a circle whose diameter is $\dfrac{2 D + d}{3}$

CENTRES OF GRAVITY.

Lines.

Circular arc. At a distance from the centre $= \dfrac{r\,c}{l}$; r being the radius, c the chord, and l the length of the arc.

Areas.

Triangle. On a line drawn from any angle to the middle of the opposite side, at two-thirds of the distance from the angle to the side.

Trapezoid. On a line a joining the middle points of the two parallel sides, B, b; distance from $B = \dfrac{a}{3}\left(\dfrac{B + 2b}{B + b}\right)$

Semicircle. Distance from the centre $= \dfrac{4\,r}{3\,\pi}$

Circular segment. Distance from the centre $= \dfrac{c^3}{12\,A}$; c being the chord of the segment, and A its area.

Circular sector. Distance from the centre $= \dfrac{2\,r\,c}{3\,l}$; c being the chord, and l the length of the arc.

Parabolic segment. Distance from the vertex = three-fifths of the abscissa.

Surface of a right cylinder, cone, or frustum of a cone. The centre of gravity is at the same distance from the base as that of the parallelogram, triangle or trapezoid, which is a right section of the same.

Surface of a spherical zone or segment. At the middle of the height.

MERLON. The space of the parapet between two embrasures.

MESNE PROCESS. Any writ issued in the course of a suit between the original process and execution. By this term is also meant the writ of proceedings in an action to summon or bring the defendant into court, or compel him to appear or put in bail, and then to hear and answer the plaintiff's claim. (*See* ARREST BY CIVIL AUTHORITY.)

MESS. The law is silent with regard to messes in the army. Executive regulations have been made on the subject, but without law it is impossible to put messes on a proper footing. In England, an allowance is granted by the king in aid of the expense of officers' messes; and every officer on appointment to a corps subscribes one month's pay to the mess fund. All the officers of the corps mess together. In France, the several grades mess separately; lieutenants and sub-lieutenants forming two tables; captains another, and field officers of different grades generally eating separately also. Colonels and general officers of

MIL.] MILITARY DICTIONARY. 419

the French service receive an allowance for table expenses, not sufficient to keep open house, but enough to enable them to entertain guests.

MIASM, MIASMATA. (*See* SANITARY PRECAUTIONS.)

MILEAGE. Travelling allowance or transportation of baggage. (*See* TRAVELLING.)

MILITARY ACADEMY. (*See* ACADEMY.)

MILITARY LAWS. (*See* GOVERNMENT, LAW (MILITARY); REGULATIONS.)

MILITIA.

GENERAL ABSTRACT OF THE MILITIA FORCE OF THE UNITED STATES, ACCORDING TO THE LATEST RETURNS RECEIVED AT THE OFFICE OF THE ADJUTANT-GENERAL.

States and Territories.	For what year.	General officers.	General staff officers.	Field officers, &c.	Company officers.	Total commissioned officers.	Non-commissioned officers, musicians, artificers, and privates.	Aggregate.
Maine	1854	10	56	13	193	272	2,345	2,617
New Hampshire	1854	11	202	119	895	1,227	32,311	33,538
Massachusetts	1856	10	46	131	521	708	154,323	155,031
Vermont	1843	12	51	224	801	1,088	22,827	23,915
Rhode Island	1854	3	39	24	49	115	1,036	1,151
Connecticut	1856	3	10	59	182	254	51,560	51,814
New York	1855	97	305	1,460	5,402	7,264	326,094	333,358
New Jersey	1852							81,984
Pennsylvania	1854							106,957
Delaware	1827	4	8	71	364	447	8,782	9,229
Maryland	1838	22	68	544	1,763	2,397	44,467	46,864
Virginia	1854	32	76	153	614	875	124,656	125,531
North Carolina	1845	28	133	657	3,449	4,267	75,181	79,448
South Carolina	1856	20	135	535	1,909	2,599	33,473	36,072
Georgia	1850	39	91	624	4,296	5,050	73,649	78,699
Florida	1845	3	14	95	508	620	11,502	12,122
Alabama	1851	32	142	775	1,883	2,832	73,830	76,662
Louisiana	1856	16	129	542	2,084	2,771	87,961	90,732
Mississippi	1838	15	70	392	348	825	35,259	36,084
Tennessee	1840	25	79	859	2,644	3,607	67,645	71,252
Kentucky	1852	43	145	1,165	3,517	4,870	84,109	88,979
Ohio	1845	91	217	462	1,281	2,051	174,404	176,455
Michigan	1854	30	323	147	2,358	2,858	94,236	97,094
Indiana	1832	31	110	566	2,154	2,861	51,052	53,913
Illinois	1855							257,420
Wisconsin	1854	15	88	125	914	1,142	48,119	49,261
Iowa								
Missouri	1853		17	4	67	88	117,959	118,047
Arkansas	1854	10	39	128	955	1,132	34,922	36,054
Texas	1847	15	45	248	940	1,248	18,518	19,766
California	1854	12	11		100	123	208,522	208,645
Minnesota Territory	1851	2	5			7	1,996	2,003
Oregon Territory								
Washington Territory								
Nebraska Territory								
Kansas Territory								
Territory of Utah	1853	2		48	235	285	2,536	2,821
Territory of New Mexico								
District of Columbia	1852	3	10	28	185	226	7,975	8,201
Grand aggregate		636	2,664	10,198	40,611	54,109	2,071,249	2,571,719

Notwithstanding the feudal military service introduced into England by William the Conqueror, ancient Anglo-Saxon laws, making it the duty of every freeman to arm himself and serve for the defence of his country against invasion, remained in full vigor. The force authorized to be raised under these conditions has from the earliest times been called the militia, and was under the command of the alderman or earl, who was at that time the governor of the county. By the 27th Henry II. (1154) this force was regulated and organized, every subject, according to his rank and means, being compelled to furnish himself with arms for the maintenance of the king's peace. A century afterwards this act was confirmed, and a fresh "Assize of arms" ordered by the statute of Wynton, by which it was enacted that every man between the ages of fifteen and sixty should be assessed, and sworn to keep armor according to the value of his lands and goods. For £15 and upwards in rent, or 40 marks in goods, a hauberk, an iron breastplate, a sword, a knife, and a horse; property of less value entailing the possession of arms of a proportionately less expensive character. Constables were also appointed to view the armor twice a year, which constables, the act says, "shall present before justices assigned such defaults as they shall see in the country about armor; and the justices assigned shall present at every parliament unto the king such defaults as they shall find, *and the king shall provide the remedy therein.* The system organized by these statutes was evidently, from the context, intended in the first place for the preservation of internal peace, by the suppression of tumults, and keeping in check the bands of robbers that infested the public ways; the sheriff, as the conservator of the public peace, had always possessed the power of calling out the posse comitatus, or assembly of liegemen of the county, to assist him on such occasions; and it is supposed that it was the object of Edward III. to confirm and extend this authority, and at the same time to organize a force readily capable of being made applicable to resist invasion. In the United States each and every free, able-bodied, white male citizen of the respective States resident therein, who is of the age of 18 years and under 45 years, (except Exempts, which see,) shall be enrolled in the militia by the captain or commanding officer of the company within whose bounds such citizen shall reside. The militia of the respective States shall be arranged into divisions, brigades, regiments, battalions and companies, as the legislature of each State shall direct. If the same be convenient, each brigade shall consist of four regiments; each regiment of two battalions; each battalion of five companies, and each company of sixty-four privates. The said militia shall be officered by the respective States as follows: to each division, one major-general and two aides-de-camp with the rank of major, one division-inspector with the rank of lieutenant-colonel, and one division-quartermaster, with the rank of major; to each brigade, one brigadier-general, one aide-de-camp with the rank of captain, one quartermaster, with the rank of captain, with one brigade-inspector, to serve also as brigade-major, with the rank of major; to each regiment consisting of two battalions one colonel, one lieutenant-colonel, and one major; where there shall be only one battalion, it shall be commanded by a major; to each regiment one chaplain; to each company one captain, one lieutenant, one ensign, four sergeants, four corporals, one drummer, and one fifer or bugler. There shall be a regimental staff, to consist of one adjutant and one quartermaster, to rank as lieutenants; one paymaster, one surgeon, and one surgeon's mate; one sergeant-major, one drum-major, and one fife-major; to the militia of each State one quartermaster-general; (*Acts* May 8, 1792, March 2, 1803, April 18, 1814, April 20, 1816.)

Out of the enrolled militia, there shall be formed for each battalion one company of grenadiers, light infantry or riflemen; and to each division there shall be at least one company of artillery and one troop of horse; there shall be to each company of artillery, one captain, two lieutenants, four sergeants, four corporals, six gunners, six bombardiers, one drummer, and one fifer. There shall be to each troop of horse, one captain, two lieutenants, one cornet, four sergeants, four corporals, one saddler, one farrier, and one trumpeter. Each troop of horse and company of artillery to be formed of volunteers of the brigade to which they belong; (*Act* May 8, 1792.)

It shall be the duty of the brigade-inspector to attend the regimental and battalion meetings of the militia, inspect their arms, ammunition, &c., superintend their exercise and manœuvres, and introduce the system of military discipline throughout the brigade agreeably to law and such orders as they shall, from time to time, receive from the commander-in-chief of the State; to make returns to the adjutant-general of the State at least once in every year, reporting the actual condition of the arms, accoutrements, and ammunition of the several corps, and every other thing which, in his judgment, may relate to their government and the general advancement of good order and military discipline; (*Act* May 8, 1792.)

Volunteer corps shall retain their accustomed privileges, subject nevertheless to all other duties required by this act, in like manner with the other militia; (*Act* May 8, 1792.)

There shall be an adjutant-general appointed in each State, whose duty it shall be to distribute all orders of the commander-in-chief of the State to the several corps; to attend all public reviews when the commander-in-chief shall review the militia; to obey all orders from him, relative to carrying into execution and perfecting the system of military discipline established by this act; to furnish blank forms of different returns that may be required, and to explain the principles on which they should be made; to receive from the several officers of the different corps throughout the State, returns of the militia under their command, reporting the actual condition of their arms, and every thing which relates to the advancement of good order and discipline; all which the several officers of the divisions, brigades, regiments, and battalions are required to make, so that the adjutant-general may be duly furnished therewith; from all of which returns he shall make abstracts and lay the same annually before the commander-in-chief of the State; and he shall also make an annual return of the militia of the State, with their arms and accoutrements, &c., to the President of the United States; and the Secretary of War shall, from time to time, give directions to the adjutant-generals of States to produce uniformity in such returns; (*Acts* May 8, 1792; March 2, 1803, and May 12, 1820.)

Whenever militia shall be called into actual service of the United States, their pay shall commence from the day of their appearance at the places of battalion, regimental, or brigade rendezvous; allowing to each non-commissioned officer and soldier a day's pay and rations for every fifteen miles from his home to such place of rendezvous, and the same allowances for travelling home from the place of discharge; (*Act* Jan. 2, 1795.)

The militia or other State troops, being mustered and in pay of the United States, shall be subject to the same Rules and Articles of War as the troops of the United States, save only that courts-martial for the trial of militia or other State troops shall be composed entirely of militia officers; (Art. 97.) All officers, serving by commission from the authority of any particular States, shall, on all detachments, courts-martial, or other duty wherein they may be employed in conjunction with the regular forces of the United States, take rank next after all officers of like grade in said regular forces, notwithstanding the commissions of such militia or State officers may be older than the commissions of the officers of the regular forces of the United States; (Art. 98.)

By the act for calling forth the militia, approved Feb. 28, 1795, militia not to serve more than three months after arrival at the place of rendezvous. Every officer, non-commissioned officer, or private of militia that shall fail to obey tne orders of the President of the United States, shall forfeit a sum not exceeding one year's pay, and not less than one month's pay, to be determined and adjudged by a court-martial; and such officer shall, moreover, be liable to be cashiered by sentence of a court-martial and be incapacitated from holding a commission in the militia for a term not exceeding twelve months, at the discretion of the said court; and such non-commissioned officers and privates shall be liable to be imprisoned by a like sentence, on failure of the payment of fines adjudged against them, for one calendar month for every five dollars of such fine.

Courts-martial for the trial of militia, shall be composed of militia officers only.

That all fines to be assessed, as aforesaid, shall be certified by the presiding officer of the court-martial before whom the same shall be assessed, to the marshal of the district in which the delinquent shall reside, or to one of his deputies, and also to the supervisor of the revenue of the same district, who shall record the said certificate in a book to be kept for that purpose. The said marshal, or his deputy, shall forthwith proceed to levy the said fines, with costs, by distress and sale of the goods and chattels of the delinquent; which costs, and the manner of proceeding with respect to the sale of the goods distrained, shall be agreeable to the laws of the State in which the same shall be, in other cases of distress. And where any non-commissioned officer or private shall be adjudged to suffer imprisonment, there being no goods or chattels to be found whereof to levy the said fines, the marshal of the district, or his deputy, may commit such delinquent to gaol, during the term for which he shall be so adjudged to imprisonment, or until the fine shall be paid, in the same manner as other persons condemned to fine and imprisonment at the suit of the United States may be committed.

That the marshals and their deputies shall pay all such fines by them levied, to the supervisor of the revenue in the district in which they are collected, within two months after they shall have received the same, deducting therefrom five per centum as a compensation for their trouble; and in case of failure, the same shall be recoverable by action of debt or information in any court of the United States of the district in which such fines shall be levied, having cognizance thereof, to be sued for, prosecuted, and recovered, in the name of the supervisor of the district, with interest and costs.

That the marshals of the several districts, and their deputies, shall have the same powers, in executing the laws of the United States, as sheriffs, and their deputies in the several States, have by law in executing the laws of the respective States.

And by a supplementary act approved in Feb. 1813, That, in every case in which a court-martial shall have adjudged and determined a fine against any officer, non-commissioned officer, musician, or private, of the militia, for any of the causes specified in the act to which this act is a supplement, or in the fourth section of an act, entitled "An act to authorize a detachment from the militia of the United States," all such fines, so assessed, shall be certified to the comptroller of the treasury of the United States, in the same manner as the act to which this act is a supplement directed the same to be certified to the supervisor of the revenue.

That the marshals shall pay all fines which have been levied and collected by them, or their respective deputies, under the authority of the acts herein referred to, into the treasury of the United States, within two months after they shall have received the same, deducting five per centum for their own trouble; and, in case of failure, it shall be the duty of the comptroller of the treasury to give notice to the district attorney of the United States, who shall proceed against the said marshal in the district court, by attachment, for the recovery of the same. (*See* CALLING FORTH MILITIA; DEFENCE, *National*.)

MINE. Powder placed in subterranean cavities, by exploding which every thing above it is overthrown. Mines are *offensive* when they are prepared by besiegers, and *defensive* when used by the besieged. The place where the powder is lodged is called the *chamber* of the mine, and it is generally made of a cubical form large enough to contain the wooden box which holds the powder necessary for the charge. The fire is communicated to the mine by means of a pipe or hose made of coarse cloth filled with powder, laid in a wooden case about an inch square, extending from the centre of the chamber to the extremity of the gallery, where a match is fixed so that the miner who applies the fire to it, may have time to retire before the flame reaches the chamber. (*See* FOUGASSE; GALLERY.)

MINORS. The Secretary of War, on demand, is required to grant the discharge from the army of any minor enlisted without the consent of parent or guardian.

MINUTE GUNS. Guns, fired at intervals of a minute, are signals of distress.

MISBEHAVIOR BEFORE THE ENEMY. Punishable with death or otherwise, according to the sentence of a court-martial; (ART. 52.)

MISNOMER. If a prisoner plead a misnomer, the court may ask

the prisoner what is his real name, and call upon him to plead to the amended charge; (HOUGH.)

MITIGATION. (*See* PARDON.)

MONEY. The embezzlement or misapplication of public money intrusted to an officer for the payment of men under his command, or for enlisting men into the service, or for other purposes, punishable with cashiering and being compelled to refund the money. In case of a non-commissioned officer, reduction to the ranks and being put under stoppages until the money is refunded, and such corporeal punishment as a court-martial shall direct; (ART. 39.)

MONTHLY RETURNS. (*See* RETURNS.)

MORTAR. The following mortars are used in the United States service: The heavy 13-inch mortar, weighing 11,500 lbs., whole length 53 inches, length of chamber 13 inches, and superior diameter of cham-

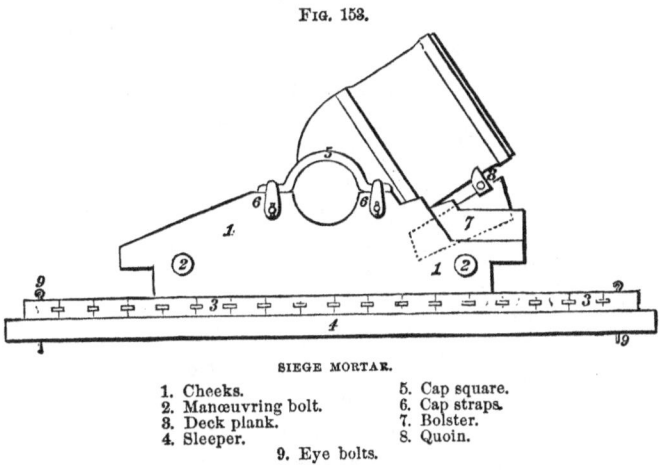

FIG. 153.

SIEGE MORTAR.

1. Cheeks.
2. Manœuvring bolt.
3. Deck plank.
4. Sleeper.
5. Cap square.
6. Cap straps.
7. Bolster.
8. Quoin.
9. Eye bolts.

ber 9.5 inches; the heavy 10-inch mortar, weighing 5,775 lbs., whole length 46 inches, length of chamber 10 inches; the light 10-inch mortar, weighing 1,852 lbs., whole length of mortar 28 inches, length of chamber 5 inches; the *light* 8-inch mortar, weighing 930 lbs., whole length of mortar 22.5 inches, length of chamber 4 inches; brass stone mortar, weighing 1,500 lbs., diameter of bore 16 inches, whole length of mortar 31.55 inches, length of chamber 6.75 inches; brass coehorn 24-pounder, diameter of the bore 5.82 inches, weight 164 lbs., whole length 16.32 inches, length of chamber 4.25 inches; iron eprouvette, diameter of the bore 5.655 inches, weight 220 lbs., length of bore exclusive of chamber, 11.5 inches, length of chamber 1.35 inch. Mortars are mounted on *beds*, and when used, siege mortars are placed on a *platform* of wood made of 6 sleepers; 18 deck planks; and 72 dowels; fastened with 12 iron eye-bolts. (Consult *Ordnance Manual* and *Instruction in Heavy Artillery for Mechanical Manœuvres*. See ARTILLERY; ORDNANCE.)

MOUNTAIN ARTILLERY. The mountain howitzer, weight 220 lbs., whole length 37.21 inches, diameter of bore 4.62 inches; length of chamber 2.75 inches, diameter of chamber 3.34; natural angle of sight, 0° 37'; RANGE 500 yards, at an elevation of 2° 30', with a charge of $\frac{1}{2}$ lb. powder and shell; time of flight, 2 seconds; with same charge and elevation, the range of spherical-case is 450 yards. At an elevation of from 4° to 5° the range with canister is 250 yards. According to elevation the range varies from 150 to 1,000 yards; at the same elevation the range with shell being greater than spherical-case. A battery of six mountain howitzers requires 33 pack-saddles and harness, and 33 horses or mules. A mountain howitzer ammunition chest will carry about 700 musket ball-cartridges, besides eight rounds for the howitzer.

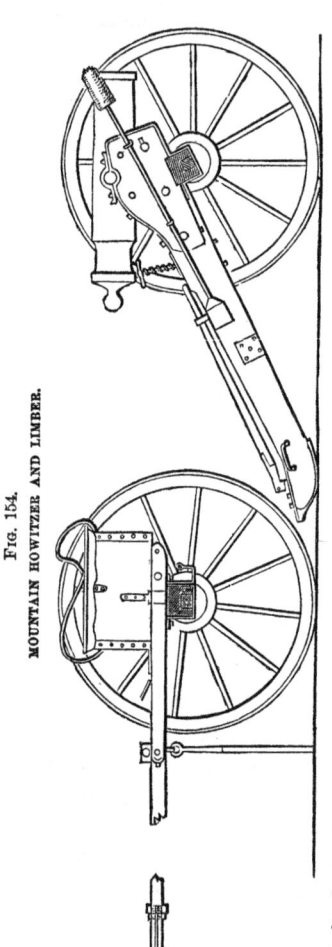

FIG. 154. MOUNTAIN HOWITZER AND LIMBER.

MOUNTED RIFLEMEN. There is one regiment of mounted riflemen in the United States army. (*See* ARMY for their organization.) The skirmish drill for mounted troops prepared by Capt. D. H. Maury, U. S. A., and used by mounted riflemen, differs from the system of cavalry exercise:

1st. *In prescribing the formation in one rank instead of in two ranks.*—Besides extending the line of front, this change develops individual instruction, and enables the officer to bring his men from column into line, and the reverse, almost as quickly as in infantry. By it a mounted company may be brought from the full gallop into fighting order *on foot*, the true order for riflemen, within *six seconds* after the command has been given.

2d. *In giving no heed to inversions.*—The effect of this change is to bring men from marching into fighting order in the simplest and most rapid manner.

3d. *The grouping together of men in sets of fours.*—This, besides being convenient for the purposes of police and guards in garrison and camp, teaches the men, when in action, to rely upon each other as near comrades. (*See* Cavalry.)

MOUNTING. The parade of marching on guard is called *guard-mounting*.

MUSKET. (*See* Arms.)

MUSTER. At every muster, the commanding officer of each regiment, troop, or company there present, shall give certificates, signed by himself, signifying how long officers who do not appear at muster have been absent, and the reason of their absence. In like manner, the commanding officer of every troop or company shall give certificates, signifying the reasons of the absence of the non-commissioned officers and private soldiers, which reasons and time of absence shall be inserted in the muster-rolls, opposite the names of the respective absent officers and soldiers. The certificates shall, together with the muster-rolls, be remitted by the commissary of musters or other officer mustering, to the Department of War, as speedily as the distance of the place will admit; (Art. 13.) Every officer, who shall be convicted of having signed a false certificate, relating to the absence of either officer or soldier, or relative to his or their pay, shall be cashiered; (Art. 14.) Every officer, who shall knowingly make a false muster of man or horse, and every officer or commissary of musters, who shall willingly sign, direct, or allow, the signing of muster-rolls wherein such false muster is contained, shall, upon proof made thereof by two witnesses before a general court-martial, be cashiered, and shall be thereby utterly disabled to have or hold any office or employment in the service of the United States; (Art. 15.) Any commissary of muster or other officer, who shall be convicted of having taken money or other things by way of gratification, on mustering any regiment, troop, or company, or on signing muster-rolls, shall be displaced from office and shall be thereby utterly disabled to have or hold any office or employment in the service of the United States; (Art. 16.) Any officer, who shall presume to muster a person as a soldier who is not a soldier, shall be deemed guilty of having made a false muster, and shall suffer accordingly; (Art. 17.) Troops are mustered every two months. (*See* Arrears of Pay; Certificate; False; Pay.)

MUTINY. Any officer or soldier, who shall begin, excite, cause, or join in any mutiny or sedition in any troop or company in the service of the United States, or in any party, post, detachment, or guard, shall suffer death, or such other puunishment as by a court-martial shall be inflicted; (ART. 7.) Any officer, non-commissioned officer, or soldier who, being present at any mutiny or sedition, does not use his utmost endeavor to suppress the same, or coming to the knowledge of any intended mutiny, does not, without delay, give information thereof to his commanding officer, shall be punished by the sentence of a court-martial with death, or otherwise, according to the nature of his offence; (ART. 8.) "Mutiny is a combined or simultaneous resistance, active or passive, to lawful military authority." The best authorities admit that a single person, without previous combination or concert with others, cannot commit mutiny. An overt act by one person, in pursuance of a combined plan or conspiracy, is, however, mutiny; and conspiracy or *intended* mutiny is, under the 8th article, punishable in the same degree as an overt act. Where an overt act, therefore, has not been committed, it is proper to base the charge on the 8th article. But all who have *conspired* in intended mutiny are alike guilty of mutiny, consisting in overt acts on the part of one or more of the conspirators.

N

NAIL BALL—is a round projectile with an iron pin projecting from it, to prevent its turning in the bore of the piece.

NATIONAL ANNIVERSARY. The 4th of July. Regulations prescribe the honors to be paid by troops to the National Anniversary.

NATIONAL DEFENCE. (*See* DEFENCE, *National.*)

NEW MATTER. It is not proper that the prosecutor should be allowed to introduce *new matter*, neither should it be admitted on the defence. There is a great difference between new matter of accusation and facts proved by evidence to mitigate the sentence. The latter are not new matter in its strict sense; (HOUGH's *Military Law Authorities.*)

NITRE. Saltpetre, or nitrate of potassa; 54 nitric acid, 48 potassa. It is spontaneously generated in the soil, and is a necessary ingredient of powder. It has occasionally been produced artificially in *nitre beds*, formed of a mixture of calcareous soil with animal matter; in these, nitrate of lime is slowly formed, which is extracted by lixiviation, and carbonate of potash added to the solution, which gives rise to the formation of nitrate of potassa and carbonate of lime; the latter is precipitated; the former remains in solution and is obtained in crystals by evaporation. Its great use is in the manufacture of gunpowder, and in the production of nitric acid. It is also employed in the curing or preservation of meat.

NOMENCLATURE. Technical designation. (*See* ARMS; ORDNANCE.)

NON-COMMISSIONED OFFICER. Grades between private and warrant officer, as corporal, sergeant, ordnance-sergeant, sergeant-major, and quartermaster-sergeant.

NOTES. Members of courts-martial sometimes take notes. They are frequently necessary to enable a member to bring the whole body of evidence into a connected view, where the case is complex.

O

OATH. " Every officer, non-commissioned officer, musician, and private, shall take and subscribe the following oath or affirmation, to wit: I, A. B, do solemnly swear or affirm (as the case may be) that I will bear true faith and allegiance to the United States of America, and that I will serve them honestly and faithfully against their enemies or opposers whomsoever; and that I will observe and obey the orders of the President of the United States, and the orders of the officers appointed over me, according to the Rules and Articles of War; (*Act March 16, 1802.*)

OATH, (COURT OF INQUIRY.) The form of the oath to be taken upon courts of inquiry by members and judge-advocate or recorder, is prescribed in ART. 93. Witnesses before courts of inquiry take the same oath as before courts-martial.

OATH, (PROFANE.) Any non-commissioned officer or soldier, who shall use any profane oath or execration, incurs the same penalties as for irreverence at divine worship. (*See* WORSHIP.) A commissioned officer shall forfeit and pay for each and every such offence one dollar, to be applied as forfeitures for irreverence at worship.

OATH OF WITNESSES. (*See* WITNESS.)

OATHS OF MEMBERS OF COURTS-MARTIAL The 69th Article of War prescribes the oath or affirmation to be taken upon courts-martial, by members, and the judge-advocate. (*See* TRIAL.)

OATS. (*See* FORAGE; WEIGHTS.)

OBEDIENCE—to "*any lawful command of his superior officer*" is exacted from all officers and soldiers under penalty of death, or such other punishment as may be inflicted by a court-martial; (ART. 9.)

Two questions, therefore, arise under this article: Who is to judge of the *legality* of the command? and, What constitutes a *superior* officer in the sense of the article?

It is evident that if all officers and soldiers are to judge when an order is *lawful* and when not, the captious and the mutinous would never be at a loss for a plea to justify their insubordination. It is, therefore, an established principle, that, unless an order is so manifestly against law that the question does not admit of dispute, the order must first be obeyed by the inferior, and he must subsequently seek such redress against his superior as the laws allow. If the inferior disputes the legality before obedience, error of judgment is never admitted in mitigation of the offence. The redress now afforded by the laws to inferiors is not, however, sufficient; for doubtful questions of the construction of statutes, instead of being referred to the Federal courts of law for their true exposition, have received variable expositions from the executive, and left the army in an unfortunate state of uncertainty as to the true meaning of certain laws, and this uncertainty has been most unfavorable to discipline.

Again, while the punishment of death is meted to officers and soldiers for disobedience of *lawful* commands, the law does not *protect* officers and soldiers for obeying *unlawful* commands. Instances have occurred in our country, where officers and soldiers have been subjected to vexatious prosecutions, simply for obeying orders, according to their oath of office. Would it not be just if the law, instead of requiring officers and soldiers thus nicely to steer between Scylla and Charybdis, should hold the superior who gives an illegal order, alone responsible for its execution?

By *superior* officer in Article 9, and every other Article of War, is meant an officer who has the right to command his inferiors in the military hierarchy. The word superior, therefore, embraces, within their appropriate circle of command, commanding generals, superior regimental and company officers, superior officers of corps or departments, and the commanding officer on guards, marches, or in quarters of whatever corps of the line of the army, marine corps or militia authorized to command the whole by the 62d Article of War, whenever different corps come together. This construction of the words "officers appointed over me, according to the Rules and Articles of War," is manifest by an attentive examination of those articles:—

See, for example, ART. 27, which gives authority to "all officers of what condition soever to part and quell all quarrels, frays, and disorders, *though the persons concerned should belong to another regiment, troop, or company.*" Here it is seen that the ordinary subordination, by grades, is found only in the same "*regiment, troop, or company.*" The power to part and quell quarrels, is, however, made an exceptional case, in favor not only of officers of *different* regiments, but the power is even extended to those of an "inferior rank." In a company, regiment, or corps, subordination by grades is established by the terms of the commission held in such regiment or corps. So also, where officers hold commissions in the army at large, their right to command when on duty is co-extensive with their commissions, except that the 61st Article of War makes such higher commissions inoperative within the regiment in which an officer is mustered. Within regiments and corps the muster-roll, then, at once determines the question of superiority of officers on duty. But when mixed corps come together, as commissions below the rank of general, excepting commissions for gallant or meritorious services, are only given in regiments and corps, and as such regimental commissions would not otherwise entitle their holders to command beyond their particular regiments, or the holders of commissions in the line of the army, marine corps, or militia, beyond the body in which they hold commissions, the 62d Article of War has provided that the officer highest in rank of the line of the army, marine corps, or militia, shall command the whole, and he likewise is thus consecrated the superior officer for the time being. (*See* BREVET; LINE; RANK.)

OBLIQUE. In tactics, *oblique* indicates a direction which is neither parallel nor perpendicular to the front, but more or less diagonal. It is a command of warning in the tactics. It is used to indicate oblique allignments, attacks, orders of battle, squares against cavalry, changes of front, fires, &c.

OBSERVATION. Army of observation; detached party of observation, &c.

OBSTACLES. The obstacles used in field-fortification are of several kinds. Their object is to render access to works more difficult.

Common harrows, picketed to the ground, with the spikes uppermost, form excellent temporary obstacles. Crows'-feet, (Fig. 155,) consisting of four iron spikes arranged at equal angles with each other, so that in any position one spike must be pointing vertically upwards, may be scattered about in front of salients or other weak points, and will render approach difficult, and for cavalry impracticable.

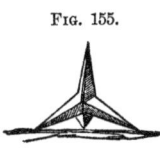

FIG. 155.

Roads or breaches, and sometimes even the restricted front of a position, my be barred by chevaux-de-frize, two forms of which are exhibited in the annexed diagrams, (Figs. 156 and 157.) Chevaux-de-frize may be formed of stout square or hexagonal beams, with iron spikes or sword blades, or even stout pointed stakes let into and standing perpendicularly from the faces; or, like Fig. 157, of stout palisades, pointed, and furnished with legs to support them, with the points towards the enemy.

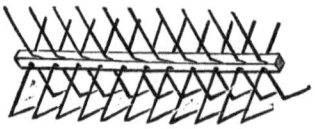

Fig. 156.

Fig. 157.

When used to close a space of any extent—indeed, where more than one length is necessary, they should be secured to each other by chains, to prevent their removal by an enemy.

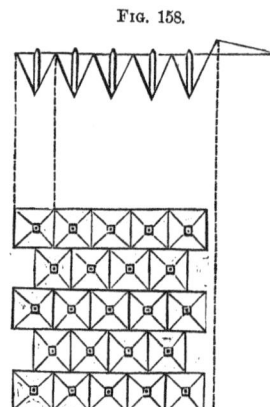

Fig. 158.

Trous-de-loup, or trap-holes, (Fig. 158,) are rows of pits in the form of inverted cones or pyramids, with a strong palisade or stake in the centre of each. They should be either too deep or too shallow to be used by riflemen, and they are, therefore, generally 8 or $2\frac{1}{2}$ feet deep.

Trap-holes, whether round or square, should always be arranged checkerwise, to prevent an enemy passing them easily. The earth from them should be formed into a glacis in front, rather than heaped up between them; as, in the latter case, they might be easily filled up again. Trous-de-loup of even two or three feet deep may be usefully employed in rendering impassable shallow, wet ditches, inundations, and fords; and, like abatis, they may be advantageously placed on the salients of works, on the weak points of lines, or in their intervals. They may thus compel the enemy to attack the strongest parts. The ardor of infantry may be much checked by unexpected obstacles within point-blank musket shot of the place attacked. (*See* ABATIS; PALISADES.)

OBSTRUCTION OF LAWS. In ordinary cases of obstruction to the laws of the United States, the powers vested in marshals are to be exercised to secure their due execution. It is only when such obstructions are too formidable to be suppressed by the ordinary course of judicial procedure or by the powers vested in the marshals, that the President of the United States is authorized to use military force. And whenever such force is employed by him, he must first, by proclamation, command insurgents to disperse, and retire peaceably to their respective abodes within a limited time; (*Acts* of Feb. 28, 1795, and *Act* March 3, 1807. See CALLING FORTH MILITIA; MARSHAL; POSSE COMITATUS.)

OCCUPY. To take or hold possession of a post or district.

OFFENCES. (*See* CRIMES; DISORDERS; NEGLECTS; ARTICLES OF WAR; JURISDICTION.)

OFFICERS. Whenever the word officer is used in the Articles of War, *commissioned* officer is understood.

OPERATIONS. *Field* operations; offensive and defensive operations; under-ground operations; siege operations, &c.

ORDER. This term, considered in its relation to the army, embraces divers subjects. It gives an idea of harmony in the accomplishment of DUTIES; a classification of corps or men; injunctions emanating from AUTHORITY; measures which regulate service, and many tactical details.

In tactics, the natural order is when troops coming upon ordinary ground are ranged in line of battle by the prescribed tactical means, and when they are formed in column right in front.

The *oblique* order is contradistinguished from the parallel, and in general means every tactical combination the aim of which is to produce an effect upon two points of an enemy's line by bringing a superior force to bear down on those two points. Such combinations constitute the oblique order, whatever manœuvres may be used to accomplish the object.

The *parallel* order operates on the contrary against the whole front of the enemy. Turenne and Condé fought habitually in parallel order, although they sometimes made a skilful use of oblique attacks. *Giubert* well says that a contiguous and regular parallel order can be of no use in war.

ORDERLIES. Non-commissioned officers and soldiers appointed to wait upon generals and other officers, to communicate orders, and carry messages.

ORDERLY SERGEANTS. The first sergeant of a company is so called. On hearing the drum beat for orders, orderly sergeants repair to the adjutant's office, and, having taken down the orders in writing, they are immediately to show them to the officers of their company, and to warn the men for duty.

ORDERLY BOOK. A book for the sergeants to insert the general and regimental orders, which are issued from time to time, is sometimes called an orderly book.

ORDERS. The principle upon which orders are to be issued is established by the 62d Article of War, which gives to the immediate commander of the troops " by commission there on duty, or in quarters," authority to " *give orders for what is needful to the service*, unless otherwise specially directed by the President of the United States, according to the nature of the case."

The President of the United States and commanding officers are, however, limited to issuing such orders as may be " according to the Rules and Articles of War ; " (*Act* March 16, 1802.) The determination of what orders are, and what are not, contrary to the Rules and Articles of War established by Congress, is a very nice question, and it is much to be regretted that Congress has not long since accurately defined the functions, rights, and duties of all officers and soldiers, and also given them some means of obtaining redress against unsound expositions of law made by the executive and military authorities. (*See* ARMY; REMEDY.)

In article INJURIES it has been shown how officers become answerable at law for their own acts or defaults occurring in the course of professional duty; but commanding officers are not legally liable for the acts of subordinates in the execution of the services confided to them.

By the general law, masters and employers of every kind are answerable for the acts or neglects of their servants or subordinate agents; but the principle of this rule is, that private individuals have the power of appointing and selecting such agents or servants as they may think proper, and are consequently bound to employ only those who are of competent skill, diligence, and ability. But this principle has no application as between superior and subordinate officers in the army, for the obvious reason that the former do not choose the latter. The rule as to military officers therefore is, that the wrong-doer alone is personally liable for the damages or injury resulting from his conduct, and the wrong-doer is he who issued the order, or otherwise gave direct occasion to the act or omission which led to the mischief.

When an officer, therefore, is employed upon a particular service, the execution of which is left to his own skill and uncontrolled judgment, the superior officer from whom he receives his orders incurs no legal responsibility for injuries occasioned to the persons or property of third parties by the conduct of the junior in executing the duty confided to him. For the senior officer has no power of appointing his subordinate officers; he is not even himself to be deemed a volunteer in that particular station merely by having voluntarily entered the army, and has no choice whether or not he will serve with the junior officers placed under his orders, but is bound to take such as he finds there, and make the best of them. He is a servant of the State, doing duty with others appointed and stationed in like manner, and by the same authority.

But the case is altered when the senior officer not only orders another to perform a particular service, but likewise prescribes the specific mode of execution. For the subordinate officer is then deprived of all exercise of his own judgment and discretion; his acts are the direct acts of his senior officer; and the latter becomes as thoroughly responsible, in a legal point of view, as if he had been personally present and assisting in the performance of the duty in question.

It frequently happens in suits at law respecting private wrongs, that the officer against whom the action is brought is the only person acquainted with some of the material facts which it may be necessary to prove against him: and though, in cases of mere debt or contract, a defendant is compellable to make a disclosure, on oath, of such facts as lie within his own knowledge, that rule does not apply to actions respecting private wrongs or injuries. An attempt, however, was made in Sir William Houston's case, by means of proceedings in the Court of Chancery, to compel that officer to produce certain military and other orders, reports, books, letters, and documents, from which the truth of the charge against him would appear. But the Master of the Rolls refused to make any order for the production; (PRENDERGAST.)

ORDNANCE DEPARTMENT. The Ordnance Department consists of one colonel, one lieut.-colonel, four majors, twelve captains, twelve first lieutenants, and six second lieutenants; master carriage-makers, master blacksmiths, master armorers, &c., &c., limited only by the judgment of the colonel of Ordnance and Secretary of War.

It shall be the duty of the colonel of the Ordnance Department to direct the inspection and proving of all pieces of ordnance, cannon-balls, shot, shells, small-arms, and side-arms and equipments, procured for the use of the armies of the United States; and to direct the construction of all cannon and carriages, and every implement and apparatus for ordnance, and all ammunition-wagons, travelling-forges, and artificers' wagons; the inspection and proving of powder, and the preparation of all kinds of ammunition and ordnance stores. And it shall also be the duty of the colonel or senior officer of the Ordnance Department to furnish estimates, and, under the direction of the Secretary for the Department of War, to make contracts and purchases for procuring the necessary supplies of arms, equipments, ordnance, and ordnance stores; (*Act* Feb. 8, 1815.)

The colonel of the Ordnance Department shall organize and attach to regiments, corps, or garrisons, such number of artificers, with proper tools, carriages, and apparatus, under such regulations and restrictions relative to their government and number as, in his judgment, with the approbation of the Secretary for the Department of War, may be considered necessary; (*Act* Feb. 8, 1815.)

The colonel of the Ordnance Department, or the senior officer of that department of any district, shall execute all orders of the Secretary for the Department of War, and, in time of war, the orders of any general, or field-officer, commanding any army, garrison, or detachment, for the supply of all arms, ordnance, ammunition, carriages, forges and apparatus, for garrison, field, or siege service; (*Act* Feb. 8, 1815.)

The costs of repairs and damages done to arms, equipments, or implements in the use of the armies of the United States, shall be deducted from the pay of any officer or soldier in whose care or use the said arms, equipments, or implements were, when the said damages occurred; *provided*, the said damages were occasioned by the abuse or negligence of the said officer or soldier. And it is hereby made the duty of every officer commanding the regiments, corps, garrisons, or detachments, to make, once every two months, or oftener if so directed, a written report to the colonel of the Ordnance Department, stating all damages to arms, equipments, and implements belonging to his command, noting those occasioned by negligence or abuse, and naming the officer or soldier by whose negligence or abuse the said damages were occasioned; (*Act* Feb. 8, 1815.)

The colonel of the Ordnance Department, under the direction of the Secretary of War, is hereby authorized to draw up a system of regulations for the government of the Ordnance Department; forms of returns and reports; and for the uniformity of manufacture of all arms, ordnance, ordnance stores, implements, and apparatus, and for the repairing and better preservation of the same; (*Act* Feb. 8, 1815.)

Regulations for the government of the Ordnance Department, &c., have been drawn up in conformity with the authority conferred by the act of 1815. (Consult *Ordnance Regulations*, 1852.) Officers and enlisted men of the Ordnance Department subject to the Rules and Articles of War; (*Act* April 5, 1832.)

ORDNANCE AND ORDNANCE STORES—comprehend all cannon, howitzers, mortars, cannon-balls, shot, and shells, for land service; all gun-carriages, mortar beds, caissons, and travelling forges, with their equipments; and all other apparatus and machines required for the service and manoeuvres of artillery, in garrisons, at sieges, or in the field; together with the materials for their construction, preservation, and repair. Also, all small-arms, side-arms, and accoutrements, for the artillery, cavalry, infantry, and riflemen; all ammunition for ordnance and small-arms; and all stores of expenditure for the service of the various arms; materials for the construction and repair of ordnance buildings; utensils and stores for laboratories, including standard weights, gauges, and measures; and all other tools and utensils required for the performance of ordnance duty. The ordinary articles of camp equipage and pioneers' tools, such as axes, spades, shovels, mattocks, &c., are not embraced as ordnance supplies. Wagons, &c., for the transport service of the army, and horse equipments, are also furnished by the Ordnance Department when practicable. Ordnance supplies are provided by open purchase, fabrication, or by contract.

The following are the kinds and calibres of cannon used in the land service of the United States:

KIND OF ORDNANCE.			CALIBRE.	MATERIAL.	WEIGHT.
					lbs.
Guns............	Field.................		6-pounder...	Bronze......	884
			12-pounder...		1,757
	Siege and garrison		12-pounder...	....	3,590
			18-pounder...	....	4,913
			24-pounder...	Iron	5,790
	Sea-coast............		32-pounder...	....	7,200
			42-pounder...	....	8,465
Howitzers....	Mountain		12-pounder...	....	220
	Field.................		12-pounder...	Bronze......	788
			24-pounder...		1,318
			32-pounder...	....	1,920
	Siege and garrison		8-inch	....	2,614
			24-pounder...	....	1,476
	Sea-coast............		8-inch	....	5,740
			10-inch	....	9,500
Columbiads			8-inch	Iron	9,240
			10-inch		15,400
Mortars.......	Light		8-inch	....	930
			10-inch	....	1,852
	Heavy		10-inch	....	5,775
			13-inch	....	11,500
	Stone mortar.........		16-inch	Bronze......	1,500
	Coehorn		24-pounder...		164
	Eprouvette...........		24-pounder...	Iron	220

A 12-inch columbiad, of cast iron, has also been made for trial; and recently Captain Rodman's 15-inch gun, now at Fort Monroe, was cast at Pittsburg, Pa. It weighs 49,100 lbs. (*See* COLUMBIAD.) For several pieces of ordnance see articles COLUMBIAD; MORTAR; MOUNTAIN ARTILLERY; RIFLED ORDNANCE. The Caisson, Travelling Forge, Seacoast Carriage, and 24-pdr. Siege Carriage, are shown in Figs. 159, 160, 161, and 162.

Cannon made of bronze are commonly called *brass* cannon.

The *cascable* is the part of the gun in rear of the base-ring; it is composed generally of the following parts: the *knob*, the *neck*, the *fillet*, and the *base of the breech*.

FIG. 159.

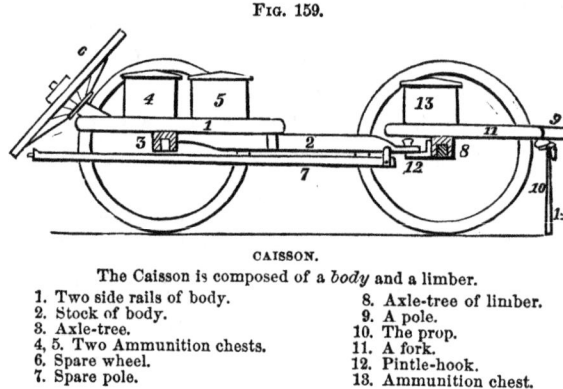

CAISSON.

The Caisson is composed of a *body* and a limber.

1. Two side rails of body.
2. Stock of body.
3. Axle-tree.
4, 5. Two Ammunition chests.
6. Spare wheel.
7. Spare pole.
8. Axle-tree of limber.
9. A pole.
10. The prop.
11. A fork.
12. Pintle-hook.
13. Ammunition chest.

The *base of the breech* is a frustum of a cone, or a spherical segment, in rear of the breech.

The *base-ring* is a projecting band of metal adjoining the base of the breech, and connected with the body of the gun by a concave moulding.

FIG. 160.

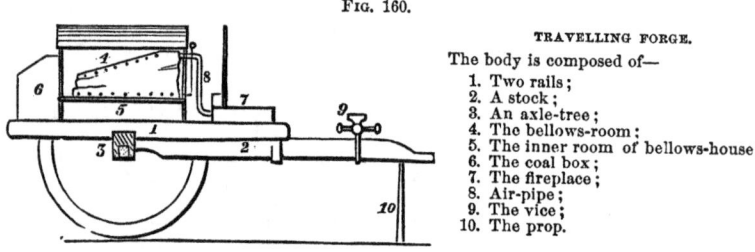

TRAVELLING FORGE.

The body is composed of—
1. Two rails;
2. A stock;
3. An axle-tree;
4. The bellows-room;
5. The inner room of bellows-house;
6. The coal box;
7. The fireplace;
8. Air-pipe;
9. The vice;
10. The prop.

The *breech* is the mass of solid metal behind the bottom of the bore, extending to the rear of the base-ring.

The *reinforce* is the thickest part of the body of the gun, in front of the base-ring; if there is more than one reinforce, that which is next to the base-ring is called the *first reinforce;* the other, the *second reinforce.* In some howitzers, instead of a reinforce, there is a *recess* in the metal around the chamber next to the base-ring.

The *reinforce band* is at the junction of the first and second reinforces in the heavy howitzers and columbiads.

The *chase* is the conical part of the gun in front of the reinforce.

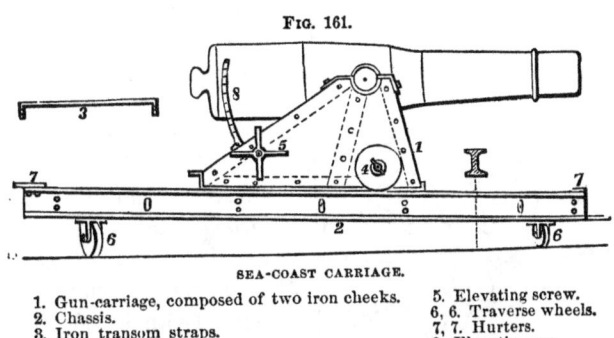

FIG. 161.

SEA-COAST CARRIAGE.

1. Gun-carriage, composed of two iron cheeks.
2. Chassis.
3. Iron transom straps.
4. Manœuvring wheels.
5. Elevating screw.
6, 6. Traverse wheels.
7, 7. Hurters.
8. Elevating arc.
I. Pintle or fixed centre.

The *astragal* and *fillets* in field-guns, and the *chase-ring* in other pieces, are the mouldings at the front end of the chase.

The *neck* is the smallest part of the piece in front of the astragal or the chase-ring.

The *swell of the muzzle* is the largest part of the gun in front of the neck. It is terminated by the muzzle mouldings, which in field and siege guns consist of the *lip* and the *fillet.* In the sea-coast guns and heavy howitzers and columbiads there is no fillet. In field and siege howitzers and in mortars a *muzzle band* takes the place of the swell of the muzzle.

The *face* of the piece is the terminating plane perpendicular to the axis of the bore.

The *trunnions* are cylinders, the axes of which are in a line perpendicular to the axis of the bore, and in the same plane with that axis.

The *rimbases* are short cylinders, uniting the trunnions with the body of the gun. The ends of the rimbases, or the *shoulders of the trunnions*, are planes perpendicular to the axis of the trunnions.

The *bore* of the piece includes all the part bored out, viz.: the cylinder, the chamber, (if there is one,) and the conical or spherical surface connecting them.

The *chamber*, in howitzers, columbiads, and mortars, is the smaller part of the bore, which contains the charge of powder. In howitzers and columbiads the chamber is cylindrical; it is united with the large cylinder of the bore by a conical surface; the angles of intersection of this conical surface with the cylinders of the bore and chamber are rounded (in profile) by arcs of circles. In the 8-inch siege howitzer, the chamber is united with the cylinder of the bore by a spherical surface, in order that the shell may, when necessary, be inserted without a sabot. A conical chamber which is joined to the cylinder of the bore by a portion of a spherical surface, (as in the 8-inch and 10-inch light mortars,) is called a *Gomer chamber*.

The *bottom of the bore* is a plane perpendicular to the axis, united with the sides (in profile) by an arc of a circle, the radius of which is one-fourth of the diameter of the bore at the bottom. In the columbiads, the heavy sea-coast mortars, the stone mortar, and the eprouvette, the bottom of the bore is hemispherical.

The *muzzle*, or mouth of the bore, is chamfered to a depth of 0.15 inch to 0.5 inch, (varying with the size of the bore,) in order to prevent abrasion, and to facilitate loading.

The *true windage* is the difference between the true diameters of the bore and of the ball.

The axis of the *vent* is in a plane passing through the axis of the bore, perpendicular to the axis of the trunnions. In guns, and in howitzers having cylindrical chambers, the vent is placed at an angle of 80° with the axis of the bore, and it enters the bore at a distance from the bottom equal to one-fourth the diameter of the bore. The diameter of the vent is *two-tenths* of an inch, in all pieces except the eprouvette, in which it is *one-tenth*. The vents of brass guns are bored in *vent pieces,* of wrought copper, which are screwed into the gun.

The *lock piece* is a block of metal at the outer opening of the vent, in some pieces of ordnance, to facilitate attaching a lock to the cannon.

The *natural line of sight* is a line drawn in a vertical plane through the axis of the piece, from the highest point of the base-ring to the highest point in the swell of the muzzle, or to the top of the *sight,* if there is one.

The *natural angle of sight* is the angle which the natural line of sight makes with the axis of the piece.

The *dispart* is the difference of the semi-diameters of the base-ring and the swell of the muzzle, or the muzzle band. It is therefore the tangent of the natural angle of sight, to a radius equal to the distance from the rear of the base-ring to the highest point of the swell of the muzzle, the sight, or the front of the muzzle band, as the case may be.

The *preponderance* of the breech of the gun is the excess of weight of the part in rear of the trunnions over that in front: it is measured by the weight which it is necessary to apply in the plane of the muzzle to balance the gun when suspended freely on the axis of the trunnions.

The *handles* of the gun are placed with their centres over the centre of gravity of the piece. The 6-pounder gun and the 12-pounder howitzer have no handles. The handle of a heavy mortar consists of a *clevis,* which is attached by a *bolt* to the *ear* of the mortar.

The *eprouvette mortar* is cast with a *sole,* which fits into a cast-iron *bed-plate,* bolted to the platform.

To designate a piece of ordnance.—State the kind, the calibre, (in inches if it be foreign ordnance,) the material, the weight, the inspector's initials, the number, the country in which it was made, the date, the place of fabrication, the founder's name, the name inscribed on it, its condition for service, the kind of chamber, if any; whether it has a vent piece, a lock piece, handles; the ornaments, and any particular marks which may serve to identify it.

There are two national armories: the Springfield Armory, Springfield, Mass., and the Harper's Ferry Armory, Harper's Ferry, Va. Their principal business is the manufacture of the rifle musket and rifle; making components, and altering other arms. The armory of James J. Ames, Chickopee, Mass., furnishes swords, sabres, and field-artillery; that of Samuel Colt, Hartford, Conn., Colt's revolving pistols, rifles, and carbines; Sharp's Manufacturing Company, Hartford, Conn., Sharp's carbines and rifles; Charles Jackson, Providence, R. I., *Burnside's* carbines; and Maynard's Arms Company, Washington, D. C., Maynard's rifles and carbines. The arms of the foregoing manufactories have been tried more or less in service and by boards, and are considered good cavalry arms. The best arms for infantry, however, are the United States rifle musket and rifle. The foundries for cannon are the South Boston, C. Alger & Co., Boston, Mass.; the West Point, R. P. Parrott, Cold Spring, N. Y.; the Tredegar, J. R. Anderson & Co., Richmond, Va.; the Bellona, J. L. Archer, Black Heath, Va., and the Pennsylvania, Knap, Rudd & Co., Pittsburg, Pa. The following are the arsenals for construction of carriages, &c., or repair: Kennebec Arsenal, Augusta, Maine; Watertown Arsenal, Watertown, Mass.; Champlain Arsenal, Vergennes, Vt.; Watervliet Arsenal, West Troy, N. Y.; New York Arsenal, New York; Alleghany Arsenal, Pittsburg, Pa.; Frankford Arsenal, Bridesburg, Pa.; Pikesville Arsenal, Pikesville, Md.; Washington Arsenal, Washington, D. C.; Fort Monroe Arsenal, Old Point Comfort, Va.; N. C. Arsenal, Fayetteville, N. C.; Charleston Arsenal, Charleston, S. C.; Augusta Arsenal, Augusta, Ga.; Mount Vernon Arsenal, Mount Vernon, Ala.; Appalachicola Arsenal, Chattahooche, Florida; Baton Rouge Arsenal, Baton Rouge, La.; Little Rock Arsenal, Little Rock, Ark.; St. Louis Arsenal, St. Louis, Mo.; Detroit Arsenal, Dearbornville, Mich.; Benicia Arsenal, Benicia, Cal.; Texas Arsenal, San Antonio, Texas.

The principal articles furnished by the Ordnance Department by fabrication at armories and arsenals and by purchase from foundries, and manufacturing establishments, are in inventories classed as follows:

PART I.

ARTILLERY, SMALL-ARMS, AMMUNITION, AND OTHER ORDNANCE STORES.

CLASS 1. *Cannon.*—The mean weight of each kind of ordnance, as well as the number of pieces, should be entered in the inventory.

CLASS 2. *Artillery Carriages* include mortar beds, different gun-carriages, battery wagons, forges, &c. "The field-carriage complete" includes the limber and ammunition chest, but no implements. The "*casemate*, or *barbette carriage complete*," includes the upper or gun-carriage, and the chassis, with all the wheels, but no implements. It is better, however, to enter the gun-carriages and the chassis separately, as above.

CLASS 3. *Artillery Implements and Equipments* include all implements and equipments used by artillerists. A set of harness for two horses includes every thing required for them except *whips* and *nose-bags*, which are reported separately.

CLASS 4. *Artillery Projectiles and their Appendages unprepared for Service.*

CLASS 5. *Artillery Projectiles with their Appendages prepared for Service.*

A *round of fixed ammunition* is used to indicate the projectile with its cartridge prepared for use, although in some cases they are not actually connected together. A *shot strapped*, or a *canister, stand of grape*, &c., indicate the projectile prepared for making fixed ammunition, or for service.

CLASS 6. *Small-arms* include muskets, rifles, carbines, pistols, swords, sabres.

CLASS 7. *Accoutrements, Implements, and Equipments for Small-arms, and Horse Equipments for Cavalry.*

CLASS 8. *Powder, Ammunition for Small-arms and Materials.*

CLASS 9. *Parts or Incomplete Sets of any of the Articles inserted in the preceding classes.*

CLASS 10. *Miscellaneous* includes gins, sling-carts, hand-carts, trucks, handspikes, rollers, &c., for mechanical manoeuvres, eprouvettes and beds, gauges, callipers, &c.

PART II.

TOOLS AND MATERIALS, CLOTHS, ROPES, THREAD, ETC., FORAGE, IRONMONGERY, LABORATORY STORES.

Lumber includes gun-carriage timber, and building materials. The number of pieces of timber for each part of a gun-carriage to be stated separately. Other plank, &c., to be stated in board measure.

LEATHER AND MATERIALS FOR HARNESS-WORK, PAINTS, OILS, GLASS, ETC., STATIONERY, TOOLS, MISCELLANEOUS ARTICLES.

To prevent the rapid decay of the wooden material of which sea-coast and garrison carriages are mainly composed, experiments have been successfully made by the Ordnance Department to ascertain whether suitable iron-carriages might not be substituted. Such carriages have been devised and fabricated even more convenient for service than those of wood, and, at the same time, fully as cheap in first cost, and of far greater durability; and more easily moved and stored. With the aid of the practical experience of officers of the different mounted corps, a new uniform model for horse equipments has also been adopted. The attention of the Ordnance Department has been given to the subject of "rifle cannon" and projectiles for the same. Many varieties of such cannon and projectiles have been devised and brought to the notice of the department for examination. Actual experiment is necessary for a comparison of the relative merits of the different devices, and the selection of the best. Such experiments have been commenced under the direction and supervision of a board of artillery and ordnance officers, who have reported their opinion "that the era of smooth-bore field artillery has passed away, and that the period of the adoption of rifle cannon for siege and garrison service is not remote. The superiority of elongated projectiles, whether solid or hollow, with the rifle rotation, as regards economy of ammunition, extent of range, and uniformity and accuracy of effect, over the present system is decided and unquestionable." Attention has been given also to experiments and tests of gunpowder with a view to ascertain the composition and manufacture of a powder which will impart a given velocity and range to a projectile, with the least strain or injury to the gun. (*See* GUNPOWDER.) Varieties of the breech-loading carbines are now on actual trial in service, either of which is probably an effective arm for cavalry. Uniformity of armament for the same kind of service is, however, essential both for tactical instruction and for adaptation to ammunition in depot, and the one arm to be adopted must be not only an effective, but the most effective of the kind. Further trials, and more extended experience, will be requisite for the selection which may yet fall on an arm not now invented. (*See* ARMS; ARTILLERY; CARBINES; FIRING; RIFLED ORDNANCE.)

ORDNANCE SERGEANTS. The Secretary of War may select from the sergeants of the line of the army, who shall have faithfully served eight years in the service, four years of which in the grade of non-commissioned officer, as many ordnance sergeants as the service may require, not to exceed one for each military post; whose duty it shall be to receive and preserve the ordnance, arms, ammunition and other military stores, at the post, under the direction of the commanding officer of the same, and under such regulations as shall be prescribed by the Secretary of War, and who shall receive for their services five dollars per month, in addition to their pay in the line; (*Act* April 5, 1832.)

ORGANIZING. "Congress shall have power to provide for organizing, arming and disciplining the militia." (*See* CONSTITUTION.)

ORILLON—is a projecting tower at the shoulder angle of a bastion, covering the flank from exterior view, frequently found in old fortresses.

ORPHANS—of officers who may die by reason of wounds received in service, to receive half pay for five years. (*See* PENSION.)

OUTLINE OR **TRACING**—is the succession of lines that show the figure of the works, and indicate the direction in which the defensive masses are laid out, in order to obtain a proper defence.

OUTPOSTS—should not only secure an army against surprise, but also be so arranged as to enable the outposts to avoid an engagement and not be enveloped by an enemy. Marshal Bugeaud has elaborated a system for outposts well worth attention. Its principal feature is the occupation at night of all avenues of approach (front, flank, and rear) to the position occupied by the detached corps, by squads of men instructed in concerted signals. These little squads do not form a continuous chain, and are each independent. It is not necessary that they should be large, for their duty is to warn. The service exacted from them is explained with clearness and precision—the signals by which the commanding officer corresponds with the posts and the posts with him, are given. These signals should be made by petards of one or two pounds of powder. About a dozen petards in all will be necessary. The posts will be made to understand what the different reports of the petard indicate, and these signals must be few in number. The little posts should be on the circumference of a circle having the village occupied by the detached corps as its centre, and at such a distance as would prevent an enemy, without warning from the night posts, from enveloping the village.

A corps detached at one or more leagues from the main body ought not only to be able to avoid a surprise, but also to prevent an enemy from cutting off its retreat. Marshal Bugeaud accordingly recommends that the chief of a detachment that is to establish itself for many days, should, on arriving in position, lose no time; but, without waiting to rest, occupy himself in reconnoitring the ground in his neighborhood, within a radius of one or two leagues. He should take with him the officers intrusted with the care of the camp for the night, some horsemen, and a few intelligent inhabitants; scan the course of the roads and pathways crossed in making his rounds, and take notes. This reconnoissance finished, he will be able to judge what will be the circle that an enemy would probably make around his post, in order to envelop it without their expectation of meeting any men in observation. This circle Marshal Bugeaud calls the supposed circle, and beyond this line he advises little posts during the night only, drawn from the posts occupied during the day. These little posts must be without fire or animals, and sometimes on the right and sometimes on the left of the pathways, and their position should be constantly varied. In the plan, (Fig. 163,) the village X is occupied by two battalions detached three leagues distant from the army. Upon arriving at the post, the chief of the detachment establishes the usual chain marked by the inner circle. This circle is about 800 metres in diameter, or 2,400 in circumference.

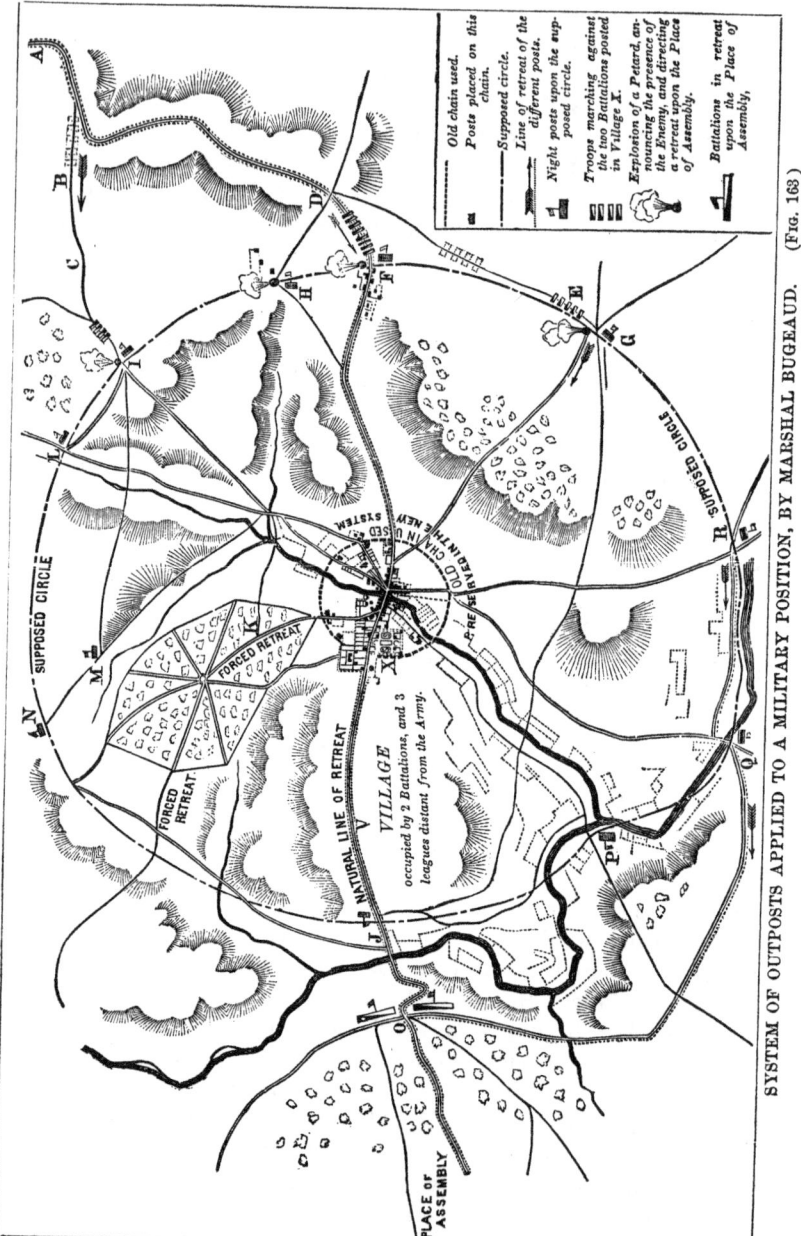

SYSTEM OF OUTPOSTS APPLIED TO A MILITARY POSITION, BY MARSHAL BUGEAUD. (Fig. 163.)

After reconnoissance, little posts for the night are sent to the points H I L M N O P Q R G. The enemy is signalled at H I G. Immediately the chief of the two battalions announces, by reports of the petard, that

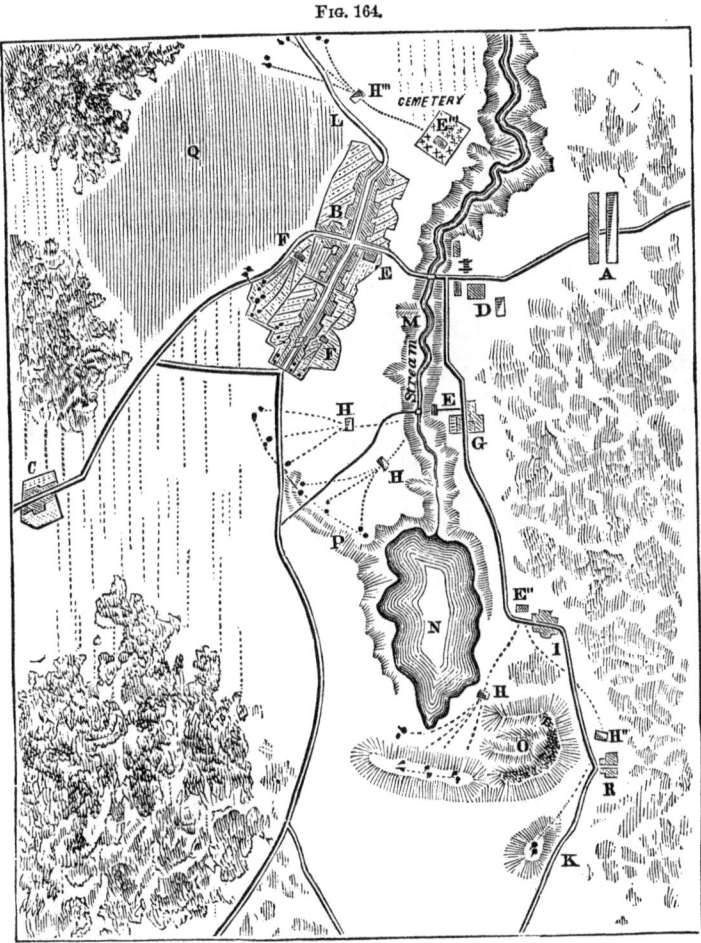

FIG. 164.

the different posts must occupy the place of assembly in rear of O. The order is given to rally, and the route followed is designated on the plan thus -------. Having re-united his men, the chief of the detachment may, according to circumstances, await daylight or continue his retreat.

Ordinary arrangement of Outposts composed of both Infantry and

Cavalry.—The best line of observation in this example, (Fig. 164,) is from the lake N and the height O, on the right flank on the road L. The advanced guard A, consisting of a brigade of cavalry and a division of infantry, is placed behind the village B, and outposts are to be posted to guard against an attack by the enemy arriving from the direction

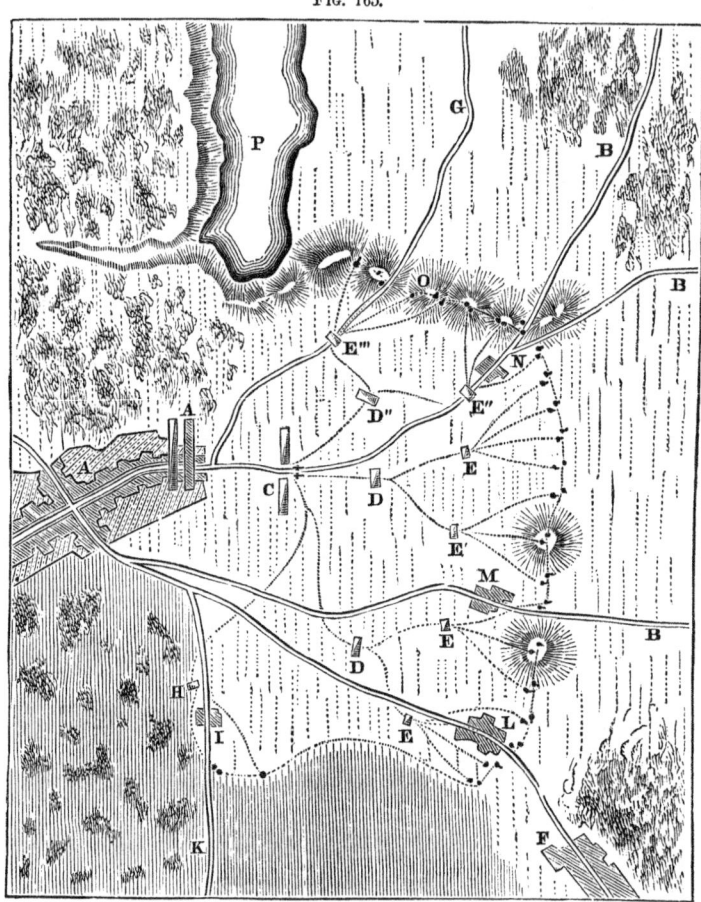

of the village C; 4 companies of cavalry, 2 regiments of infantry, and 2 pieces of foot artillery are detailed to furnish the outposts and reserve.

D is the reserve, consisting of 1½ companies of cavalry, 11 companies of infantry, and 2 pieces of foot artillery.

E is a mainguard of 3 companies of infantry, which furnishes the two pickets F, each of which posts 5 pairs of sentinels.

E′ is an infantry main guard of 3 companies, which supports the two cavalry pickets H, each of which consists of a platoon, and posts 3 pair of videttes. E″ is an infantry main guard of two companies, to support the cavalry picket H′, which posts 4 videttes, and H″, which posts 1 vidette. E‴ is an infantry main guard of 2 companies, which holds the cemetery and supports the cavalry picket H‴, posting 3 videttes.

Ordinary arrangement of Outposts composed of Cavalry alone.—In this example, (Fig. 165,) the most advantageous line of observation is that proceeding from the village L, through the villages M and N, thence following the ridge O to the lake P. The extent of this line is a little more than 5 miles. The advanced guard A, composed of a brigade of cavalry and a division of infantry, is in the village A, and it is necessary to post outposts to guard it against the enemy, expected by the roads B; 10 companies of cavalry and 2 pieces of horse artillery are detailed for the outposts and reserve. C is the reserve, consisting of 4 companies and 2 guns. D, D′, and D″ are the main guards, consisting of one company each. E, E, E′, E′, E″, E‴, are the pickets, of a platoon each. H is an independent picket of one platoon, observing the road K. (Consult Bugeaud and McClellan.)

OUTWORKS—are such works as are constructed between the enceinte and the glacis, of whatever kind.

OVEN. Ovens are always provided in garrisons, so that the troops may bake their own bread. A large saving of flour is thus made, which is the most considerable element of the post fund. A brick oven, 3m. 33 in breadth, 4m. 50 in depth, and 0m. 75 in height, contains 500 rations. It may be constructed in less than 24 hours. The cylindrical form is greatly to be preferred, as it is more easily made and requires less material than the ordinary form. The want of brick for the arch and fireplace of ovens may be supplied in the field by two gabions of semi-circular or semi-elliptical form

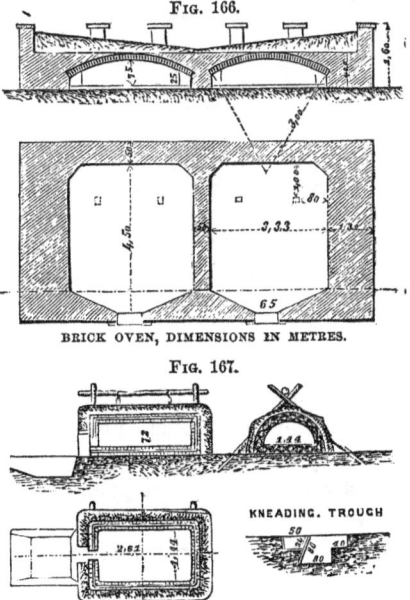

Fig. 166.

BRICK OVEN, DIMENSIONS IN METRES.

Fig. 167.

KNEADING TROUGH

OVEN OF GABIONS, DIMENSIONS IN METRES.

1m. 44 in diameter; the basket work is not so close as the ordinary gabion, and is 1m. 32 in height. The two gabions, resting one over the other upon the flat side, make a cradle 2m. 64 long, 1m. 44 broad, and 0m. 72 high. The interior and exterior is then plastered with clay, which must penetrate the interstices of the basket work. The front and back part is shut in the same manner, or with sods. The cradle is then covered with earth to retain the heat, and in order that the superincumbent weight may not cause it to give way. Withes are attached to the top of the basket work, and passed vertically through the embankment, and then fastened to the longitudinal beam of a wooden horse straddled against the exterior curve. Eight of these furnaces may be made in 24 hours. Ovens may also be made of wood or of earth.

The wooden oven (Fig. 168) is made by digging an excavation of 3m. 20 in length by 2m. 40 in breadth, and 0m. 50 in depth, making the fireplace slightly descending towards the mouth. This trench is covered with pieces of wood of 0m. 15 to 0m. 25 square, placed close to-

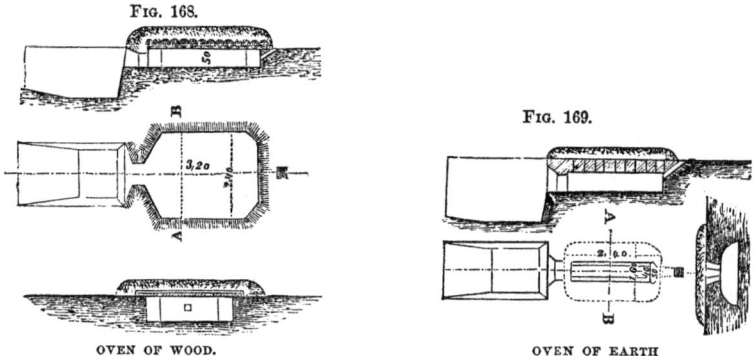

Fig. 168.   Fig. 169.

OVEN OF WOOD.   OVEN OF EARTH

gether; the wood is covered with earth carefully packed, the chimney-place is sodded. The fireplace is dried by heating for 7 or 8 hours, and subsequent heatings require two hours. Such ovens resist very well five or six bakings. They require only two hours for construction with prepared wood, and if the wood should burn, the fire may be extinguished by closing the chimney and mouth, and in half an hour the wood may be replaced, if consumed. To construct rapidly an earthen oven dig a slope with a step, and on its prolongation, dig the length of the oven in a trench separated from the step by a mass of earth to be pierced later as the mouth of the oven. The trench, when finished, to be 0m. 80 in depth, 0m. 40 in breadth, and 2m. 40 in length. Then dig laterally portions of an elliptical arch in such a manner as to make the arch 1m. 70 in breadth. This work finished, pierce the mouth and cover the trench with from three to five sods as arch stones, leaving a chimney-place to the bottom. Ovens for from 100 to 250 rations may be thus made. Dough is kneaded with 6 parts of wheat, 4 of water, and a little salt, a half part to the hundred parts. (Consult *Memorial des Officiers d'Infanterie et de Cavalerie*.)

OVER. Authority; command; above. (*See* OATH OF OFFICE.)

P

PACES. The length of each pace of the infantry soldier is 28 inches from heel to heel; which he must be trained to take in proper cadence and in perfect steadiness.

PALISADES—are strong palings six or seven inches broad on each side, having about one foot of their summits sharpened in a pyramidal form. They are frequently placed at the foot of slopes, as an obstacle to the enemy. A large beam or lintel, sunk about 2 or 3 feet, is often used to unite them more firmly. Their tops should be a foot above the crest of the parapet behind which they stand, and in field-fortifications they form a very good obstruction, if protected from artillery. An expeditious mode of planting them, is to sink a small ditch, about 2 feet 6 inches deep, and the same breadth, and to nail the ends of the palisades to a piece of timber, or the trunk of a tree, laid on the bottom of it, and then fill in the earth, and ram it well. (Fig. 170.)

The palisades should be 9 or 10 feet long, so that when finished, the ends shall be at least 7 feet above the ground. They may be made out of the stems of young trees of 6 or 8 inches diameter; but stout rails, gates with the ends knocked off, planks split in half, cart shafts, ladders, and a variety of such things, will come into play, where more regular palisades are not to be had. If the materials are weak, a crosspiece must be nailed to them near the top, to prevent their being broken down, and they must not be placed so close together as to cover an enemy. (Fig. 171.)

FIG. 170.   FIG. 171.

PARADE. An assemblage of troops in a regular and prescribed manner, for guard-mounting, field-exercises, or dress parade.

PARADOS—is a traverse, covering the interior of a work from reverse fire.

PARALLELS—in the attack of a place, are wide trenches, which afford the besieged troops a free covered communication between their various batteries and approaches, and a secure position for the guards of the trenches. (*See* SIEGE.)

PARAPET. (*See* FORTIFICATION.) In field works, while the height is fixed at about seven feet, the thickness of the parapet varies according to the kind of fire it is intended to resist. Should the ground in front be inaccessible to artillery at 800 yards, the parapet is constructed of dimensions sufficient only to resist musketry, or from two to two and a half feet thick. To resist field-artillery, a thickness of from six to ten feet is required.

PARBUCKLES—are 4-inch ropes, 12 feet long, with a hook at one end and a loop at the other. To parbuckle a gun, is to roll it in either direction from the spot in which it rests. To do this, place the gun on skids, and if it is to be moved up or down a slope, two $4\frac{1}{2}$-inch ropes are made fast to some place on the upper part of the slope, the ends are carried under the chase and breech of the gun respectively round it, and up the slope. If the running ends of these ropes are hauled upon, the gun ascends; if eased off, it descends. If the ground is horizontal, handspikes only are necessary to move the gun.

PARDON. Every officer authorized to order a general court-martial, shall have power to pardon or mitigate any punishment ordered by such court, except the sentence of death; or of cashiering an officer, which, in cases where he has no authority (by ART. 65) to carry them into execution, he may suspend, until the pleasure of the President of the United States can be known, which suspension, together with copies of the proceedings of the court-martial, the said officer shall immediately transmit to the President for his determination. And the colonel or commanding officer of the regiment or garrison where any regimental or garrison court-martial shall be held, may pardon or mitigate any punishment ordered by such court to be inflicted; (ART. 89.)

PARK—is literally an inclosed space. In military language it means the space occupied by the animals, wagons, pontoons, and material of all kinds, whether of powder, ordnance stores, hospital stores, and provisions when *parked*. The meaning is also extended to embrace not only the space occupied, but also the whole of the objects occupying the space. We say park of wagons, park of artillery; reserve park; division park, &c.; camp park; engineer park.

PAROLE. Word distinguished from the countersign. The latter is given to all sentinels; the former only to officers of the guard, and those authorized to inspect guards or give orders to guards. Giving a different parole from that received punishable with death, or according to the discretion of a court-martial; (ART. 53.)

Parole is also a pledge of honor required of prisoners when they are liberated on *parole*.

PARTISAN. The name given to small corps detached from the main body of an army, and acting independently against the enemy. In partisan warfare, much liberty is allowed to partisans. Continually annoying the flanks and rear of columns, they intercept convoys, cut off communications, attack detachments, and endeavor to spread terror everywhere. This kind of warfare is advantageously pursued only in mountainous or thickly-wooded districts. In an open country, cavalry very readily destroys partisans. The Spanish race make active partisans. The party is called a *guerilla*, the partisan a *guerillero*.

PASS. A straight, difficult, and narrow passage, which, well defended, shuts up the entrance to a country. A short permission to be absent given to a soldier.

PASSAGE OF RIVERS. The passage is effected by surprise or by main force, and detachments are thrown by one means or the other upon the enemy's bank of the river before proceeding to the construction of bridges. The passage by force ought always to be favored by diversions upon other points. Infantry cross bridges without keeping step. Cavalry dismount in crossing, leading their horses. Wagons, heavily loaded, pass at a gallop. (*See* BRIDGE; DEFILE; DISTANCES; FORDS.)

PASSAGES—are openings cut in the parapet of the covered way, close to the traverses, in order to continue the communication through all parts of the covered way. (*See* TRAVERSE.)

PASSPORTS. Foreigners going into the Indian territory without passports subject to a penalty of $1,000. (*See* INDIAN; WAR.)

PATROL. A small party detached from a guard to gain information from a neighboring post, to scour a village or wood, or to supply the place of an insufficiency of sentinels by making constant rounds.

PAWL. The click or detent which falls into the teeth of a ratchet-wheel to prevent its motion backward.

PAY.

MILITARY DICTIONARY.

Rank and Classification of Officers.	PAY. Per month.	SUBSISTENCE. 30 cents per ration.—Act, Feb. 21, 1857, Sect. 1.		FORAGE. 8 00 per mo. for each horse—Act, April 24, 1816, Sect. 12.		SERVANTS. Pay, subsistence & clothing of a private soldier.—Act, April 24, 1816, Sec. 12.		TOTAL MONTHLY PAY.
		No. of rations per day.	Monthly commutation value.	No. of horses allowed.	Monthly commutation value.	No. of servants allowed.	Monthly commutation value.	
GENERAL OFFICERS.								
Lieutenant-general................................	$270	40	$360	..	$50	4	$78 00	$758 00
Aids-de-camp....................................	80	5	45	3	24	2	45 00	194 00
Secretary.......................................	80	5	45	3	24	2	45 00	194 00
Major-general...................................	220	15	135	3	24	4	78 00	457 00
Senior Aid-de-camp to General-in-chief........	80	4	36	3	24	2	41 00	181 00
Aid-de-camp, in addition to pay, &c., of Lieut..	24	1	9	1	8	..	......	41 00
Brigadier-general	124	12	108	3	24	3	58 50	314 50
Aid-de-camp, in addition to pay, &c., of Lieut..	20	..	..	1	8	..	......	28 00
ADJUTANT-GENERAL'S DEPARTMENT.								
Adjutant-general—Colonel........................	110	6	54	3	24	2	41 00	229 00
Assistant Adjutant-general—Lieut.-colonel.......	95	5	45	3	24	2	41 00	205 00
Assistant Adjutant-general—Major................	80	4	36	3	24	2	41 00	181 00
Assistant Adjutant-general—Captain..............	70	4	36	1	8	1	20 50	134 50
Judge-advocate—Major............................	80	4	36	3	24	2	41 00	205 00
INSPECTOR-GENERAL'S DEPARTMENT.								
Inspector-general—Colonel.......................	110	6	54	3	24	2	41 00	229 00
QUARTERMASTER'S DEPARTMENT.								
Quartermaster-general—Brigadier-general.........	124	12	108	3	24	3	58 50	314 50
Assistant Quartermaster-general—Colonel.........	110	6	54	3	24	2	41 00	229 00
Deputy Quartermaster-general—Lieut.-colonel.....	95	5	45	3	24	2	41 00	205 00
Quartermaster—Major.............................	80	4	36	3	24	2	41 00	181 00
Assistant Quartermaster—Captain.................	70	4	36	1	8	1	20 50	134 50
SUBSISTENCE DEPARTMENT.								
Commissary-general of Subsistence—Colonel.......	110	6	54	3	24	2	41 00	229 00
Ass't Commissary-general of Subsistence—Lieut.-col..	95	5	45	3	24	2	41 00	205 00
Commissary of Subsistence—Major.................	80	4	36	3	24	2	41 00	181 00
Commissary of Subsistence—Captain...............	70	4	36	1	8	1	20 50	134 50
Assistant Commissary of Subsistence, in addition to pay, &c., of Lieut..........................	20	..	..	..	..	..	......	20 00
PAY DEPARTMENT.								
Paymaster-general, $2,740 per annum.............	..	..	..	..	..	..	......	228 33
Deputy Paymaster-general........................	95	5	45	3	24	2	41 00	205 00
Paymaster.......................................	80	4	36	3	24	2	41 00	181 00
MEDICAL DEPARTMENT.								
Surgeon-general, $2,740 per annum...............	..	..	..	..	..	..	......	228 33
Surgeons of ten years' service..................	80	8	72	3	24	2	41 00	217 00
Surgeons of less than ten years' service........	80	4	36	3	24	2	41 00	181 00
Assistant Surgeons of ten years' service........	70	8	72	1	8	1	20 50	170 50
Assistant Surgeons of five years' service.......	70	4	36	1	8	1	20 50	134 50
Assistant Surgeons of less than five years' service..	53 33	4	36	1	8	1	20 50	117 83
OFFICERS OF THE CORPS OF ENGINEERS, CORPS OF TOPOG. ENG., AND ORDNANCE DEPARTMENT.								
Colonel...	110	6	54	3	24	2	41 00	229 00
Lieutenant-colonel..............................	95	5	45	3	24	2	41 00	205 00
Major...	80	4	36	3	24	2	41 00	181 00
Captain...	70	4	36	1	8	1	20 50	134 50
First Lieutenant................................	53 33	4	36	1	8	1	20 50	117 83
Second Lieutenant...............................	53 33	4	36	1	8	1	20 50	117 83
Brevet Second Lieutenant........................	53 33	4	36	1	8	1	20 50	117 83
OFFICERS OF MOUNTED DRAGOONS, CAVALRY, RIFLEMEN, AND LIGHT ARTILLERY.								
Colonel...	110	6	54	3	24	2	41 00	229 00
Lieutenant-colonel..............................	95	5	45	3	24	2	41 00	205 00
Major...	80	4	36	3	24	2	41 00	181 00
Captain...	70	4	36	2	16	1	20 50	134 50
First Lieutenant................................	53 33	4	36	2	16	1	20 50	125 83
Second Lieutenant...............................	53 33	4	36	2	16	1	20 50	125 83
Brevet Second Lieutenant........................	53 33	4	36	2	16	1	20 50	125 83
Adjutant and Regimental Quartermaster, in addition to pay, &c., of Lieut...................	10 00	..	..	..	..	..	......	10 00
OFFICERS OF ARTILLERY AND INFANTRY.								
Colonel...	95	6	54	3	24	2	39 00	212 00
Lieutenant-colonel..............................	80	5	45	3	24	2	39 00	188 00
Major...	70	4	36	3	24	2	39 00	169 00
Captain...	60	4	36	..	..	1	19 50	115 50
First Lieutenant................................	50	4	36	..	..	1	19 50	105 50
Second Lieutenant...............................	45	4	36	..	..	1	19 50	100 50
Brevet Second Lieutenant........................	45	4	36	..	..	1	19 50	100 50
Adjutant and Regimental Quartermaster, in addition to pay, &c., of Lieut...................	10	..	..	1	8	..	......	18 00
MILITARY STOREKEEPERS.								
Attached to the Quartermaster's Department, at armories, and at arsenals of construction; the storekeeper at Watertown Arsenal, and storekeepers of ordnance serving in Oregon, California, and New Mexico, $1,490 per annum............................	..	..	..	..	..	..	......	124 16
At all other arsenals, $1,040 per annum.........	..	..	..	..	..	..	......	86 66

DATES OF THE ACTS OF CONGRESS ESTABLISHING THE PRESENT RATES OF PAY, ETC.

Act of May 28, 1798, Sec. 5—Feb. 15, 1855—Feb. 21, 1857—March 3, 1859.
" " " " " " " " " "
" Jan. 11, 1812, Sec. 6—Mar. 30, 1814, Sec. 9—April 24, 1816, Sec. 9 & 12—Mar. 3, 1845, Sec. 1—Feb. 21, 1857, Sec. 1.
" Sept. 26, 1850, Sec. 2—March 3, 1845, Sec. 1—Feb. 21, 1857, Sec. 1.
" Jan. 11, 1812, S c. 6—April 24, 1816, Sec. 9 & 12—March 3, 1845, Sec. 1—Feb. 21, 1857, Sec. 1.
" April 12, 1808, Sec. 4—Mar. 30, 1814, Sec. 9—Apr. 24, 1816, Sec. 9 & 12—Mar. 3, 1845, Sec. 1—Feb. 21, 1857, Sec. 1.
" " " 6—April 24, 1816, Sec. 9 & 12—March 3, 1845, Sec. 1—Feb. 21, 1857, Sec. 1.

Act of March 3, 1813, Sec. 3—March 30, 1814, Sec. 9—March 3, 1845, Sec. 1—Feb. 21, 1857, Sec. 1.
" " 1847, Sec. 2—March 3, 1845, Sec. 1—Feb. 21, 1857, Sec. 1.
" " 1813, Sec. 3—April 24, 1816, Sec. 9 & 12—March 3, 1845, Sec. 1—Feb. 21, 1857, Sec. 1.
" July 5, 1838, Sec. 7—March 3, 1845, Sec. 1—Feb. 21, 1857, Sec. 1.
" Mar. 2, 1849, Sec. 4—Mar. 3, 1813, Sec. 3—Apr. 24, 1816, Sec. 9 & 12—Mar. 3, 1845, Sec. 1—Feb. 21, 1857, Sec. 1.

Act of April 14, 1818, Sec. 5—March 3, 1845, Sec. 1—Feb. 21, 1857, Sec. 1.

Act of Mar. 28, 1812, Sec. 2—Mar. 30, 1841, Sec. 9—Apr. 24, 1816, Sec. 9 & 12—Mar. 3, 1845, Sec. 1—Feb. 21. 1857, Sec. 1.
" July 5, 1838, Sec. 9—March 3, 1845, Sec. 1—Feb. 21, 1857, Sec. 1.
" " " " " " " "
" March 2, 1851, Sec. 7— " " " " " "
" July 5, 1838, Sec. 9— " " " " " "

Act of April 14, 1818, Sec. 6—July 5, 1838, Sec. 13—March 3, 1845, Sec. 1—Feb. 21, 1857, Sec. 1.
" July 5, 1838, Sec. 11—March 3, 1845, Sec. 1—Feb. 21, 1857, Sec. 1.
" March 2, 1829, Sec. 2— " " " " " "
" July 5, 1838, Sec. 11— " " " " " "
" March 2, 1821, Sec. 8— " " " " " "

Act of April 24, 1816, Sec. 3—Feb. 21, 1857, Sec. 1.
" March 3, 1847, Sec. 13—March 3, 1845, Sec. 1—Feb. 21, 1857, Sec. 1.
" April 24, 1816, Sec. 3—July 5, 1838, Sec. 24—March 3, 1845, Sec. 1—Feb. 21, 1857, Sec. 1.

Act of April 14, 1818, Sec. 2—Feb. 21, 1857, Sec. 1.
" June 30, 1834, Sec. 2 & 3—July 5, 1838, Sec. 24—March 3, 1345, Sec. 1—Feb. 21, 1857, Sec. 1.
" " " " 2 — " " " " " " " " " "
" " " " 2 & 3— " " " " " " " " " "
" " " " 2 — " " " " " " " " " "
" " " " 2 — " " " " " " " " " "

Act of July 5, 1838, Sec. 2, 5 & 13—March 3, 1845, Sec. 1—Feb. 21, 1857, Sec. 1.
" " " " " " " " " "
" " " " " " " " " "
" " " " " " " " " "
" " " " " " " " " "
" April 29, 1812, Sec. 4—Feb. 21, 1857, Sec. 1.

Act of April 12, 1808, Sec. 4—March 30, 1814, Sec. 9—March 3, 1845, Sec. 1—Feb. 21, 1857, Sec. 1.
" " " " Apr. 24, 1816, Sec. 9 & 12— " " " " " "
" " " " " " " March 2, 1827, Sec. 1—March 3, 1845, Sec. 1—Feb. 21, 1857, Sec. 1.
" " " " " " " " Feb. 21, 1857, Sec. 1.
" April 29, 1812, Sec. 4—Feb. 21, 1857, Sec. 1.
{ " April 12, 1808, Sec. 4.
{ " May 30, 1796, Sec. 12—Feb. 11, 1847, Sec. 4—Feb. 21, 1857, Sec. 1.

Act of Mar. 16, 1802, Sec. 4, 5—Mar. 30, 1814, Sec. 9—Apr. 24, 1816, Sec. 9, 12—Mar. 3, 1845, Sec. 1—Feb. 21, 1857, Sec. 1.
" " " " April 24, 1816, Sec. 9, 12—Feb. 21, 1857, Sec. 1.
" " " " " " " March 2, 1827, Sec. 1—Feb. 21, 1857, Sec. 1.
" " " " " " " " " "
" April 29, 1812, Sec. 4—Feb. 21, 1857, Sec. 1.
{ " March 16, 1802, Sec. 4—April 24, 1816, Sec. 9, 12—March 3, 1845, Sec. 1.
{ " May 30, 1796, Sec. 12—Feb. 11, 1847, Sec. 4—Feb. 21, 1857, Sec. 1.

Act of August 23, 1842, Sec. 2—March 3, 1849, Sec. 2—March 3, 1853, Sec. 1—Feb. 21, 1857, Sec. 1.

Act of August 23, 1842, Sec. 2—Feb. 21, 1857, Sec. 1.

1. The officer in command of a company is allowed $10 per month for the responsibility of clothing, arms, and accoutrements; (*Act* March 2, 1827; Sec. 2.)

2. Subaltern officers, employed on the *General Staff*, and receiving increased pay therefor, are not entitled to the additional or fourth ration provided by the Act of March 2, 1827; Sec. 2.

3. Additional rations allowed to officers while commanding separate armies, divisions, departments, posts, armories, and arsenals; (*Act* March 3, 1797, Sec. 4; *Act* March 16, 1802, Sec. 5; *Act* August 23, 1842, Sec. 6; *Act* March 3, 1849, Sec. 1.)

4. Every commissioned officer of the line or staff, exclusive of *general* officers, receives an additional ration per diem for every five years' service; (*Acts* July 5, 1838; July 7, 1838.)

5. The allowances for forage and servants are contingent.

6. The following is the monthly pay of non-commissioned officers and soldiers: Each ordnance-sergeant, twenty-two dollars, and each sergeant-major, quarter-master sergeant, and chief musician, twenty-one dollars; to each first sergeant of a company, twenty dollars; to all other sergeants, seventeen dollars; to each artificer, fifteen dollars; to each corporal, thirteen dollars; to each musician and private of artillery or infantry, eleven dollars—one dollar per month of each private's pay being retained to the expiration of his term of service; (*Acts* July 7 and 8, 1838, and *Act* Aug. 4, 1854.)

Sec. 2. *And be it further enacted*, That every soldier, who, having been honorably discharged from the service of the United States, shall, within one month thereafter, re-enlist, shall be entitled to two dollars per month in addition to the ordinary pay of his grade, for the first period of five years after the expiration of his previous enlistment, and a further sum of one dollar per month for each successive period of five years, so long as he shall remain continuously in the army; and that soldiers now in the army, who have served one or more enlistments, and been honorably discharged, shall be entitled to the benefits herein provided for a second enlistment.

Sec. 3. *And be it further enacted*, That soldiers who served in the war with Mexico, and received a certificate of merit for distinguished services, as well those now in the army as those that may hereafter enlist, shall receive the two dollars per month to which that certificate would have entitled them, had they remained continuously in the service.

Sec. 4. *And be it further enacted*, That non-commmissioned officers, who, under the authority of the seventeenth section of the act approved March third, eighteen hundred and forty-seven, were recommended for promotion by brevet to the lowest grade of commissioned officer, but did not receive the benefit of that provision, shall be entitled, under the condition recited in the foregoing section, to the additional pay authorized to be given to such privates as received certificates of merit; (*Act* Aug. 4, 1854.)

Non-commissioned officers, musicians, and privates are also allowed one ration per day, and an allowance of clothing, both to be prescribed by the President of the United States; (*Act* April 24, 1816, and *Act* April 14, 1818.)

Troops shall be paid in such manner that the arrears shall, at no time, exceed two months, unless the circumstances of the case shall render it unavoidable; (*Act* March 16, 1802, and March 3, 1815.)

No assignment of pay made by a non-commissioned officer or private shall be valid; (*Act* May 8, 1792.)

Brevet officers shall be entitled to, and receive, pay and emoluments according to their brevet rank " when on duty, and having a command according to their brevet rank, and at no other time;" (*Act* April 16, 1818.)

No money shall be paid to any person for his compensation, who is in arrears to the United States, until such person shall have accounted for, and paid into the treasury, all sums for which he may be liable. Provided, however, that the officers of the treasury shall, upon demand of the party, forthwith report the balance due, and it shall be the duty of the solicitor of the treasury within sixty days thereafter to order suit to be commenced against such delinquent; (*Acts* Jan. 15, 1828, and May 29, 1830.)

PAY DEPARTMENT. (See ARMY for its organization.) It is the duty of paymasters to pay all the regular and other troops in the service of the United States; and, to insure punctuality and responsibility, correct reports shall be made to the paymaster-general once in two months, showing the disposition of the funds previously transmitted, with accurate estimates for the next payment of such regiment, garrison, or department, as may be assigned to each; and whenever any paymaster shall fail to transmit such estimate, or neglect to render his vouchers to the paymaster-general for settlement of his accounts, more than six months after receiving funds, he shall be recalled and another appointed in his place; (*Acts* April 24, 1816, and July 14, 1832.) (*See* ACCOUNTABILITY; DISBURSING OFFICERS.)

When volunteers or militia are called into service, so that the paymasters authorized by law shall not be deemed sufficient to enable them to pay the troops with proper punctuality, the President may assign to any officer of the army the duty of paymaster, who shall perform the same duty, give the same bond, and receive the same pay and emoluments as are provided for the paymasters of the army; but the number of officers so assigned shall not exceed one for every two regiments of militia or volunteers; (*Act* July 4, 1836.)

PAYMASTER-GENERAL. Under the direction of the Secretary of War, the paymaster-general assigns paymasters to districts; (*Act* April 24, 1816.) He receives " from the treasurer all the moneys which shall be intrusted to him for the purpose of paying the pay, the arrears of pay, subsistence, or forage due to the troops of the United States; he shall receive the pay abstracts of the paymasters of the several regiments or corps, and compare the same with the returns or muster-rolls, which shall accompany the said pay abstracts. He shall certify accurately to the commanding officer the sums due to the respective corps, which shall have been examined as aforesaid, who shall thereupon issue his warrant on the said deputy paymaster for the payment accordingly; (*Act* May 8, 1792.)

The paymaster-general may, in his discretion, allow to any paymaster's clerk, in lieu of the pay now allowed by law, an annual salary of $700. The paymaster-general shall have the rank of colonel; the deputy paymaster-general the rank of lieutenant-colonel, and in addition to paying troops, shall superintend the payment of armies in the field. Paymasters have the rank of major; but it is provided that paymasters, in virtue of such rank, shall not be entitled to command in the line or other staff departments of the army; (*Act* March 3, 1847.)

PENDULUM. The times of vibration of pendulums are proportional to the square roots of their lengths.

$$T = \pi \sqrt{\frac{l}{g}}$$

Therefore, if l be the length of a pendulum vibrating seconds, and l' the length of any other simple pendulum, or the distance from the point of suspension to the centre of oscillation of a compound pendulum, vibrating in the time t at the same place, then: $l' = l\,t^2$

The length of a pendulum vibrating seconds is in a constant ratio to the force of gravity:

$$\frac{g}{l} = 9.8696044.$$

Length of a pendulum vibrating seconds at the level of the sea, in various latitudes.

At the Equator . . . . . . 39.0152 inches.
    Washington, Lat. 38° 53′ 23″ . . . 39.0958 "
    New York, Lat. 40° 42′ 40″ . . . 39.1017 "
    London, Lat. 51° 31′ . . . 39.1393 "
    Lat. 45° . . . . . 39.1270 "
    Lat. L. . . 39.1270 in.—0.09982 cos. 2 L

PENDULUM HAUSSE—is a tangent-scale, the graduations of which are the tangents of each quarter of a degree of elevation, to a radius equal to the distance between the muzzle-sight of the piece, and the axis of vibration of the hausse, which is one inch in rear of the base-ring. At the lower end of the scale is a brass bulb filled with lead. The *slider* which marks the divisions on the scale is of thin brass, and is clamped at any desired division on the scale by means of a screw. The scale passes through a slit in a piece of steel, with which it is connected by a screw, forming a pivot on which the scale can vibrate laterally. This piece of steel terminates in pivots, by means of which the pendulum is supported on the *seat* attached to the gun, and is at liberty to vibrate in the direction of the axis of the piece. The *seat* is of metal, and is fastened to the base of the breech by screws, so that the centres of the steel pivots of vibration shall be at a distance from the axis of the piece equal to the radius of the base-ring.

A MUZZLE-SIGHT of iron is screwed into the swell of the muzzle of guns, or into the middle of the muzzle-ring of howitzers. The height of this sight is equal to the dispart of the piece, so that a line joining the muzzle-sight and the pivot of the tangent-scale is parallel to the axis of the piece.

PENETRATION. The penetration of a solid shot, other circumstances being the same, varies with its diameter, and with the distance and material of the substance penetrated.

In the subjoined table are given the penetrations of a 24-pounder shot, whence a tolerably accurate estimate may be formed of the penetrations of shot of other calibres.

Substance penetrated.	RANGE.			
	100 yards.	400 yards.	1,200 yards.	
Good Masonry	2 ft.	1¼ ft.	¾ ft.	Penetration in feet.
Oak	4 "	3 "	1½ "	
Firm Earth	6¼ "	5 "	2¼ "	
Fresh dug Earth	12 "	9 "	4½ "	

Sand, sandy earth mixed with gravel, small stones, chalk and tufa resist shot better than the productive earths. Shells may be considered as round shot of a lower specific gravity, and their penetrations are therefore proportionately less. A bank of earth, to afford a secure cover from heavy guns, will require a thickness from 18 to 24 feet. In guns below 18-pounders, if the number of the feet in thickness of the bank be made equal to the number of lbs. in the weight of the shot by which it is to be assailed, the requisite protection will be obtained. Earth possesses advantages over every other material. It is easily obtained, regains its position after displacement, and the injury done to an earthen battery by day can be readily repaired at night. Where masonry is liable to be breached, it should be covered with earth. Wrought-iron plates $4\frac{1}{2}$ inches in thickness will withstand the effects of 32-pound shots, and of all inferior calibres at short ranges as 400 yards. Plates of this thickness, however, are soon destroyed by 68-pound shots, and afford little protection from the elongated shots of the new rifled ordnance. (*See* IRON PLATES.)

To resist successfully the fall of heavy shells, buildings must be covered with arches of good masonry, not less than 3 feet thick, having bearings not greater than 25 feet, and these must be again protected by a covering of several feet of earth. Iron plates half an inch thick, oak planks 4 inches thick, or a nine-inch brick wall, are proof against musketry or canister at a range of 100 yards. Iron plates 1 inch thick, oak from 8 to 10 inches thick, a good wall a foot thick, or a firm bank of earth 4 feet thick, will afford secure cover from grape shot, from any but the largest guns at short ranges. The common musket will drive its bullet about a foot and a half into well-rammed earth, or it will penetrate from 6 to 10 half-inch elm boards placed at intervals of an inch. The penetration of the rifled musket is about twice that of the common musket. A rope matting or mantlet $3\frac{1}{2}$ inches thick is found to resist small-arm projectiles at all distances; it may therefore be employed as a screen against riflemen.

Experiments were made in 1848 at Portsmouth against the "Leviathan," to ascertain whether a round shot fired at a depression into the water close to a ship would continue its course, and passing through the water, can maintain force sufficient to penetrate into the ship considerably below the water-line; for this a 32-pounder gun of 56 cwt., with a charge of 10 lbs., was fired at a depression of 7 degrees from a dockyard "lump," 16 yards distant from the "Leviathan." The shot struck the water 4 feet from the ship's side, rose immediately, passed through the orlop, and was found on the lower deck. Another shot, fired under the same circumstances, only indented the wood 18 inches below the water line. But elongated rifle-shot fired into the water have the faculty of entering and passing through the fluid in the direction of their axes, and, after passing through many feet of water, retain force sufficient to penetrate any ship's side below the water-line. This was proved by firing Whitworth's hexagonal shot under circumstances nearly similar to the preceding experiments against the "Leviathan," when a flat-headed hexagonal shot fired from a 24-pounder passed through 33 feet of water, and then penetrated into the ship through 12 or 14 inches of oak and planking; (Sir Howard Douglas; Hyde and Benton.) (*See* Rifled Ordnance.)

PENSION. No person in the army, navy, or marine corps, shall be allowed to draw both a pension as an invalid and the pay of his rank or station in the service, unless the alleged disability for which the pension was granted, be such as to have occasioned his employment in a lower grade, or in some civil branch of the service; (*Act* April 30, 1844.) Any officer, non-commissioned officer, or soldier of the army, including militia rangers, sea-fencibles and volunteers, disabled by wounds or otherwise, while in the line of his duty in public service, shall be placed on the list of invalids of the United States, at the following rates of pay: No officer shall receive more than the half pay of a lieutenant-colonel; half the monthly pay of inferior grades; or, for a first lieutenant, seventeen dollars; a second lieutenant, fifteen dollars, a third lieutenant fourteen dollars, an ensign thirteen dollars; and a non-commissioned officer, musician, or private, eight dollars per month for the highest disability, and for less disabilities a sum proportionably less; (*Act* March 16, 1802, and April 24, 1816.)

The widow of an officer dying of wounds received in military service, or if the officer have no widow, any child or children left by the officer, is entitled to his half pay for five years; provided that the pension to the widow shall cease upon her death or intermarriage, and shall also cease upon the death of such child or children; (*Act* March 16, 1802.)

In an elaborate opinion given by Mr. Attorney-general Cushing, published by the War Department in General Orders, No. 11 of 1855, he draws the conclusion that "the phrase 'line of duty' is an apt one, to denote that an act of duty performed must have relation of causation, mediate or immediate, to the wound, the casualty, the injury, or the disease producing disability or death." "Every person" (says Mr. Cushing) who enters the military service of the country—officer, soldier, sailor, or marine—takes upon himself certain moral and legal engagements of duty, which constitute his official or professional obligations. While in the performance of those things which the law requires of him as military duty, he is in the line of his duty. But at the same time, though a soldier or sailor, he is not the less a man and a citizen, with private rights to exercise and duties to perform; and while attending to these things he is not in the line of his public duty. In addition to this, a soldier or sailor, like any other man, has the physical faculty of doing many things which are in violation of duties either general or special; and in doing these things he is not acting in the line of his duty. Around all those acts of the soldier or sailor which are official in their nature the pension laws draw a legislative line, and then they say to the soldier or sailor: If, while performing acts which are within that line, you thereby incur disability or death, you or your widow or children, as the case may be, shall receive pension or allowance; but not if the disability or death arise from acts performed outside of that line; that is, absolutely disconnected from, and wholly independent of, the performance of duty. Was the cause of disability or death a cause within the line of duty or outside of it? Was that cause appertaining to, dependent upon, or otherwise necessarily and essentially connected with, duty within the line; or was it unappertinent, independent, and not of necessary and essential connection? That, in my judgment, is the true test-criterion of the class of pension cases under consideration."

PERCUSSION. Twelve percussion caps are issued to ten cartridges. (*See* ARMS and ACCOUTREMENTS.)

PERCUSSION BULLETS—are made by placing a small quantity of percussion powder, inclosed in a copper envelope, in the point of an ordinary rifle musket bullet. The impact of the bullet against a substance no harder than wood is found to ignite the percussion charge, and produce an effective explosion. These projectiles can be used to blow up caissons and boxes containing ammunition at very long distances; (BENTON.)

PETARD. An engine made of gun-metal, fixed upon a board, and containing about nine pounds of powder. Sometimes attached to gates, &c., to burst them open. In an attack upon a fortification, leathern bags containing fifty pounds of powder have been found more useful.

PICKER. A small pointed brass wire, which is supplied to every infantry soldier for the purpose of cleaning the vent of his musket.

PICKET. Sharp stakes used for securing the fascines of a battery. To picket horses in camp. STOCKADES, which see, are also sometimes called picket works.

Also a detachment composed of cavalry or infantry, whose principal duty is to guard an army from surprise, and oppose such small parties as the enemy may push forward for the purpose of reconnoitring. (*See* OUTPOSTS.)

PIECE—designates any gun, large or small.

PIERRIER—was a term originally applied to an engine for casting stones; then to a small kind of cannon; now to a mortar for discharging stones, &c.

PIERS. The columns upon which a bridge is erected.

PIKE. A military weapon formerly used as a bayonet. The pike had a shaft from ten to fourteen feet long, with a flat pointed steel head called the spear.

PILE. A beam of wood driven into the ground to form a solid foundation for building. Also a heap, as a pile of balls. To *pile arms*, is to stack arms in the prescribed manner, that they may remain steady on the ground. Balls are piled according to kind and calibre, under cover if practicable, in a place where there is a free circulation of air, to facilitate which the piles should be made narrow if the locality permits; the width of the bottom tier may be from 12 to 14 balls, according to the calibre. Prepare the ground for the base of the pile by raising it above the surrounding ground so as to throw off the water; level it, ram it well, and cover it with a layer of screened sand. Make the bottom of the pile with a tier of unserviceable balls buried about two-thirds of their diameter in the sand; this base may be made permanent: clean the base well and form the pile, putting the fuze holes of shells downwards, in the *intervals*, and not resting on the shells below. Each pile is marked with the number of serviceable balls it contains. The base may be made of bricks, concrete, stone, or with borders and braces of iron. Grape and canister shot should be oiled or lackered, put in piles, or in strong boxes, on the ground floor, or in dry cellars; each parcel marked with its kind, calibre, and number.

PILLAGE. (*See* PLUNDER.)

PIONEERS. Soldiers sometimes detailed from the different companies of a regiment and formed under a non-commissioned officer, furnished with saws, felling axes, spades, mattocks, pickaxes, and billhooks. Their services are very important, and no regiment is well fitted for service without pioneers completely equipped.

PISTOL. Horsemen have one or two pistols furnished them. General, field and staff officers also carry pistols in their holsters. Colt's pistol is a revolver composed of a cylinder containing six charges, a rifled barrel, and a handle or stock. The length of bore (navy) 9 in.; weight 2.40 lbs.; weight of projectile 125 grs.; weight of powder 14 grs.; initial velocity 760 feet. (*See* ARMS for Pistol-Carbine.)

PIVOT. That officer or soldier upon whom the company wheels. The *pivot flank* in a column is that which, when wheeled up, preserves the proper front of divisions of the line in their natural order. The opposite flank of the column is called the reverse flank.

PLACE. Town or city is but little used in military parlance. A strong *place* is a fortified city.

PLACES OF ARMS—are enlargements in the covered-way, at the re-entering and salient angles of the counterscarp; hence the terms re entering places of arms, and salient places of arms; the latter space is formed simply by rounding the counterscarp; and the former by setting off demi-gorges of thirty yards, (more or less,) and making the faces form angles of 100° with the adjoining branches of the covered-way.

PLAN. A plan of campaign (says Napoleon) should anticipate all that an enemy may do, and combine within itself the means necessary to baffle him. Plans of campaign are modified by circumstances, the genius of the chief, the nature of the troops, and topography. There are good and bad plans of campaign, but sometimes the good fail from misfortune or mismanagement, while the bad succeed by caprices of fortune.

PLAN OF A WORK. A plan shows the tracing; also the horizontal lengths and breadths of the works; the thickness of the ramparts and parapets; the width of the ditches, &c.: it exhibits the extent, division, and distribution of the works; but the depth of the ditches and the height of the works are not represented in a plan.

PLANE OF COMPARISON—is a plan of a fortress, and of the surrounding country, on which are expressed the distances of the principal points from a horizontal plane, imagined to pass through the highest or lowest points of ground, in the survey This imaginary plane is called a plane of comparison.

PLANE OF DEFILADE—is a plane supposed to pass through the summit or crest of a work, and parallel to the plane of site.

PLANE OF SITE. The general level of the ground, or ground line, upon which the works are constructed, is called the plane of site, whether that plane be horizontal or oblique to the horizon.

PLATFORM. There are six sleepers, 18 deck planks, 72 dowels, and 12 iron eye-bolts, used for the platform of siege mortars. The weight of the platform made of yellow pine is 837 lbs.

PLATOON. The half of a company.

PLONGÉE. The dip or declension of the superior slope of the parapet, is called the plongée. The amount of it is regulated by the distance of the nearest spot, to which the fire of musketry is to be directed; that is, generally, the exterior edge of the ditch in front of it.

PLUMMET. A leaden or iron weight suspended by a string, used by artificers to sound the depth of water, or to regulate the perpendicular direction of any building. Pendulums, called also plummets, which vibrate the required times of march in a minute, are of great utility; they must be in the possession of, and be constantly referred to by, each instructor of a squad. (*See* PENDULUM.)

PLUNDER. Every officer or soldier, who shall quit his post or colors to plunder and pillage, shall suffer death or such other punishment as may be ordered by a general court-martial; (ART. 52.)

POINT-BLANK. The point-blank is the second point at which the line of sight intersects the trajectory of the projectile. The *natural point-blank* is when the natural line of sight is horizontal. The point-blank made by the use of the *hausse*, is called an artificial point-blank.

In the British service, the point-blank distance is the distance at which the projectile strikes the level ground on which the carriage stands, the axis of the piece being horizontal. This definition conveys a better idea of the power of the piece than the French and American definition. For the same piece, the point-blank distance increases with the charge of powder; for the same initial velocity, a large projectile has a greater point-blank distance than a small one; a solid shot than a hollow one; and an oblong projectile than a round one. (*See* FIRING.)

POINTING. To point a gun is to give it such direction, and elevation or depression, that the shot may strike the object. The general rule is, first give the direction, and then the elevation or depression. In pointing mortars, the elevation is first given and then the direction. The direction of a gun or howitzer is given by directing the line of metal upon the object. The elevation or depression depends upon the charge, the distance and the position of the object above or below the battery, and it is ascertained by reference to tables of fire, or by experiment; and the proper angle is given by means of instruments—the gunner's quadrant or tangent-scales. In the absence of tangent-scales or quadrant, the gunner may point his gun by placing one or more fingers of the left hand upon the base-ring perpendicularly to the axis, and using them as a breech-sight.

In pointing a mortar, the elevation is given by applying the quadrant to the face of the piece, and adjusting the quoin until the required number of degrees is indicated. The direction is given by determining practically, two fixed points which shall be in a line with the piece and object, and sufficiently near to be readily distinguished by the eye. These points being covered by the plummet, determine a vertical plane which, when including the line of metal, becomes the plane of fire. Various methods are given for the accomplishment of this object in Roberts's Handbook of Artillery. (Consult *Instructions for Field and Heavy Artillery*, published by the War Department.)

POLYGON OF FORTIFICATION. Every piece of ground to

be fortified, is surrounded by a polygon, either square, pentagonal, hexagonal, &c., according to the number of its sides, which are called exterior sides; upon these the fronts of fortifications are constructed.

PONTONIERS. (*See* SAPPERS.)

PONTOON. Vulcanized India rubber pontoons, consisting of three cylinders connected together, have been made in the United States. The three cylinders weigh 260 lbs., and with their flooring of three chesses can be packed in a box 5 feet × 3½ feet × 1 foot. The India rubber pontoons are made of India rubber cloth, and consist each of three tangent cylinders, peaked at both extremities like the ends of a canoe; the ends are firmly united together by two strong India rubber ligaments which extend along their lines of contact and widen into a connecting web towards the ends in proportion as these diminish, the three thus forming a single boat 20 feet long by 3 feet broad, of great buoyancy and stability, and from its form and lightness presenting but trifling resistance to the water. Each cylinder, including its peaked extremities, is 20 inches in diameter, and is divided into three distinct air-tight compartments, each of which has its own inflating nozzle. The middle compartment occupies the whole width of the roadway of the bridge. The inflating nozzles are made of brass, with stopple and tube, the former screwing into the latter to open or close the nozzle. The frame lies on the top of the pontoon to which it is lashed, and serves as a means of attaching the baulks to the pontoon and preventing their chafing it: the baulks are of white pine or spruce 19 feet long; the chesses are also of white pine or spruce 13 feet 9 inches long. The equipment and management of these pontoons are nearly similar to the means employed for bridges of a different kind. The floating portion constitutes the essential difference, and this, being light and compact when folded up, may be easily transported. (Consult *Papers published by United States Engineers in* 1849.)

The chief engineer, with the approbation of the Secretary of War, regulates and determines the number, quality, forms, dimensions, &c., of the necessary vehicles, pontoons, tools, implements, and other supplies for the use of the company of sappers, miners, and pontoniers; (*Act* May 15, 1846.)

PORT-FIRE. A composition of nitre, sulphur, and mealed powder driven into a case of strong paper used to fire guns previous to the introduction of the friction primer.

POSSE COMITATUS. A sheriff or marshal, for the purpose of keeping the peace and pursuing felons, may command all the people of his county above 15 years old to attend him, which is called the *posse comitatus*, or power of the county; (BLACKSTONE.)

Can United States troops stationed in any county be employed as a *posse comitatus?* Their service does not give them residence where they are employed, and moreover the Acts of Congress of 1795, and March 3, 1807, restrict the employment of the United States military forces in civil commotions to clearly defined cases, and then authorize the President of the United States alone to use such force after he shall have by proclamation commanded the insurgents to disperse and retire peaceably to their homes within a reasonable time. (*See* CALLING FORTH MILITIA; OBSTRUCTION OF LAW.)

These enactments of Congress would seem to make inapplicable to United States troops the doctrine of English judges, that the soldier, being still a citizen, acts only in preservation of the public peace as another citizen is bound to do. *See* EXECUTION OF LAWS, for the learning on the subject of using troops in civil commotions where the common law is not changed by legislation.

POST. It is synonymous with position. Thus a post is said to be good or not tenable. Post is also the walk or position of a sentinel. Any officer or soldier, who shall shamefully abandon any fort, post, or guard which he may be commanded to defend, or speak words inducing others to do the like, shall suffer death or such other punishment as a court-martial may direct; (ART. 52.)

Any sentinel, who shall be found sleeping upon his post, or shall leave it before he shall be regularly relieved, shall suffer death or such other punishment as shall be inflicted by a court-martial; (ART. 66.) (*See* PAY.)

POSTERN OR SALLY-PORT—is a passage usually vaulted, and constructed under the rampart, to afford a communication from the interior into the ditch. The passages from the covered way into the country, are likewise called sally-ports; as they afford free egress and ingress to troops, engaged in making a sally or sortie.

POWDER. (*See* GUNPOWDER.)

PRESIDENT. The President of the United States is commander-in-chief of the army, navy, and militia, called into service. His functions as such are assigned by Congress, but embrace of course whatever authority may be assigned to any military commander, on the principle that the authority of the greater includes that of the less. For the command, government, and regulation of the army, however, Congress has created a military hierarchy or range of subordination in the army with rights and duties regulated by Congress, and the commander-in-chief cannot make use of any other agents in exercising his command; and all orders issued by him must be according to the rules and articles made by Congress for the government of the army. In his capacity of chief-magistrate of the Union, Congress has also invested the President with many administrative functions relating to military affairs; and for the performance of the latter duties the Secretary of the Department of War has been made his minister, upon matters connected with *materiel*, accounts, returns, the support of troops, and the raising of troops. (*See* ARMY REGULATIONS; CONGRESS; DEPARTMENT OF WAR; ORDERS; REGULATIONS; SECRETARY OF WAR.)

PRESIDENT, (COURT-MARTIAL.) The President of a court-martial is the senior member. He preserves order in court; administers the oath taken by the judge-advocate, and the proceedings of the court are authenticated by his signature and that of the judge-advocate.

PRINTING. The following explanation of the marks which are in general use by printers for correcting proofs, with the annexed specimen, will enable an officer, who has to superintend a work through the press, to correct the proof sheets in a way that will be clearly understood by the printer, and thus promote its accuracy.

If it is desired to change any word to capitals, small capitals, Roman text, (the ordinary letter,) or italics, draw a line beneath it, and write in the margin, *Caps.*, *S. caps.*, *Rom.*, or *Ital.*, as the case may be. See corrections 1, 2, 14, and 8, on the proof-sheet.

When it is necessary to expunge a letter or word, draw a line through it, and place in the margin a character resembling a *d* of current hand, which stands for the Latin word *dele* (*erase*); as in No. 3.

When a wrong letter or word occurs in the proof-sheet, draw a line through it, and place what must be substituted for it in the margin, with a vertical line at the right; as in the corrections marked 4.

Attention is drawn to an inverted letter by underscoring it, and writing opposite the character used in No. 5.

An omitted word, letter, comma, semicolon, colon, exclamation-point, or interrogation-point, as well as brackets and parentheses, are written in the margin with a vertical line at the right; as in the various corrections marked 6 : a caret shows where to introduce what is thus marked in. When there is so much omitted that there is not room for it in the margin, it is written at the top or bottom of the page, and a line is used to show where it is to be introduced; as at the bottom of the proof-sheet.

A period is marked in by placing it in the margin inside of a circle, as in No. 9.

Apostrophes and quotation-points are introduced in a character resembling a V, and a caret is placed in the text to show where they are to be inserted. This is illustrated in No. 11.

No. 22 shows how the dash and hyphen are introduced.

When a letter or word should be transposed, a line is drawn around it and carried to the place where it should stand, and the letters *tr.* are placed opposite, as in No. 7.

No. 10 shows how to mark out a quadrat or space which improperly appears.

If a broken or imperfect letter is used, draw a line through or beneath it, and make an inclined cross in the margin, as in No. 12.

Sometimes a letter of the wrong size will be used by mistake; in such a case, underline it and place the letters *w. f.* (*wrong font*) in the margin as in 13.

If the letters of a word stand apart from each other, draw a curved line beneath the space which separates them, and two curves in the margin, as in No. 15. If the proper space is wanting between two contiguous words, place a caret where the space should be, and opposite to them make a character like a music sharp, as shown in No. 16.

Two parallel horizontal lines, as in No. 17, are used when the letters of a word are not all in the same level, and a horizontal line is also drawn under such as are out of place.

When a new paragraph has been improperly begun, a line is drawn from its commencement to the end of the previous paragraph, and the words *no break* are written in the margin; see No. 18. When it is desired to commence a new paragraph, the paragraph mark (¶) is introduced at the place, and also in the margin.

MILITARY DICTIONARY.

When letters at the commencement of a line are out of the proper level, a horizontal line should be drawn beneath them, and a similar one placed in the margin; as in No. 21. When any portion of a paragraph projects laterally beyond the rest, a vertical line should be drawn beside it, and a similar one must stand opposite to it in the margin; see No. 23.

When a lead has been improperly omitted, the word *Lead* is written at the side of the page, and a horizontal line shows where it is to be introduced, as in No. 25. If a lead too many has been introduced, the error is corrected, as in 24.

When uneven spaces are left between words, a line is drawn beneath, and *space better* is written opposite; see 26.

If it is desired to retain a word which has been marked out, dots are placed beneath it, and the word *stet* (*let it stand*) is written in the margin; as in 27.

MARKS USED IN CORRECTING PROOF-SHEET.

WILLIAM FALCONER.

William Falconer was the son of a barber in Edinburgh, and was born in 1730. He had very few avantages of education, and went to sea in early life in the merchant service. He afterwards became mate of a vessel that wrecked in the Levant, and was saved with only two of his crew: this catastrophe formed the subject of his poem entitled "The Shipwreck, on which his reputation as a writer chiefly rests. Early in 1769, his "*Marine Dictionary*" appeared, which hasbeen highly spoken of by those capable of estimating its merits.

In this seam year, he embarked on the AURORA but the vessel was never heard of after she passed the Cape; the poet of the Shipwreck is therefore supposed to have perish'd by the same disaster he had himself so graphically described. ¶ The subject of the "Shipwreck" and its authors fate demand our interest and sympathy. If we pay respect to the ingenuous scholar who can produce agreeable verses in leisure and retirement, how much more interest must we take in the "shipboy on the high and giddy mast' cherishing the hour which he may casually snatch from danger and fatigue.

refined visions of fancy at

PRISONER OF WAR. Agreements are made between governments at war; or, when governments do not make such agreements, opposite commanding generals, during a campaign, regulate mutual exchanges of prisoners, and also determine the allowances to be made to prisoners while they are held in captivity.

PRISONERS. Whenever any officer shall be charged with a crime, he shall be arrested and deprived of his sword by the commanding officer; (ART. 77.) Non-commissioned officers and soldiers charged with crimes shall be confined until tried by a court-martial, or released by proper authority; (ART. 78.) (*See* PROVOST-MARSHAL; REFUSAL.) When brought into court, a prisoner should be without irons or any manner of shackles or bands; unless there is danger of an escape, and then he may be secured with irons; (BLACKSTONE.) (*See* ARREST; COUNSEL; JUDGE-ADVOCATE.)

PRIVATE. The term applied to the rank of a common soldier.

PRIZE-MONEY. (*See* BOOTY.)

PROCEEDINGS. The proceedings of courts-martial of the previous day are usually read over each day by the judge-advocate. Much time is lost by adopting this measure, and there is no rule directing the court to read them; (HOUGH's *Military Law Authorities.*) (*See* PRESIDENT.)

PROJECTILES. The projectiles for unrifled ordnance are solid shot and shells. (*See* CANISTER; CARCASSES; GRAPE; GRENADES; LIGHT and FIRE BALLS; SHELLS; SPHERICAL CASE; STONES.)

PROJECTILES, (CYLINDRO-CONOIDAL.) Sir Isaac Newton has given, in the "Principia," (lib. ii., schol. to prop. 34,) an indication of the form of a solid body which, in passing through a fluid, would experience less resistance than a body of equal magnitude and of any other form. He imagined that this might be of use in ship-building, and it is evident that the principle is equally applicable in the theory of projectiles. Investigations of the differential equations of the curve may be seen in the writings of mathematicians. The body is a solid of revolution, and the differential equation is—

$$y = C \frac{d z^4}{d y^3 d x}$$

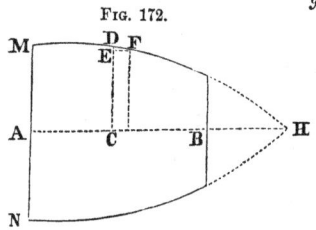

FIG. 172.

in which C is a constant. The form of a section through the axis of the solid is given in the annexed diagram, (Fig. 172.) A B is the axis, and in the direction of that line the solid is to move; y is any ordinate, as D C; and dx, dy, dz, are elementary portions, E F, E D, D F, respectively. The end B, as well as A, of the solid is a plane surface; for the numerator of the fraction in the above equation will evidently be always greater than the denominator, and therefore y, the ordinate to the curve, can never be zero. It is plain, however, that the minimum of resistance would not be obtained with a shot of an elongated form, when discharged from a musket or piece of ordnance, unless the axis A B can be kept in the direction of the trajectory. This may be accomplished if the shot be caused to have a rotatory motion on that axis by being discharged from a rifled bore; and without such rotation, not only will the axis perpetually deviate from the direction of the path, but the projectile will even turn over. The advantages of this form of shot are, that when rotating on their longitudinal axes, and moving with their smaller extremities in front, they experience less resistance from the air than spherical projectiles of the same diameter. To this form alone are to be referred the long range with the great momentum and penetrating power of the projectiles for rifle-muskets and other rifled ordnance now used; (Sir HOWARD DOUGLAS.) The elongated bullet was first experimented upon by M. Tamisier. It had a groove around the bottom or cylindrical part designed to attach the cartridge. A change having been made in the manner of attaching the cartridge to the projectile this groove was omitted as useless. The accuracy of the fire was thereupon diminished. The groove being replaced, it was found that the slightest change in its shape or position had much influence on the accuracy of fire. M. Tamisier made experiments with a ball, the point of which, instead of being curved, was a cone and the rest a cylinder; he varied the length of each part, and determined that these variations always produced variations in the accuracy of fire. These researches brought him to results of the greatest importance, and led, with the idea of M. Minié of causing the ball to expand by the explosion of the charge, to the adoption of the Minié projectiles now used, which however are not identical in different countries. (*See* RIFLED ORDNANCE.)

FIG. 173.

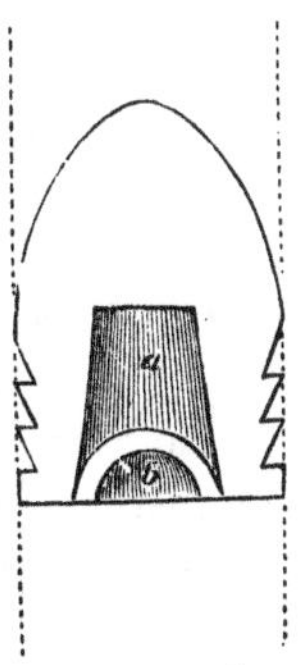

PROLONGE—is a stout hempen rope, sometimes used to connect the lunette of a field-carriage with the limber when the piece is fired; it has a hook at one end and a toggle at the other, with two intermediate rings, into which the hook and toggle are fastened to shorten the distance between the limber and carriage.

PROMOTION. " Congress may fix the rules for promotions and

appointments; and, in the reduction of the army and navy, determine from whom such promotions and appointments shall be made. Every promotion is a new appointment, to be confirmed by the Senate;" (*Report of Committee of Senate*, April 25, 1822.) (*See* CONSTITUTIONAL.)

"Promotions may be made through the whole army in its several lines of light artillery, light dragoons, artillery, infantry, and riflemen, respectively;" (*Act* March 30, 1814) "Promotions by brevet may be conferred for gallant actions or meritorious conduct;" (*Act* July 6, 1812.) "All promotions in the staff departments or corps shall be made as in other corps of the army;" (*Act* March 3, 1851.)

The French army has the most democratic organization of any army in the world. The following rules regulate promotions in that army; (*Law of* April 14, 1832; and *Law of* March 16, 1838.)

ART. 1. No person can be corporal, until he has served at least six months as a private soldier in some one of the corps of the army.

2. No one can be sergeant until he has served at least six months as corporal. All vacancies of corporal or sergeant on campaign, in any battalion, belong exclusively to those present in the field where the vacancies occur.

3. No one can be *sous*-lieutenant, unless he is at least 18 years of age, and has either served at least two years as a non-commissioned officer in one of the corps of the army; or has been two years a pupil of a military school, and has passed a satisfactory examination upon leaving the school. The first vacancy occurring on campaign, is given to some sergeant present. The 2d and 3d from those eligible, according to a fixed rule adopted at the beginning of the year. But when a non-commissioned officer has merited, for distinguished conduct mentioned in the orders of the army, a nomination for the grade of *sous*-lieutenant, and no vacancy exists in his regiment for the promotion of a non-commissioned officer, he is named for promotion, either in his own corps or in other regiments of his *arm*, to a vacancy belonging to the 2d and 3d classes.

4. All soldiers of the army, until the age of 25, may be received to undergo an examination for the polytechnique school.

5. No one can be lieutenant, unless he has served two years as *sous*-lieutenant.

6. No one can be captain, unless he has served two years in the grade of lieutenant.

7. No one can be chief of battalion, chief of squadron, or major until he has served four years as captain.

8. No one can be lieutenant-colonel, until he has served three years the grades of chief of battalion, chief of squadron, or major.

9. No one can be colonel, until he has served two years in the grade of lieut.-colonel.

10. No one can be promoted to a grade superior to that of colonel, until he has served three years in the grade immediately inferior.

11. One-third of the vacancies in the grade of *sous*-lieutenant of the different corps of troops of the army, shall be given to the non-commissioned officers of the respective corps in which the vacancies occur. (*See* ART. 3.)

12. Two-thirds of the grades of lieutenant and captain shall be given by seniority, to wit: in the infantry and cavalry, to the officers of the respective regiments; in the staff corps, to the officers of the corps; in the artillery and engineers, to the officers among themselves who stand in competition. Promotions to the grades of lieutenant and captain are made as follows: Half of the vacancies in the battalions, squadrons, or detachments which form an active army, and two-thirds of those occurring elsewhere, are given to *sous*-lieutenants, and lieutenants by seniority in their respective corps. All officers, whether with that portion of their corps in campaign or not, may be selected to fill vacancies in their corps belonging to the class of selections. But when, from distinguished conduct duly mentioned in army orders, a sous-lieutenant or lieutenant merits promotion to the next superior grade, and there is no vacancy among the class of selections in his own regiment, he may be promoted to a vacancy in some other regiment of his arm. When so many vacancies in the grades of lieutenant and captain of a regiment occur in war, that there is not a sufficient number of the inferior grade with the exacted qualifications to fill them, they will be filled from other regiments of the same arm.

13. Half of the grades of *chef-de-bataillon* and chief of squadron will be given by seniority of grade, as follows: In the infantry and cavalry and staff corps, to the captains of each arm; in the artillery and engineers to the captains among themselves, who stand in competition. The employment of major (a regimental administrative officer) will be given by selection from those eligible.

14. All the grades superior to that of chief of battalion, chief of squadron, or major, will be by selection from those eligible.

15. Seniority of grade will be determined by date of commission, or in cases of similar date by the date of the commission of the inferior grade.

16. When an officer is no longer borne on the list of some one of

the active corps of the army, the time that he thus passes out of service shall be deducted from his seniority, except in cases of mission, disbandment, or suppression of employment. There shall also be deducted from his seniority the time passed in a foreign service; but not the time passed upon detached service in the national guard, in the navy, or upon a diplomatic mission. Officers who cease to be borne on the list of corps of the army, in consequence of suppression of employment or disbandment of regiments, will nevertheless be entitled to promotion in the regiments of the same arm to which they belong, and which may be retained or subsequently created.

17. Officers, prisoners of war, will retain their rights of seniority for promotion; but they can only be promoted to the grade immediately superior to that which they had when made prisoners.

18. The term of service exacted for passing from one grade to another, may be reduced one half by service in war or in colonies.

19. The conditions exacted by the preceding articles for passing from one grade to another, can be departed from only in the following cases: 1st. For distinguished conduct duly set forth and published in the general orders of the day to the army; and 2d, when it is not otherwise possible to fill the vacancies of corps in the presence of the enemy.

20. In time of war, and in corps in presence of the enemy, there shall be given by seniority half the grades of lieutenant and captain. All the grades of chief of battalion and chief of squadron shall be made by selection from those eligible.

21. In no case shall any one be appointed to a grade without command, nor be granted an honorary grade, nor shall a rank be given superior to that of actual command.

22. All promotions of officers shall be immediately made public, with an indication of the vacancy filled, and the cause of promotion, whether by seniority, by selection, or distinguished action.

23. No officer admitted to the retired list can resume his position upon the active list.

24. Command is distinct from grade. No officer can be deprived of his grade, except in the cases and under the forms determined by law.

25. All the provisions of the present law are applicable to marines.

26. All provisions repugnant to the present law are abrogated.

Selections by the law of March 16, 1838, are made as follows:—Recommendations for appointment of non-commissioned officers are to be made to the colonel of the regiment by captains, accompanied by remarks of the chiefs of battalions, squadrons, and lieutenant-colonel. The colonel appoints from this list those who are to fill vacancies. He may also, besides this list, select from those distinguished by an action of eclat. For promotion to the grades of *sous*-lieutenant, lieutenant and captain, the chief of the corps *recommends*, after taking the advice of the chiefs of battalions or squadrons, and also of the lieutenant-colonel, when he is present. For promotion to the grade of chief of battalion or squadron, the general of brigade *recommends*, after taking the advice of the chiefs of corps of his brigade. For promotion to the grade of lieutenant-colonel, the general of division *recommends*, after taking the advice of the chiefs of corps and that of the generals of brigade. For promotion to the grades of colonel or general of brigade, the general in chief *recommends*, after taking the advice of the generals of brigade and division for the promotion of a colonel, and that of generals of division for the promotion of a general of brigade. These propositions for the different grades of officers are addressed through the regular channels of communication, and transmitted with his opinion to the Minister of War. The chiefs of corps and the general officers to whom this right of nomination is given, designate for each vacancy three candidates taken from among the non-commissioned or commissioned officers under their orders, who have been presented for promotion in the form indicated. The number of candidates for the grades of lieut.-colonel, colonel, and general of brigade may be reduced.

PROMULGATION. (*See* COURT-MARTIAL.)

PROOFS. (*See* PRINTING.)

PROSECUTOR. The judge-advocate is the prosecutor, usually; but if an officer prefers a charge, he sometimes appears to sustain the prosecution. No person can appear as prosecutor not subject to the Articles of War, except the judge-advocate; (HOUGH.)

PROVOST-MARSHAL. An officer appointed in every army in the field to secure prisoners confined on charges of a general nature. In the British army he is intrusted with authority to inflict summary punishment on any soldier, follower, or retainer of the camp, whom he sees commit the act for which summary punishment may be inflicted. (*See* CONFINEMENT; PRISONER; REFUSAL TO RECEIVE PRISONER.)

PULLEY. FIXED PULLEY. The power is equal to the weight. The pressure Q on the axis is to the power or weight as the chord c of the arc enveloped by the rope is to the radius r of the pulley. $P = w = \dfrac{Qr}{c}.$

MOVABLE PULLEY.—The power is to the weight, as the radius of the pulley is to the chord of the arc enveloped by the rope. The pressure on the fixed end of the rope is equal to the power: $P = Q = \dfrac{wr}{c}$.

In a system of n movable pulleys, the power is to the weight, as the product of the radii of the pulleys is to the product of the chords of the arcs enveloped by the rope: $P = w \dfrac{r\, r'\, r''\, ..\, n}{c\, c'\, c''\, ..\, n}$.

If the ropes are parallel, $c = 2\,r$, and $P = \dfrac{w}{2n}$.

PUNISHMENT. It is often necessary to punish to maintain discipline, and the Rules and Articles of War provide ample means of punishment, but not sufficient rewards and guards against errors of judgment. In the French army degrading punishments are illegal, but soldiers may be confined to quarters or deprived of the liberty of leaving the garrison; confined in the guard-room, in prison, or in dungeon; required to walk or to perform hard labor; and officers may be subjected to simple or rigorous arrests. Every officer who inflicts a punishment, must account for it to his superior, who approves or disapproves, confirms, augments, or diminishes it. If an inferior is confined to the guard-room, he cannot be liberated except upon application to a superior. An officer who has been subjected to punishment, must, when relieved, make a visit to him who ordered it. The French code has, in a word, been careful to provide for both the security of its citizens, and the strength of authority. The punishments established by law or custom for U. S. soldiers by sentence of court-martial, according to the offence, and the jurisdiction of the court, are: death; stripes for desertion only; confinement; hard labor; ball and chain; forfeiture of pay and allowances; and dishonorable discharge from service, with or without marking. It is regarded as inhuman to punish by solitary confinement, or confinement on bread and water exceeding 14 days at a time, or for more than 84 days in a year at intervals of 14 days.

PURCHASING—from any soldier his arms, uniform, clothing, or any part thereof, may be punished by any civil court having cognizance of the same by fine in any sum not exceeding three hundred dollars, or by imprisonment not exceeding one year; (*Act* March 16, 1802.)

PURVEYOR. A person employed to make purchases, or to provide food, medicines, and necessaries for the sick.

PYRAMID. A pyramid is a solid whose base is any right-lined plane figure, and its sides are triangles having all their vertices or tops meeting together in one point, called the vertex of the pyramid.

PYROTECHNY. Artificial fire-works and fire-arms, including not only those used in war, such as cannon, shells, grenades, gunpowder, wildfire, &c.; but also those intended for amusement, as rockets, St. Catherine's wheels, &c.

Q

QUARRELS. All officers of what condition soever have power to part and quell all quarrels, frays, and disorders, though the persons concerned should belong to another regiment, troop, or company, and either to order officers in arrest, or non-commissioned officers or soldiers into confinement, until their proper superior officers shall be acquainted therewith; and whosoever shall refuse to obey such officer, (though of an inferior rank, and of a different regiment, troop, or company,) or shall draw his sword upon him, shall be punished at the discretion of a general court-martial; (ART. 27.)

QUARTERMASTER'S DEPARTMENT. (See ARMY for its organization.) This department provides the quarters and transportation of the army, except that, when practicable, wagons and their equipment are provided by the Ordnance Department; storage and transportation for all army supplies; army clothing; camp and garrison equipage; cavalry and artillery horses; fuel; forage; straw and stationery. The incidental expenses of the army (also paid through the quartermaster's department) include per diem to extra duty men: postage on public service; the expenses of courts-martial; of the pursuit and apprehension of deserters; of the burials of officers and soldiers; of hired escorts, of expresses, interpreters, spies, and guides; of veterinary surgeons and medicines for horses; and of supplying posts with water; and, generally, the proper and authorized expenses for the movements and operations of an army not expressly assigned to any other department. (Consult *Regulations of the War Department for the Quartermaster's Department.*)

These regulations derive their validity from the following acts of Congress: " It shall be lawful for the Secretary of War to cause to be provided, in each and every year, all clothing, camp utensils and equipage, medicines and hospital stores, necessary for the troops and armies of the United States for the succeeding year, and for this purpose to make purchases, and enter or cause to be entered into all necessary contracts or obligations for effecting the same; (*Act* March 3, 1799.) The Secretary of War shall be authorized and directed to define and prescribe the species, as well as the amount of supplies to be respectively purchased by the commissary-general's and quartermaster-general's departments, and the respective duties and powers of the said departments respecting such purchases. And the secretary aforesaid is also authorized to fix and make reasonable allowances for the store rent, storage, and salary of storekeepers necessary for the safe keeping of all military stores and supplies; (*Act* March 3, 1813.) The acts of March 3, 1813, and April 24, 1816, make it also the duty of the Secretary of the War Department to prepare general regulations better defining and prescribing the duties and powers of the several officers of the quartermaster's department, and other *staff* officers; which regulations, when approved by the President of the United States, shall be respected and obeyed until altered or revoked by the same authority.

An essential element, in all services of supply, is the means of transportation; and its formation, maintenance, and management call for the exercise of unremitting intelligence and activity on the part of the quartermaster. The most important want is the carriage of provisions, to which a very large portion of all military transport must be devoted. The next in importance is the hospital transport service. (*See* AMBULANCE.) The carriage of ordnance and engineer stores requires a large number of wagons; and the conveyance of camp equipage, regimental and staff baggage, as also of reserve small-arm ammunition, is also indispensable. In most foreign armies the nucleus of a trained transport corps is maintained in times of peace, organized with especial view to its easy extension for the purposes of war, so that when a force takes the field it carries with it the means of conveying its most essential supplies; while whatever transport can be drawn from the country under occupation, whether by hire or purchase, by requisition or by seizure, can at once be united to the trained and organized corps, and brought under the influence of military order and discipline. In our own army we have in this, as in other respects, too much neglected to prepare in peace for the exigencies of war. Relying upon our financial resources, and believing that while money abounds the *matériel* of war will not be wanting, we have overlooked the necessity which exists in every branch of the military service for preliminary practice and training, in order to turn our means to good account. Transport, to be effective, must be organized and trained to a systematic performance of duty, and this cannot be the work of a day. Whatever the nature or organization of the transport, however, a quartermaster should devote his best exertions to maintaining it in a state of efficiency. The men, whether soldiers or natives of the scene of operations, should be as much as possible encouraged to attach themselves to the service. Exposed, as they necessarily are, to so many fatigues and hardships in all weathers, they should be suitably clothed and well fed, and be rendered as comfortable when off duty as circumstances may allow. In the case of native drivers, their peculiar habits should be consulted as far as may be practicable; and while a strict discipline shoud be maintained, and misconduct immediately and severely punished, good behavior, steadiness, and attention to duty should be noticed and rewarded. *Esprit du corps* is to masses of men what self-respect is to individuals, and should be fostered by all possible means, since it tends to impress men in every position with a sense of their duty. A quartermaster, who fully understands the importance of his functions, will not find it unworthy of his attention to study the character and disposition of the most humble individual under his orders, with the view of developing his good qualities and abilities to the greatest advantage of the public service. In dealing with people of different nations this becomes peculiarly necessary, and as a large portion of the *personnel* belonging to the transport of armies is generally drawn from the local population, care should be taken not to offend unnecessarily feelings or even prejudices which, if properly directed, may be used to our advantage.

Another error to be avoided is unnecessary interference in the attempt to improve indiscriminately upon local practices and habits. Both men and animals will work best in the way they have been accustomed to, and even the most obvious improvements should be effected gradually and cautiously, lest in endeavoring to teach a new method before the old has been unlearnt, only the worst features of each should be the result. As a rule the practice in force, however opposed to our notions, is founded upon some sufficiently valid reasons. In this respect we have generally more to learn than to teach, and a little careful observation will probably serve to convince us that practices which at first sight we are disposed to deride or condemn are, under the peculiar circumstances of the case, preferable to any thing we could substitute.

But while unnecessary interference is to be deprecated, the importance of attending to the conditions of transport animals cannot be too strongly insisted upon. A quartermaster in charge should satisfy himself by frequent personal inspection that the animals are properly stabled, fed, cleaned, and shod; the state of saddlery and harness should be carefully attended to, and on the march no halt should be made without the wagons being examined, and, if necessary, repaired. The break-down of a single wagon may, on a narrow road, seriously obstruct the whole line of march, besides causing the loss of its load. Every cart or wagon should be required to carry the necessary tools for effecting repairs, as also the means of greasing the wheels, by which the draught is greatly diminished, and much wear and tear saved.

These are trivial details, but nothing is unimportant that tends to maintain the efficiency of army transport.

In loading, the greatest care should be taken to adapt the weight to the capability of the animal or vehicle, and full allowance must be made for the chances of heavy roads or forced marches. Mules, which for mountainous roads are by far the best pack animals, can carry continuously 2 cwt. for long marches; they are moreover more hardy and less dainty in their food than horses, and, with common care, can withstand any weather. Mules also work well in draught when no great speed is required; but whenever supplies are expected to keep up with cavalry or artillery, light wagons with two horses are preferable to any other kind of transport. A good horse should, over even roads, be able to draw 10 cwt., vehicle included; but over mountainous or heavy roads 12 cwt. (including the carriage) is more than a full load for a pair of horses. For the baggage and supplies required to accompany armies *en masse* on their ordinary marches, common country wagons drawn by oxen do excellent service; they are slow, but can carry large loads, and the beasts get through a great deal of work upon small quantities of food. A well-organized train of pack animals, though a greater number is requisite than would suffice for draught, is the most manageable transport that can be devised, and for rapid marches far preferable to any other.

The transport required for carriage of the ordinary material of war, and for hospital purposes, can always be computed with tolerable accuracy, since its extent is little affected by local circumstances. But it is different as regards consumable stores. In a country rich in resources, and with a friendly population, a small train suffices even for continuous marches; but if the scene of operations yield little or nothing, if the progress of the army be through a wilderness or a desert of ruined fields and burning villages, it would be necessary to provide transport for the carriage of provisions and forage, and perhaps even wood and water, for the full number of days that the march is calculated to last. The quartermaster must in these cases exercise his own judgment, in concert with the officer commanding the expedition.

It must be borne in mind that every additional transport animal calls for a corresponding addition of supplies. It was computed, during the organization of the British Land Transport in the Crimea, that it would require about 9,000 men and 12,500 animals to carry the rations, ammunition, and hospital establishments for 58,000 men and 30,000 horses for three days. At this rate, additional provision would require to be made for one-third as much forage and one-fifth as many rations as may be requisite for the actual combatant force in order to subsist the transport establishment. In other words, every three horses would have to be calculated as four, and every five soldiers as six, to cover the additional demands of the transport attached to the force. (*See* TRANSPORTATION.)

In most foreign armies, ships of war are as much as possible used for the transport of troops; and although the presence of soldiers may, to a certain extent, interfere with the economy and discipline of a vessel—this objection, particularly in time of peace, is not so powerful as to justify the employment, at a large cost, of private ships, while numbers of our own are making objectless cruises over all the oceans of the globe or lying idle in harbor. A naval officer very naturally dislikes to be encumbered with some hundreds of soldiers with their wives and children, or to have a number of idle officers lounging about his quarter-deck; but there are interests to be consulted beyond even the most praiseworthy professional *amour propre*, and it ought to be considered whether economy and good policy do not require that a more frequent use should be made of ships of war as transports, and also whether general regulations might not be adopted for the transportation of the articles of supply from the places of purchase to the several armies, garrisons, posts, and recruiting places, and for the safe keeping of such articles, and for the distribution of an adequate and timely supply of the same to the regimental quartermasters, and such other officers as may, by virtue of such regulations, be intrusted with the same. (*See* ADMINISTRATION; ARMY REGULATIONS; CAMP; CLOTHING; SUPPLIES; TRAIN; WAGON. Consult FONBLANQUE.)

QUARTERMASTER-GENERAL—has the rank, pay, and emoluments of brigadier-general. He is not liable for any money or property that may come into the hands of subordinate agents of the department; (*Act* May 22, 1812.) He accounts as often as required, and at least once in three months, with the Department of War, in such manner as shall be prescribed, for all property which may pass through his hands, or the hands of the subordinate officers in his department, or that may be in his or their possession, and for all moneys which he or they may expend in discharging their respective duties; he shall be responsible for the regularity and correctness of all returns in his department, and he, his deputies, and assistant deputies, before they enter on the execution of their respective offices, shall severally take an oath *faithfully to perform the duties thereof*; (*Act* March 28, 1812.) The quartermaster-general is authorized to frank and receive letters and packets by post, free of postage; (*Act* March 2, 1827.)

Each quartermaster-general attached to any separate army, command, or district shall be authorized, with the approbation and under the direction of the Secretary of War, to employ as many artificers, mechanics, and laborers as the public service may require; (*Act* March 3, 1813.)

QUARTERS. "No soldier shall, in time of peace, be quartered in any house, without the consent of the owner; nor in time of war, but in a manner to be prescribed by law;" (*Constitution*, 3d *Amendment.*) The law not having made any provision for quartering soldiers in time of war, troops of the United States at home would be subjected to exorbitant demands for the hire of quarters. (*See* BARRACKS; BILLETS.)

QUESTIONS. (*See* EVIDENCE; TRIAL.)

QUICK-MATCH. It is made of threads of cotton or cotton wick, steeped in gummed brandy or whiskey, then soaked in a paste of mealed powder and gummed spirits, and afterwards strewn over with mealed powder. It is used to fire stone and heavy mortars, in priming all kinds of fireworks, such as fire-balls, light-balls, carcasses, priming tubes, &c. A yard burns in the open air in 13 seconds.

QUOINS. In gunnery, a quoin is a wedge used to lay under the breech of a gun to elevate or depress it.

R

RACK-STICK AND LASHING—consist of a piece of two-inch rope, about 6 feet long, fastened to a picket about 15 inches long, having a hole in its head to receive the rope. Rack-lashings are used for securing the planks of a gun or mortar platform, between the ribbons and the sleepers.

RAFT. (*See* BRIDGE.)

RAFTERS. (*See* CARPENTRY.)

RAISE. To raise a siege is to abandon a siege. Armies are *raised* in two ways: either by voluntary engagements, or by lot or conscription. The Greek and Roman levies were the result of a rigid system of conscription. The Visigoths practised a general conscription; poverty, old age, and sickness were the only reasons admitted for *exemption.* "Subsequently, (says Hallam,) the feudal military tenures had superseded that earlier system of public defence, which called upon every man, and especially upon every landholder, to protect his country. The relations of a vassal came in place of those of a subject and a citizen. This was the revolution of the 9th century. In the 12th and 13th another innovation rather more gradually prevailed, and marks the third period in the military history of Europe. Mercenary troops were substituted for the feudal militia. These military adventurers played a more remarkable part in Italy than in France, though not a little troublesome to the latter country." A necessary effect of the formation of mercenaries was the centralization of authority. Money became the sinews of war. The invention of fire-arms caused it to be acknowledged that skill was no less essential for warlike operations than strength and valor. Towards the end of the middle ages, the power of princes was calculated by the number and quality of paid troops they could support. France first set the example of keeping troops in time of peace. Charles VII., foreseeing the danger of invasion, authorized the assemblage of armed mercenaries called *compagnies d'ordonnance*. Louis XI. dismissed these troops, but enrolled new troops composed of French, Swiss, and Scotch. Under Charles VIII., Germans were admitted in the French army, and the highest and most illustrious nobles of France regarded it as an honor to serve in the *gens d'armes*.

Moral qualifications not being exacted for admission to the ranks, the restraints of a barbarous discipline became necessary, and this discipline divided widely the soldier from the people. The French revolution overturned this system. "Now (says Decker) mercenary troops have completely disappeared from continental Europe. England only now raises armies by the system of *recruiters*. The last wars of Europe have been wars of the people, and have been fought by nationalities. After peace armies remain national, for their elements are taken from the people, and are returned to the people by legal liberations. The institution of conscription is evidently the most important of modern times. Among other advantages, it has bridged the otherwise impassable gulf between the citizen and soldier, who, children of the same family, are now united in defence of their country. Permanent armies have ceased to be the personal guard of kings, but their sympathies are always with the people, and their just title is that of skilful warriors maintained as a nucleus for the instruction of their countrymen in the highest school of art. (*See* CONSCRIPTION; DEPOT; DEFENCE, *National*; ENLISTMENT; MILITIA; RECRUITING; RE-ENLISTING; VOLUNTEERS.)

RALLY. To re-form disordered or dispersed troops.

RAMP. A ramp is a road cut obliquely into or added to the interior slope of the rampart, as a communication from the town to the terre-plein.

RAMPART. A broad embankment or mass of earth which surrounds a fortified place, and forms the *enceinte* or body of the place. On its exterior edge the parapet is placed, while towards the place it is terminated by the interior slope of the rampart, on which *ramps* are made for the easy ascent of the troops and material.

RAMROD. The rod of iron used in loading a piece to drive home the charge.

RANGES. The extreme ranges of smooth-bored guns firing solid shot may be considered to vary, according to their size, from 2,000 to 4,000 yards. These great ranges are only attained by firing at great elevations, and the practice at such distances is consequently uncertain. Ranges of 1,000 to 1,200 yards for field-guns and of 1,500 to 2,000 yards for heavy guns are as great as can be secured with any thing like accuracy. It seems, however, more than probable, that smooth-bored guns will, before long, be altogether superseded by rifled ordnance, and reasoning from what has been already accomplished, we may at least expect to double the present ranges, and greatly to increase the accuracy of fire. The ranges of grape-shot are equal only to the ranges of the individual balls of which the grape-shot is composed; they are, therefore, subject to considerable variation, according to the dimensions of the gun from which the grape is discharged. The most effective ranges for grape-shot may be considered to lie between 300 and 600 yards. The range of canister-shot is very limited. From the small size of the bullets they rapidly lose their initial velocity. At ranges below 300 yards canister-shot against bodies of troops is very destructive. Spherical-case shot is effective at much greater ranges than canister or grape shot. It may be employed with good effect at any distance between 600 and 1,500 or even 1,800 yards. The ranges of shells vary according to their size from 1,000 to 4,000 yards. They are fired either from mortars or guns. With the method of firing them from mortars at an elevation of 45°, with a charge of powder proportioned to the range desired, any great accuracy of practice is not to be expected. (*See* ARTILLERY; COLUMBIAD; FIRING; RIFLED ORDNANCE; SPHERICAL CASE.)

RANK. A range of subordination; a degree of dignity. *Rank* also means a line of soldiers, side by side. *Ranks* in the *plural*, the order of common soldiers. Questions as to the positive or relative rank of officers may often be of the greatest importance at law, in consequence of the rule, that every person who justifies his own acts on the ground of obedience to superior authority must establish, by clear evidence, the sufficiency of the authority on which he so relies. There may also be many occasions on which the propriety of an officer's assumption of command, or his exercise of particular functions, or his right to share with a particular class of officers in prize-money, bounties, grants, and other allowances, may depend on the correctness of the view taken by himself or others of his right to a specific rank or command; and an error in this respect may expose him to personal loss and damage in suits before the civil tribunals.

The regulation of military rank is vested absolutely in Congress, which confers or varies it at pleasure. The will of Congress in this respect is signified by the creation of different grades of rank; by making rules of appointment and promotion; by other rules of government and regulation; or is by fair deduction to be inferred from the nature of the functions assigned to each officer; for every man who is intrusted with an employment, is presumed to be invested with all the powers necessary for the effective discharge of the duties annexed to his office.

Rank and Grade are synonymous, and in their military acceptation indicate rights, powers, and duties determined by laws creating the different degrees of rank, and specifying fixed forms for passing from grade to grade; and when rank in one body shall give command in another body; and also when rank in the army at large shall not be exercised. Rank is a right of which an officer cannot be deprived, except through forms prescribed by law. When an officer is on Duty, his rank itself indicates his relative position to other officers of the body in which it is created. It is not, however, a perpetual right to exercise command, because the President may, under the 62d Article of War, at any time relieve an officer from duty; or an officer may be so relieved by arrest duly made according to law; or by inability to perform duty from sickness, or by being placed by competent authority on some other duty. But whenever an officer is on duty his rank indicates his command.

During the Mexican war, an attempt was made to procure the passage of a law creating the rank of lieutenant-general, in order that Mr. Senator Benton might be placed in command of the army with that rank. Congress, however, refused to create the rank. The President then sought to obtain the passage of a law authorizing him to put a junior major-general in command of a senior. Congress likewise refused him that power. On the 9th of March, Mr. President Polk, in a letter to Mr. Senator Benton, thus writes:

"Immediately after your nomination as major-general had been unanimously confirmed by the Senate, I carefully examined the question, whether I possessed the power to designate you, a junior major-general, to the chief command of the army in the field. The result of the examination is, I am constrained to say, a settled conviction in my mind, that such power has not been conferred on me by existing laws."

Struggle as commentators may, who desire to subject rank to executive caprice, rather than have its powers and duties defined by law, as the constitution requires in giving to CONGRESS the power to make rules for the government and *regulation* of the army, the rights of rank cannot, without usurpation, be varied at the will of the President. The law has created rank. Rank means a range of subordination in the particular body in which it is created. It is, therefore, effective in that body, without further legislation, and its effect, when the officer is present for duty, is extended beyond that particular portion of the army in which the officer holds rank, or its exercise is restricted within a corps only by legislation. Executive authority cannot make rank vary at will, but whatever authority the executive has over rank must be determined by law. A reference to the 62d Article of War will show that the President is given the authority to limit the discretion of commanding officers, in special cases, in respect to what is needful for the service, and also to relieve the senior officer from any command, so that the command may fall upon the next officer in the line of the army, marine corps, or militia, " by commission there on duty or in quarters," or assign some senior to duty with troops, in order that such officer may become entitled to command under the 62d Article of War. Any power of *assignment* claimed for the President beyond this is not and ought not to be sanctioned by law. The 62d Article extends the validity of commissions in any part of the line of the army, marine corps, or militia, and thus enables the senior officer of the line of the army present for duty to command the whole when different corps come together—while the 61st Article provides that in the regiment, troop, or company, to which officers belong, although they may also hold higher commissions in the army at large, they shall nevertheless do duty and take rank both in courts-martial and on detachments, which shall be composed only of their own corps, according to the commissions by which they are mustered in said corps.

The legislation on the subject of rank is thus complete. Officers, when serving only with their own regiment, serve according to their regimental rank; but when with other corps, the senior by commission in the line, whether by brevet or otherwise, is entitled to command. (*See* ASSIGNMENT.)

RASANTE—is a French term, applied to a style of fortification, in which the command of the works over each other, and over the country, is kept very low, in order that the shot may more effectually sweep or graze the ground before them.

RATCHET-WHEEL. A wheel with pointed and angular teeth, against which a ratchet abuts, used either for converting a reciprocating into a rotatory motion on the shaft to which it is fixed, or for admitting of its motion in one direction only.

RATION. The President may make such alterations in the component parts of the ration as a due regard to the health and comfort of the army and economy may require; (*Act* April 24, 1818.) The allowance of sugar and coffee to the non-commissioned officers, musicians, and privates, in lieu of the spirit or whiskey component part of the ration, shall be fixed at six pounds of coffee and twelve pounds of sugar to every one hundred rations, to be issued weekly, when it can be done with convenience to the public service, and when not so issued, to be paid for in money; (*Act* July 5, 1838.)

Women not exceeding four to a company, and such matrons and nurses as may be necessarily employed in the hospital, one ration each; (*Act* March 16, 1802.) The President may authorize rations to be issued to Indians visiting military posts; (*Act* May 13, 1800.) (*See* Pay; Wagon.)

TABLE, SHOWING THE WEIGHT AND BULK OF 1,000 ARMY RATIONS.

One thousand rations of	Nett weight in pounds.	Gross weight in pounds.	Bulk in barrels.	100 rations consist of
Pork	750	1,218.75	3.75	75 lbs., or
Bacon	750	903.19	4.90	75 lbs.
Flour	1,125	1,234.06	5.74	112.5 lbs., or
Pilot Bread	750	921.69	9.03	75 lbs., or
"	1,000	1,228.91	12.05	100 lbs. In the field.
Beans	155	177.32	0.71	8 quarts, or
Rice	100	114.50	0.46	10 lbs.
Coffee	60	70.90	0.35	6 lbs.
Sugar	120	135.62	0.50	12 lbs.
Vinegar	92.5	107.50	0.83	4 quarts.
Candles	15	17.50	0.09	1¼ lbs.
Soap	40	46.89	0.19	4 lbs.
Salt	33.75	38.63	0.16	2 quarts.

Forage.

14 lbs. hay or fodder, } per horse { When pressed 11 lbs. to cubic foot.
12 qts. oats, or } per day. { 40 lbs. to bush., 33.14 lbs. cub. foot.
8 qts. corn } { 55 lbs. to bush., 45.65 lbs. cub. foot.

Average mule pack, New Mexico, 175 lbs.

Average load to mule team across the prairies, 2,000 lbs.

RAVELIN—is the work constructed beyond the main ditch, opposite the curtain, composed of two faces, forming a salient angle, and two demi-gorges, formed by the counterscarp. It is separated from the covered way by a ditch which runs into the main ditch.

RAVELIN, (REDOUBT OF THE)—is a work constructed within the ravelin, but separated from it by a ditch.

RAZED. Works or fortifications are said to be razed, when they are totally demolished.

READINESS. A state of alertness or preparation; thus, to *hold a corps in readiness*, is to have it prepared in consequence of some previous order to march at a moment's notice.

REAR, REAR RANK. The hinder rank.

REAR GUARD. A detachment of troops in the rear of an army.

RECEIPT. A voucher or acknowledgment, which should always be given when official papers are received. When flags of truce are the bearers of a parcel or a letter, the officer commanding at an outpost should give a receipt for it, and require the party to depart forthwith.

RECOIL. The motion which a cannon takes backward when fired.

RECOMMENDATIONS. All members of a court who concur in recommendations to mercy sign. The recommendation is introduced after the finding and sentence are closed and authenticated. The recommendation should distinctly set forth the reasons which prompt it; (HOUGH.)

RECOMPENSE. (*See* ALLOWANCE; GRATIFICATION; INDEMNITY; PAY.)

RECONNOISSANCE, RECONNOITRE, RECONNOITRING, —may be distinguished into reconnoissance of the enemy, and topographical reconnoissances.

Reconnoissances are warlike operations for the purpose of procuring information of the positions and strength of corps of the enemy. Without such knowledge, no well-concerted measures of attack or defence can be made. First of all, notes of information are gained from spies, deserters, and travellers, and the position of the different corps of the enemy is marked out upon a good map. But when the opposing armies are more nearly approximated, it becomes necessary to ascertain, every day, what changes and movements have taken place, whether for purposes of concentration or withdrawal to other points. Reconnoissances by force result from this necessity, and lead sometimes to bloody actions.

The custom is almost universal to cover an army by outposts, and to detach clouds of light troops to mask the camp and prevent an enemy from seeing what dispositions are made for attack or defence. To gain information, it is therefore necessary to *push* a reconnoissance through the curtains of light troops, by which the enemy has enveloped himself, and drive back or cut off outposts, so as to enable the officer charged with the reconnoissance clearly to see the *army* of the enemy, note the advantages and disadvantages of his positions, count his battalions, and judge of his means of resistance: whether he is intrenched, what artillery he has; whether the ground is or is not favorable for cavalry; where the cavalry is encamped, &c. These different objects ought to be seen rapidly and by a practised eye, for the reconnoissance will have called to arms a greatly superior force, and it is necessary as soon as possible to fall back. But the aim will have been attained, for the enemy having been compelled to unmask and deploy his forces, the reconnoitring officer will know all that he desires, and consequently hastens his return to camp, in order that his party may not be exposed to have its retreat cut off.

Similar reconnoissances ordinarily precede battles. By their means a general is assured of the true state of the enemy, before giving his last orders. On a march, the advance guard reconnoitres the enemy. Sometimes a reconnoissance has for its object to discover if a point is solidly occupied; if a bridge over which an army is to pass has been broken; whether a defile is fortified; whether the enemy has guns in any particular position; whether he is in a certain city, or whether he has followed such and such routes after losing a battle, &c., &c. Such reconnoissances are often made by small parties of cavalry alone to ensure rapidity; but if resistance is anticipated or foreseen, the party must consist of all arms, or be constituted according to circumstances, and the command be given to an experienced officer.

The commander of a reconnoissance ordinarily receives written instructions. He should well understand the object before him, and demand such explanations as he may require. He is furnished with a good map, a telescope, writing materials, and means of making field-sketches of the positions of the enemy. He secures two or three inhabitants of the country to serve as guides, and to answer his inquiries relative to the names and populations of villages, the nature of the roads, the extent of woods, the condition of water-courses, ground, &c. He ought to be accompanied by an officer who knows the language of the country, and he should, before commencing his march, inspect the troops intrusted to him to satisfy himself of the good condition of their arms, ammunition, and provisions.

The detachment charged with pushing a reconnoissance marches with its advance guard and flankers; stops all persons who would precede it, and might give information of its march; questions inhabitants of villages, and, if necessary, takes hostages to secure true information. The attention of the commander is particularly directed to the ground over which he passes, to determine, in advance, points where a stout resistance may be made in the event of his being obliged to fight when making his retreat. He frequently consults his map to ascertain its fidelity to the country over which he passes, and notes its variations. The detachment pushes forward, using all necessary precautions, without fear of compromising itself, attacking boldly such antagonists as present themselves, until the information has been gained for which it was despatched.

There are other reconnoissances made by small detachments, which employ stratagem rather than force, and which consequently ought to shun any engagement that can be avoided. In strong reconnoissances or reconnoissances by force, on the contrary, the aim is to penetrate to the positions of the enemy, and the design must not be permitted to fail by an accidental meeting with troops; but, profiting by such good fortune, the opposing troops must be overthrown, prisoners made who will give useful information, and the fugitives rapidly followed to the outposts, which will probably be in confusion at the repulse of the detachment. The line of the enemy is then soon pierced, and his corps will be soon seen deployed to repulse the attack. The commandant of the reconnoissance ought now to seek some elevated point from which he can gain a good knowledge of the force and positions of the enemy, and make, or have made by officers who acccompany him, a rapid sketch of the ground and the positions of the enemy. When once this object has been gained, a retreat must be sounded even in the middle of the combat. And it is under such circumstances that skill and prudence guide courage; and *sang-froid* is absolutely indispensable. The object of the reconnoissance is to gain information. Boldness must be employed to attain that end; but, if in the hope of surprising a post, carrying off a convoy, or destroying troops, the commander forsakes his route and loses time, it is a violation of duty; he is blamable, even if success attends his enterprise.

Secret reconnoissances are conducted on different principles. They are ordinarily composed of a single kind of troops; of cavalry in flat, open districts, and of infantry in mountainous or intersected countries. The detachment marches with caution. If the *eclaireurs* announce the approach of an enemy, it endeavors to avoid observation by the shelter furnished by woods or any accident of ground at hand; or else escaping by a prompt retreat if necessary; or, if near its own outposts, and the enemy is in strength, sending back information, and retarding the column of the enemy as much as possible, by simulating strength.

When the commandant of such a reconnoitring party has reached his destination without hindrance, he holds his men concealed behind some curtain, such as a clump of trees, an old wall or ditch, and followed only by a few men in echelons, he takes some elevated position with his guide and two or three soldiers, whence he can observe the enemy. He notes what he sees, with the explanations of his guide. If the positions of the enemy are well seen, he makes sketches, which are always valuable even when very rough. He must not be imposed on by first appearances, but examining with *sang-froid*, he endeavors to seize exact ideas, and exposes himself when necessary to attain his aim. Inexact knowledge or lies are worse than total ignorance. Montluc well says that discretion must be exercised in selections for such expeditions, for an inexperienced man may soon take alarm, and even imagine "bushes to be battalions of the enemy." Send always some fearless and skilful officer, and if you would do better go yourself.

When the reconnoissance is finished, the commanding officer makes a written report to the general, when his verbal account is not sufficient. This report ought to be clear, simple, and as brief as possible. The officer will state only facts of which he is perfectly sure. His conjectures will be presented with great reserve, and always as conjectures. He will guard against flights of imagination, but confine himself to realities, and will avoid speaking much of himself; but, knowing the satisfactory result of his mission must do him honor, he will bestow just praise upon his troops. (*See* SURVEY, *Military*.)

There are many *signs* which, if reported to a general and his staff, enable them to judge of what they wish to know, as clearly as if a detailed picture of the enemy were spread before them. It is necessary, therefore, that every officer and soldier should know how to mark and collect these signs. They consist, when a camp, bivouac, or cantonment is observed, in the color of coats and pantaloons; other distinctive marks, the numbers of videttes, sentinels, fires, and tents of the enemy; the frequency and direction of rounds, patrols, and reconnoissances; the nature and time of signals by trumpet or drum; the placing of signal posts; measures of straw; boughs broken off; the arrival of reinforcements; new uniforms; collections of fascines, beams, joists, ladders, boats. When a corps is watched on the march, the signs to observe are the depth and front of columns; the number of subdivisions; the sort of troops, infantry, cavalry, artillery, trains; the quickness and direction of the march; the height of the dust; the reflection of arms; the number of the flankers and the eclaireurs. When an army ready for battle is observed, we should particularly note the number of its lines, their extent, the composition of the troops in column or in line of battle; the calibre of pieces; their position relative to cavalry and infantry; the number of skirmishers; their manœuvres; the concentration of forces or artillery on such a point; flank marches of one or many corps. If troops are followed on their march, we note the tracks of men and horses, those made by wheels, cattle, and beasts of burden; the relative positions of these tracks: whether they are regular and preserve an invariable order; whether the places where they stop have little or much space between them; whether the route passed over is covered with remains of animals; whether the skeletons of the horses are lean and sore; whether the ground is bloody; if graves have been freshly made, whether some indications may not show them to be for superior officers; whether the country has been devastated; whether the entrails of beef, mutton, or horses are seen; whether the fires are recent; whether they are numerous, and show much or little ashes; whether bridges are broken, and in what parts; whether the inhabitants of the country are anxious, sad, humble, animated, or satisfied.

Topographical reconnoissances are not less important than reconnoissances of the enemy. It is necessary to know the distances of places to combine the march of different columns, and without a knowledge of the difficulties of a route, necessary measures to overcome them cannot be prescribed. It is by special reconnoissances that such knowledge is gained, for maps are never sufficient. They do not give the nature of the soil, the quality of the roads, the condition of rivers or bridges, the thickness of forests, or the slope of mountains, &c., &c., but it is necessary to know all these things before undertaking any important enterprise. If this detailed information has not been collected in time of peace through special corps, officers of the staff, in presence of the enemy, and protected by troops, commonly make sketches, representing more or less exactly the most essential localities. Those officers, also, on the march of an army, make out itineraries, survey positions, fields of battle, and not unfrequently great extents of country.

Officers of all arms, however, are liable to be placed in situations which require them to explore localities and give correct descriptions. The following means may be employed for that purpose without becoming an expert in the art of drawing. The system of showing upon plans the levels of the ground by means of contour lines is one of some utility, but it is the most difficult representation in a topographical map. The art is only acquired by study and practice, and even with skill there is not always time for its display in the field. Instead of attempting lines to represent slopes, the contour of hills may be marked by two curves, one for the top and one for the foot of the slope, and these contour lines naturally present themselves to the eye, and are at once put upon paper, to indicate the general form of the hill. The space between these two lines is sufficient to write a few words indicating the slope, &c. Whether, for instance, the slope is gentle or steep, accessible or not to cavalry, its approximative height. In order that the lines of *circumscription* representing heights may not be confounded with other conventional signs, they must be long dots. Ciphers in parenthesis give the heights of points of the superior curve above corresponding points of the inferior curve.

Other objects, as water-courses, ponds, marshes, woods, vines, towns, villages, large farms, and other isolated constructions which may play an important part in battle, embankments, ferries, fords, stone and wooden bridges, all may be represented as in Fig. 174.

Water-courses.—Two lines, one heavier than the other, are sufficient to represent them. It is usual to add other lines between the two first.

FIG. 174.

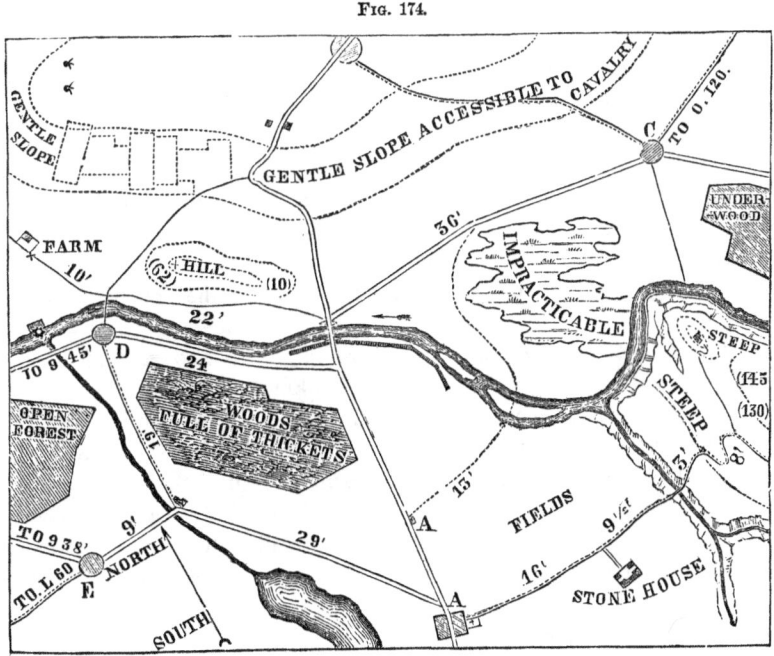

Sometimes a blue shade advantageously takes the place of the intermediate threads. An arrow indicates the direction of the current. A mill is seen in the lower part of the river. Smaller streams empty into the river.

Means of crossing.—A ferry boat. A stone bridge, distinguished from a wooden bridge by being wider and having wings on the opposite banks. A ford, marked by dotted lines across the river.

Ponds or lakes are designated by lines of contour, and by threads or a blue tint.

Marshes.—By a line of contour, and horizontal lines in the interior, with some points representing grass in the interior. *Practicable* or *impracticable*, &c., is written.

Woods and vines.—These objects are designated by tracing the contour. If colors are used India ink will designate woods, and violet vines. Write, in the interior, the nature and characteristic circumstances of the wood; whether it is undergrowth or forest, thickset or open, &c.

Rocks.—Endeavor to imitate them, but if they present themselves in prolonged walls, the crest and foot may be designated as in the sketch. Or a few written words may give a better idea.

Habitations.—A village is represented by a circle filled with parallel lines. A town in the same manner, except that a square is substituted for the circle. A red tint may replace the parallel lines in habitations. Isolated houses are designated merely by their form, without regard to the scale.

Communications.—A great route is represented by two parallel lines. A wagon road in the same manner, except that the lines are nearer together. Roads practicable only for light carriages by the same means, except that one of the lines is dotted. Distances being essential in a plan of this kind, they must be written along the routes between the objects.

Levees and Embankments are represented by two parallel lines, with cross lines in the interior. *See* embankment near stone bridge.

The sketch is completed by a meridian line.

However rapidly such a sketch as Fig. 174 may be made, there are circumstances in which it is not possible to give that time, and a reconnoissance must be made at a gallop. In the latter case, the reconnoitring officer confines himself to taking rapid notes, and afterwards making his sketch from recollection. This is a most useful talent, and officers should be exercised in noting the prominent features of localities, and tracing their recollections upon paper. Reconnoissances are

MILITARY DICTIONARY. 495

Fig. 175.

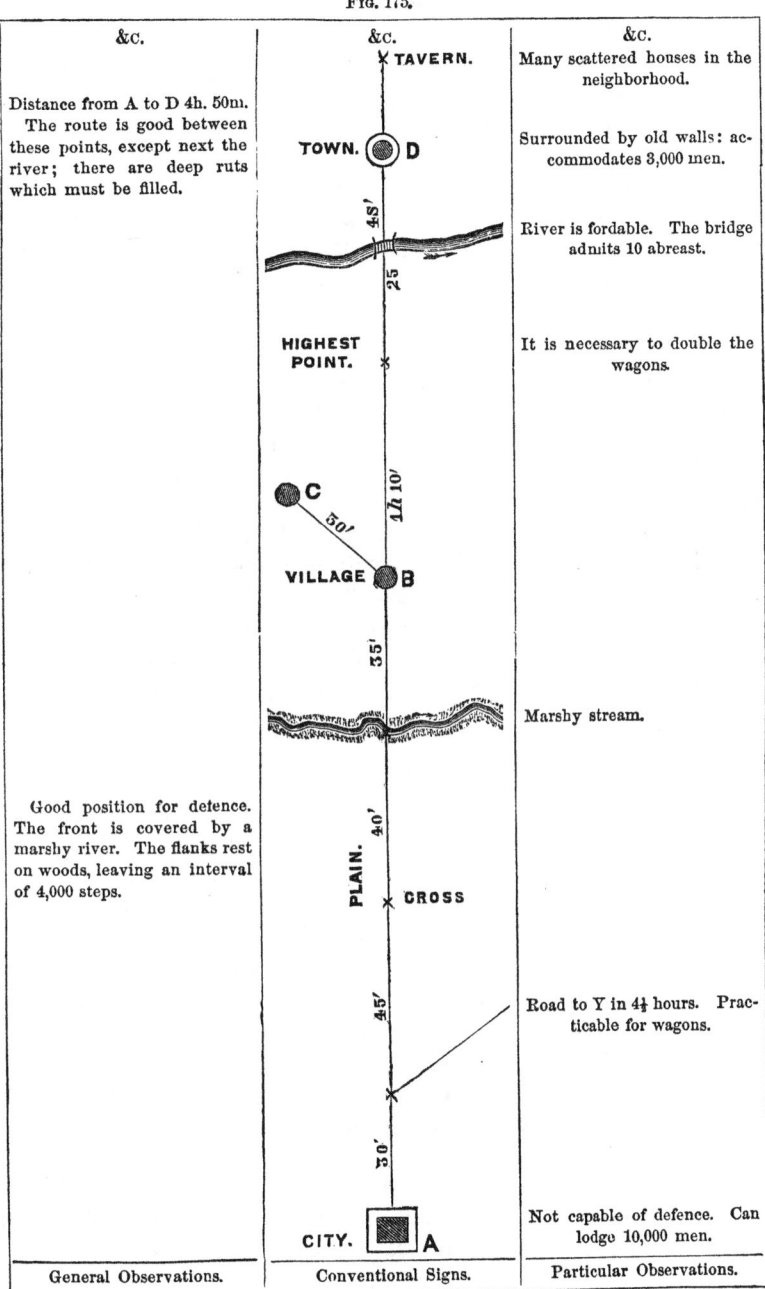

Itinerary from A to X.

much simplified when confined to noting circumstances along a route, and are then called *Itineraries.* All particularities of the route are noted, whatever is remarkable on the right or left, the breadth of defiles, military positions, the steepness of slopes, what is necessary to improve a road, the distances between points in time; covers, that is, houses of all kinds are given according to their capacity of containing soldiers, &c. In itineraries, conventional signs as well as written notes are used. Itineraries are made of leaves of paper five or six inches in breadth. Leaves are subsequently united, and represent entire routes. Notes begin at the foot of the leaf, and are continued above, as in Fig. 175. (*See* also *article* JOURNAL.)

Details concerning the resources of a country must be embodied in statistical tables. The itinerary would be too much complicated by embracing them. Such information is most important, however, in supplying an army; but statistical tables, prepared with that view, should be confined to necessary objects. They should embrace details of the population of towns, inhabited houses, workmen, mills, ovens, grain, wagons, boats, horses, mules, beef cattle, with general observations which would aid the departments of supply in the performance of their duties. (Consult DUFOUR; BUGEAUD; *Aide Memoire d'Etat Major.* See SURVEYS, *Military or Expeditious.*)

RECRUITING. The system of recruiting armies practised in England and the United States by voluntary enlistments, is vicious. In continental Europe, the obligation is acknowledged that every subject or citizen of a certain age owes military service to his country, either personally or by substitute. The government consequently annually calls for as many men as are needed for the military service. In answer to this call, lots are drawn by the whole class liable to service, and those upon whom the lot falls become soldiers for a fixed period, varying in different countries from three to eight years. The military have but little to do with such a system of recruiting. There is in France simply a council for recruiting, in each department, instituted to pronounce upon the fitness for service of those men desigated by lot. It is composed of a prefect, a commanding general, a field-officer designated by the minister of war, a councillor of the prefect, and an officer of the *gendarmerie.* Those upon whom the lot has fallen, who think that they have good reasons for being exempted, present their cases before this committee, who examine such applications, and pronounce what exemptions shall be made, and in what cases substitutes shall be admitted. With such a system of recruiting, the ranks of an army are composed of all classes of the community. Promotion from the ranks is of ordinary occurrence. The soldier has a career before him. He is proud of his profession. The army is a national army, or an army of the people. Its sympathies are all with the people, and it is ever, as in France, a true representative of the popular sentiment.

In England, where it is the policy of the government to keep the army under the control of the aristocracy, they are logical in rejecting a system of conscription, and adhering to a system of recruiting which divides an army into two castes: the officer and the soldier. What possible reason can be given for adopting that system in the United States, is unknown. (*See* DEPOT; RAISE.)

REDAN. Small work with two faces terminating in a salient angle, used to cover a camp, the front of a battle-field, advanced posts, avenues of a village, bridge, &c. Fig. 176 exhibits a bridge-head, composed

FIG. 176.

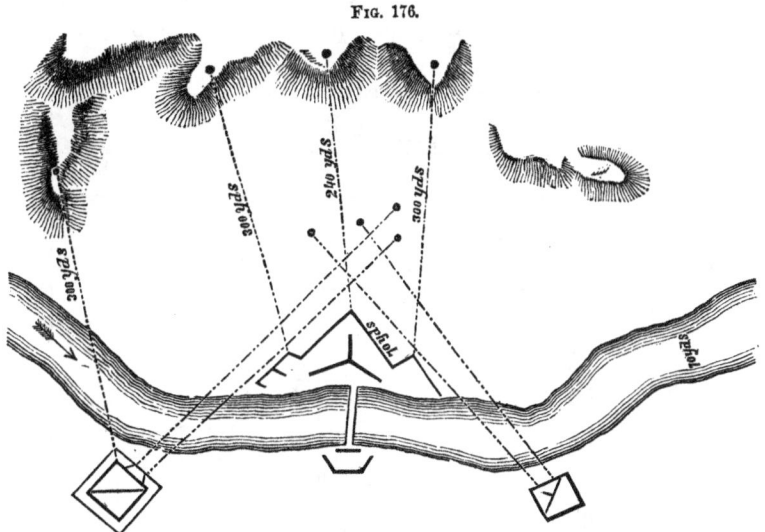

of a redan with flanks, flanked by two redoubts on the opposite bank of the river. These works are supposed to be in the neighborhood of hills, from which it is necessary that they should be defiladed. This is effected by traverses to cover the bridge, and by a traverse across the centre of each redoubt. (*See* FIELD-WORKS.)

REDOUBTS—are works inclosed on all sides of a square, polygonal, or circular figure. The latter form is rarely used, being unsuitable to ground in general, and from the impossibility of giving any flanking defence to the ditch. Redoubts on level ground are generally square or pentagonal. On a hill or rising ground their outline will, in most cases, follow the contour of the summit of the hill. Their dimensions should be proportioned to the number of men they are to contain. One file, that is, two men, are required for the defence of every lineal yard of parapet; the number of yards in the crest line of any redoubt should not, therefore, exceed half the number of men to be contained in it. Again, as every man in an inclosed work requires 10 square feet of the interior space, that space clear of the banquette must not contain less than ten times as many square feet as the number of men to be contained in it. From these considerations it follows: 1st. To find the least number of men sufficient to man the parapet of an inclosed work, multiply the number of yards in the crest line by two. 2d. To find the greatest number of men that an inclosed work can contain, find the area, clear of the banquette, in square feet, and divide this number by 10.

When the redoubt contains guns, 324 square feet must be allowed for each gun, and this quantity, multiplied by the number of guns,

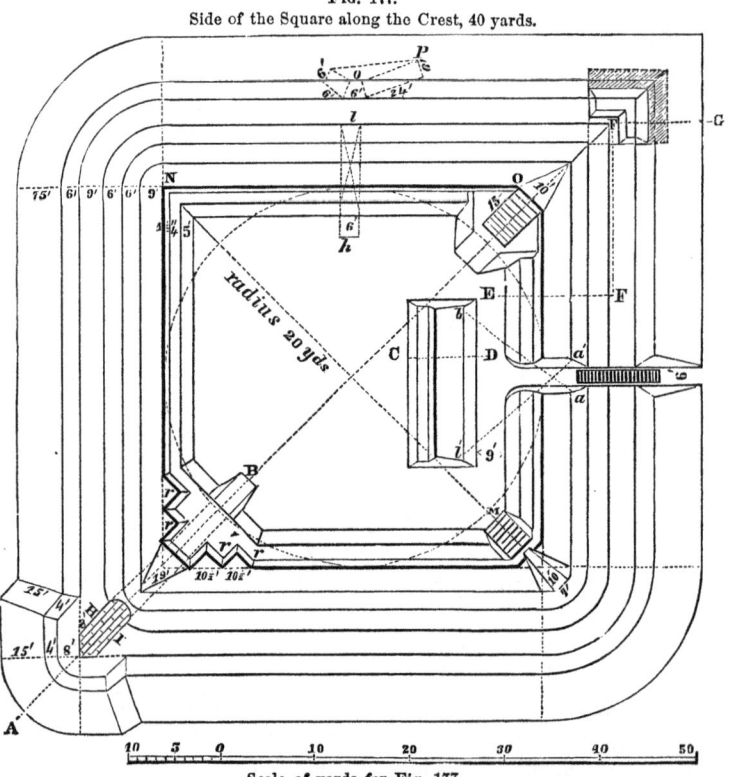

Fig. 177.
Side of the Square along the Crest, 40 yards.

Scale of yards for Fig. 177.

must be subtracted from the whole interior space. The remaining number of square feet, divided by 10, will give the number of men which the redoubt can hold. The side of a square redoubt should, under no circumstances, be less than 50 feet.

The great objections to small inclosed works are: 1st, the liability of their faces to be enfiladed from without; 2d, the difficulty of providing an effective flanking defence for their ditches; 3d, the weakness of their salient angles, the ground in front of them being undefended by a direct fire. In the preceding diagram (Fig. 177) is shown a square redoubt, having a side of 40 yards, and capable of holding four pieces of artillery, and one hundred and twenty men.

In tracing redoubts and all inclosed field works, care must be taken to direct as much as possible their faces upon inaccessible ground, so as to reduce to a minimum the effects of an enemy's enfilade, while approach on the salients must be rendered difficult by abatis, trous-de-loup, and obstacles of all available descriptions. It will henceforward be very difficult to guard the interior of inclosed works from the effects of distant musketry. Well-trained troops from a distance of 900 yards could throw with certainty every shot into the interior of even a small redoubt; while the angle at which they fall, some 15° to 20°, would enable them to sweep the whole interior and make every part of the redoubt too hot. It seems to be a question whether such a work can be protected by traverses from such a plunging fire; (HYDE's *Fortification*.) (*See* ATTACK AND DEFENCE of field-works.)

REDRESSING WRONGS. If any officer shall think himself wronged by his colonel, or the commanding officer of the regiment, and shall, upon due application being made to him, be refused redress, he may complain to the general, commanding in the State or territory where such regiment shall be stationed, in order to obtain justice; who is hereby required to examine into the said complaint, and take proper measures for redressing the wrong complained of, and transmit, as soon as possible, to the Department of War, a true state of such complaint, with the proceedings had thereon; (ART. 34.) If any inferior officer or soldier shall think himself wronged by his captain or other officer, he is to complain thereof to the commanding officer of the regiment, who is hereby required to summon a regimental court-martial for the doing of justice to the complainant; from which regimental court-martial, either party may, if he thinks himself still aggrieved, appeal to a general court-martial. But if, upon a second hearing, the appeal shall appear vexatious and groundless, the person, so appealing, shall be punished at the discretion of the said court-martial; (ART. 35.) (*See*

INJURIES, for liability for private injuries, personal injuries, and criminal liabilities; REMEDY.)

REDUCE. To *reduce a place*, is to oblige the garrison to surrender. To *reduce to the ranks*, is when a sergeant or a corporal, for any misconduct, has his rank taken from him, and is obliged to return to the duty of a private soldier. Non-commissioned officers cannot be reduced to the ranks except by the sentence of a court-martial, or by the order of the colonel of the regiment.

RE-ENLISTING. Every able-bodied non-commissioned officer, musician, or private soldier, who may re-enlist into his company or regiment, within two months before or one month after the expiration of his term of service, shall receive three months' extra pay; (*Act* July 5, 1838.) (*See* ENLISTMENT.)

RE-ENTERING ANGLE—is an angle pointing inwards, or towards the work.

RE-ENTERING ANGLE OF THE COUNTERSCARP—is that formed by the intersection of the two lines of the counterscarp, opposite the curtain.

REFUSAL TO RECEIVE PRISONERS. No officer commanding a guard, or provost-marshal, shall refuse to receive or keep any prisoner committed to his charge by an officer belonging to the force of the United States; provided the officer committing shall, at the same time, deliver an account in writing, signed by himself, of the crime with which the prisoner is charged; (ART. 80.) No officer commanding a guard, or provost-marshal, shall release any prisoner committed to his charge without proper authority for so doing, nor shall he suffer any person to escape on penalty of being punished at the discretion of a court-martial; (ART. 81.) Every officer or provost-marshal to whose charge prisoners are committed, shall, within twenty-four hours after such commitment or as soon as he shall be relieved from guard, report in writing to the commanding officer their names, crimes, and the names of the officers who committed them, on penalty of being punished for disobedience or neglect, at the discretion of a court-martial; (ART. 82.) (*See* CONFINEMENT; PROVOST-MARSHAL.)

REGIMENT. (Lat. *rego*, I rule.) A body of troops organized by law, subject to the same administration, discipline, and duties, having a legal head and members, and composed according to arm of companies, battalions, squadrons, or batteries. (*See* ARMY for the organization of the several regiments of infantry, cavalry, and artillery.)

REGIMENTAL COURT-MARTIAL. (*See* COURT-MARTIAL.)

REGIMENTAL NECESSARIES. (*See* NECESSARIES.)

REGIMENTALS. The uniform clothing of regiments, such as coats, trousers, caps, &c.

REGULATIONS. Under the Constitution of the United States, rules for the government and *regulation* of the army must be made by Congress. Regulation implies *regularity*. It signifies fixed forms; a certain order; method; precise determination of functions, rights, and duties. (*See* ARMY REGULATIONS.) Rules of Regulation also embrace, besides rules for the administrative service, systems of tactics, and the regulation of service in campaign, garrison, and quarters. In the case of the staff departments, legislative authority has been delegated *jointly* to the President and Secretary of War. But in relation to the powers, rights, and duties of officers and soldiers in campaign, garrison, and quarters, Congress has not delegated its authority to the President, nor have such matters been *precisely* determined by military laws. Even rights of rank, command, and pay, concerning which Congress has legislated, are subjects of dispute, and variable expositions of laws regulating those essentials of good government have been given by different executives, with an increasing tendency to invalidate rank created by Congress. There can be no remedy for these encroachments, unless Congress should pass a law to enable cases to be brought before the Federal civil courts, in order that the true exposition of military statutes and authorities in dispute may be determined. With such a remedy, laws, however defective they may be, would at least be known, and rights powers, and duties established by law would be well determined.

But it may be said in relation to such rules of regulation, how can a body like Congress determine upon systems of tactics, &c.? Their constitutional duty might easily be performed as follows:—1. By clearly declaring, in a manner not to be misunderstood, that the general-in-chief is charged with the discipline and military control of the army under the rules made by Congress and the orders of the President. 2. The Secretary of War is charged with the administrative service of the army under the rules made by Congress and the orders of the President. 3. By directing the general-in-chief, with the advice of properly constituted military boards, to report to the President rules for the government and regulation of the army in campaign, garrison, or quarters, including systems of tactics for the different arms of the service. 4. By directing the Secretary of War, with the advice of properly constituted boards, to report to the President rules for raising and supporting armies; including regulations for the administrative service. 5. By directing the President to submit the rules made in accordance with provisions 3 and 4, to another board organized by the President, with directions to harmonize the details of the several reports; which last report shall be submitted to Congress for confirmation or orders in the case. 6. By directing that each year, previous to the meeting of Congress, the following boards be assembled under the orders of the general-in-chief, viz.: a board of general staff officers; a board of artillery officers; a board of cavalry officers; and a board of infantry officers. The Secretary of War to assemble the following boards, viz.: a board of engineer officers; a board of ordnance officers; a board of medical officers; and a board of quartermasters, commissaries, and paymasters. Each of the boards so assembled to report to the general-in-chief or Secretary of War, such suggestions of improvements in their respective services as it may be desirable to adopt. 7. The repeal of all laws delegating legislative authority to the President and Secretary of War. (*See* ADMINISTRATION, and references; ARTICLES OF WAR; COMMAND; CONGRESS; GOVERNMENT, and its references; LAWS, (*Military*;) OBEDIENCE; ORDERS; ORDNANCE DEPARTMENT; SECRETARY OF WAR; SERVICE, and references; STAFF, and references.)

REJOINDER. The weight of authority is against permitting a rejoinder on the part of the prisoner, unless evidence has been adduced in the reply of the prosecutor. But such evidence should not be permitted in reply, and there should be no rejoinder; (HOUGH's *Military Law Authorities*.)

RELEASE OF PRISONERS. (*See* REFUSE.)

RELIEF. A guard is usually divided into three reliefs. Relief is also the height to which works are raised. If the works are high and commanding, they are said to have a bold relief; but if the reverse, they are said to have a low relief. The relief should provide the requisite elevations for the musketry and artillery, to insure a good defence.

RELIEVING THE ENEMY. Whosoever shall relieve the enemy with money, victuals, or ammunition, or shall knowingly harbor or protect an enemy, shall suffer death, or such other punishment as shall be ordered by the sentence of a court-martial; (ART. 56.)

REMBLAI—is the quantity of earth contained in the mass of rampart, parapet, and banquette.

REMEDY. The rules and articles for the government of the army are defective in not providing sufficient remedies for wrongs. The army of the United States is governed by law. The law should therefore provide a sufficient remedy for cases in which the rights of officers are wrested from them by illegal regulations, purporting to interpret the true meaning of acts of Congress. In cases arising in the land and naval forces of the United States, where the true construction of any act of Congress is in dispute, legislation is wanted to enable an officer, who thinks himself wronged by an illegal executive decision, to bring the matter before the federal civil courts to determine the true exposition of the statute or authority in dispute. (*See* REDRESSING WRONGS; SUIT.)

REPAIRS OF ARMS. (*See* DAMAGE.)

REPLY. It is the duty of a court to prevent new matter from being introduced into the prosecution or defence, but a prisoner may urge in his defence mitigating circumstances, or examine witnesses as to character or services, and produce testimonials of such facts, without its being considered new matter. If any point of law be raised, or any matter requiring explanation, the judge-advocate may explain. No other reply to be admitted; (HOUGH.)

REPORTING PRISONERS. (*See* REFUSE.)

REPRIEVE. The President of the United States has power to grant reprieves and pardons for offences against the United States, except in cases of impeachment; (*Constitution.*)

REPRIMAND. It is earnestly recommended to all officers and soldiers diligently to attend divine service; and all officers, who shall behave indecently or irreverently at any place of divine worship, shall, if commissioned officers, be brought before a general court-martial, there to be publicly *reprimanded* by the President; (ART. 2.)

REPRISALS. Acts of war to obtain satisfaction for losses or acts of retaliation. (*See* WAR.)

REPROACHFUL or provoking speeches or gestures, used by one officer to another, are punished by the arrest of the officer; in the case of a soldier, he is to be confined and ask pardon of the party offended, in the presence of the commanding officer; (ART. 24.)

REQUISITIONS. Forms prescribed for the demand of certain allowances, as forage, rations, &c. (*See* ADMINISTRATION.)

RESERVE. A select body of troops kept back to give support when needed, or to rally upon.

RESIGN; RESIGNATION. The voluntary act of giving up rank or an appointment. (*See* DISCHARGE.)

RETAINERS. All sutlers and retainers to the camp, and all persons whatsoever, serving with the armies of the United States in the field, though not enlisted soldiers, are to be subject to orders according to the rules and discipline of war; (ART. 60.)

RETREAT. Retrograde movement before an enemy; by retreat is also understood the drum-beat at sunset.

RETRENCHMENT—is an inner defensible line, either constructed in the original design, or executed on the spur of the occasion, to cut off a breach, or other weak point; so that the capture of the latter shall not involve that of the retrenched post.

RETURNS. Every officer who shall knowingly make a false return to the Department of War, or to any of his superior officers, authorized to call for such returns, of the state of the regiment, troop, company, or garrison, under his command; or of the arms, ammunition, clothing, or other stores, thereunto belonging, shall on conviction thereof before a court-martial be cashiered; (ART. 18.) The commanding officer of every regiment, troop, independent company, or garrison of the United States, shall, in the beginning of every month, remit, through the proper channels, to the Department of War, an exact return of the regiment, troop, independent company, or garrison under his command, specifying the names of the officers then absent from their posts, with the reasons for, and the time of, their absence. And any officer who shall be convicted of having, through neglect or design, omitted sending such returns, shall be punished according to the nature of his crime, by the judgment of a general court-martial; (ART. 19.) Disbursing agents shall make monthly returns, in such forms as may be prescribed by the treasury department, of the moneys received and expended during the preceding month, and of the unexpended balance in their hands; (*Act* March 3, 1809. *See* ACCOUNTABILITY; ORDNANCE DEPARTMENT.)

REVEILLE. Drum-beat and roll-call at daybreak.

REVERSE. The reverse flank in a column is the flank at the other extremity of the pivot of a division.

REVETMENTS. The interior slopes of the parapets of permanent and field-works, as well as in some cases the sides of the ditches of the latter, require revetments to enable them to stand at that slope which is necessary, and to endure the action of the weather. The materials made use of in the construction of field-revetments are: fascines, gabions, hurdles, sod, sand-bags, and timber. In siege operations, and in fact in all operations in active warfare, vast quantities of these materials are required, and are daily consumed, in the construction of breastworks, parapets, batteries, magazines, and a variety of miscellaneous purposes. Large quantities, then, must be prepared or manufactured by the ordinary troops of the line, superintended by their own officers, who should be acquainted with all the details necessary for their production.

Fascines are strong, close, regular fagots, carefully and compactly made, generally of green brushwood. They should be straight, cylindrical, and pliant; bound round with good thick, unbroken gads or withes, of pliant wood, at equal distances, the knots well tied, and all in one line; no variation in girth exceeding 1 inch to be allowed.

Fascines are of several kinds and various dimensions, according to the purposes for which they are intended. The most common are the long fascines or saucissons, 18 feet long, 9 inches in diameter, about 140 lbs. in weight; such a fascine can be made by five men in one hour, including the cutting of the wood when at hand. Water fascines, 18 inches in diameter, 6 to 9 feet long. Trench fascines, 4 or 5 feet long, 6 inches in diameter. Sap fagots, 3 feet long, 9 inches in diameter, having a sharp-pointed stake, passed longitudinally through the centre, and projecting a foot or so beyond the extremity of the fascine. To make good fascines requires considerable practice and much care and attention, (Fig. 178.) The process is this: Stakes are driven into the

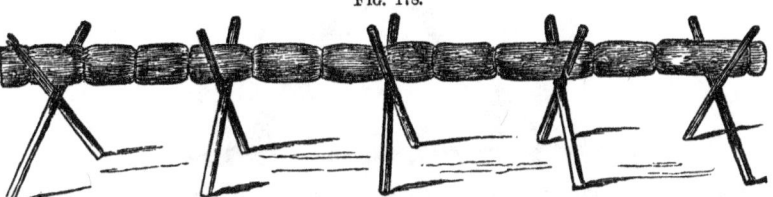

Fig. 178.

ground, obliquely, in pairs, so that the stakes in each pair cross at the same height above the ground about 3 feet, where they are firmly bound together, forming a row of trestles each in shape like the letter X. These trestles should be placed about 4 feet apart when the brushwood is good; closer together when it is bad. Thus 5 trestles at least will be requisite to prepare 18-feet fascines.

A choker must now be prepared. This is made by fastening, by an iron ring, each extremity of a chain about 4 feet long, to an ash stake. Each stake is 4 feet long, and the point where the chain is fastened is about 18 inches from the thicker end. Two small rings are attached to the chain $28\frac{1}{2}$ inches apart, (equal to the circumference of the fascine,) and equidistant from its middle point. In choking the fascine, the middle of the chain is placed under it, and the ends brought over and crossed as in Fig. 179. Two men, one on each side, then bearing on the longer arms of the levers tighten the chain, and compress the fascine to the proper dimensions, that is, until the rings on the chain meet. A third man now binds the fascine as close as possible to the choker, with a strong gad, or with stout spun yarn, when the choker may be removed and the operation repeated at the proper intervals, generally 18 inches. For withes or gads to bind fascines, very straight rods must be selected; they should be 5 feet long, not thicker at the thickest part than the thumb, nor thinner at the thinnest than the little finger. To prepare them for use, place the thick end under the foot, and twist the

506 MILITARY DICTIONARY. [REV.

rod from the top downwards, by which the rod will become flexible and capable of being securely knotted without fracture. The knot to

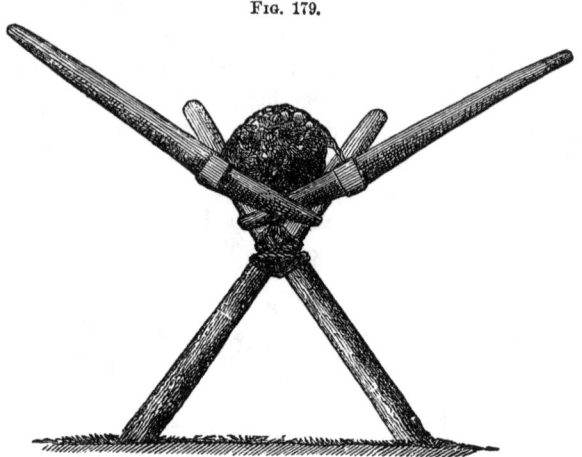

FIG. 179.

be formed in fastening the gad round the fascine is shown in Fig. 180. To make the fascine, the brushwood is laid in the trestles, the longest and straightest rods being kept round the outside, the inferior material in the middle. The proper quantity of brushwood having been thus carefully arranged, the choker is applied near the extremity of the fascine, and subsequently at intervals of 18 inches as already mentioned. The ends and exterior are now neatly trimmed by the hand saw and billhook, and the fascine is complete. When good gads or withes cannot be procured, stout, well-tarred spun-yarn may be substituted for them. With fascines are prepared bundles of stakes, called fascine pickets, in the proportion of six to each fascine; they should be 4 feet long, $\frac{1}{2}$ inch in diameter, and be cut to triangular points.

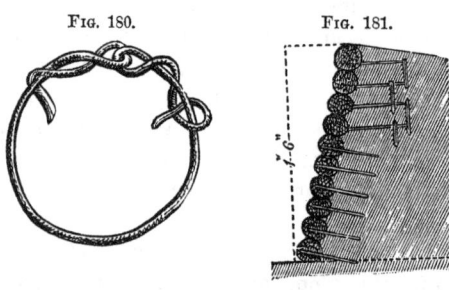

FIG. 180.   FIG. 181.

Slopes, to be revetted with fascines, have usually a base equal to one-fourth their height. The fascines are placed horizontally one over another, as the work is built, until the whole slope is covered by one layer of fascines. Pickets are driven through each fascine to secure it to the work, and these are sometimes fastened to other pickets, buried vertically in the mass of parapet, as shown in Fig.

181. To find the number of fascines required to revet any slope, divide the length of the slope by the length of the fascine, and the height of the slope by the diameter of the fascine: these two quotients multiplied together will be the requisite number.

Gabions are stout, rough, cylindrical baskets, open at top and bottom; they are made of various dimensions according to their intended use. Those for revetting the interior slopes of parapets are usually 3 feet high and 2 feet in diameter; strongly and somewhat coarsely made. Those used in sapping (called sap gabions) have about the same dimensions, but are carefully finished. To construct a gabion, a circle of 22 inches diameter must be traced on a clean, hard, level piece of ground, each quarter of this circle is then divided into four or five equal parts, and small holes made at the points of division, to receive straight uprights of $3\frac{1}{2}$ feet in length, around which the withes are interwoven. Gabions may be made with one, two, or three rods woven together about the uprights; when two rods are woven together, the work is called pairing; when three, waling. The last gives the strongest gabions. The method of working will be best understood by reference to Fig.

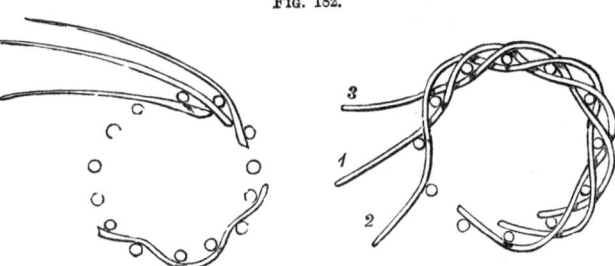

Fig. 182.

182. Each rod passes outside two, and inside one, upright, and the three are twisted together like a rope.

In revetting with gabions, a base is first made for them at right angles to the slope, so that when standing upon this, their surfaces will be coincident with the slope, (Fig. 183). When one row of gabions has been thus placed, and the parapet has risen as high as their upper surfaces, a row of fascines is laid horizontally upon the tops of the row of gabions. Above these again another row of gabions is placed at the same inclination with the former, and finally another row of fascines completes the whole. Two rows of gabions and two of fascines are required for the revetment of an interior slope, of the usual height, without a banquette, and one row of gabions

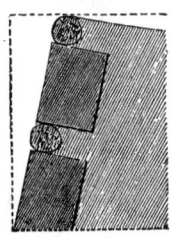

Fig. 183.

and two of fascines with a banquette; therefore, in the former case, the number of gabions required, will be equal to the number of feet of crest to be revetted, and in the latter case to half that number. The number of fascines, in either case, will be equal to twice the length of the slope divided by the length of a fascine.

Hurdles (Fig. 184) are the common coarse wicker hurdles made for farming, and other purposes, usually 3 or 4 feet high and 6 to 9 feet long. They are useful in temporary works, to retain earth at a steep slope, for a short time. When thus used, they should be secured by anchoring pickets. Hurdles are moreover useful, to form a dry footing in trenches, during wet weather; in the passage of wet ditches, and for many similar purposes. Sods or turfs are used for the formation of the interior slopes of parapets, and the cheeks of embrasures. Sods should be cut from fine close turf, with thickly matted roots, previously mown, and if possible, watered, to make the earth adhere more closely to the roots of the grass. The sods are laid, with the grass downwards, alternately headers and stretchers, like bricks in a wall. Their under or upper surfaces should be perpendicular to the slope of the parapet, and not horizontal, except in a vertical revetment, and each sod should be fastened to those beneath, by two or three wooden pegs. Sod work can be made with great perfection, and is very durable. The arrangement of the sods is shown in plan and in rear elevation in Fig. 185, and in side elevation in Fig. 186. In meadows, the dimensions of sods may be from 12 to 18 inches long, 12 inches wide, and 4 to 6 inches thick.

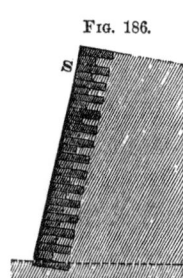

Fig. 184.

Fig. 185.

Fig. 186.

In heath, having large roots, they may be 2 feet long, 12 or 18 inches wide, and 8 to 10 inches thick. To find the number of sods required to revet any given length of slope, the revetment being one sod thick:

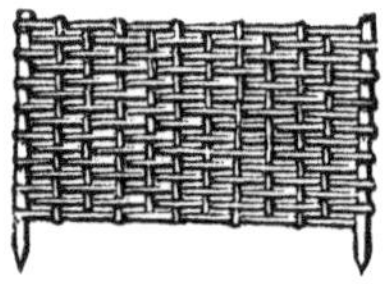

Divide the height of slope by thickness of sods, for the number of rows. Divide twice the length of the slope by the sum of the length and breadth of a sod for the number in one row. Multiply these two quotients together, for the whole.

Sand-bags are coarse canvas bags, of a capacity sufficient to hold about a bushel of earth; when empty they occupy only a small space, and are frequently of great use. A good field-revetment can be built with filled sand-bags, laid as sods; such a revetment, however, is only fit for temporary purposes, as the sand-bags soon rot; they are unfit for lining the cheeks of embrasures, as the flash of the guns speedily destroys them. In rocky positions, it is sometimes necessary to construct entire batteries and parallels with filled sand-bags. In Figs. 187 and 188, are shown a section of a parapet revetted with sand-bags, and an enlarged plan of the same. Many of the British trenches and batteries before Sebastopol, owing to the rocky nature of the ground, were formed of sand-bags, baskets, casks, &c., filled with earth brought from a distance. Sand-bags are used in great numbers, laid on the superior slopes of parapets, to form loop-holes for riflemen.

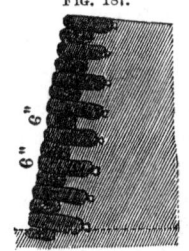

FIG. 187.

FIG. 188.

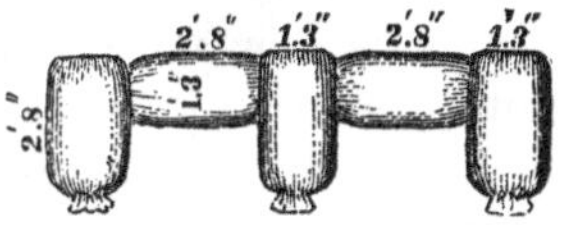

SAND-BAGS SHOULD BE TARRED, AND HOLD ONE CUBIC FOOT OF EARTH.

Timber is used for revetments, in particular cases only, as where it may be considered advisable, in important field-works, to retain the escarp of the ditch at a steep slope. In this case, a revetment is necessary, which may be constructed of beams or the trunks of small trees, planted 3 or 4 feet deep, vertically in the ground and touching each other, or by lining the surface of the slope with planks secured by stout posts, 3 or 4 feet apart, planted several feet in the ground, and there fastened to heavy horizontal beams. The strength of the revetment may be still further increased, by connecting the upper extremities of the posts to others buried under the mass of the rampart; (HYDE's *Fortification*.)

REVIEW. Prescribed form of passing troops before a general officer, an inspector, or other reviewing personage.

REVISION. Where an officer, who orders a court-martial, does not approve their proceedings, he may, by the custom of war, return them to the court for revision, and no additional evidence can be taken on such revision; (HOUGH.)

REWARD. Thirty dollars are paid for the apprehension of deserters.

RICOCHET. Guns fired with a small charge and a low elevation, project ricochet shot, which merely clear a parapet, and thence bound along a rampart, destroying gun-carriages, &c. (*See* FIRING.)

RIFLED ORDNANCE. Rifle-muskets are wholly indebted to the elongated projectile for their efficiency and celebrity. Elongated shot possess, when their axes are coincident with the path they describe, the properties of being less resisted by the air, having longer ranges and greater penetrating power than spherical projectiles of the same diameter. To obviate the difficulty and loss of time in loading ordinary rifles, by forcing the ball into the barrel by repeated blows of the ramrod or a mallet, on account of which that arm had been little used, M. Delvigne proposed that the bullet should have sufficient windage to enter freely into the barrel, in order that, when stopped by the contraction of the chamber with which this arm was furnished, it might be forced to expand and enter into the grooves, on receiving a few smart blows; thus the piece being fired, the bullet would come out a forced, or rifle ball, without having been forced in. But this ingenious contrivance was not found to answer. The edge of the chamber on which the ball lodged, not being opposite to the direction of the blow, did not form a sufficient support upon which to flatten the ball when struck by the ramrod, and thus cause the bullet to expand; whilst portions of the charge of powder previously poured in, having lodged on the contraction, cushioned and still further impeded the expansion of the shot; and as, obviously, no patch could be used, the grooves were liable to get foul, and to become leaded, to an extent which could not be effectually obviated. To remedy this defect, Colonel Thouvenin proposed in 1828 to suppress the chamber, and substitute a cylindrical tige or pillar of steel, screwed into the breech in the centre of the barrel, so that the bullet, when stopped by, and resting upon the flat end of the pillar, directly opposite to the side struck, might more easily be flattened and forced to enter the grooves. But here another defect appeared. The pillar occupying a large portion of the centre of the barrel, and the charge being placed in the annular space which surrounds it, the main force of the powder, instead of taking effect in the axis of the piece, and on the centre of the projectile, acted only on the spherical portion of the bullet which lies over this annular chamber, and thus the ball, receiving obliquely the impulse of the charge, was propelled with diminished force. The next improvement, which was proposed by M. Delvigne, was to make the bottom of the projectile a flat surface; the body cylindrical, and to terminate it in front with a conical point, thus diminishing the resistance of the air comparatively with that experienced by a solid of the same diameter having a hemispherical end. The form of the projectile was, therefore, an approximation to that of Newton's solid of least resistance. (*See* PROJECTILE.) In 1841 a patent was obtained by Captain Tamisier for his method of giving steadiness to the flight of cylindro-conical shot, by cutting three sharp circular grooves each .28 inches deep, on the cylindrical part of the shot, by which the resistance of the air behind the centre of gravity of the projectile being increased, the axis of rotation was kept more steadily in the direction of the trajectory; the grooves being to this projectile what the feathers are to the arrow, and the stick to the rocket.

But the tige musket having been found inconvenient in cleaning, the pillar liable to be broken, and, after firing some rounds, the operation of ramming down so fatiguing to the men as to make them unsteady in taking aim, M. Minié, previously distinguished as a zealous and able advocate for restoring the rifle to the service in an improved form, proposed to suppress the tige, and substitute for it an iron cup, *b* (Fig. 189,) put into the wider end of a conical hollow, *a*, made in the shot: this cup being forced further in by the explosion of the charge, causes the hollow cylindrical portion of the shot to expand and fix itself in the grooves, so that the shot becomes forced at the moment of discharge. A slip of cartridge-paper is wound twice round the cylindrical part of the projectile, so that, as the latter does not become forced or rifled till the charge is fired, it fits so tightly to the barrel as to be free from any motion which would be caused by the carriage of the rifle on a march, or by its being handled before the shot is fired. But unless the cup *b* (Fig. 189) be driven, by the first action of the explosion of the charge, so far into the conical space in which it is placed, as to cause the lead to enter into the grooves of the rifle before the shot moves, there will be no rotation—the paper wrapped round the shot not sufficing for this purpose. In the experiments of 1850 it was found that the hollow part of the Minié cylindro-conical shot was very frequently separated entirely from the conical part by the force with which the cup was driven into the hollow part of the shot, and sometimes remained so firmly fixed in the barrel that it could not be extracted; but in the more recent trials

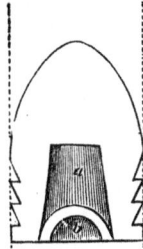

FIG. 189.

with shot made by compression and with better lead, no such failure occurred.

While efforts were being made in France to augment the power and accuracy of small-arms, loaded at the muzzle, as already described, M. Dreyse, of Sommerda, in Thuringia, was led to test whether the inconvenience of ramming down and flattening the shot might not be got rid of by loading the barrel at the breech—an old project; and he suggested a plan for this purpose, which has been adopted to a great extent in the Prussian army. The Prussian rifled musket for firing cylindro-conical shot is loaded at the breech, and is designated "zundnadelgewehr," from the ignition of the charge being produced by passing a needle through the cartridge to strike the percussion-powder placed in the wooden bottom, or spiegel. The escape of gas at the junction of the chamber and barrel is considered by all as a great objection to the needle-prime musket: it is stated that the point of the igniting-needle soon becomes furred, so that it is difficult, and, after a time, impossible, to draw it back by the thumb. The Prussians, however, appear to be quite confident of the superiority of the latter over other rifle-muskets; their government is said to have caused 60,000 stand of these arms to be executed, and at least half as many more are ordered. Their fusiliers, who are armed with the needle-prime musket, have also a short sword, with a cross hilt: this they plant in the ground; and, lying down, they use the hilt as a rest for the purpose of taking a steady aim.

It is, no doubt, in some respects, an important advantage in the Prussian rifles, that they may be loaded more quickly than the ordinary musket or rifle; but rifle actions are generally decided, not by mere rapidity of fire, but by each soldier taking time to use his arm in the most efficient manner possible. Although the use of the rifle was suspended in the French armies throughout the whole of the general war (1794–1815,) yet the French infantry, armed with the common musket, were well trained to act *en tirailleur*, and showed great aptitude for that kind of service. Good patterns having been obtained of the Delvigne carabine à tige, the French and the Belgian Minié rifles, experiments were made at Woolwich in 1851 with these three arms and with Lancaster's pillar-breech rifle, in order to test their relative merits in firing at a target 6 feet square, at 400 yards' distance. The results of these experiments fully established the peculiar advantages of M. Minié's method of quick loading, and forcing the shot into the rifled state, and a large supply of what has been called the regulation Minié musket was ordered. The form of its projectile, which is simply conoidal, is given in Fig. 190 annexed.

Mr. Lancaster, who invented the ordnance with an elliptical bore, spirally formed, and the pillar-breech rifle, proposed also a description of musket having a bore of a similar kind. No grooves are cut in the interior surface of the barrel; but in a transverse section, the bore has the form of an ellipse of small eccentricity, being *freed* at the breech: the projectile is cylindro-conoidal, with a circular base, and, when heated by the fired gunpowder, it expands so far as to take a form corresponding to the elliptical section of the bore. The bore, being a continuous spiral, fulfills the object of grooves, and causes the shot, in passing along it, to acquire a rotatory motion on its axis. The spiral is not uniform in its whole length, but has what is called by Americans a *gaining twist* or *an increasing spiral*. The advantages of this rifle are supposed to be—greater accuracy of practice, less recoil than other muskets have, and no tendency to cause the rifle to turn over sideways.

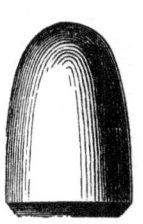
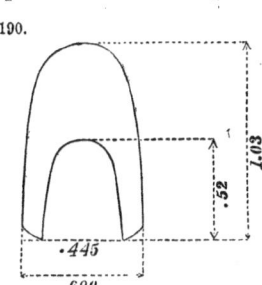

Fig. 190.

In December, 1853, a trial was made at Hythe of Mr. Lancaster's elliptically-bored muskets freed at the breech, in order to compare their shooting with that of a rifle-musket of .577 bore, having three grooves regularly spiral of one turn in 6 ft. 6 in., which was manufactured at Enfield in the same year; the report of this trial was in favor of the Enfield rifle, Lancaster's muskets evincing a strong tendency to *strip*, and at the longer ranges this defect was very marked.

In 1858, Mr. Whitworth of Manchester produced a musket having a hexagonal bore of a spiral figure, making one turn in 20 in., by which the projectiles—either of hexagonal or cylindro-conoidal form—in passing along the barrel acquire a swift and steady rotation on their axes. This species of rifle has been found considerably superior in accuracy of shooting to the Enfield rifle, which has been adopted in England.

In order to test the relative merits of these two kinds of weapons, a series of trials were made at Hythe, under the direction of Colonel Hay, the able superintendent of the school of musketry at that place, and the results are stated in the following table. The rifles were fired from rests, and ten or twenty rounds were fired from each at the several distances. The numbers in the fourth column express, in feet and decimals, the means and the distances of the ten or twenty points of impact on the target, from a nearly central point of the group in each trial.

TABLE SHOWING THE RESULTS OF EXPERIMENTS WITH THE "WHITWORTH" AND "ENFIELD" RIFLES.

Description of Rifle.	Distance in yards.	Angle of elevation.	Mean radial deviation.	Remarks.
		° ′	Feet.	
Enfield	500	1 32	2.24	
Whitworth	......	1 15	.37	
Enfield	800	2 45	4.20	
Whitworth	......	2 22	1.00	
Enfield	1,100	4 12	8.04	
Whitworth	......	3 8	2.62	
Enfield	1,400	......	......	Shooting so wild, no diagram taken.
Whitworth	......	5 0	4.62	
Enfield	1,880	......	......	Not tried.
Whitworth	......	6 40	11.62	

The superiority of the Whitworth rifle in accuracy of fire is hence manifest; and it may be added that, from its form, the bore is less liable to be worn than that of any grooved rifle. As the projectile may be made harder, it will, consequently, have greater penetrating power; and, in fact, the Whitworth projectile went through 35 half-inch planks of elm wood, and remained in a bulk of solid oak beyond, while the Enfield projectile went through only 12 such planks.

Till within the last twenty years, no *sight* was considered necessary for a common musket—the stud at the muzzle being sufficient for the purpose of taking aim. When percussion-arms were first introduced, a fixed block-sight for 120 yards was adopted; and subsequently a block-sight for 200 yards and a leaf for 300 yards were affixed to the two-grooved rifle. At present every English rifled musket is furnished with a complicated and delicate sight. The rifles used by the Russians at the battle of the Alma were of good construction; they have two grooves, and carry conoidal shot, each weighing 767 grains, equivalent in weight to a spherical bullet of 9 to the pound. They are flat at the base, and have projections at the sides corresponding with the grooves of the musket. The great weight of these projectiles is very objectionable; the soldiers who carry them must be very much distressed by the loads in their pouches, or these must contain a smaller number of shot than are usually carried. The Russian missile is more pointed than the English Minié shot, and no part being cylindrical, it must be liable to irregular movements in the barrel, and, consequently, to unsteadiness in its flight.

It has the designation of a Minié shot, a term now generally but improperly applied to all elongated shot for musketry, since they differ from one another both in form and weight.

The rifle used in the French service up to the commencement of the late Italian war consisted only of the carabine à tige, and these were given only to special corps of riflemen. However eminent the authority of Colonel Minié on the subject of rifles, his method of rifling was never introduced into the French service. Throughout the Crimean war, the French infantry of the line were armed with the smooth-bored regulation musket. Some time previous to the Italian campaign the whole of the French infantry had their old muskets rifled, and conical shot introduced—the rifling principle being a triangular hollow cut in the bottom of the shot, without any cup, as in the Minié system. The efficient range did not exceed 600 yards, and was very inaccurate beyond 400 yards. This imperfect measure, as admitted by the French authorities, hardly kept pace with the general improvement in small-arms; but they were restricted by considerations of economy, which did not admit of any general alteration of the muskets in store. Thus all the French infantry during the Italian campaign used these defective rifled muskets, with the exception of the chasseurs, who retained the carabine à tige, the range of which was far superior to other French musket rifles.

In 1846, iron rifled cannon, loaded at the breech, were invented by Major Cavalli and Baron Wahrendorff, for the purpose of firing cylindro-conical and cylindro-conoidal shot. In these guns the mechanical contrivances for securing the breech, are very superior to the rude processes of earlier times; yet it appears doubtful whether or not, even now, they are sufficiently strong to insure safety when high charges are used in long continued firing. The length of the Cavalli gun is 8 feet 10.3 inches; it weighs 66 cwt., and its calibre is $6\frac{1}{2}$ inches. Two grooves are cut spirally along the bore, each of them making about half a turn in the length, which is 6 feet 9 inches. The chamber, which is cylindrical, is 11.8 inches long and 7.008 inches diameter.

In the summers of 1853 and 1854, trials were made at a spot between Leiny and Cirie, in Piedmont, of a rifled Cavalli gun, loaded at the breech, and with various improvements in the apparatus for loading and pointing. The gun carried cylindro-ogivale shells, each weighing 30 kilogrammes, (66 lbs. 3 oz. English,) and provided with a metal fuze. The shells were fired with charges equal to one-tenth of the weight of the projectile, at elevations varying from 5 to 25 degrees. The firing was directed against a target about 10 feet square, and placed at the distance of 3,050 yards from the gun. In ten trials, at an elevation of 10 degrees, the mean of the ranges obtained was 3,058 yards; the means of the deviations were to the right 3.4 yards, and to the left 3.39 yards. After one rebound the shot went to the distance of 4,096 yards from the gun, with a deviation to the right equal to 126 yards. The mean time of flight was 11 seconds. In fifteen trials, at an elevation of 15 degrees, the mean of the ranges was 4,128 yards; the mean deviations were, to the right 11 yards, and to the left 1 foot 11 inches. The time of flight was 16 seconds. In fifteen trials, at an elevation of 20 degrees, the mean of the ranges was 4,917 yards; while the mean deviations were, to the right 6 yards 2 feet, and to the left 10 yards. The time of flight was 19 seconds. Lastly, in ten trials, at an elevation of 25 degrees, the mean of the ranges was 5,563 yards, while the deviations were, to the right 3 yards, and to the left 4 yards. These trials were considered highly satisfactory; and no less so were some experiments also made with metal fuzes, and with a charge equal to one-thirtieth of the weight of the projectile; the first shell so fired struck against one of the beams of the target, and tore away splinters of the wood varying in length from 1 ft. 9 in. to 1 ft. 11 in. The bursting-charge appeared to be fired a little before the moment of the shell falling.

Baron Wahrendorf invented a 24-pounder gun, which is also to be loaded at the breech. It is mounted on a cast-iron traversing carriage; and, taking little room, it appears to be very fit for casemates. The upper part of the carriage has, on each side, the form of an inclined plane, which rises towards the breech, and terminates near either extremity in a curve whose concavity is upwards. Previously to the gun being fired the trunnions rest near the lower extremity; and on the discharge taking place, the gun recoils on the trunnions, along the ascending plane, when its motion is presently stopped. After the recoil, the gun descends on the plane to its former position, where it rests after a few short vibrations. The axis of the gun constantly retains a parallel position, so that the pointing does not require readjustment after each round. The gun was worked easily by eight men, apparently without any strain on the carriage, With a charge of 8 lbs., and with solid shot, the recoil was about 3 feet, and the trunnions did not reach the upper extremity of the inclined plane, though the surface was greased.

THE ARMSTRONG GUN.—In the latter part of the year 1854, Mr. William George Armstrong (now Sir William George Armstrong) submitted to the Duke of Newcastle, then Minister at War, a proposal for a rifled field-piece on a new principle, and undertook, with his grace's authority, to construct a gun upon the plan he had suggested. This gun was completed early in the following year, (1855,) and became the subject of a long course of experiments, which ultimately led to the general introduction of the weapon into the British service. Fig. 191 shows the exterior of a 12-pounder Armstrong gun, such as is now used for field artillery, and also an end view of the same, showing the hole through the breech-screw for loading and sponging the gun. These guns can be fired with careful aim twice in a minute, and fully three times per minute without aim.

The following description of the Armstrong gun, as now manufactured, was given by Sir William in the discussion which recently took place at the Civil Engineers' Institute.

"The gun is composed wholly of wrought iron, and the prominent feature in its manufacture is the application of the material in the form of long bars, which are coiled into spiral tubes, and then welded by forging. For the convenience of manufacture, these tubes are made in lengths of from 2 to 3 feet, which are united together, when necessary, by welded joints. From the muzzle to the trunnions the gun is made in one thickness, and is therefore, so far as that portion is concerned, strictly analogous to the barrel of a fowling-piece.

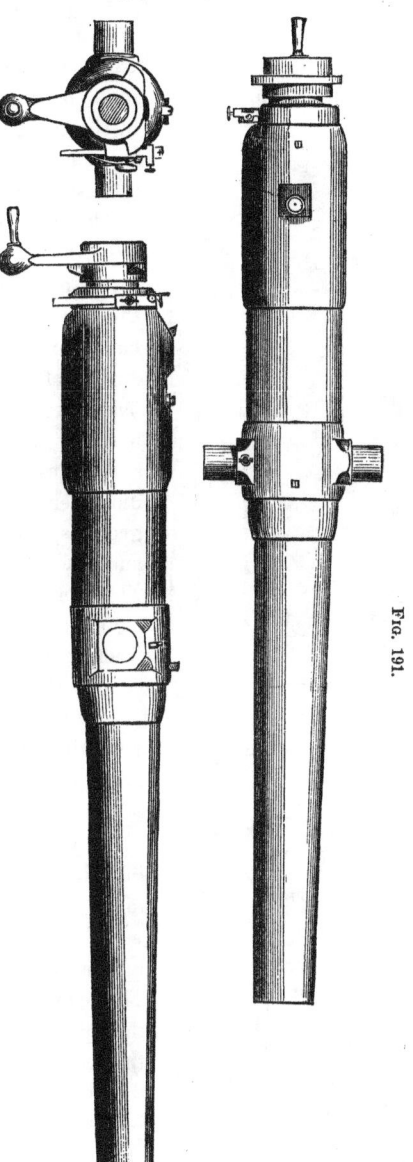

Fig. 191.

Behind the trunnions two additional layers of material are applied. The external layer consists, like the inner tube, of spiral coils; but the intermediate layer is composed of iron slabs bent into a cylindrical form and welded at the edges. The reason for this distinction is, that the intermediate layer has chiefly to sustain the thrust on the breech, and it is therefore desirable that the fibre of the iron should be in the direction of the length, while elsewhere in the gun it is more advantageously applied in the transverse direction. The back end of the gun receives the breech-screw, which presses against a movable plug, or stopper for closing the bore. This screw is hollow, and when the stopper is removed, the passage through the screw may be regarded as a prolongation of the bore. The screw is turned by means of a handle, which is free to move through half a circle before it begins to turn the screw. It has thus a certain amount of run, which enables it to act as a hammer, both in tightening and slackening the screw. The bore is 3 inches in diameter, and is rifled with thirty-four small grooves, having the driving side rectangular and radial, and the opposite side rounded. The bore is widened at the breech end one-eighth of an inch, so that the shot may enter freely and choke at the commencement of the grooves.

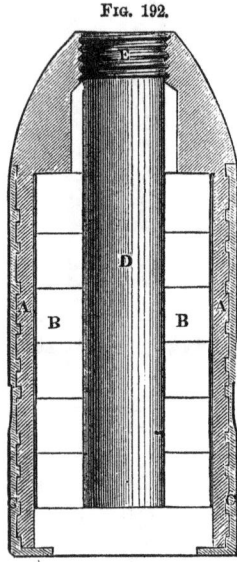

Fig. 192.

"The projectile (Fig. 192) consists of a very thin cast-iron shell, the interior of which is composed of forty-two segment-shaped pieces of cast iron, built up in layers around a cylindrical cavity in the centre, which contains the bursting-charge, and the concussion arrangement. The exterior of the shell is thinly coated with lead, which is applied by placing the shell in a mould, and pouring melted lead around it. The lead is also allowed to percolate among the segments, so as to fill up the interstices, the central cavity being kept open by the insertion of a steel core. In this state the projectile is so compact that it may be fired through six feet of hard timber without injury; while its resist-

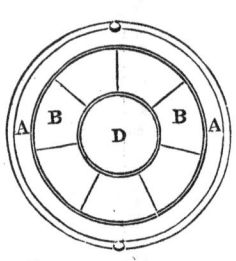

12-PDR. SEGMENT SHELL.
A A. The cast-iron case or shell.
B B. The segment shot in layers.
C C. The lead covering.
D. The central cavity for bursting-tube, and concussion-fuze.
E. Screw for time-fuze.

ance to a bursting force is so small, that less than one ounce of powder is sufficient to break it in pieces. When this projectile is to be used as a shot, it requires no preparation, but the expediency of using it in any case otherwise than as a shell, is much to be doubted. To make it available as a shell, the bursting-tube, the concussion arrangement, and the time-fuze, are all to be inserted; the bursting-tube entering first and the time-fuze being screwed in at the apex. If then the time-fuze be correctly adjusted, the shell will burst when it reaches within a few yards of the object; or, failing that, it will burst by the concussion arrangement, when it strikes the object, or grazes the ground near it. Again, if it be required to act as "canister," upon an enemy close to the gun, the regulator of the time-fuze must be turned to zero on the scale, and the shell will then burst at the instant of quitting the gun. In every case the shell on bursting spreads into a cloud of pieces, each having a forward velocity equal to that of the shell at the instant of fracture. The explosion of one of these shells in a closed chamber, where the pieces could be collected, resulted in the following fragments:—106 pieces of cast iron, 99 pieces of lead, and 12 pieces of fuze, &c.; making in all 217 pieces. The construction of the time-fuze and the concussion arrangement are described as follows:—The body of the time-fuze (Fig. 193) is made of a mixture of lead and tin, cast to the required form, in a mould. The fuze-composition is stamped into a channel forming nearly an entire circle round the body of the fuze, and is afterwards papered and varnished on the external surfaces. As the shell fits accurately into the gun, there is no passage

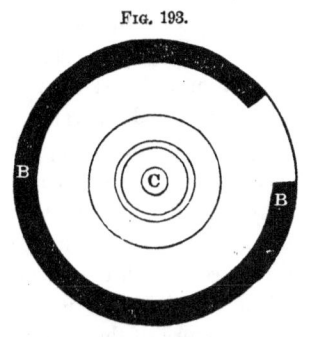

Fig. 193.

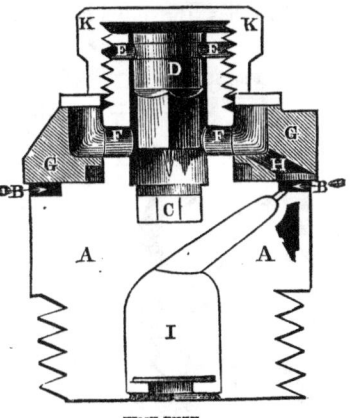

TIME-FUZE.

A A. The body of the fuze.
B B. Groove containing fuze-composition.
C. The detonator.
D. The striker.
E E. The holding pin.
F F. The flame passage.
G G. Revolving cover, or regulator.
H. Igniting aperture.
I. Chamber for priming-powder.
K K. Tightening cap.

of flame by which the fuze could be ignited. That effect is therefore produced in the following manner: A small quantity of detonating composition is deposited at the bottom of the cylindrical cavity in the centre of the fuze, and above this is placed a small weight, or striker terminating in a sharp point presented downwards. This striker is secured in its place by a pin, which, when the gun is fired, is broken by reason of the vis inertiæ of the striker. The detonator is then instantly pierced by the point, and thus fired. The flame thus produced passes into an annular space, formed within the revolving cover, which rests on the upper surface of the fuze-composition, and from this annular space, it is directed outwards, through an opening, so as to impinge on and to ignite the fuze-composition, at any required part of the circle. The fuze, thus ignited, burns in both directions, but only takes effect at one extremity, where it communicates with a small magazine of powder in the centre. The fuze is surrounded by a scale-paper, graduated to accord with the elevation of the gun, so that when the range of a distant object is found by trial, it is only necessary to turn the igniting aperture of the cover to the point on the fuze-scale corresponding with the degrees and minutes of elevation on the tangent-scale. This fuze has the advantage of being capable of adjustment and readjustment any number of times, before entering the gun, and the officer in command has the opportunity of seeing that it is correctly set, at the moment of being used.

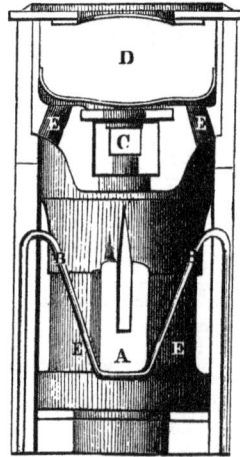

Fig. 194.

CONCUSSION-FUZE
A. The striker.
B B. The holding wire.
C. The detonator.
D. The chamber for priming-powder.
E E. Flame passages.

"The concussion-fuze (Fig. 194) is on nearly the same principle. A striker with a point, presented upwards, is secured in a tube by a wire fastening, which is broken on the firing of the gun; the striker, being then liberated, recedes through a small space, and rests at the bottom of the tube, but as soon as the shell meets with any check in its motion, the striker runs forward and pierces the detonator in front, by which means the bursting-charge is ignited. The process of loading is effected by placing the projectile, with the cartridge and a greased wad, in the hollow of the breech-screw, and thrusting them either separately or collectively, by a rammer, into the bore opposite; (Fig. 195.) The stopper is then dropped into its place, and secured by half a turn of the screw. The gun is fired by the ordinary friction-tube, the vent being contained in the stopper. The whole operation is simple, and can be very rapidly performed.

"In the early guns it was necessary that the portion of the bore which was occupied by the shot should be perfectly clean, otherwise the shot would not always enter its place. A wet sponge had therefore to be used; but in the new guns, now issued for service, a slight alteration in the bore has enabled a greased wad to be employed with perfect effect, in substitution of the wet sponge. The gun can now be fired with great rapidity, and apparently for any length of time, without being sponged at all. The reason for making the vent in the stopper is, that, since the chief wear of the gun always takes place at the vent, it is better to make it in a part which can be easily replaced, than in the body of the gun itself. The breech-screw being internal is never exposed to injury, nor can drifting sand, or dust, ever reach the oiled surfaces, so as to impede the action of the screw by adhering to the lubrication. The screw is of small diameter, and the few inches of extra length in the gun, required for its reception, cannot be of any importance, considering that any further reduction of weight is prohibited by recoil. The stopper is secured from falling by a chain, but in practice it is preferred to leave it loose. The man who fires the gun lifts the stopper after each round, and in so doing only occupies time that would otherwise be vacant. A duplicate stopper accompanies each gun. The form of carriage which was originally used, is represented in the following diagram, (Fig. 196.) It was fitted with a recoil slide, which was afterwards abandoned for field guns; but it has been decided that the principle should be retained in ship guns, (Fig. 197.) It is a point of great importance, that a breech-loading gun should be self-acting, in recovering its position after recoil, so as to obviate the employment of

FIG. 195.

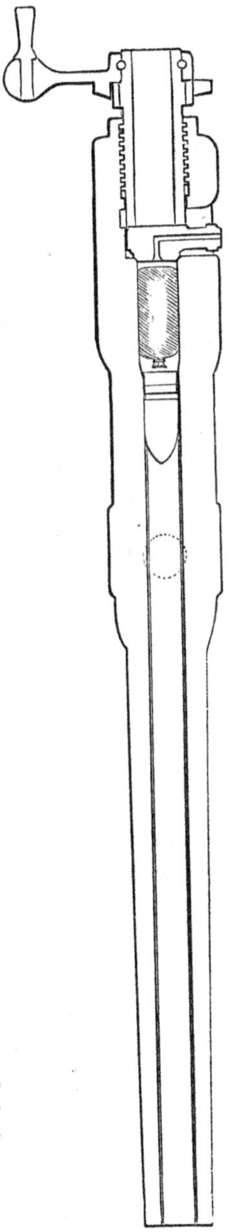

Sectional view of the Armstrong Gun; exhibiting the interior, containing the elongated projectile, the greased wad, and the cartridge, ready for discharge.

Fig. 196.

so many men to run out the gun. A traversing movement was originally applied to the field-carriages, as shown in the diagram, and was found to afford great facility in laying the gun. A very neat modification of this traversing movement has recently been contrived in the Royal Carriage Department, and adopted for the field carriages."

The greatest range which has yet been attained with the Armstrong gun is 9,175 yards, or nearly $5\frac{1}{4}$ miles. The conditions which are chiefly conducive to an extended range are, a small bore and a very lengthened projectile; but the more a projectile assumes the character of a bolt, the less suitable it becomes for a shell. Sir William Armstrong, therefore, deprecates any further increase of range at expense of efficiency in the shell; and, indeed, it may well be doubted whether an extension of range beyond a distance of five miles would prove of any practical utility. The following is an example of practice with the Armstrong 12-pounder field-gun of 8 cwt., at an angle of 5° and with a charge of 1 lb. 8 oz.

No.	Range.	Deflection.	
		Left.	Right.
	Yards.		
1	1920		1 ft.
2	1910		1 ft.
3	1909	In line.	
4	1923	1 ft.	
5	1945	3 ft.	
6	1923	3 ft.	
7	1906	3 ft.	
8	1911	3 ft.	
9	1903	2 ft.	
10	1921	4 ft.	
11	1918	2 ft.	
12	1924	6 ft.	

THE ARMSTRONG GUN MOUNTED FOR SEA SERVICE.

A. The breech-stopper.
B. The upper carriage, which recoils on the incline C.
P. The pivot bolt, which connects the Armstrong carriage with the common slide.

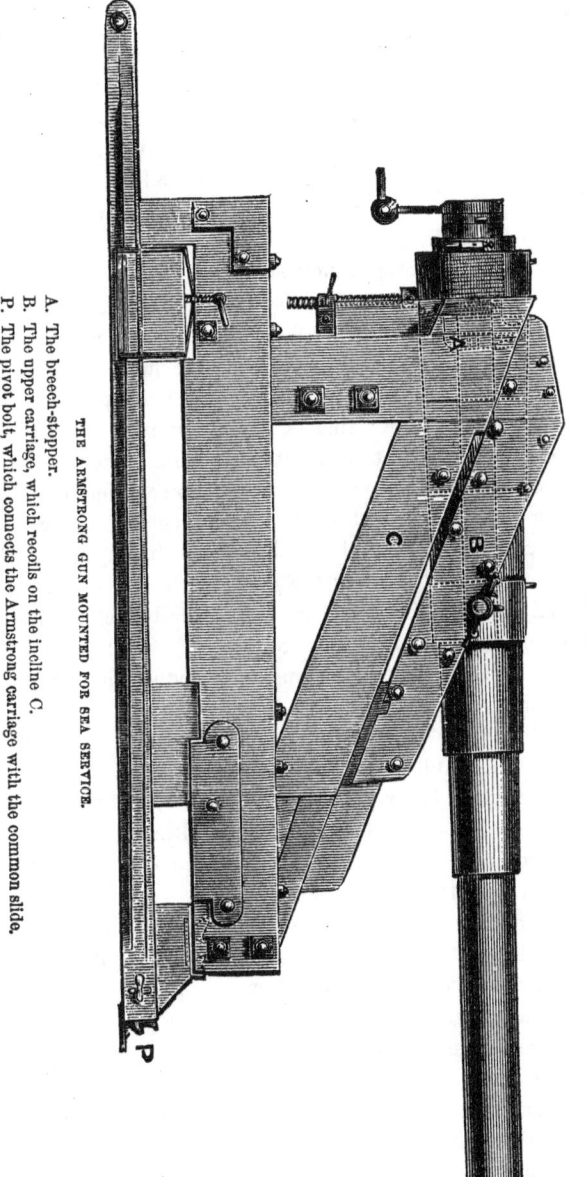

Fig. 197.

The above practice was made with the ordinary shell adapted for this gun, and the minimum charge. By increasing the charge, and using a longer projectile, the same range is attained with less elevation, but the recoil becomes too severe upon the carriage for long continued firing. The projectiles, as now used for these guns, are in all cases made of cast iron, thinly coated with lead, and, being of somewhat larger diameter than the bore, the lead is crushed into the grooves; by means of which the necessary rotation is given, while all shake and windage are prevented. The projectile for field-service admits of being used indifferently as solid shot, shrapnell shells, or canister shot. It is composed of separate pieces, so compactly bound together that it has been fired through a mass of oak timber 9 feet in thickness without sustaining fracture. When used as a shell it divides into the number of pieces of iron, lead, and fuze, stated in p. 519. It combines the principle of the shrapnell and of the percussion shell: that is, it may be made to explode either as it approaches the object or as it strikes it. The shock which the projectile receives in the gun puts the percussion arrangement as it were from half-cock to full-cock, and it then becomes so delicate that it will burst by striking even a bundle of shavings. It may also be made to explode at the instant of leaving the gun, in which case the pieces produce the usual effect of grape or canister. For breaching purposes or for bursting in the side of a ship, a different construction of shell is adopted. The object in that case being to introduce the largest possible charge of powder, the projectile used is simply a hollow shot, and from its length and form is capable of containing a much larger bursting charge than is compatible with a spherical form of the same diameter. The largest gun which has yet been completed upon Sir William Armstrong's principle is one of 65 cwt., which, although only designed to throw a projectile of 80 lbs., has been frequently tried with a shot weighing upwards of 100 lbs.

Early in the course of his experiments, Sir William Armstrong's attention was directed to the improvement of the sights, as the means of aiming guns previously employed were obviously not sufficiently delicate for a gun having 57 times their accuracy. The sights which he has introduced present many peculiarities. The eye-piece of the tangent-scale is in the form of a cross slit, and has a traversing movement for correcting the effect of side wind. The vertical and lateral movements of the sight are each regulated by means of a vernier which enables the scale to be read off to one minute of a degree both for elevation and deflection. With regard to the strength of the Armstrong guns to resist explosion, the 12-pounders have been proved by filling the chamber with powder (about 2¼ lbs.), and using a shot of double the service-weight. In the case of the 40-pounders, it is intended to apply double charges and single shot. To provide for a large charge of powder, it is only necessary to reduce the lead on the shot, so as to allow it to enter further into the bore. Sir W. Armstrong believes the strength of his guns to be enormously in excess of these charges, the object of the proof being rather to detect defects in the surface of the bore than the resistance to bursting, which he considers to be almost uniform in all guns constructed on his principle.

The Whitworth Gun.—Mr. Whitworth, of Manchester, has succeeded in constructing several rifled breech-loading cannon of various calibres: his 3-pounder gun, 208 lbs. in weight, with a calibre of 1½ inches, a charge of 8 oz. of powder, and an elevation of 35°, projects its shot to a distance of more than 5½ miles, and this with remarkable accuracy. He applies the same principles to his guns which have been so successful in his small-arms—using a very long projectile, 3½ diameters in length, that the resistance of the air may be as small as possible, (Fig. 198.) To overcome the tendency of so long a projectile to turn over in its flight, a rapid spin or rotation is impressed upon

Fig. 198.

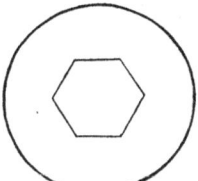

it, by a more than usually rapid twist in the grooves of the rifle. The bore of the barrel is described by its hexagonal section moving parallel to itself from breech to muzzle, and at the same time rotating uniformly about its centre with such a velocity, that it completes one whole rotation while its centre is moving over a space of 20 inches in the small-arms and 3 feet 8 inches in the 3-pounder gun. So that the barrel may be considered as a rifle with six grooves, making one turn in 20 inches in the one case, and in 3 feet 8 inches in the other. The bullets are made of a hard metal, an alloy of 9 parts lead with 1 part tin, and they are shaped to fit accurately the interior of the bore. Experiments made to test the penetrating powers of Whitworth's hexagonal 80-pounder shot, have established its superiority to any other gun or projectile yet produced in penetrating power. The hexagonal bore is also the best for communicating a rapid rifle motion to the projectile, but experiments in the United States have not shown it to be safe for ordinary cast-iron cannon.

MILITARY DICTIONARY. [RIF.

TABLE I.—*Ranges obtained at Southport, February 15th and 17th, 1860, of a 3-pounder Whitworth Gun, length 6 ft., weight 208 lbs., diameter of bore 1½ in., charge 7½ oz., at the undermentioned angles of elevation.*

Angles of Elevation.	Yards Range.	Deviation from Line of Fire.	Angles of Elevation.	Yards Range.	Deviation from Line of Fire.
3°	1,607	¼ yard to the right.	20°	6,784	12 yards to the left.
"	1,593	Line.	"	6,720	14¼ " "
"	1,589	Line.	"	6,910	2 " "
"	1,588	1 yard to the right.	35°	8,907	22 yards to the right.
"	1,577	¼ " "	"	8,930	10 yards to the left.
"	1,575	¼ " "	"	9,059	11 yards to the right.
"	1,573	¼ " "	"	9,164	23¼ " "
"	1,568	2 " "	"	9,688	34 " "
"	1,552	¼ " "	"	9,645	31 " "
10°	4,171	6 yards to the left.	"	9,611	89 " "
"	4,179	4 " "	"	9,547	57 " "
"	4,224	5 " "	"	9,503	72 " "
"	4,122	2 " "	"	9,463	58 " "
20°	6,760	5 " "			

TABLE II.—*Ranges of a 3-pounder Whitworth Gun, at 20° Elevation. Charge 7½ oz. of Powder.*

Yards Range.	Deviation from Line of Fire.	Yards Range.	Deviation from Line of Fire.
6,818	26 yards to the left.	6,561	20 yards to the left.
6,749	27 " "	6,316	20 " "
6,602	54 " "	6,469	11 " "
6,556	35 " "	6,389	12 " "
6,511	34 " "		

TABLE III.—*Ranges of a 12-pounder Whitworth Gun; length 7 ft. 9 in., weight 8 cwt., diameter of bore 3½ in., with a charge of 1¼ lbs. of powder, at elevations of 2°, 5°, and 10°.*

Angles of Elevation.	Yards Range.	Deviation from Line of Fire.	Angles of Elevation.	Yards Range.	Deviation from Line of Fire.
2°	1,280	¼ yard to the right.	5°	2,333	2 yards to the left.
"	1,270	⅞ yard to the left.	"	2,298	1 yard to the left.
"	1,257	½ " "	10°	3,942	15 yards to the right.
"	1,254	1¼ yards to the right.	"	4,120	13 " "
"	1,208	½ " "	"	4,011	7 " "
5°	2,342	4 yards to the left.	"	4,002	16 " "
"	2,321	On the line.	"	4,059	9 " "
"	2,326	1 yard to the right.			

TABLE IV.—*Ranges of an 80-pounder Whitworth Gun; weight 4 tons, with a charge of 10 lbs. of powder, and a solid shot of 90 lbs. weight, at elevations of 5°, 7°, and 10°.*

Angles of Elevation.	Yards Range.	Deviation from Line of Fire.	Angles of Elevation.	Yards Range.	Deviation from Line of Fire.
5°	2,544	5 yards to the right.	7°	3,487	6¼ yards to the right.
"	2,604	2 " "	"	3,482	6¼ " "
7°	3,503	4¼ " "	10°	4,700	5 " "
"	3,498	6 " "	"	4,409	6 " "

All serviceable cannon, whether of bronze or iron, may be rifled for the use of General James's projectile. It is, therefore, an invention of the greatest practical utility, and the author is much indebted to Major W. A. Thornton, U. S. Ordnance Department, for the following description and experiments made by a board of officers of the U. S. Army:

General James's Projectile—is a cylindro-conoidal missile of cast-iron, having a compound envelop of canvas—sheet tin, and lead, called packing, encircling nearly the entire length of the body of the cylinder. The canvas, being the external portion of the packing, is well saturated with a tallow lubric, which renders the loading easy, and cleans the gun at each discharge. The head of the projectile may be solid, or, if it has a prepared cavity, the missile then becomes a shell. The average weight of the projectile for a 42-pounder gun is, if a solid, $81\frac{1}{4}$ lbs., if a shell, $64\frac{1}{4}$ lbs., of which in either case $6\frac{1}{2}$ lbs. is the weight of the packing. Its length is 13 inches, of which $6\frac{1}{4}$ inches is the measurement of the conical head, and $6\frac{3}{4}$ inches is the length of its cylindrical body. The diameter of the cylinder is $6\frac{3}{4}$ inches, or $\frac{1}{4}$ of an inch less than the bore of a 42-pounder gun. It retains its full diameter for $\frac{3}{4}$ of an inch of its length at each end:—then for the intermediate space, the diameter is shortened half an inch, thereby forming a recess round the body of the cylinder, between the ends; (Fig. 199.) The shortening of the diameter, and consequent loss of iron to the circumference of the body of the cylinder, is replaced by the before-named

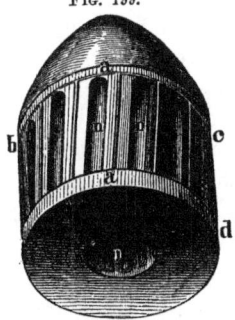

Fig. 199.

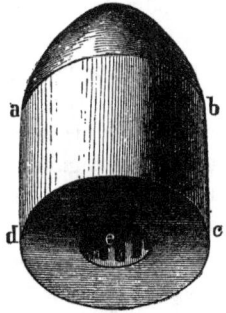

Fig. 200.

JAMES'S SHELL, BEFORE THE APPLICATION OF THE PACKING.
    a. Band $\frac{3}{4}$ inch wide at ends of cylinder.
b, c, d, e. Recess round body of cylinder.
    m. Rectangular openings through to recess.
    n. Orifice in base, leading to the recess.

JAMES'S SHELL, AFTER THE APPLICATION OF THE PACKING, AND READY FOR USE.
a, b, c, d. Belt of canvas, tin, and lead, called packing.
    e. Orifice in base, leading to recess.

packing, when the projectile is prepared for use, (Fig. 200.) The solidity of the conical head is continued into, and forms the solid end of the cylinder. The base, or opposite end of the projectile, has a central orifice, of $3\frac{1}{4}$ inches in diameter, which extends $2\frac{1}{2}$ inches into the cylinder; and from which *ten* rectangular openings diverge, (like the mortises for spokes in the hub of a wheel,) through the body, to the periphery of the cylinder, in the recess of its circumference. The packing is formed by a plate of sheet tin, of the length of the greatest circle of the cylinder; and in width, equal to the length of the recess caused by the shortening of the diameter. This plate of tin is laid on a piece of strong canvas, which is two inches wider, but of the same length of the plate; and the canvas is folded over the side edges of the plate and firmly secured by cross sewing. The tin plate, when so prepared, or half covered, is folded round the body of the missile in the recess, and retained in position by an iron collar clamp; (Figs. 199 and 200.) The space between the inner surface of the envelop and the body of the cylinder is filled with melted lead, which, adhering to the tin and iron, forms a compact mass round the body of the projectile. When the charge is fired, the power or gas generated by the burning of the powder, in its effort to expel the projectile and to escape from the gun, is forced into the orifice in the base of the missile, and through the *ten* openings against the packing, which is thereby pressed into the grooves, in the gun's bore, and by its firm hold in them the rifle motion is imparted to the projectile. The packing has not been known to strip from the projectile while in the gun; and the certainty that it compels obedience on the part of the missile to the rifling, is demonstrated in direct hits, by the perfect circular orifice cut by the shot in entering targets; and when the projectiles are obtained after firing, their head and body are frequently found cut in furrows, conforming to the rifling of the gun, by stones, against which the missile impinged in entering the ground. All serviceable guns, either of bronze or iron, can be made available by rifling, for the use of the said projectiles. The rifling should be of the gain twist nature. It should be shallow; say, for field-guns $\frac{1}{20}$, and for siege-guns $\frac{1}{13}$, and $\frac{1}{17}$ of an inch in depth. The lands and grooving should be of the same width, and about 18 of each, for the bore of a 42-pounder gun. The ordinary grained cannon powder does not appear to act too violently in projecting these heavy missiles from field-guns; but there can be no doubt that the coarse-grained $\frac{6}{10}$ inch powder is far the best for service, in firing James's projectiles from long-bored guns.

When the projectile is a shell, (Fig. 201,) its fuze-orifice is in its head and axis. The length of the orifice for a 42-pounder shell is $2\frac{1}{2}$ inches. For two inches of its length, its diameter is 1 inch, and for the remainder of the length, the diameter is reduced to $\frac{3}{4}$ of an inch; so forming a shoulder in the fuze-orifice, to prevent the fuze-plug from being driven into the cavity of the shell, when, by firing, the missile is expelled from the gun. The threads of a female screw are cut in the head of the fuze-orifice for the reception of the body of the fuze-orifice cap. This cap is of brass. Its diameter is an inch, its length half an inch; its head is convexed, and has a slot cut in it for the reception of a screw-driver; the base end is deeply cupped, to admit the nipple of a musket cone, and to give more play to the fuze-plug.

FIG. 201.

JAMES'S SHELL.
Section through the axis.
a. Brass fuze-orifice screw-cap.
b. Fuze slide-plug.
c. Cone to fuze-plug—musket size.
d. Lead portion of packing.
e. Canvas and tin portion of packing.
m. Rectangular openings to periphery in recess.

The fuze-plug is of wrought iron, surmounted by a musket cone; and its action in the fuze-orifice is like the ordinary working of a piston. Its length is 1¼ inches, of which the quarter is the length of its shoulder. The diameter of its shoulder and body, is very nearly the same as the two diameters of the fuze-orifice. Its vent is in its axis, and in size to receive the male screw of the musket cone. The threads of a female screw are cut in the head end of the vent, of sufficient length to receive the screw end of the said cone.

When the shell is loaded, care should be taken not to overfill its cavity, and thereby prevent the working of the fuze-plug. The powder should be cleaned from the fuze-orifice; the plug should be oiled to ensure its free and sure action. Its cone should be capped, but before the application the percussion cap should be carefully examined to see that it is perfect, and of the best quality. The fuze-plug, when so prepared, is then inserted into the fuze-orifice, and it should enter freely but not by its own weight, until the shoulders of the fuze-plug and orifice are in contact. The cap for the fuze-orifice should be then firmly screwed in, which completes the charging of the shells. If after the shell is loaded the fuze-plug should be disturbed by handling; that is, if the plug has slidden forward, it will be forced back to its proper position by the impulse given to the missile, by the firing of the gun charge; and it will so remain during the flight, until the shell impinges against any hard substance; as ground, wood, &c., which, by obstructing the progress of the missile, causes the fuze-plug to slide forward with violence, and by the collision of the cone's point against the bottom of the fuze-orifice cap-plug, the percussion cap on the cone will be exploded, and the bursting charge of the shell fired.

GENERAL C. T. JAMES'S PROJECTILE. SUMMARY OF TARGET-FIRING, WATCH HILL, R. I., 1860.

42-pdr. Service Gun, Rifled.

81¼ lbs. averaged weight of projectile, of which 6½ lbs. packing,
2 " of powder, the loading charge of shell,
8 " averaged weight of charge of powder,

2° ⅛″ elevation,
3⅛″ time of flight to target,
45 projectiles fired,
31 hits direct,
8 hits ricochet,
68.8 proportional direct hits per 100 shots
17.7 " ricochet " "

target 20 by 40 feet.
distance 1,000 yards.

5° 2′ elevation,
6 ⅖″ time of flight to target,
65 projectiles fired,
15 hits direct,
7 hits ricochet,
23 proportional direct hits per 100 shots
10.7 " ricochet " "

target 20 by 40 feet.
distance 2,000 yards.

Remarks.—The averaged weight of the projectile at rest in the gun was 81¼ lbs.; averaged weight of packing thrown off was 6½ lbs.; weight of projectile when it impinged, 74¾ lbs. Penetration, through 45 inches of the best well-seasoned oak, at 2,000 yards; weight of oak target 17 tons, well bedded and firmly braced by back timbers; forced back 10 inches by impact of shot; range, at 15° elevation 4,346 yards, or nearly 2½ miles; ricochet on water, in prolongation of line of fire, but the projectile does not bound as often as round balls. When the missile is a shell loaded, it bursts by percussion, in penetrating earth, or other denser material.

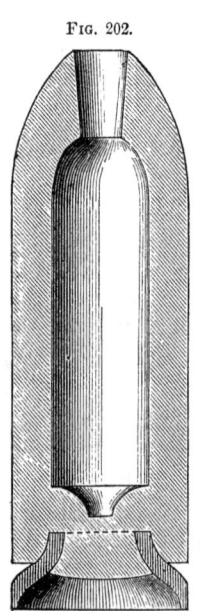

FIG. 202.

The Reed projectile is also an American invention. Its peculiarity, whether shot or shell, consists in its having a base or cup of wrought iron connected by casting in, or in any other mode of attachment, to the cast-iron projectile, (Fig. 202.) The object is to obtain a material pliable enough to be forced by the expansive action of the powder into the grooves of the gun, and strong enough to give the

SYNOPSIS OF EXPERIMENTS WITH RIFLED GUNS AT FORT MONROE, TARGET 40' BY 20', FROM ROBERTS' HAND-BOOK OF ARTILLERY.

Name of Gun	Calibre	Bore. Diameter	Bore. Length	Grooves. No.	Grooves. Width	Grooves. Depth	Twist	Weight of Gun	Weight of Projectile	Weight of charge	1,000 Yards. No. of shots fired	1,000 Yards. No. of direct hits	1,000 Yards. No. of ricochet hits	1,000 Yards. Angles of Elevation	1,000 Yards. Time of flight	2,000 Yards. No. of shots fired	2,000 Yards. No. of direct hits	2,000 Yards. No. of ricochet hits	2,000 Yards. Angle of Elevation	2,000 Yards. Time of flight	Average range	Corresponding Elevation	Time of flight
		in.	in.		in.	in.		lbs.	lbs.	lb.				° '					° '			° '	
Sawyer	24-pdr.	5.862	110	6	1.5	0.25	Uniform, one turn in 34⅓ feet.	8,822	45	5¼	15	13	2	2 2		119	32	17	4 30	6	4,859	18¾	
Dimick	32-pdr.	6.4	101	6	2.0	0.2 rectan.	Increasing from 0 to one turn in 62½ ft. at muzzle; twist to the right.	9,300	51	6	7	5	1	2 15		58	21	6	5	6¼			
Dr. Reed	12-pdr. Siege.	4.854	109	7	1-14th circum.	.03 to .08	Increasing from 0 at commencement to one turn in 50 feet at muzzle.	5,000	22	3	26	14	9	2 15		30	5	8	4 30				
Do.	12-pdr. Field.	4.636	74	7	do.	do.	Do. do.	1,900	15	2	48	16	3	2	3								
Do.	32-pdr.	6.425	110	8	1-6th circum.	.085 to .12 circular.	Uniform, one turn in 40 feet.	8,500	50	6	10	8	2	2 15	3	84	19	8	5	6¼	3,665	11 30	
Do.	6-pdr.	3.69	103.4	8	do.	.077 to .111 circular.	Uniform, to the right, one turn in 25 feet.	1,200	12	1¼	28	18	4	2 10		52	9	5	4 45	7			
Capt. Dyer	3-pdr.	2.9	44.5	8	0.4	.05	Uniform, one turn in 16 feet.	250	9	1	28	16	5	2 25		18	4	2	5¼		3,270	13 30	15
Do.	6-pdr. bronze.	3.67	57.5	16	0.5	.025	Uniform, one turn in 19 feet.	880	14	1¼	22	11	4	2 15									

The following is a description of the several projectiles, viz.:—

Sawyer's.—Flanged projectile; elongated; entire shell coated with an alloy chiefly of lead, and has a percussion cup on small end.

Dimick's.—Expanding shell; elongated; cup of soft metal cast on rear end of projectile.

Reed's.—The body is of cast iron, and the expanding portion is a cup of wrought iron, which is fastened to the body by inserting it in the mould and pouring the melted metal around it.

Dyer's.—Description nearly the same as that of Dimick's.

necessary rotative movement to the projectile resulting from the twist of these grooves. The action is in fact similar to that of the common elongated bullet for the rifle musket, or the application of the Minié ball to cannon. The projectile is 2.9 inches. R. P. Parrott, Esq., West Point Foundry, has produced a field-gun for firing this elongated bullet reinforced by wrought iron, the idea of which is not novel, but which he claims to have arranged in proper proportions, and otherwise to have brought into practical shape so as to make a safe, cheap, and good rifled cannon. The gun has, in reference to the projectile, three grooves and a twist of one turn in 10 feet. It has not yet been before a board, but has been successfully tried before officers of the army. (Consult Sir Howard Douglas; Hyde; Wilcox. *See* Ammunition; Arms; Bullet; Carbine; Firing; Percussion; Projectile.)

RIFLE PITS—are holes or short trenches, about four feet long and three feet deep, forming, with the earth thrown out in front of them, cover for two men. There is generally a loophole on the top of the breastwork, made, by placing two sand-bags across the parapet, and a third resting on these, in the direction of it, to cover the head and shoulders of the riflemen. A rifle pit of this construction is shown in plan, section, and elevation in Fig. 203.

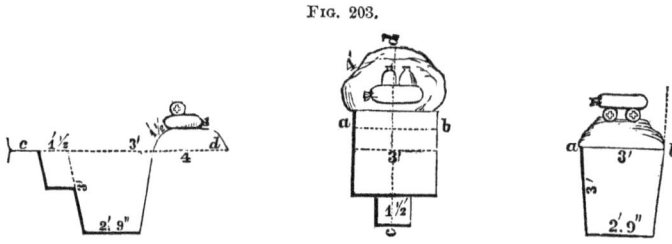

Fig. 203.

RIOT. (*See* Execution of Laws.)

ROADS. When it is proposed to construct a line of road, extending between two places, the officer upon whom such duty devolves, first makes himself well acquainted with the surface of the country lying between the two places; he is then to select what he thinks, all circumstances being taken into consideration, the best general route for the proposed road. But previously to laying it out with accuracy, it is necessary to make an instrumental survey of the country, along the route thus selected; taking the levels from point to point throughout the whole distance, and making borings in all places where excavations are required, to determine the strata through which such cuttings are to be carried, and the requisite inclinations of the slopes or slanting sides as well of the cuttings as of the embankments to be formed by the material thus obtained. It is also requisite, in the selection of the route for the proposed road, to have regard to the supply of materials, not only for first constructing it, but for maintaining it in repair. The results of such an investigation should be reduced to plan and section; the plan of the road being on a scale not less than 66 yards to an inch, and the section not less than 30 feet to an inch. The loss of *tractive* power and consequent danger produced by steep acclivities, render it necessary that a proper limitation should be imposed on the acclivities or inclinations on every line of road. As, however, this reduction of hills in a country where much inequality of surface exists, is attended with great labor and expense, greater rates of inclination must be allowed to hills or roads where the traffic is not sufficient to repay the expense of excavations. A dead level, even where it can be obtained, is not the best course for a road; a certain inclination of the surface facilitates the drainage, and keeps the road in a dry state. There is a certain inclination or acclivity, which causes, at a uniform speed, the traces to slacken, and the carriages press on the horses, unless a drag or break is used; the limiting inclination within which this effect does not take place is called the *angle of repose*. On all acclivities less steep than the angle of repose, a certain amount of tractive force is necessary in the descent, as well as in the ascent; and the means of the two drawing forces, ascending and descending, is equal to the force along a level road. The exact course of the road, and the degree of its acclivities being determined, the next thing to be considered is the formation of its surface. The qualities which ought to be imparted to it, are twofold: first, it should be smooth; secondly, it should be hard; and the goodness of the road will be exactly in proportion as these qualities can be imparted to it, and permanently maintained upon it. The means resorted to accomplish these objects are: 1. *Gravel Roads.* A coating of four inches of gravel should be spread over the road bed, and vehicles allowed to pass over it, till it becomes tolerably firm—men being required to rake in the ruts as fast as they appear; a second coating of 3 or 4 inches of gravel should be then added and treated like the first, and finally a third coating. 2. *Broken Stone Roads*, or McAdam roads. French engineers value uniformity in size of the broken stone less than McAdam. They use all sizes from $1\frac{1}{2}$ inches to dust. McAdam considers from 7 to 10 inches of depth of stone on the road sufficient for any purpose. He earnestly advocates the principle, that the whole science of road-making consists in making a solid dry path on the natural soil, and then keeping it dry by a durable waterproof coating. 3. Broken stone roads with a paved bottom or foundation, or *Tilford Roads*; a road thus constructed will, in most cases, cost less than one entirely of broken stone. 4. *Roads of Wood.* The abundance, and consequent cheapness of wood, renders its employment in road-making of great value. It has been used in the form of logs, of charcoal, of planks, and of blocks. When a road passes over soft swampy ground it is often made passable by felling straight young trees, and laying them side by side across the road at right angles to its length. This is the primitive *corduroy* road. A very good road has been lately made through a swampy forest, by felling and burning the timber, and covering the surface with charcoal thus prepared. Timber from 6 to 18 inches through is cut 24 feet long, and piled up lengthwise in the centre of the road about five feet high, and then covered with straw and earth in the manner of coal pits. The earth required leaves two good ditches, and the timber, though not split, is easily charred; and when charred the earth is removed to the side of the ditches, and the coal raked down to a width of 15 feet, leaving it two feet thick at the centre and one at the sides. 5. *Plank Roads.* Two parallel rows of small sticks of timber (called sleepers) are imbedded in the road three or four feet apart. Planks, 8 feet long and 3 or 4 inches thick, are laid on these sleepers across them. A side track of earth to turn out upon is carefully graded. Deep ditches are dug on each side to ensure perfect drainage; and thus we have the plank road. 6. *Roads of Earth.* These roads are deficient in the important requisites of smoothness and hardness, but they are the only roads usually made in the field to carry on military operations. Its shape, when well made, is properly formed with a slope of 1 in 20 each way from the centre. Its drainage should be made thorough by deep and capacious ditches, sloping not less than 1 in 125. Trees should be removed from the borders of the road, so as not to intercept the sun and wind. The labor expended upon it, will, however, depend upon circumstances. Every hole or rut in the road should, however, be at once filled up with good materials, for the wheels fall into them like hammers, deepening them at each stroke and thus increasing the destructive effect of the next wheel. (Consult GILLESPIE, *Roads and Road-making.*) The cross-section of a road embraces: 1. *The width of the road*—from $16\frac{1}{2}$ to 30 feet, according to its importance, and the amount of travel upon it. 2. *The shape of the road-bed.* The best shape of the transverse profile for a road on level ground is two inclined planes meeting in the centre of road, and having their angle slightly rounded. On a steep hill, the transverse profile should be a single slope inclining inwards to the face of the hill. 3. *Footpaths, &c.* 4. *Ditches.* The ditches should, if possible, lead to the natural water-courses of the country. 5. *The side slopes of the cuttings and fillings.* These vary with the nature of the soil.

ROCKET, (War.) A projectile set in motion by a force within itself. It is composed of a strong case of paper or wrought iron, inclosing a composition of nitre, charcoal, and sulphur; so proportioned as to burn slower than gunpowder. The head is either a *solid shot*, shell, or spherical-case shot. The base is perforated by one or more vents, and in the case of the Congreve rocket, with a screw hole to which a guide-stick is fastened. The rockets used in the United States service are Hale's, in which steadiness is given to the flight of the rocket by rotation, as in the case of the rifle ball, around the long axis of the rocket. This rotation is produced by three small vents placed at the base of the head of the rocket. Fig. 204 shows Hale's rocket now used in the United States. Mr. Hale's last improvement (Fig. 205) consists in

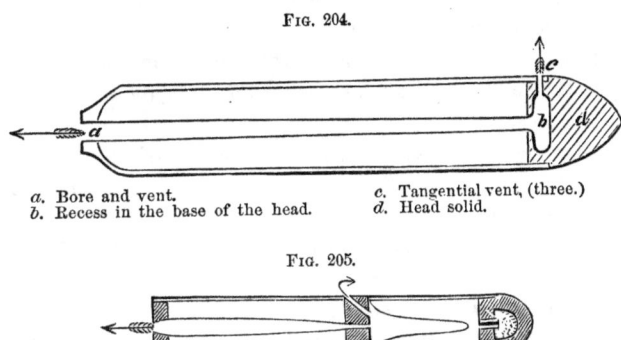

Fig. 204.

a. Bore and vent.
b. Recess in the base of the head.
c. Tangential vent, (three.)
d. Head solid.

Fig. 205.

placing three tangential vents in a plane passing through the centre of gravity of the rocket, and at right angles to the axis. This is accomplished by dividing the case into two distinct parts, or rockets, by a perforated partition. The composition in the front part furnishes the gas for rotation, and that in the rear the gas for propulsion. The two sizes of Hale's rockets in use, are the

$2\frac{1}{4}$ inch, (diameter of case,) weighing 6 lbs.; and
$3\frac{1}{4}$ inch   "   "   "   16 lbs.

Under an angle of from 4° to 5° the range of these rockets is from 500 to 600 yards, and under an angle of 47° the range of the former is 1,760 yds., and the latter 2,200 yards. War rockets are usually fired from tubes or troughs, mounted on portable stands, or on light carriages.

The following rules concerning the length of rocket-fuzes, the ranges and elevations, for Congreve's rockets, may be useful, though they have not been confirmed by an extensive course of practice:—

For 24-pounder rockets; if the whole length of the fuze is left in the shell of the 4-pounder rocket, it may be expected to burst at about 3,700 yards, elevation 47 degrees.

If the whole of the fuze-composition be bored out, and the rocket-composition left entire, the shell may be expected to burst at about 2,000 yards, elevation 27 degrees.

If the rocket-composition be bored into, to within 1.5 inch of the top of the cone, the shell may be expected to burst at about 700 yards, elevation 17 degrees.

For 12-pounder rockets; if the whole length of fuze be left in the shell of the 12-pounder rocket, it may be expected to burst at about 3,000 yards, elevation 40 degrees.

If the whole of the fuze-composition be bored out, and the rocket-composition left entire, the shell may be expected to burst at about 1,500 yards, elevation 20 degrees.

If the rocket-composition be bored into, to within one inch of the top of the cone, the shell may be expected to burst at about 420 yards, elevation 10 degrees.

For 6-pounder rockets; if the whole length of fuze be left in the shell of the 6-pounder rocket, it may be expected to burst at about 2,300 yards, elevation 37 degrees.

If the whole of the fuze-composition be bored out, and the rocket-composition be left entire, the shell may be expected to burst at about 1,100 yards, elevation 15 degrees.

If the rocket-composition be bored into within one inch of the top of the cone, the shell may be expected to burst at about 20 yards, elevation 10 degrees.

For 3-pounder rockets; if the whole length of the fuze be left in the shell of the 3-pounder rocket, it may be expected to burst at about 1,800 yards, elevation 25 degrees.

If the whole of the fuze-composition be bored out, and the rocket-composition be left entire, the shell may be expected to burst at about 850 yards, elevation 12 degrees.

If the rocket composition be bored into within one inch of the top of the cone, the shell may be expected to burst at about 420 yards, elevation 8 degrees; (Sir HOWARD DOUGLAS.)

ROLL. A uniform beat of the drum, without variation for a certain length of time.

Long-roll.—A beat of the drum, as a signal for the assembling of troops at any parade.

Muster-roll.—A return, forwarded every two months from every company in the service to the adj.-general and paymaster. It contains a list of the officers, non-commissioned officers, and privates, specifying their pay, and the casualties arising from deaths, promotions, &c.

ROSTER or ROLLSTER. Lists of officers for duty. The principle which governs details for duty is from the eldest down; longest off duty first on. If an officer's tour of duty for armed service, court-martial, or fatigue happen when he is upon either duty, he is credited with both duties. A regiment, or detachment, detailed for any duty, receives credit for the duty when it marches off parade to perform the duty, but not if it is dismissed on parade. Officers on inlying pickets are subject to all details.

ROUNDS. Visiting rounds; grand rounds; visiting small posts, guards, and sentinels by commanders or staff officers. He who makes the round is alone, or accompanied according to grade and circumstances.

ROUT. To put to *rout* is to defeat and throw into confusion. It is not a retreat in good order, but also implies dispersion.

ROUTE. An open road; the course of march of troops. Instructions for the march of detachments, specifying daily marches, means of supply, are given from the head-quarters of an army in the field, and are called marching *routes*.

RUFFLE. A low, vibrating sound beat upon a drum not so loud as a roll.

RULES and ARTICLES of WAR. (*See* ARTICLES of WAR.)
RUN; RUNNING. (*See* MANŒUVRES of INFANTRY IN COMBATS.)
RUNNING FIRE. Rapid and successive fire by troops.

S

SABOT. In *field-guns*, when firing solid shot, the charge is usually about $\frac{1}{5}$ the weight of the shot. For spherical case and canister, the charge is less. These projectiles are always fixed to a block of wood, called a *sabot*, (Fig. 206,) to which the cartridge is also attached; forming what is called a round of *fixed ammunition*; (Fig. 207.) In the 12-pdr. field-howitzer, also, the ammunition used is fixed, A, (Fig. 206;) but with the other howitzers the projectile and charge are separate; the latter being attached to a block of wood called a *cartridge-block*, (Fig. 208,) the object of which is to give a finish to the cartridge and fill the chamber, the dimensions of the block being so calculated for each different charge as to reach to the mouth of the chamber. The sabots used with these heavy howitzers are conical in shape to fit the connecting surface between the chamber and bore. Care should be taken in loading to put the seam of the cartridge to the sides, so that it will not come under the vent. In loading the 32 and 24-pdr. howitzer, the cartridge is first pushed carefully into the chamber without ramming, and the shell is then sent home, also without ramming.

FIG. 206. FOR GUNS.

Shot.

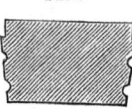
Canister.

12-PDR. HOWITZER.

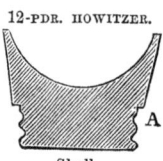
Shell.

Canister.

FIG. 207.

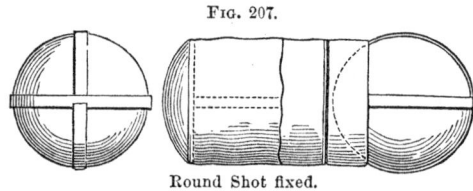
Round Shot fixed.

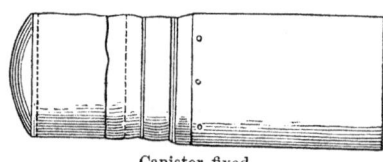
Canister fixed.

FIG. 208.
Cartridge Block.

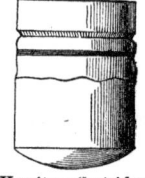
Howitzer Cartridge.

When sabots cannot be obtained, place upon the powder a layer of tow, about 0.2 in. thick, forming a bed for the shot; tie the bag over the shot and around the tow; the bag requires to be one inch longer than for strapped shot; (GIBBON.)

SABRE. The cavalry sabre blade has shoulder, back, edge, bevel point, curvature, large groove, small groove, tang reveting. The HILT has a brass surmounting (gilt for officers) guard, and steel scabbard. The blade of the mounted artillery sabre has but one groove; the guard but one branch, (cavalry sabre guard has three;) steel scabbard. Officers of mounted artillery, and mounted officers of artillery and infantry use the sabre for mounted artillery with gilt mounting. (*See* SWORD.)

SABRETASCHE. From the German, *Sabel*, a sabre, and *Tasche*, a pocket. The sabretasche is part of the accoutrements of a cavalry or staff officer, consisting of a leathern case or pocket, suspended at the left side from the sword belt by three slings, corresponding with the belt.

SACK. An expression used when a town has been taken by storm, and given up to pillage.

SADDLER. All acts of Congress previous to the Act of March 2, 1833, allowed one saddler to each company of dragoons. The omission to provide for saddlers in the present cavalry organization would seem to be accidental.

SAFEGUARD. Whosoever, belonging to the armies of the United States, employed in foreign parts, shall force a safeguard, shall suffer death; (ART. 55.) The men left with a safeguard may require of the persons for whose benefit they are so left, reasonable subsistence and lodging; and the neighboring inhabitants will be held responsible by the army for any violence done them.

The bearers of a safeguard left by one corps, may be replaced by the corps that follows; and if the country be evacuated, they will be recalled; or they may be instructed to wait for the arrival of the enemy, and demand of him a safe conduct to the outposts of the army. The following form will be used:—

SAFEGUARD.

By authority of Major-gen. ———, (or *Brig'r-gen.* ———.)

The person, the property, and the family of ———, (or such a college, and the persons and things belonging to it; such a mill, &c.,) *are placed under the safeguard of the United States. To offer any violence or injury to them is expressly forbidden; on the contrary, it is ordered that safety and protection be given to him, or them, in case of need.*

Done at the head-quarters of ———, this ——— day of ———, 18—.

Forms of safeguards ought to be printed in blank, headed by the article of war relative thereto, and held ready to be filled up, as occasions may offer. A duplicate, &c., in each case, might be affixed to the houses, or edifices, to which they relate.

SALE. The President is authorized to cause to be sold unserviceable ordnance or stores of any kind, but the inspection or survey of unserviceable stores shall be made by an inspector-general, or such other officer or officers as the Secretary of War may appoint for that purpose; and the sales shall be made under such rules and regulations as may be prescribed by the Secretary of War; (*Act* March 3, 1825.) In all cases where lands have been, or shall hereafter be, conveyed to or for the United States, for forts, arsenals, dock-yards, light-houses, or any like purpose, or in payment of debts due the United States, which shall not be used, or necessary for the purposes for which they were purchased, or other authorized purpose, it shall be lawful for the President of the United States to cause the same to be sold, for the best price to be obtained, and to convey the same to the purchaser by grant or otherwise; (*Act* April 28, 1828.)

SALIENT. The salient angle of a fortification is an angle projecting towards the country.

SALLY. A sally or sortie is a movement made by strong detachments from a besieged place to attack the besiegers or destroy their works.

SALLY-PORTS. Openings to afford free egress to troops for a sortie. They are cut in the faces of the re-entering places of arms, and in the middle of the branches of the covered-ways. When sally-ports are not in use, they are closed by strongly constructed gates of timber supported by bars of iron.

SALTPETRE. (*See* GUNPOWDER; NITRE.)

SALUTE. A discharge of artillery in compliment to some individual; beating of drums and dropping of colors for the same purpose; or by carrying or presenting arms according to the rank and position of an officer.

SAND-BAGS. Bags filled with earth, usually from 12 to 14 inches wide, and about 30 inches long. They are employed sometimes in constructing batteries, and in repairing breaches and embrasures when damaged by the enemy's fire. (*See* REVETMENT.)

SANITARY PRECAUTIONS. Send troops where we may, they are destroyed by fevers. Is there any safeguard? None, but in the good keeping, good condition, physical and moral, of the troops. After a fever has been established, physic does little, but the battle is fought by the nurse; let that attendant be sagacious and vigilant, and the patient is saved; the contrary, and he dies. The most successful treatment (the necessary evacuations always being premised) is cold water, or, in other words, the regulation of the temperature. Fever, when once it has gained entry, is the most tenacious of all pre-occupants. Rhythm, the rule of number counting by day, as if it played upon the nervous chords, paroxysm, remission and crisis, proclaim its sway. Let the practitioner obviate evil tendencies whenever he can, but if he turn to his medical books he will find in the medical records of two thousand years always the same results, viz.: the futility of interfering with medicines of specific power, and the deaths of a given number, almost always the same, when the air is pure, and the patient has had any thing like fair play. *Quinine* is a specific in intermittent fever, but it is as futile as all other specifics in continued fevers. The practitioner must content himself with taking for his guides depletion at the outset, refrigeration during all the middle stages, and stimulation with support at the close of the disease. This course may be taken with very little aid from medicine, and the event will be more successful than if the patient had been drugged with all the stuff of an apothecary's shop.

Disinfectants.—The best disinfectants are caloric, light, ventilation, and the operation of water, and a bountiful Providence has placed them all at our disposal. It is a matter of experiment that even the concentrated matter of small-pox, cow-pox, and the fomites of scarlatina are deprived of all infecting power on being subjected to a heat of 140° of Fahrenheit's thermometer. It may then be fairly inferred that if these could be so neutralized, gaseous factitious infectants, such as that of typhus fever, would be dissipated under a much inferior degree of heat, and it is accordingly found that typhus will not readily cross the tropic of cancer, and the plague of the Levant goes out at the same boundary. *Boiling water*, then, must be all-sufficient for the purification of whatever it can be made to touch; and a portable iron stove, filled with ignited charcoal, will infallibly disinfect any building or apartment. The infection constantly given out from a living body cannot, while it continues diseased, be so disposed of; but all that it has inhabited is easily rendered harmless.

Light is another sure disinfectant; the strongest poisons, as prussic acid, when exposed to its influence lose their power.

Ventilation comprehends all that the atmosphere can bring to the process of disinfection; and *water* is only a more concentrated application of the same principle. *Chlorine fumigation* is utterly *useless*, " but the burning of a few handfuls of charcoal, with the aid of clean linen, will certainly disinfect the most saturated lazar that ever came out of a pest-house; but until that ceremony, or an equivalent to it, such as a hot bath, be performed, no one can answer for his being otherwise than dangerous."

Dysentery is truly an army disease. In some services the soldiery in the field may escape fever, but never dysentery if they lie on the ground. Atmospherical vicissitudes, cold of the night, chill of the morning, after heat of preceding day, will cause it to spread. Heat is, however, uniformly the remote cause. The disease is purely inflammatory in the beginning; yet, because the acid and sub-acid fruits sometimes occasion griping when in health, these and vegetables of every kind are sometimes strictly prohibited. They are, however, amongst the best remedies. For the peculiar inflammation which dysentery sets up in the mucous linings of the intestines, there has been no remedy yet discovered at all comparable to mercury, (calomel.) The specific inflammations, such as the iritic, the hepatic, the pneumonic, the syphilitic, &c., all fall before its peculiar superseding stimulus. The habitual use of mercury is not fitted to all constitutions, and it has often been abused; but the discovery of its power to supersede inflammation is one of the happiest of the uncertain art of medicine.

Miasmata or marsh poisons, it has been supposed, are exhalations produced by the agency of vegetable or aqueous putrefaction. More general knowledge has, however, established the fact, that one condition only is necessary to the production of miasma on all surfaces capable of absorption, and that is, the *paucity of water where it has previously and recently abounded.* The greatest danger may exist, where there is no evidence of putrefaction, as every one can testify who has seen pestilence steam forth, to the paralyzation of armies, from the barren sands of the Alentyo in Portugal, the arid burnt plains of Estremadura in Spain, and the recently flooded table-lands of Barbadoes, which have seldom more than a foot of soil to cover the coral rock, and are therefore, under the drying process of a tropical sun, brought almost immediately after the rains into a state to give out pestilential miasmata. It is not known whether miasma is lighter or heavier than air, but it is established that the inhabitants of ground floors are affected by it in a greater proportion than those of upper stories; and that this is caused by its attraction by the earth's surface is proved by its creeping along the ground, and concentrating and collecting on the sides of adjacent hills, instead of floating directly upwards in the atmosphere. Miasma is certainly lost and absorbed *by passing over a small surface of water.* The rarefying heat of the sun, too, certainly dispels it, and it is only during the cooler temperature of the night that it acquires body, concentration, and power. All regular currents of wind have also the same effect. The leeward shore of Guadaloupe, for a course of nearly thirty miles, under the shelter of a very high steep ridge of volcanic mountains, never felt the sea breeze, nor any breeze but the night land wind from the mountains; and though the soil is a remarkably open, dry, and pure one, being mostly sand and gravel, altogether and positively without marsh in the most dangerous places, it is inconceivably pestiferous throughout the whole tract, and in no spot more so than the bare sandy beach near *the high water mark.* The colored people alone ever venture to inhabit it, and when they see strangers tarrying on the shore after nightfall, they never fail to warn them of their danger.

The chief predisposing causes of every epidemic, and especially of cholera, are: damp, moisture, filth, animal and vegetable matters in a state of decomposition, and in general, whatever produces atmospherical impurity; which always have the effect of lowering the health and vigor of the system, and of increasing the susceptibility to disease. Attacks of *cholera* are uniformly found to be most frequent and virulent in low-lying districts, on the banks of rivers, in the neighborhood of sewer mouths, and wherever there are large collections of refuse, particularly amidst human dwellings. The practical precautions given in Russia are " to keep the person and dwelling-place clean, to allow of no sinks close to the house, to admit of no poultry or animals within the house, to keep every apartment as airy as possible by ventilation, and to prevent crowding wherever there are sick." Next to perfect cleansing of the premises, *dryness* ought to be carefully promoted, by keeping up in damp and unhealthy districts sufficient fires, and this agent will promote ventilation as well as warmth and dryness. If, notwithstanding these precautions, cholera break out, the premonitory symptom of looseness of the bowels almost universally precedes the setting in of the more dangerous state of the disease. This looseness of the bowels may be accompanied with some degree of pain, but in many cases pain is wholly absent, and for some hours or even days the bowel complaint may appear so slight, without previous knowledge of the importance of its warning, as to escape notice altogether. But when the Asiatic cholera is epidemic, never neglect the slightest degree of looseness of the bowels. If neglected only a few hours, it may suddenly assume the most fatal form. The most simple remedies will suffice, if given on the first manifestation of the premonitory symptom, and the following, which are within the reach and management of every one, may be regarded as among the most useful, namely: twenty grains of opiate confection, mixed with two tablespoonfuls of peppermint water, or with a little weak brandy and water, and repeated every three or four hours, or oftener, if the attack is severe, until the looseness is stopped; or an ounce of the compound chalk mixture, with ten or fifteen grains of the aromatic confection, and from five to ten drops of laudanum repeated in the same manner. From half a drachm to a drachm of tincture of catechu may be added to the last, if the attack is severe. Half these quantities should be given to young persons under 15, and still smaller doses to infants. It is recommended to repeat these remedies night and morning for some days after the looseness of the bowels has been stopped, *and in all cases to have recourse to medical advice as soon as possible.* Next in importance to the immediate employment of such remedies, is attention to proper diet and clothing. The most wholesome articles of vegetable diet are *well-baked but not new bread, rice, oatmeal, and good potatoes. The diet should be solid rather than fluid,* and with the means of choosing, it is better to live principally upon animal food, as affording the most concentrated and invigorating diet—avoiding salted and smoked meats, pork, salted and shell-fish, cider, perry, ginger beer, lemonade, acid, liquors of all description, and ardent spirits. If, notwithstanding these precautionary measures, a person is seized suddenly with *cold, giddiness, nausea, vomiting,* and *cramps,* under circumstances in which *instant* medical assistance cannot be procured, the concurrent testimony of the most experienced medical authority shows that the proper course is to get as soon as possible into a warm bed; to apply warmth by means of heated flannel, or bottles filled with hot water, or bags of heated camomile flowers, sand, bran, or salt, to the feet and along the spine; to have the extremities diligently rubbed; to apply a large poultice of mustard and vinegar over the region of the stomach, keeping it on fifteen or twenty minutes; and to take every half hour a teaspoonful of *sal volatile* in a little hot water, or a dessert-spoonful of brandy in a little hot water, or a wine glass of hot wine whey, made by pouring a wine glass of sherry into a tumbler of hot milk; in a word, to do every thing practicable to procure a warm, general perspiration, until the arrival of the physician whose immediate care under such circumstances is indispensable.

(This article is an abstract from an article in the British *Aide Memoire to the Military Sciences,* under the head of *Sanitary Precautions,* and that article is taken entirely from the works of Dr. W. Ferguson, Inspector-general of Military Hospitals, and Reports of the General Board of Health, London, 1849.)

SAP. The sap is an apparently slow means of constructing trenches, but being continued by night as by day without cessation, its progress is soon felt. The work is executed by sappers rolling before them a large gabion, which shelters the workmen from musketry. In this manner one gabion after another is filled with earth and rolled in advance of its predecessor *after* that part of the trench already made has been well consolidated. A trench thus formed is called a sap. When the fire of the enemy is slack, so that many gabions may be placed and filled at the same time, it is called a flying sap. If two parapets, one on each side of the trench, be formed, it is then called a double sap.

SAP-FAGOTS—are fascines three feet long, placed vertically between two gabions, for the protection of the sappers before the parapet is thrown over.

SAPPERS. There is attached to the corps of engineers a company of sappers, miners, and pontoniers, called engineer soldiers The company is composed of ten sergeants or master workmen, ten corporals or overseers, two musicians, thirty-nine privates of the first class or artificers, and thirty-nine privates of the second class or laborers. The said engineer company shall be subject to the Rules and Articles of War, be recruited in the same manner and with the same limitation, and are entitled to the same provisions, allowances, and benefits, as are allowed to other troops constituting the present military peace establishment. The said company shall be officered by officers of the corps of engineers, shall perform all the duties of sappers, miners, and pontoniers, and shall aid in giving practical instructions in those branches at the Military Academy; and shall, under the orders of the chief engineer, be liable to serve by detachments in overseeing and aiding laborers upon fortifications or other works under the engineer department, and in supervising finished fortifications as fort-keepers, preventing injury and applying repairs; (*Act* May 15, 1846.) In marches near an enemy, every column should have with its advance guard a detachment of sapppers, furnished with tools to open the way or repair the road. It would be well if these sappers, as suggested by General Dembinski, were mounted, in order rapidly to regain the advance guard, after having finished their work.

SAP-ROLLER—consists of two large concentric gabions, six feet in length, the outer one having a diameter of four feet, the inner one a diameter of two feet eight inches, the space between them being stuffed with pickets or small billets of hard wood, to make them musket-shot-proof. Its use is to protect the squad of sappers, in their approach, from the fire of the place.

SASH. A mark of distinction, worn by officers round the waist, and composed of silk.

SAW-MILL, (PATENT, UPRIGHT, PORTABLE.) It is composed of eight pieces of timber, from five to eight feet long; four pieces of plank, from four to six feet long; and about fifteen hundred pounds of iron; besides two long bed-pieces, a carriage, some small wooden fixtures, pulleys, etc. The common up-and-down saw, six and one-half or seven feet long, is used without sash-gate or muley, and will saw timber of the largest or smallest size. It is so very simple in its construction that it has but few bearings, and consequently but little friction, and will therefore require much less power to drive it than the more complicated mills now in general use. As much of the cumbrous machinery of other mills, such as large, heavy frames, sash-gates, etc., is dispensed with in this, it is much less liable to get out of order; while its simplicity enables any one of ordinary mechanical ability to repair or build it. The amount of repairs required with fair usage is of insignificant import. The great advantage of such a mill for military purposes is its portability. The engines and boilers furnished with these mills are constructed specially for it. The first size is a boiler 10 feet long, 24 tubes $2\frac{1}{2}$ inches in diameter, and $7\frac{1}{2}$ feet long, shell over the fire-box 44 inches in diameter, shell over the tubes 34 inches in diameter, and engine of 7-inch cylinder and 15-inch stroke. This is a large eight-horse power, and is sufficient to drive the mill with any rapidity in the hardest and heaviest timber. It is sold with the mill—the whole establishment weighing about 6,500 pounds—for $1,250. The second size is a boiler $11\frac{1}{2}$ feet long, 25 tubes $2\frac{1}{4}$ inches in diameter, and $7\frac{1}{2}$ feet long, shell over the fire-box 44 inches in diameter, shell over the tubes 34 inches in diameter, engine same as that described above, (7-inch cylinder and 15-inch stroke,) excepting that it has extra connections. It may be rated as good ten-horse, and is capable of driving the mill, together with some other machinery at the same time, such as circular-saw for sawing slabs, lath, and other light work. This power is recommended. It is sold with the mill—the whole weighing about 7,500 lbs.—for $1,400. In these prices smoke pipes, connections, and every thing necessary for running are included. The mill may be put up and at work in two or three days after its receipt at any given place. It is said to saw three thousand feet a day, and has been made to saw nine hundred feet per hour. With an exhaust pipe on the smoke stack the sawdust may be used for fuel.

SCALING LADDERS. (*See* ESCALADE.)

SCARFED. (*See* CARPENTRY.)

SCARP. (*See* ESCARP.)

SCARP (To.) To cut down a slope, so as to render it inaccessible.

SCHOOL. (*See* ACADEMY, *Military*.)

SCOUTS. Horsemen sent in advance, or on the flanks to give an account of the force and movements of the enemy.

SCREWS. In screws the parts are—the stem, the head, the slit, and the thread. The bottom of the slit of the larger screws of small-arms is concave; the base screw of the rear sight has two holes in the head instead of a slot, in order that it may not be removed by the ordinary screw-driver. The *Screw* is also a mechanical power. The power applied perpendicular to the axis, is to the weight, as the *pitch* of the screw s, or the distance between the two threads, is to the circumference described by the point to which the power is applied. Thus, if the power is applied by means of a lever l,

$$P = \frac{w\ s}{2\ \pi\ l}$$

SECANT. (*See* TRIGONOMETRY.)

SECRETARY OF WAR. The principal officer of the Executive Department of War. (*See* DEPARTMENT OF WAR.) Mr. Attorney-general Wirt, in an opinion, dated Jan. 25, 1821, says, the Secretary of War " does not compose a part of the army, and has no duties to perform in the field." The duties assigned by law for the Secretary of War are the following : 1. The act creating the new department (*Act* Aug. 7, 1789) gives to the Secretary, besides the custody of records, books, and papers of the old department, the record of military commissions, the care of warlike stores and other duties clearly ministerial. 2. Section 5, *Act* March 3, 1813, continued in force by the 9th section of the *Act* of April 24, 1816, delegates jointly to the President and Secretary of War the power to make regulations better defining and describing the respective powers and duties of staff officers. 3. Articles of War, 13, 18, and 19, intrust the Secretary of War with muster-rolls and returns, and give him authority over the forms of such papers, and to require stated returns. 4. The 11th Article of War authorizes him to grant discharges to non-commissioned officers and soldiers ; and the 65th of the same articles makes him the medium in passing proceedings of certain courts-martial, and the organ of the President's orders thereon ; 5. Another Article of War (the 95th) charges the Secretary with receiving accounts of the effects of deceased officers and soldiers. 6. *Act* May 18, 1826, section 1, respecting clothing, &c., charges certain duties upon the Quartermaster-general " under the direction of the Secretary of War." 7. Several acts authorize the Secretary to purchase sites for arsenals. 8. The Ordnance Department and its *materiel* are made subject to the Secretary by the *Act* February 8, 1815. 9. Under the *Act* March 2, 1803, Section 1, the Secretary of War is authorized to give direction to the State Adjutants-general, in order " to produce uniformity " in returns, and to lay abstracts of the same, &c. 10. The Secretary shall lay before Congress on the 1st of February in each year a statement of the appropriations of the preceding year showing the amount appropriated, and the balance remaining unexpended on the 31st of December preceding. He shall estimate the probable demands which may remain on each appropriation, and the balance shall be deducted from the estimates of his department for the service of the current year ; (*Act* May 1, 1820.)

11. He shall render annually accounts exhibiting the sums expended out of such estimates, together with such information connected therewith as may be deemed proper; (*Act* May 1, 1820.) 12. The Secretary of War shall cause to be collected and transmitted to him at the seat of Government all flags, standards, and colors, as may be taken by the army of the United States from their enemies; (*Act* April 18, 1814.) 13. The Secretary may employ for the office of the War Department one chief clerk, and such other clerks as may be authorized by law; (*Acts* April 20, 1818, and May 26, 1824.) 14. The Secretary of War may furnish to persons who design to emigrate to Oregon, California, or New Mexico, such arms and ammunition as may be needed to arm them for the expedition at the actual cost of such arms and ammunition; (*Resolution* March, 2, 1849.) 15. All purchases and contracts for supplies or services for the military service of the United States, shall be made by or under the Secretary of War; (*Act* July 16, 1798.) 16. He shall annually lay before Congress a statement of all contracts, with full details; (*Act* April 21, 1808.)

Not one of the numerous acts of Congress relative to the War Department gives him authority to command troops. His lawful duties are all purely administrative, and as " he does not compose a part of the army," the President, in the exercise of his office of commander-in-chief, can of course only use the military hierarchy created by Congress. The English, from whom our system is borrowed, opposed to centralization of authority as adverse to freedom, have judiciously recognized the fact, in practice as well as theory, that the War Department is not of such a nature that it can be directed as other departments of the cabinet, or even be made to work by the simple play of constitutional changes in the ministry. They have consequently separated the *action* of the public force from the *direction* of financial matters. But as the safety of the state depends upon the stability of its .military institutions, the steadfastness of the means at work, and the skilful direction of all details, the Minister of War, who is changed by every triumph of opposite opinion, is not a military officer, and not charged with military authority. The permanent military institutions of the country do not depend upon him. The army does not look to him for nominations to office, discipline, or military control. He is simply the great provider, the superintendant of accounts, the financier, the interpreter of the plans of the cabinet for exterior and politico-military operations. He is aided by under-secretaries, who do not go out of office with the cabinet, and who are charged with the administration and payments for *materiel*.

The commander-in-chief, on the contrary, is the conservator of discipline, the centre of nominations, the life-spring which animates and directs the army, the source of orders, the regulator of tactics. He occupies himself with improvements of all kinds, and with the destination of *materiel.* It is to him that the Minister of State for War has recourse when he communicates to parliament or the cabinet the condition of the army, details of organization and other military information. Military finance and the support of armies are thus left with the Secretary of War, while command, discipline, and improvements are regulated by the commander-in-chief. The Minister of War thus follows the fortunes of a cabinet without the military institutions of the country being in any manner affected by party changes. Practice in the United States has widely diverged from this theory. (Consult BARDIN, *Dictionnaire de l'Armée de Terre; Milice Anglaise; Debates in Parliament.*)

SECTION, PROFILE, GROUND-PLAN. If a plane pass through work in any direction, the cut made by it is a section; if the cut be vertical and perpendicular to the face of the work, it is a ground-plan: thus, when the foundation of a house appears just above the ground, it shows the ground-plan of the building.

SELLING. (*See* AMMUNITION.)

SENIOR. Superior rank.

SENTENCE. (*See* COURT-MARTIAL.)

SENTRY OR SENTINEL. Any sentinel sleeping on post or leaving it before being regularly relieved, shall suffer death, or such other punishment as may be inflicted by sentence of a court-martial.

SERGEANT. Non-commissioned officer above corporal. There are various grades of sergeants: 1st. Sergeant-major, the first non-commissioned officer of a regiment, whose principal office is to assist the adjutant; 2d. Quartermaster-sergeant, assistant to the regimental quartermaster; 3d. Principal musicians of a regiment; 4th. Ordnance sergeant; 5th. First sergeant, or orderly sergeant of a company, and 6th. Sergeants, without prefix.

SERVANTS. (*See* PAY, for the number allowed to officers.) Company officers only can take soldiers from the line as servants; (*Act* April 24, 1816.)

SERVICE. The military art is the art of serving the state in war. All studies, acts, and efforts of the profession of arms have this end in view. To belong to the army and to belong to the land service, are the same thing. In a more restricted sense, service is the performance of military duty. In its general sense, service embraces all details of the military art. But in its restricted sense, actual service is the exercise of military functions. We say the Military Service; Cavalry, Artillery, or Infantry Service; Active Service; Regimental Service; Detached Service; Service on the Staff; Garrison Service; Camp Service; Campaign Service; Service in peace; Service in war; Daily Service; Service abroad; Service at home; Frontier Service; Service as captain, &c.; Armed Service; Actual Service. *To see service* implies actual combat with an enemy. *Service in campaign*, is service in the field; and in the French army, service in war or in colonies counts double, in estimating length of service, for promotions, pensions, retreat, and other remunerations. (*See* ABATIS; ADJUTANT-GENERAL; AIDE-DE-CAMP; ARMS, (*Small*;) ARTILLERY; ASSAULT; ATTACK AND DEFENCE; BARRICADES; BARRIER; BATTERIES; BATTLE; BAYONET; BLACKING; BLINDAGE; BLOCK-HOUSE; BOMBARDMENT; BRIDGES; CAMP; CAMPAIGN; CAPITULATION; CARPENTRY; CAVALRY; CHARGE; CONVOYS; COOKING; COUP D'ŒIL; DEFENCE, (*Coast*;) DEFILE; DEFILEMENT; DISEMBARKATION; DRAGOONS; EMBARKATION; ENGINEERS; ESCALADE; FASCINES; FIELDWORKS; FIRING; FLAGS OF TRUCE; FLANK; FORAGING; GABIONS; GUNNERY; INFANTRY; LANCE; LAW, (*Martial*;) LODGMENT; MANŒUVRES IN BATTLE; MARCH; MINE; OBSTACLES; OVEN; OUTPOSTS; PARTISAN; RECONNOISSANCE; RIFLEMEN; ROADS; SANITARY PRECAUTIONS; SAW-MILL; SIEGES; SQUARES; STADIA; STRATEGY; SURVEYS, (*Military*;) TACTICS; TARGET; TELEGRAPH; TOOLS; VETERINARY; WAGON; WAR; and Alphabetical list generally.

SEXTANT. An instrument for measuring the angular distances of objects by reflection. It is a segment of a circle of 60°. The quadrant and reflecting circle are instruments which depend on the same principle of optics, viz.: if an object be seen by reflection from two mirrors which are perpendicular to the same plane, the angular distance of the object from its image is double the inclination of the mirrors. The purpose, then, of the sextant, quadrant, and reflecting circle, is the adaptation of a convenient method for measuring the angle between two mirrors perpendicular to the same plane, and thus ascertaining the angle between two objects. This is accomplished by a contrivance which enables the mirrors to be so arranged that an object seen directly is brought to coincide with the *image* of another object seen by reflection, and the angle is shown by an index.

SHAFT—in mining, is a perpendicular excavation.

SHEERS. (*See* DERRICK; GIN.) By removing the pry pole of the gin, it may be used as sheers. When thus used, a block of wood of the same dimensions as the head of the pry pole with a hole in it large enough to receive the clevis blot, must be inserted in place of the pry pole.

SHELLS. A shell is a hollow shot with a hole to receive the fuze. They are usually fired from mortars and howitzers, and are charged with a sufficient quantity of powder to burst them, when they reach the end of their range. When fired at troops, the shells should be prepared to burst over their heads; or if the ground be favorable, to ricochet in front and plunge into the column. When fired at works or buildings, the shells should burst after penetration. (*See* AMMUNITION; FUZE; RIFLED ORDNANCE; SABOT; SPHERICAL CASE.)

SHOT, (SOLID OR ROUND.) Made of cast iron and used as projectiles when great accuracy, range, and penetration are required. (*See* BREACH; SABOT.)

SIEGES. An army, to undertake the siege of a fortress, must have superiority in the field, so that while some of the corps are occupied in besieging the place, others are employed in *covering* this operation, or in repulsing the enemy whenever he endeavors to succor the place. The army covering the siege is called an *Army of Observation*, and that which endeavors to give aid to the place is called the *Succoring Army*. The *Besieging Army* is that which, protected by the army of observation, throws up all the works necessary to take the place, such as trenches, batteries, &c. It begins its operations by investing the fortress; that is, it will advance with the greatest secrecy and rapidity, and occupy positions on every side, to cut off all communication with the adjacent country, and confine the garrison entirely to their own resources. The positions thus occupied are strengthened by field-works, and a sure communication is kept up between them.

It is absolutely necessary to invest the fortress attacked, so as to prevent the garrison holding any intercourse with the neighboring country; for if this precaution be not taken, the defenders will be able to draw fresh supplies of men, provisions, and ammunition from the country, increasing greatly the duration of the siege, and reducing the chances of ultimate success. At the late siege of Sebastopol, the ground being intersected by the inlet of the harbor of Sebastopol, the allied army was unable to complete the investment. Thus the fortress on the northern side was left open to receive all the reinforcements of men and materiel which could be furnished by the resources of Russia. Fresh officers, fresh troops, fresh provisions were continually poured in; the defences were enlarged and multiplied; and the besiegers, attacked in their own lines, held at one period a very critical position. The siege was thus prolonged beyond that of any other of modern times, and success was ultimately attained by a loss of men and materiel altogether unprecedented. Ground was broken on the 10th

October, 1854, and on the 10th September, 1855, the Russians, having sunk their ships, retreated from the southern to the northern side of the harbor, leaving the works on the southern side in the hands of the allies, exactly eleven months after the commencement of their attack.

A place may sometimes be reduced by investment or blockade alone, and where it is possible suddenly to blockade a place ill provisioned and filled with a numerous garrison and population, it may be the most ready and bloodless mode of proceeding. Indeed, many other circumstances may render it desirable to endeavor to reduce a place by blockade. When the defenders have been driven within their works, and the place invested, the ground before the fronts to be attacked is carefully examined, and the most suitable situations selected for the park of artillery, and the engineer's park: the former to receive all the ordnance stores and ammunition; the latter all the engineers' stores and materials to be used in the construction of the trenches, batteries, &c. These parks should be placed in secure localities, behind the slopes of hills or in ravines, beyond the general range of the guns of the fortress, but with a ready access to the trenches and batteries of attack, for the use of which they are formed.

The artillery and engineer parks having been duly established, and an adequate supply of ordnance, ammunition, and materials collected in them, for a week's or ten days' consumption, the actual work of the siege begins. The objects of the besiegers are three: 1st. By a superior fire of artillery to dismount the guns and subdue the artillery fire of the place. 2d. To construct a secure and covered road by which his columns may march to assault the defensive works, so soon as they are sufficiently destroyed to justify the attempt. 3d. To breach or batter down the escarp revetments of the fortress in certain spots, causing the fall of the rampart and parapet supported by them, and thus exposing the interior of the place to the assaulting columns.

Now, before any means can be taken to attain any one of these objects, a strong force must be placed under cover, close at hand to the spots on which the necessary operations are to be commenced, whose duty it is to repel any sortie of the enemy, and drive back any parties which issue from the place to destroy or interrupt the works of the attack. The cover provided for this guard of the trenches is usually a trench and parapet called the first parallel, formed around the whole of the fronts attacked: its distance from the advanced works has usually been between 600 and 700 yards. In the late siege of Sebastopol, the first parallel was opened at a distance of 1,200 yards; and doubtless, in future sieges, owing to the increased range of fire-arms, the first parallel will seldom be less, and may probably be considerably more distant. This parallel is formed by approaching the place secretly in the night with a body of men; part carrying intrenching tools, and the remainder armed. The former dig a trench in the ground parallel to the fortifications to be attacked, and with the earth excavated from the trench raise a bank on the side next the enemy, while the latter remain under arms, usually in a recumbent posture, in readiness to protect the working party, should the garrison sally out. During the night, this trench and bank are made of sufficient depth and extent to cover from the missiles of the place the number of men requisite to cope with the garrison, and the besiegers remain in the trench throughout the following day, in despite of the fire or of the sorties of the besieged. This trench is afterwards progressively widened and deepened, and the bank of earth raised till it forms a covered road, called a parallel, embracing all the fortifications to be attacked; and along this road, guns, wagons, and men securely and conveniently move, equally sheltered from the view and the missiles of the garrison. So soon as the first parallel is established, the engineers select positions for the batteries to silence the defensive artillery. In the positions of these batteries lies one of the principal advantages of the besiegers.

Batteries of guns and mortars are now constructed a little in advance of this parallel, in positions, such that their guns enfilade all the faces of the works attached. The crest lines of these batteries are therefore made perpendicular to the prolongations of the faces of the ravelins and bastions of the fronts attacked, and so great is the advantage to the besieger arising from such positions of his batteries, that with an equal or sometimes smaller number of guns he is able speedily to subdue the artillery fire of the defence. These enfilading batteries on the first parallel should be completed and ready to open fire on the third morning after breaking ground.

After the fire of the defensive artillery has been sufficiently subdued, the approaches are commenced. These, like the first parallel, are trenches dug in the ground and protected by a parapet formed of the excavated earth, thrown up on the side of the enemy's works. The approaches are made on the capitals of the ravelins and bastions attacked, but not in a straight line directly towards the salients, as in that case they could be enfiladed from end to end, but in a zigzag direction, alternately to the right and to the left of the capitals, in such a manner that their prolongations fall clear of the fortress, and the possibility of enfilading them is entirely removed.

The heads of these approaches are pushed forward by small parties of men, who, from their great numerical inferiority, are quite unable to contend with sorties issuing from the place. To prevent the repeated destruction of the approaches, and the continual loss of the working parties engaged in their construction, a guard of sufficient strength must always be stationed within a distance from these works not exceeding the distance of these works from the covered-way of the place: so that a sortie issuing from the place for the purpose of destroying the approaches may be met and repulsed by the guard of the trenches before they can have time to carry their object into effect; and as the approaches themselves, from their limited dimensions, afford no accommodation for a guard of the trenches, a parallel must always be established at least as near to the head of the approaches as the heads of approaches to the covered-way of the place.

It may then be considered a general principle of the attack that a new parallel or place of arms becomes necessary when the approaches have advanced half way between the last formed parallel and the covered-way of the fortress. So soon, therefore, as the approaches have advanced half the distance between the first parallel and covered-way of the fortress, a second parallel must be established to accommodate a guard of the trenches, or the working parties at the heads of the approaches will be liable to be swept off by parties of cavalry issuing from the covered-way, before aid can reach them from the first parallel. The approaches are then pushed forward, parallels being made according to the principles just laid down, wherever required, until they reach nearly the crest of the covered way. Here a trench of greater magnitude is formed, and in it batteries of heavy guns are constructed to silence the remaining artillery of the defence, and to breach in certain selected spots the escarp revetment wall, thus destroying the formidable obstacle to assault presented by the high perpendicular sides of the ditches of the fortress.

The order for the assault is given when the breach has been rendered practicable by the overthrow of the parapet upon the ruins of its walls; and after a gallery has been opened for descending into the ditch, across which a good epaulement has been made joining the breach to the gallery. The troops for the assault are held in the ditch, in the crowning of the covered-way, and in the third parallel. These detachments are to sustain each other and to do it with strong arms. At the concerted signal, the first detachment mounts the breach, driving back the defenders, and seeking to establish themselves firmly upon the height by constructing with gabions a lodgement in the angle of the bastion. This is a little intrenchment, called by the French *nid de pie*, which crowns the breach, and under shelter of which the soldiers fire upon all who present themselves. The sappers are charged with its construction, and in sufficient numbers for this purpose, accompany the assaulting party, each carrying a shovel, a pick-axe, and a gabion. The second detachment aids the first in surmounting the breach, and relieves it if the struggle is obstinate. The third detachment lines the trenches upon the glacis, and sweeps with its fire the parapets and top of the breach, and wherever else there is resistance, but care must also be taken, before coming to close quarters, to facilitate the assault by directing upon the work attacked, as many pieces of artillery as possible. When the close combat begins, the artillery ceases, as it would otherwise fire upon friend and foe.

Frequently the taking of the first works brings about the surrender of the place, but again it often happens that their resistance is but a foretaste of the obstinate defence to be made, and it is necessary to grasp, step by step, the fortifications of the besieged. Sometimes, again, the possession of the ramparts does not put an end to the fighting, but courageous citizens, willing to sacrifice their property to the honor and independence of their country, dispute inch by inch the possession of the streets and houses. The defence of Saragossa in 1808 is a heroic instance of such devotedness. The Spaniards, after losing their fortifications, sustained during twenty-three days attacks in streets and from houses. They capitulated for want of powder, and only after the enormous loss of fifty-four thousand persons of all ages and sexes.

A commanding officer, *defending the approaches of a fortress* threatened by armed enemies, declares it in a *state of siege*, and from that moment martial law prevails; or, in other words, the military authority alone governs. Every thing is brought into the place necessary for defence, in the shape of wood, fascines, gabions, animals, grain, and eatables of all kinds. All useless mouths are sent out of the place, and those inhabitants who remain are required to provide themselves with wheat, dried vegetables, oil, salt meats, &c., for many months, in order that the garrison may not be obliged to share their provisions with them. The place is put in a state of defence by arming and repairing the fortifications, planting palisades, clearing away the incumbrances in the communications, &c., &c.

When the garrison is sufficiently numerous, and that is the case here supposed, it guards against being entirely shut up in the place, by disputing all approaches. Positions are taken in advance of the suburbs, and far from destroying the suburbs as a smaller garrison must do, they should be covered by intrenchments, in the double aim of preserving them, and sparing the rear as long as possible.

Besides the preceding intrenchments, advantageous points are selected for solid redoubts and small posts. The most exposed passages are closed by abatis or deep cuts. Walls are pierced with embrasures, the different stories of houses made defensible, and all means whatever resorted to that can prolong the defence.

Upon a field of battle thus prepared, a long resistance may be expected, and the attacking force will experience great losses before they can open their trenches and begin the ordinary labors of the siege. Perhaps even during this exterior struggle, political events or other warlike operations may extricate the garrison from the impending siege, and its glorious struggle will then have freed the place committed to it from many horrors.

If the moment at last comes when it is necessary for the garrison to shut itself up, then follows that series of operations properly called a siege. The defence has a thousand means of prolonging its duration, because his exterior defence has given time to prepare them. Knowing the point of attack indicated by the first operations, the defence will have redoubled his intrenchments. The garrison will have been made warlike by frequent combats. It occupies, it is true, a post hard pressed, but its force is the more concentrated from that cause, and is still imposing notwithstanding the losses that it has experienced.

It is by *sorties* that we retard the operations of the besiegers. Large sorties are executed by numerous corps, and are generally made by day to avoid confusion. Small sorties are made at night, and consist of but few men. The first are designed to overthrow the trenches, fire the batteries, and spike the pieces, and they are consequently always followed by a sufficient number of workmen, provided with the necessary instruments. The smaller sorties are only directed against the workers of the sap; they present themselves unexpectedly and frequently drive away the workmen, and break up the gabions. The sap thus interrupted progresses but slowly.

Defensive mines are also a powerful means of prolonging the defence, as they force the besieger to make works that require much time in their preparation. As soon as the point of attack is known the besieged prepare under the glacis chambers of mines, which threaten the batteries of the besieger and constrain him to dig under the ground. The defence has in this subterranean war a great advantage, as he expects the attack in galleries previously prepared. The attack has no other resource than to prepare his chambers at a great distance in order to destroy those of the defenders, and for this purpose *globes* of *compression* are employed. These overcharged chambers, however, require a great deal of powder, and also much time for their preparation.

The besieged has also an advantage in the defence of breaches, because the attacking force may be surrounded, and can only reach their object by a narrow and difficult ascent. In defending a breach, therefore, all the energies of the defence should be brought into action. Preparations should be made in advance for this period of the siege, and some pieces of artillery should be carefully preserved, to arm at the moment of the assault these works which take in flank and reverse the columns of attack. At the top of the breach loaded shells are kept ready to roll down upon the assailants; a large fire should be lighted at the foot of the breach, and kept up by fagots. Or, if the enemy has only partially beaten down the wall, the foothold may be cleared away during the night in such a manner as to make the breach impracticable. Mines may be dug under the ruins by which the assailants may be overthrown. Long arms, as pikes, may be given to the soldiers who defend the breach, and those in the front ranks may be protected by cuirasses. If the work attacked has much capacity, reserves may be held in the interior to charge the enemy when he shows himself, and cavalry may also be brought up at this decisive moment.

Such are, in general, the steps to be taken to defend a work; but success will at last depend upon the character, firmness, and skill of the governor, and upon the intrepidity of his soldiers.

The army of observation ought not to be too far from that engaged in the siege, because it may be necessary to call for reënforcements from the latter, and they should be able to return to their camps after the action. Such aid furnished at the opportune moment is precious, and may contribute powerfully to defeat or repulse an enemy. When Napoleon covered the siege of Mantua he did not confine himself to drawing battalions from the besieging army, in order to fight the numerous troops striving to surround him, but he marched the whole besieging army, and uniting it with the army of observation, he gained the celebrated battle of Castiglione.

Besides, if the army of observation be too far off, there is nothing to prevent the enemy from unexpectedly attacking the besieging army, which, occupying a long line of investment, is rarely in a condition to repulse such an attack, and may therefore, without aid, be compelled to raise the siege, with the loss of ordnance and other *materiel*. General rules cannot be laid down for the position to be taken by an army of observation. It must possess mobility of action, and seek concentration as much as circumstances admit. It must not consider itself tied to the besieging army, and yet be always ready to succor the latter as well as repel a succoring army: conditions which demand much consideration, and which will be fulfilled only by varying dispositions according to circumstances.

In 1640, Prince Thomas of Savoy and the Spaniards held the city of Turin, whose citadel was defended by a French garrison. At the beginning of May, the Count d'Harcourt, celebrated for his courage and his military talents, set himself down before the place, but it was scarcely invested before the Marquis de Lénages arrived with heavy forces to blockade him in his lines. Turin, invested in this manner, presented the singular spectacle of a citadel besieged by the city, of the city besieged by a French army, and the latter surrounded by a Spanish army. In this position the Prince of Savoy corresponded with Lénages by means of shells without fuzes, in which letters were enclosed. The same means were used to introduce into the city a little salt and medicine, of which they were much in want. Count d'Harcourt, obliged to defend himself from continual sorties made by Prince Thomas, as well as from reiterated attacks of the Spanish army, covered himself by double lines as a protection against both. At length after a siege of four months and a half, after having endured in his camp every privation, he forced the city to capitulate. The Prince of Savoy marched out with the honors of war, and the Spanish army retired; (*Memoirs of the House of Savoy, par le* Marquis DE COSTA.)

Marshal Soult, at the siege of Badajoz, being occupied with an army larger than his own, nevertheless found an occasion by which he ably profited to establish an equilibrium of forces. Ten thousand Spaniards, to avoid being an encumbrance, went out from Badajoz to encamp upon the heights separated from the French army by the Gaudiana, and covered by the Gebora. Howitzers of long range were fired by the French army upon the Spanish camp, in order to drive it as far as possible from the works of Badajoz, from which it was separated by a valley of 600 toises in breadth. An hour before day, the Gaudiana was crossed in boats, the torrent of Gebora forded, and while Marshal Mortier directed a front attack upon the heights and sent his cavalry to turn the right, two or three thousand infantry placed themselves in the valley between the fortress and the camp, and facing both ways cut off all communication. Complete success crowned these beautiful dispositions. Eight thousand Spaniards grounded their arms, five or six hundred were killed, and the remainder escaped. Such was the brilliant combat of Gebora fought Feb. 19, 1811; (*Victoires et conquêtes.*)

These dissimilar instances, with hundreds that might be cited, show that no rules can determine the conduct of an army in the field, but genius in war may derive instruction from the memoirs of able commanders. (*Consult* DUFOUR; HYDE. *See* BATTARDEAU; BATTERY; DITCH; FIELD-WORKS; FORTIFICATIONS.)

SIGHT. A small piece of brass or iron, fixed on a gun at its muzzle, to serve as a point of direction, and also to hold fast the bayonet on the firelock. (*See* HAUSSE.)

SIGNAL. To attract the notice of a division of your party, five or even ten miles off, glitter a bit of looking-glass in the sun towards where you expect them to be, (Fig. 209.) It is quite astonishing at how great a distance its flashes will catch the sharp eyes of a bushman who has learnt to know what it is. It is now a common signal in American prairies. The sparks from a well-struck flint and steel can be seen at an equal distance.

FIG. 209.

If, instead of flashing with the mirror, the glare be steadily directed to where the party are, it will be seen at a far greater distance, and appear as a brilliant star; but it requires some practice to do this well. The rays from the mirror, whatever its size may be, form a cone whose vertical angle is no greater than that subtended by the diameter of the sun, and it is therefore necessary that the signaller should be satisfied that he throws his flash within that degree of accuracy. Moreover, a rapidly passing flash has far less brilliancy than one that dwells steadily for a fraction of a second.

An instrument, called a "hand heliostat," has been contrived by Galton for ascertaining the direction of the flash. Mr. Galton says: The instrument is perfectly easy to manage, and letters can be signalled by a combination of flashes, which I need not here describe. Its power is perfectly marvellous. On a day so hazy that colors, on the largest scale—such as green fields and white houses—are barely distinguishable at seven miles' distance, a looking-glass no larger than the finger-nail, transmits signals clearly visible to the naked eye.

The result of several experiments in England showed that the smallest mirror visible (under atmospheric conditions such that the signaller's station was discernible, but dim) subtended an angle of one-tenth of a second. It is very important that the mirror should be of truly parallel glass, such as instrument-makers procure. There is loss of power in more than one way from a slight irregularity. A plane mirror only three inches across, reflects as much of the sun as a globe of 120 feet diameter, and looks like a dazzling star at 10 miles' distance.

There are makeshift ways of directing the flash of the mirror; as, by observing its play on an object some paces off, nearly in a line with the station it is wished to communicate with. In doing this, be careful to bring the eye to the very edge of the mirror; there should be as little "dispart" as possible, as artillerymen would say. The aim must be a very true one, or the flash will never be seen. An object, in reality of a white color but apparently dark, owing to its being shaded, shows the play of a mirror's flash better than any other. The play of a flash, sent through an open window, on the walls of a room, can be seen at upwards of 100 yards. It is a good object by which to adjust the above-mentioned instrument. Two bits of paper and a couple of sticks, arranged as in Fig. 210, serve pretty well to direct a flash. Sight the distant object through the holes in the two bits of paper, A and B, at the ends of the horizontal stick; and, when you are satisfied that the stick is properly adjusted and quite steady, take your mirror and throw the shadow of A upon B, and further endeavor to throw the white speck in the shadow of A, corresponding to its pin-hole in it, through the centre of the hole in B. Every now and then lay the mirror aside, and bend down to see that A B continues to be properly adjusted.

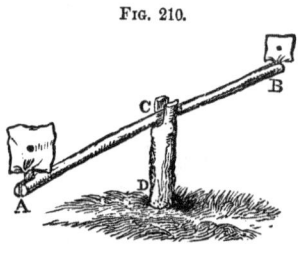

FIG. 210.

In short reconnoitring expeditions with a small detachment of a party, the cattle or dogs are often wild, and certain to run home to their comrades on the first opportunity; and, in the event of not being able to watch them, owing to accident or other cause, advantage may be taken of their restlessness, by tying a note to one of their necks, and letting them go and serve instead of postmen or carrier-pigeons.

Fire-beacons, hanging up a lantern, setting fire to an old nest high up in a tree—make night-signals; but they are never to be depended on without previous concert, as bushes and undulations of the ground may often hide them entirely. The smoke of fires by day is seen very far, and green wood and rotten wood make the most smoke. It is best to make two fires 100 yards apart. In the old-fashioned semaphores, or telegraphs, with arms to them, it is a common rule to allow, for the length of the arms, one foot for every mile it is intended to be seen from, and the eye is supposed to be aided by a telescope.

A line of men can be turned into a line of semaphores, by making them each hold a cap or something black and large in their hands, and mimic the movements of one another. Only a few simple signals could be transmitted in this way with any certainty. There are four elementary signals, which deserve general adoption. I fear the use of more would perplex. Men should be practised at these four, (Fig. 211.)

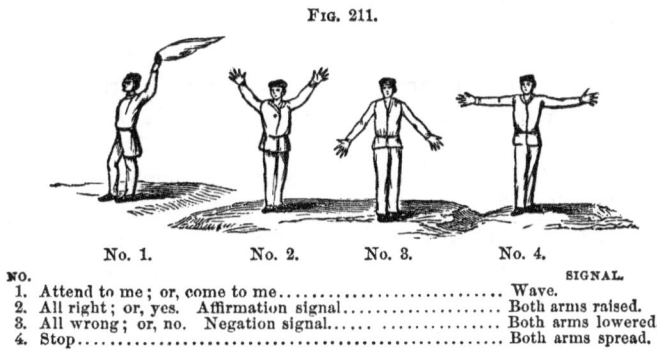

FIG. 211.

No. 1.　　No. 2.　　No. 3.　　No. 4.

NO.		SIGNAL.
1.	Attend to me; or, come to me	Wave.
2.	All right; or, yes. Affirmation signal	Both arms raised.
3.	All wrong; or, no. Negation signal	Both arms lowered.
4.	Stop	Both arms spread.

Energetic movements, of course, intensify the meaning. To use the signals, wave until you are answered; then make your signal while you count five, and wait five. Continue this till your friend does the same, then make a rapid "all right;" he does the same, and all is concluded. In order that you may be seen, try and stand in a position where your friend would see you against the sky; (GALTON's *Art of Travel.*)

A kite has been suggested as a day-signal; and also a kite with some kind of squib let off by a slow-light and attached to its tail, as one by night; (Col. JACKSON.)

A common signal for a distant scout is, that he should ride or walk round and round in a circle from right to left, or else in one from left to right. "At other times they will lie concealed near a road, with scouts in every direction on the look-out; yet no one venturing to speak, but only making known by signs what he may have to communicate to his companions or leader. Thus he will point to his ear or foot on hearing footsteps, to his eyes on seeing persons approach, or to his tongue if voices be audible; and will also indicate on his fingers the number of those coming, describing also many particulars as to how many porters, beasts of burden or for riding, there may be with the party; (PARKYNS.) Balloons, rockets, flags, &c., may be used to signal. (*See* TELEGRAPH; RECONNOISSANCE.)

SIGNAL OFFICER. By *Act* approved June, 1860, there was added to the staff of the army "one signal officer with the rank, pay, and allowances of a major of cavalry, who shall have charge, under the direction of the Secretary of War, of all signal duty, and of all books and papers, and apparatus, connected therewith." (*See* SIGNALS; TELEGRAPH.)

SINE. In trigonometry the sine of any arc of a circle is the straight line drawn from one extremity of the arc perpendicular to the radius passing through the other extremity. The sine of an arc is half of the chord of the double arc. It is positive in the first and second quadrants and negative in the third and fourth. (*See* TRIGONOMETRY; TABLES.)

SKETCHING. (*See* RECONNOISSANCE and SURVEYING.)

SKIRMISH. A loose, desultory engagement. Light infantry are the troops usually employed for such service; (*Infantry Tactics*.)

SLEEPERS. Small joists of timber, which form the foundation for the platform of a battery, and upon which the boards for the flooring are laid.

SLING-CARTS. A wooden sling-cart is composed of two wheels, 8 feet in diameter, an *axle-tree*, a *tongue*, and the *hoisting apparatus*, and is used to transport cannon and their carriages. The hoisting apparatus is a screw, which passes through the axle-tree, and is worked by a nut with long handles. The lower part of this screw is terminated with two hooks, to which are fastened the chains and trunnion rings; the breech of the piece being supported by the cascable chain. Or, if a chain be passed around the piece to be raised, it may be fastened to the hooks. The iron sling-cart is smaller than the wooden, and is used to transport cannon in the siege trenches.

SLUICE—is a strong vertical sliding door to regulate the flow of water.

SOLDIER. Whoever belonging to the military service of the State receives pay is a soldier. The term is derived from *solde*. It is an appellation, however, which a soldier proudly claims; and it is on the contrary an outrage to a brave man to say to him, "you are no soldier." "Every means (says Napoleon) should be taken to attach a soldier to his colors. This is best accomplished by showing consideration and respect to the old soldier. His pay likewise should increase with his length of service. It is the height of injustice not to pay a veteran more than a recruit." "There are five things a soldier should never be without—his musket, his ammunition, his knapsack, his provisions, (for at least four days,) and his intrenching tool."

SOLID SHOT. (*See* SHOT.)

SORTIE. An attack by a besieged garrison. (*See* SIEGE.)

SPHERICAL-CASE SHOT. A spherical-case shot consists of a thin shell of cast iron, containing a number of musket balls, and a charge of powder sufficient to burst it; a fuze is fixed to it as in an ordinary shell, by which the charge is ignited and the shell burst at any particular instant. A spherical case-shot, when loaded ready for use, has about the same specific gravity as a solid shot, and therefore, when fired with the service charge of powder, its range, and its velocity at any point in its range, is about equal to that of a solid shot of the same calibre. The spherical case mostly used for field-service is the 12-pdr., and contains, when loaded, 90 bullets. Its bursting charge is 1 oz. of powder, and it weighs 11.75 lbs. Its rupture may be made to take place at any point in its flight, and it is therefore superior to grape or canister. The attrition of the balls with which it is loaded, formerly endangered the firing of the bursting charge. This is now obviated, in making one mass of the balls, by pouring in melted sulphur. It is also prevented by Captain Boxer's improved spherical-case shot, two forms of which are shown in Fig. 212.

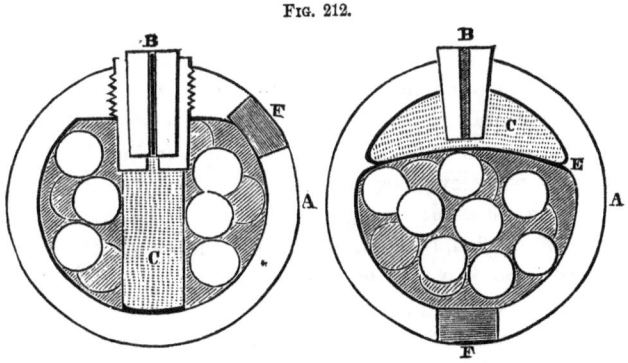

FIG. 212.

In either case, it is evident that the bursting charge of powder is kept separate from the balls. In one fig., it is contained in a cylindrical tin box, attached to a brass socket which receives the fuze, and which is screwed into the shell. In the other, the part of the shell containing the bursting charge is separated from that containing the bullets, by a diaphragm of sheet iron, E, cast into the shell; (*i. e.*, the shell is cast on to the diaphragm which is inserted into the core.) The bullets are introduced into the shell by a second orifice F, and are kept in their places by a composition afterwards poured in. The present 12-pounder spherical-case shot, fired with a charge of 2½ lbs. of powder, is effective at 1,500 yards. The proper position of the point of rupture varies from 50 to 130 yards in front of, and from 15 to 20 feet above, the object. The mean number of destructive pieces from a 12-pdr. spherical-case shot, which may strike a target 9 feet high and 54 feet long, at a distance of 800 yards, is 30. The spherical-case shot from rifle cannon is said to be effective at over 2,000 yards. Spherical case should not be used at a less distance than 500 yards.

SPIES. In time of war all persons not citizens of, or owing allegiance to the United States, who shall be found lurking, as spies, about the fortifications or encampments of the armies of the United States, shall suffer death by sentence of a general court-martial; (*Act* April 10, 1806, Sec. 2.)

SPIKE. To spike guns, is to drive large nails, or a piece of small rod, into the vent, so as to render guns unserviceable. To do this effectually drive into the vent a jagged and hardened steel spike with a soft point, or a nail without a head; break it off flush with the outer surface, and clinch the point inside by means of the rammer. Wedge a shot in the bottom of the bore by wrapping it with felt or by means of iron wedges, using the rammer to drive them in; a wooden wedge would be easily burnt by means of a charcoal fire lighted with a bellows. Cause shells to burst in the bore of brass guns, or fire broken shot from them with high charges. Fill a piece with sand over the charge to burst it. Fire a piece against another, muzzle to muzzle. Light a fire under the chase of brass guns and strike on it to bend it. Break off the trunnions of iron guns, or burst them by firing with heavy charges and full of shot, at great elevations.

To unspike a piece.—If the spike is not screwed in or clinched, and the bore is not impeded, put in a charge of ½ the weight of the shot, and ram junk wads over it with a hard spike, laying on the bottom of the bore a strip of wood with a groove on the under side containing a strand of quick-match by which fire is communicated to the charge; in a brass gun take out some of the metal at the upper orifice of the vent, and pour sulphuric acid into the groove for some hours before firing. If this method, several times repeated, is not successful, unscrew the vent piece, if it be a brass gun, and if an iron one, drill out the spike or drill a new vent.

To drive out a Shot wedged in the Bore.—Unscrew the vent piece if there be one, and drive in wedges so as to start the shot forward, then ram it back again in order to seize the wedge with a hook; or pour in powder and fire it, after replacing the vent piece. In the last resort, bore a hole in the bottom of the breech, drive out the shot, and stop the hole with a screw; (*Ordnance Manual.*)

SPLINTER-PROOF. Strong enough to resist the splinters of bursting shells.

SPRING. (*See* ARMS for the springs in the musket lock.)

SQUAD. A small party of men. A company should be divided into squads, each under a responsible officer or non-commissioned officer; the whole under the superintendence of the captain or company commander.

SQUADRON. Two companies or troops of cavalry.

SQUARES. My opinion (says Marshal Bugeaud) is that a large square has not proportionally a greater fire than a small one, and that it is no stronger. In a charge of cavalry, that portion only which attacks the face of a square is to be feared. In extending the face of a square, therefore, if its fire is augmented, the number of cavalry that can bear down against it is augmented in the same proportion. A square of three thousand men is not then any stronger than a square of one thousand. It would therefore be absurd to form three thousand men in one square, because they can be more readily formed into three or four squares, which will mutually protect each other, and form, as it were, a system of redoubts. And if one of these combined squares is broken by cavalry, the cavalry becomes disordered in the act, and the remaining squares are left intact. Besides, in presenting a small front to the attack of cavalry, horses, fearing to charge against the shower of balls which welcome them, are apt to oblique to the right or to the left. If the face of the square is extended they cannot do so, and the shock must fall on some part of the face, but the smaller the faces of combined squares the greater will be the intervals, and the more certain the success of the defence.

From these considerations, it is apparent that large squares ought not to be used, but that squares of a single battalion are worthy of all commendation. The formation of troops in two ranks is the prescribed order of the United States infantry tactics. Marshal Marmont says: "Nothing can be said in favor of a third rank. Persons of experience know that if one can, at a review, fire a volley in three ranks, it is impossible in war. It is better, therefore, to adopt the two-deep formation, and to render it permanent." The tactics direct that the divisions, as a general rule, shall always be formed before forming square. Marshal Bugeaud is of opinion that the square formed from the column by company, which would give a depth of four or six men to the different faces of the square, is greatly to be preferred. Apart from the fact that such squares are more expeditiously formed, the face of the square is reduced one-half, and the square is strengthened by the reduction.

STABLES AND STABLE DUTIES. The following arrangement of stables is recommended:

As far as possible, the horses of the same squadron should be placed in the same building, divided by partition walls or staircases into stables of equal capacity. When windows can be arranged in both long walls, place the horses head to head, separating the two rows of stalls by a longitudinal partition, which should not be more than 1' higher than the top of the hay rack, between the pillars which support the roof. The interior width of a stable, for 1 row of stalls, is 20'; for 2 rows, it is 40', when they are head to head; 34' 8'', when they are tail to tail; height of ceiling, 16' 8''. Doors should be pierced in the gable ends, and in the transverse partition walls, to secure a longitudinal ventilation during the absence of the horses. The doors for ordinary use should be pierced in the long walls; width, 6' 8''; height, at least 8' 8''.

There should be a window, with an area of about 16 square feet, for every 3 stalls; the sill 10' above the floor; the sash revolving around a horizontal axis at the bottom, and opening by the simplest mechanism; wooden shutters to be provided, if necessary. The recesses for the windows should extend to the floor, and be provided with hooks and racks for suspending the horse equipments; in these recesses openings 3' 4'' × 2' 4'' should be made through the wall, for throwing out the litter. If necessary, ventilators may be cut through the roof in the middle of the passage ways behind the stalls; ventilators near the floor should be employed only in cases of absolute necessity.

The floor ought to be of hard stones, laid on a firm foundation, and the joints filled with hydraulic mortar, cement, or asphalt; slope of floor of stall from two to three-tenths of an inch in ten inches. Mangers of wood, stone, or cast iron, placed on a mass of masonry, the front surface of which, as well as that of the manger, has a reversed slope of $\frac{1}{5}$. The wooden mangers are divided by partitions; those of stone or iron are hollowed out to the length of 2' for each horse, being solid between the hollows; depth 8″, width at top 1′, at bottom 9″.6; top of manger 3′ 8″ above the floor. The hay racks of wood and continuous, 3′ 4″ high, and placed 5′ 4″ above the floor. The bars round and capable of turning in their sockets, each bar 1″.2 in diameter, and placed 4″ apart; racks of iron may be authorized. The system of securing the horse consists of: 1st, a bar of round iron bent at both ends, placed up and down, parallel to the face of the manger, the upper end secured to the manger, the lower built into the masonry; 2d, a ring sliding on this bar, and having a chain 2′ long, with a T at the free end, attached to it; this T toggles to the halter ring. Fig. 213 shows this arrangement.

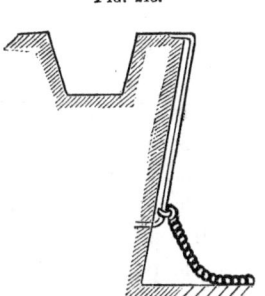

FIG. 213.

Each horse is allowed a width of 4′ 10″, never less than 4′ 8″, so that he may have the allowance of 70 cubic feet, and the space necessary for stable guards, utensils, &c., may be preserved. Stables which are less than 29′ wide and 12′ high can be used for two rows of horses only as a temporary arrangement.

The French have stables of all dates and varieties; one recently completed at Saumur, and the new ones at Lyons are justly regarded as models of excellence. Their dimensions and general arrangements are in conformity with the regulations given above; there are, however, some details worthy of notice; that at Saumur being the most perfect will be described in preference. The stalls are 4′ 10″ wide in the clear, and 10′ long to the heel posts; they are separated by suspended swinging planks.

The floors are of cubical blocks of stone, laid in cement. A shallow gutter in the rear of each row of stalls allows the stale to drain off. The longitudinal partition is of masonry, and about 10′ high. The interior of the stable is plastered; the woodwork painted oak color. In the window recesses there are racks, on which to hang the horse equipments when saddling and unsaddling. The equipments are kept in rooms in the loft, where the saddles are placed on horizontal wooden pins, the bridles hung on hooks. The racks are continuous, and of wood; the string-pieces, and each bar, are bound with narrow strips of sheet iron. The lower string-piece rests upon iron hooks, let into the wall, the upper one is held firm by iron bars, also let into the wall. The manger is a continuous mass of stone, with an excavation for each animal; these excavations are 22″ long, 12″ deep, and 12″ wide at top.

The building is divided into apartments, for about 20 horses each, by transverse partitions and stairway halls; there are large doors in these partitions. In a central hall there are water tanks.

The openings mentioned in the regulations for removing the litter do not exist. The halter bars are arranged as described in the regulations; but there is another ring and chain, above the manger, for use in the day time. Forage for 3 or 4 days is kept in the loft, where there are also rooms for a few non-commissioned officers. In the floor of the loft there are trap doors, so that hay and straw may be thrown down into the halls below. The oats run down from the bin, through a wooden pipe, into a large box on wheels. On the outside of the walls there are rings for attaching the horses while being groomed. At Lyons, some of the stables had quarters in the second story; this is stated by many officers to be an admirable arrangement, and attended with no inconvenience whatever; there are a few who object to it. The hospital stables are always separate from the others, and have box stalls.

Stable Duty.—In each squadron, the stable guard generally consists of a corporal and 1 man for every 20 horses. It is their duty to feed the horses, watch over their safety during the night, and attend to the general police of the stables, being assisted by an additional detail at the hours of stable call.

About one-half the litter is usually kept down during the day. The oats are given in two feeds: one-half at morning stable call, the rest in the evening. The hay is divided into three equal portions—at morning, noon, and night; in the forage magazine it is put up in trusses of 1 ration each, and thus received in the stable loft; at each feed the stable guard receive these trusses, and divide each one among three horses. If straw is fed, it is given either just before or just after the hay, always in the same order. The horse is watered twice a day, either just before or after his grain. The horse is cleaned principally with a *bouchon* of straw and with the brush; the comb is used only to clean the brush.

In the Crimea, the cavalry usually encamped in line, with two rows of picket ropes and a line of shelter tents in front of and behind the picket ropes; the arms and equipments between the shelters and the picket ropes.

The picket rope is stretched on the ground, and the horses secured to it by a hobble on the right fore-foot; the hobble is of leather, and about 3′ long; it buckles around the pastern joint; sometimes the hobble is attached to a picket pin, instead of a picket rope. Fig. 214 shows this arrangement; it is spoken of by the French officers as being the best manner of securing the horses. Officer's horses are on the flanks of the squadron picket ropes; those of the field and staff are near the tents of their owners. For the latter, rude stables are usually formed, by excavating to the depth of a couple of feet, banking up the earth around three sides, and then forming a roof and walls of brush.

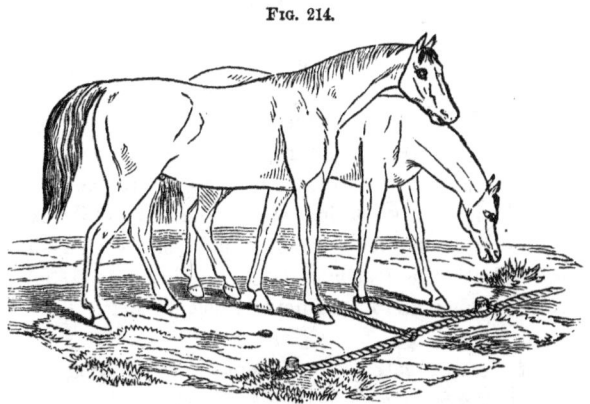

Fig. 214.

When time and circumstances permitted, the same was done for the horses of the men, especially in the winter. It was stated that a very slight protection of this kind produced very marked beneficial results. In this connection, it may be said, that companies of cavalry ought always to be provided with a sufficient number of tools to enable them to improvise some such shelter in any camp at all permanent; any thing which partially protects the horses from the cold winds is of great service. The French horses were blanketed in camp. (Consult McClellan.)

STADIA. A very simple aid in estimating distances, consists of a small stick, held vertically in the hand at arm's length, and bringing the top of a man's head in line with the top of the stick, noting where a line from the eye of the observer to the feet of the man cuts the stick, or *stadia*, as it is called.

To graduate the stadia, a man of the ordinary height of a foot-soldier, say 5 ft. 8 in., is placed at a known distance, say 50 yards; and the distance on the stick covered by him when it is held at arm's length is marked and divided into 8 equal parts. If the distance is now increased, until the man covers only one of these divisions, we know he is at a distance equal to 50 yds. $\times$ 8 = 400 yards. This instrument is not very accurate, except for short distances.

A much more accurate stadia is constructed by making use of a metal plate, having a slit in it in the form of an isosceles triangle, the base of which, held at a certain distance from the edge, subtends a man, (5 ft. 8 in.) say at the distance of 100 yards. A slider, *ab*, (Fig. 215,) moves along the triangle, being always parallel to the base, AB, and

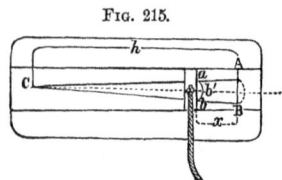

Fig. 215.

the length of it comprised between the two sides of the triangle, represents the height of men at different distances, which are marked in yards on the side of the triangle, above or below, according as the object looked at is a foot soldier or horseman. In order to keep the stadia always at the same distance from the eye, a string is attached to the slider, the opposite end having a knot tied in it, which is held between the teeth while using the instrument, which is held in the right hand, the slider being moved with the left-hand finger. The string should always be kept stretched when the instrument is used, and the line AB in a vertical position.

It must be graduated experimentally, by noting the positions in which the slider *a b* represents the height of the object. The instrument used is not, however, reliable. Its uncertainty increases in an equal ratio with the distance of the object observed. At the extreme ranges it is quite useless. At the school for firing at Vincennes, therefore, they rely entirely on the eye alone for the judgment of distances, and great pains by careful practice and instruction is taken to perfect that judgment. A simple instrument by which distances can be determined is, therefore, still a great desideratum. The prismatic teliometer of M. Porro, of the Sardinian army, is however the best measurer of distances that has been yet invented. It is described in Wilcox's *Rifle Practice*.

STAFF. The staff of an army may be properly distinguished under three heads:—

1. The *General Staff*, consisting of adjutants-general and assistant-adjutants-general; aides-de-camp; inspectors-general and assistant-inspectors-general. The functions of these officers consist not merely in distributing the orders of commanding generals, but also in regulating camps, directing the march of columns, and furnishing to the commanding general all necessary details for the exercise of his authority. Their duties embrace the whole range of the service of the troops, and they are hence properly styled general staff-officers.

2. *Staff Corps*, or staff departments. These are special corps or departments, whose duties are confined to distinct branches of the service.

The engineer corps and topographical engineers are such staff corps. The ordnance, quartermasters', subsistence, medical and pay departments are such staff departments.

3. The *Regimental Staff* embraces regimental officers and non-commissioned officers charged with functions, within their respective regiments, assimilated to the duties of adjutant-generals, quartermasters and commissaries. Each regiment has a regimental adjutant, and a regimental quartermaster, appointed by the colonel from the officers of the regiment. Ideas concerning the utility, organization, and duties of the staff may be found in many writers. Until the end of the reign of Louis XIV., feudal manners and arbitrary notions accommodated themselves badly to written rules; but about this period more wholesome ideas began to prevail; mathematics made some progress; its application spread; the military art felt its effects; it was admitted that a single head was not sufficient for all the details necessary to conduct an army. It was agreed that the general-in-chief should have assistants to perform various duties. Hence certain military grades and financial employments were created. Those thus invested with authority were associated under the same designation. But this STAFF was far from being a special permanent corps. It was only a temporary assemblage of officers, and later took the name of staff, to indicate that they were AIDES of the general in regulating and supplying troops. Frederick the Great and Bonaparte undertook and gloriously terminated more than one war with the aid of staff-officers, but without a staff corps. At the beginning of the last century there existed in regard to the staff a few traditions, or customs, which differed in different armies. Neither laws, regulations, nor instructions had yet been established defining the rights, powers, and duties of the staff. Staff-officers were principally employed in reconnoissances; and on duties connected with lodging troops as aides of the quartermaster-general; in the preparation and distribution of the orders of the day, &c.; and as bureau officers. The war of the French Revolution was finished by the French army without a staff corps. The French army had staff-officers under the names of adjutants-general, commandants, adjunct-captains, and orderly officers; but such officers were rather a momentary aggregation of officers of divers corps, than a special and permanent corps. Officers of cavalry, artillery, engineers, and infantry, if they had not the title, often exercised the functions of general staff-officers, and made reconnoissances. But in 1818, upon the return of the Bourbons, in imitation of the Austrians, Prussians, and Russians, a staff corps was formed in France. The corps was recruited from pupils leaving the school of

St. Cyr; after study they were admitted to the school of the staff; they subsequently served in regiments of infantry and cavalry. After having been advanced a grade, they were definitely admitted as lieutenants of the staff, and became entitled to cavalry pay, with the title of aide-major.

It is necessary that a general staff-officer should have a knowledge of horsemanship—that he should not be ignorant of the sword exercise; he should have some knowledge of topography; he should be familiar with foreign languages, should have studied military administration and castrametation; but above all, he should possess a complete knowledge of tactics, and be able to judge skilfully of military positions. An officer grown old in the silence of a bureau would hardly in the tumult of battle, or under critical circumstances, second his general by aiding him intelligently concerning warlike operations. Can he interrogate spies, watch over the observance of order in military trains; draw up orders and instructions, mark out military positions; improvise a fortification; organize and conduct foraging parties, direct markers for grand manœuvres? Open the march of armies? Vault at the head of the light cavalry? Stimulate and enlighten the troops by his interpretation of the orders he carries, by his intuitive knowledge of their tactical position, by his coup d'œil, by the propriety of his counsels, and by the vigor of his impulsions? None, but officers whose experience has been gained by service with troops, can do these things with promptitude and effect; but these are the important duties of the general staff, and service with troops therefore is the true criterion of merit in such staff-officers. In organizing a permanent general staff corps, it consequently becomes necessary either to employ in peace that large body of officers necessary in war for staff duties, upon duties entirely foreign to their functions in war, or else leave them in idleness. Either course must unfit them for the services required of them on campaign, and it therefore follows, that a permanent general staff involves a useless number of officers in time of peace, and a deficiency of experience, instruction, and aptitude for their duties in time of war. It is impossible to avoid this vicious circle with a permanent general staff. The only true system of staff organization, then, is that which admits of supernumerary general and regimental officers, selected temporarily for staff duties by commanders of troops, as provided by the Act of Congress of 1799, drawn by Alexander Hamilton. "The leading qualifications which should distinguish an officer selected for the head of the staff (says Napoleon) are: to know the country thoroughly; to be able to conduct a *reconnaissance* with skill; to superintend the transmission of orders promptly; to lay down the most com-

plicated movements intelligibly, but in few words, and with simplicity."
(*See* ADJUTANT-GENERAL; AIDES-DE-CAMP; ENGINEERS; INSPECTOR-GENERAL; LINE; MEDICAL DEPARTMENT; ORDNANCE DEPARTMENT; PAY DEPARTMENT; QUARTERMASTER'S DEPARTMENT; SECRETARY OF WAR; SUBSISTENCE DEPARTMENT.)

STANDARDS. Flags, standards, and colors, taken by the army and navy of the United States from their enemies, to be delivered with all convenient despatch to the President of the United States, for the purpose of being, under his direction, preserved and displayed in such public place as he shall deem proper; (*Act* April 18, 1814.)

STATE TROOPS. (*See* MILITIA.)

STOCKADE. A work which may be substituted with advantage for earthen works of very small profile, if it can be covered from the fire of artillery; (Fig. 216.) The stockades or picket works usually

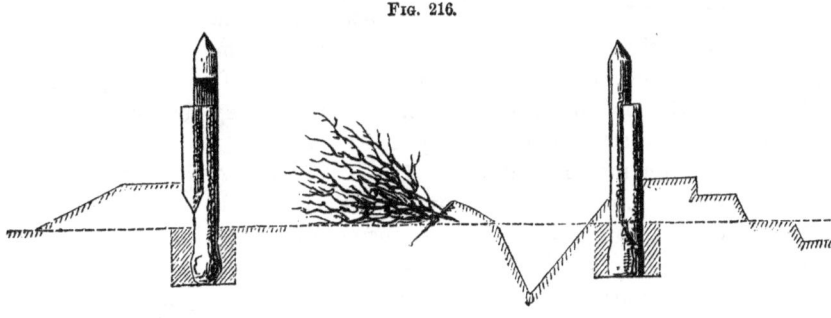

FIG. 216.

employed against Indians are composed of rough trunks of young trees cut into lengths of 12 or 14 feet, and averaging 10 or 12 inches in diameter. They should be firmly planted close together. A banquette or step will generally be required, and the loopholes so arranged that they cannot be used from the outside. If necessary, such a work can be strengthened by ditch and abatis, and flanked by block-houses. The figures show the manner of planting the pickets.

STOPPAGE OF PAY. Where pay is stopped on account of arrears to the United States, the party whose pay is stopped may demand a suit, and the agent of the treasury is required to institute a suit within sixty days thereafter; (*Act* Jan. 25, 1828.) (*See* DEFAULTER; SUIT.)

STORE-KEEPERS. (See ARMY ORGANIZATION for the number.) Military store-keepers and paymasters receive twelve hundred dollars per annum; other military store-keepers receive eight hundred dollars per annum; (*Act* Aug. 2, 1842.)

STORES. All public stores taken in the enemy's camp, towns, forts, or magazines, whether of artillery, ammunition, clothing, forage, or provisions, shall be secured for the service of the United States; for the neglect of which the commanding officer is to be answerable. (*See* BOOTY; EMBEZZLEMENT; SALE.)

STORM. To storm is to make a vigorous assault on any fortified place, or on its outworks. The storming party is a select body of men, who first enter the breach, and are, of course, imminently exposed to the fire of the enemy.

STRAGGLERS. Individuals who wander from the line of march. It is part of the rear guard's duty to pick up all stragglers.

STRATAGEM—is a scheme or plan devised to cover designs during a campaign, or to deceive and surprise the enemy.

STRATEGY—is the art of concerting a plan of campaign, combining a system of military operations determined by the end to be attained, the character of the enemy, the nature and resources of the country, and the means of attack and defence. The *theatre of operations* selected, embraces the territory we seek to invade or that to be defended. It comprehends a *base of operations*; the *objective point* of the campaign; the *front of operations*, that is, the extent of the line occupied by the army in advance of its *base*; *lines of operations*, the routes followed to reach the objective point or end proposed; *lines of communication* which unite the different lines of operation together; *obstacles*, natural or artificial, and places of refuge.

STREET-FIGHTING. In an enemy's country the case is much simplified: a town so occupied is all inimical, and under the most desperate state of opposition; consequently in the attack there is no respect to person or property. If the houses are combustible, a ready means of subduing the place is within reach; and if not, it is forced in different directions by siege operations, as practised by the French at Saragossa.

On occasions of internal dissensions and insurrectionary movements, the case is different; the efforts of the troops and of the well-disposed citizens are greatly impeded by the difficulty of distinguishing between friend and foe, or of the premises or property with which it may be justifiable to interfere. This, and the very natural and proper anxiety to avoid bloodshed and injury to one's own countrymen, frequently lead to a habit of temporizing with the circumstances, and by this indication of timidity and weakness give such confidence to the rebels as to enable them, and perhaps with comparatively insignificant numbers, to gain in moral effect as the others lose; by degrees the wavering and the timid are led to join them; the troops themselves imagine that there is a declared power manifested that is not to be opposed, and thus the former obtain a complete ascendency, which the exertion of more firmness and system at first would effectually have prevented.

The best institutions of any country become endangered by such a state of things; but a remedy may be found in a more systematic manner of proceeding. The troops should never be brought into the presence of the insurrectionists until fully authorized to act—the consequence would be that the very appearance of the soldiers would be a warning to every one of the immediate consequences of prolonged opposition, which would prevent further conflict, or make it very short. In order to promote the power of vigorous action by the military, and to prevent the innocent from suffering, the most solemn warning should be issued, in case of tumult, against the presence in the streets of women, children, and persons who do not join in the troubles, intimating that the consequences of any bad result from their being thus incautiously exposed must rest on themselves. These are necessary preliminaries to the consideration of the means of attacking an insurrectionary force. When disturbances are to be quelled in a town, cavalry, artillery, and infantry can act with full effect, and with every advantage of organization, so long as their opponents occupy the open streets. If barricades are constructed across them, the cavalry become unserviceable; the infantry, however, have still full force; for one side of an ordinary barricade is as good as the other, and the infantry can cross any of them without difficulty.

But when it is found that the insurgents have had recourse to the most determined means of resistance, by occupying the interior of houses in support of barricades, the mode of attack must be adapted to the circumstances. The operation should be conducted under due deliberation, nor would any triumph be conceded by care being taken that the use of cover shall not give the impression of defeat. It will be readily ascertained what part or parts of the town are so occupied as to render the movement of the troops through the open streets unadvisable. An endeavor should be made to isolate those portions by detachments of troops posted at all the approaches to them. This of itself would throw the rioters into a most uncomfortable and false position: they would find themselves shut up without any internal organization to enable them to act to any useful purpose, or to make any combined forcible effort for their release; or, indeed, if they could do so, it would have all the effect of *an escape* instead of a victory.

Nor would it be necessary, under such circumstances, that these detachments should be at all large, numbers of them being supported by some general reserve. Active measures, however, might at the same time be carried on against any portions of the houses that it may be considered advisable to force, for the purpose of confining the resistance within narrower limits, or for subduing it at once altogether. Although in towns the attack of a mass of houses is formidable, and almost impracticable to troops unprepared for such an operation, it will not present much difficulty to a systematic proceeding. One great defect for defence in a house or street is its want of a flanking fire, although every part may obtain a support from the opposite houses in the same street. If, therefore, only one side of the street is occupied, individuals or parties moving close along that side are in security, except from the chance missiles that may be blindly thrown down from the windows. Nothing of that kind could prevent two or three soldiers, under cover of a partial fire on the windows, from passing up and breaking open the doors; by which means, the troops being admitted, possession of the entire building would soon be obtained.

When, however, from any peculiarity of the building, or of others contiguous, or from the circumstance of both sides of the street being occupied in force, such a mode of proceeding would be too hazardous, the soldiers might make an entrance into the nearest available house in the same block of buildings, and, supported by detachments of troops, work their way, through the partition walls, from one house to another; or by the roofs or the back premises, where the defenders will be quite unprepared to oppose them, or, if they make the attempt, would not have the same advantages as in front: small parties, if necessary, keeping up a fire on the windows from the walls of the back yards, or from the opposite houses, would effectually cover these advances of the troops. To carry on such approaches, the men should be provided with an assortment of crowbars, sledge-hammers, short ladders, and, above all, some bags of powder not less than 5 or 6 lbs. weight.

In these desultory operations in the defiles of streets and houses, the troops should not be in heavy columns, but in small detachments well supported; and by acting thus in order, and on system, the effect will be the more certain, as a popular movement is, necessarily, without subordination or unity of action, and peculiarly subject to panics at any proceeding differing from what had been anticipated; (*Aide Mémoire.*)

STRIPES AND LASHES—infliction of, allowed only in case of desertion; (*Act* May 16, 1812, and March 2, 1833.)

SUBALTERN. Commissioned officer below captain.

SUBSCRIBING. Every officer must subscribe the Articles of War; (ART. 1.)

SUBSISTENCE DEPARTMENT. (*See* ARMY for its organization.) Provides subsistence stores for the army, either by contract or purchase. Assistant commissaries subject to do duty as assistant-quartermasters. The President, under authority of law, has fixed the ration at $\frac{3}{4}$ lb. of pork or bacon, or $1\frac{1}{4}$ lbs. of fresh or salt beef, 18 oz. of bread or flour, or 12 oz. of hard bread, or $1\frac{1}{4}$ lbs. of corn meal; and at the rate to 100 rations of 8 qts. of peas or beans, or 10 lbs. of rice; 6 lbs. of coffee, 12 lbs. sugar, 4 qts. of vinegar, $1\frac{1}{2}$ lbs. of tallow, $1\frac{1}{4}$ lbs. adamantine or 1 lb. of sperm candles, 4 lbs. of soap and 2 qts. of salt. In different climates and on different kinds of service, soldiers require different articles of diet; some latitude should therefore be given to commanders of armies and military departments in making variations from the prescribed ration.

A conscientious administrator should acquaint himself with the peculiar properties of different kinds of food, their relative nutriment, and the differences of food best suited to promote health under the various circumstances incident to field-service. The following extract from some observations made by an eminent Scotch chemist, is worthy of every attention on the part of the student of military administration:

"In consequence of the advances made in physiology and chemistry the nutritive value of any dietary, deduced from practical experience, may be tested with care and certainty by reference to its chemical composition. As this fact is little known to practical men, it may be well to explain the principles on which the method is founded.

"1. All articles of food used by man consist of one or more, and generally several nutritive principles; and most of them contain water and an indigestible cellular tissue. The two latter must, of course, be deducted in estimating nutritive value.

"2. The nutritive principles consist of two sets, one of which maintains respiration, and the other repairs the waste constantly incurred by the animal textures in the exercise of their functions. As the respiratory principles commonly abound in carbon, they are sometimes called carboniferous, while the reparative principles, because they all contain nitrogen, are termed nitrogenous.

STANDARD TABLE OF NUTRIMENT.

NAME OF ARTICLES.	Percentage of Nutriment.		
	Carboniferous.	Nitrogeneous.	Total.
Wheat flour...............................	71·25	16·25	87·5
Bread.......................................	51·5	10·5	62·0
Oatmeal....................................	65·75	16·25	82·0
Barley (pearl)...........................	67·0	15·0	82·0
Pease.......................................	55·5	24·5	80·0
Potatoes..................................	24·5	2·5	27·0
Carrots....................................	8·5	1·5	10·0
Turnips....................................	5·7	0·3	6·0
Cabbage...................................	6·7	0·3	7·0
Lean of beef and mutton...........	0·0	27·0	27·0
Fat of meat...............................	100·0	0·0	100·0
Average beef and mutton...........	15·0	20·25	35·25
Bacon.......................................	62·5	8·36	70·86
Skimmed milk cheese................	0·4	64·6	65·0
White fish.................................	0·0	21·0	21·0
New milk..................................	8·0	4·5	12·5
Skimmed milk...........................	5·5	4·5	10·0
Butter milk...............................	1·0	6·0	7·0
Beef tea (strong).......................	0·0	1·44	1·44
Beef tea, and meat decoction of broth......	0·0	0·72	0·72

" 3. Experience has shown that the most successful dietaries for bodies of men, deduced from practical observation, contain carboniferous and nitrogenous food in the proportion of about three of the former to one of the latter, by weight. During two-and-twenty years that my attention has been turned to the present subject, not a single exception has occurred to me.

" 4. Hence it is obvious that the least weight of food in the rough state will be required, first, when there is least moisture and cellular tissue in it; and secondly, when the carboniferous and nitrogenous principles are nearest the proportion of three to one.

" 5. Of the various nutritive principles belonging to each set, some may replace one another; some are better than others; some are probably essential. This branch of the science of the subject is unfortunately still imperfect.

" 6. Two things, however, are certain, that nitrogenous may replace carboniferous food, for supporting respiration, though at a great loss; but that carboniferous food (without nitrogen) cannot replace nitrogeneous food, for repairing textural waste.

" 7. The daily amount of nutritive principles of both sets must increase with exercise and exposure, otherwise the body quickly loses weight, and ere long becomes diseased. If the above proportion between the two sets be maintained, the weight of real nutriment per day varies, for adults at an active age, between seventeen and thirty-six ounces; the former being enough for prisoners confined for short terms, the latter being required for keeping up the athletic constitution, or that which is capable of great continuous muscular efforts, as in prize-running and other similar feats.

"8. Dietaries ought never to be estimated by the rough weight of their constituents, without distinct reference to the real nutriment in these, as determined by physiological and chemical inquiry.

"Keeping these principles in view, and with the help of a simple table, it is not difficult to fix the dietary advisable for any body of men, according to their occupation. It is, also, in general, easy to detect the source of error in unsuccessful dietaries. For example, any scientific person conversant with the present subject could have foretold, as a certain consequence, sooner or later, of their dietary, that the British troops would fall into the calamitous state of health which befell them last winter in the Crimea.

"Soldiers in the field will be the more efficient the nearer they are brought to the athletic constitution. But as the demand for protracted, unusual exertion occurs only at intervals, the highly nutritive athletic dietary is not absolutely necessary. *On the whole, from experience in the case of other bodies of men somewhat similarly circumstanced,* 28 *ounces of real nutriment, of which* 7 *are nitrogeneous or reparative, will probably prove the most suitable.* Any material reduction below 28 ounces will certainly not answer; and under unusual exertion kept up for days continuously, as in forced marches, or forced siege labor, the quantity should for the time be greater, if possible."

Biscuit, particularly when salted meat is the principal article of diet, is very apt to produce dysentery and scrofulous complaints; it becomes, moreover, unpalatable when continuously used; and so eager were English soldiers in the Crimea for soft bread that they used to exchange 5 lbs. of biscuit for 1 lb. of bread with the French soldiers, whose first work, after pitching their camps, was generally to construct field-bakeries, and whose supply of soft bread seldom failed. Sallust tells us (De bello Jugurth. 44) that the Roman soldiers used to sell their ration of *grain* for a trifle in order to purchase bread, which at that time they had not the means of manufacturing. Mills and ovens exist in some form or other in all countries, and they should be made available whenever an army halts for a sufficiently long period to admit of their being worked; but as the enemy frequently destroys these means of contributing to the soldier's comfort, the use of hand mills and field-ovens must under such circumstances be resorted to; and to construct these in the most rapid and at the same time the most effectual manner, should always be done where circumstances permit. The description of camp ovens must necessarily depend upon the permanency of the encampment. If the army be likely to remain in position for any length of time, they should be constructed of durable materials, such as bricks; but for hurried operations a mere excavation of the earth suffices in the course of a very short time to produce an oven capable, with a little care, of baking bread. The impromptu ovens used by the American backwoodsmen, as described by Sir Randolph Routh, are usually raised upon a platform about 3 feet high, and 5 or 6 feet long, by 4 feet broad, and on this they construct the circular form of the oven by means of forest twigs and boughs of sufficient strength to receive and support the cement, which is made of common clay soil and water, mixed to a proper consistence, and put on in successive layers until it acquires the necessary thickness. An opening is left to introduce the bread, and a common piece of wood with a handle supplies the place of a door until it is baked. (*See* also article OVEN for the ovens made by French soldiers; and TRAVELLING KITCHEN, for a suggested improvement for field-service.) It is very important that soldiers should be instructed in making field-ovens.

Nothing is more important in the field than to keep up the supply of fresh meat. It is the only article of the soldier's ration that provides its own transport, and though a supply of salted provisions is indispensable as a reserve in case of accidents, and to provision fortified places in the event of a siege or blockade, it should be economized as much as possible, and issued only in cases of necessity, not only as being more expensive and absorbing a great deal of transport, but because the frequent use of salted provisions is invariably detrimental to the health of the troops.

The importance of providing the soldier with vegetables is now universally admitted. When salted provisions are much used, it is essentially desirable to counteract the tendency to scrofulous complaints induced by such diet by means of vegetable food; to obtain fresh vegetables in the field is, however, a matter of considerable difficulty, their liability to spoil and their bulk are obvious objections to their use by an army in movement; but the process of compressing vegetables, which has now been brought to perfection, enables a commissariat to keep up this supply at the cost of but little transport, and in the most convenient form for immediate use. Rice is an admirable article of diet, more particularly when there is any tendency to bowel complaints.

It contains more nutriment than wheat flour, is easily conveyed and cooked, and is not liable to suffer from exposure. There would, probably, be no difficulty in making a preparation of rice which would greatly reduce its bulk and still further facilitate its cooking. Corn meal and pease are likewise excellent articles of food; but the latter should, if possible, be issued in a ground state, as it otherwise requires more soaking to render it fit for use than there is time for on the march. The supply of coffee or tea should never be allowed to fail. Dr. Christison says :—" It is difficult to over-value the proposed addition of tea and coffee to the men's rations. They possess a renovating power, in circumstances of unusual fatigue, which is constantly experienced in civil life, and which I have often heard officers, who served in the Spanish campaigns, as well as in the late Burmese war, describe in the strongest terms. This, however, is not all, for it has been recently shown by a very curious physiological inquiry, that both of them, and especially coffee, possess the singular property of diminishing materially the wear and tear of the soft textures of the body in the exercise of its functions in an active occupation."

The object of accounts is to insure the application of public resources to their prescribed ends, and within regulated limits. This is perfectly feasible under ordinary circumstances; but on active service it is not always possible to procure vouchers and receipts according to the established forms, and it is far better to establish, by means of a well-organized department of control, a strict and efficient local supervision over the conduct of supply duties in the field than to exact accounts, which, however correct in their outward form, can but rarely represent the actual transactions as conducted during the hurried and ever-changing events of active warfare. A judicious system of musters and inspections would do more to check waste or malversation in the field than the most ingenious accountability that could be devised; and if a commissariat officer were simply required to furnish the head of his department with a periodical " state of supplies," showing where and how obtained and issued; and officers commanding corps a return of the number of men fed, noting any deficiency of supply; both reports being subject to verification by means of personal inspections and musters, the object in view would be attained with far greater certainty than under the present complicated system of returns, abstracts, and vouchers, the preparation of which occupies much of the time of a commissariat officer that might be more profitably employed for the benefit of the troops, while their subsequent examination, probably after a lapse of one or two years, answers no possible purpose except to find employment for a large number of clerks. (Consult FONBLANQUE.)

SUIT. In all cases where the pay or salary of any person is withheld, in consequence of arrears to the United States, (and salary can be legally withheld from no other cause except by sentence of court-martial,) it shall be the duty of the accounting officers, if demanded by the party, his agent, or attorney, to report forthwith to the agent of the Treasury Department, the balance due; and it shall be the duty of the said agent, within sixty days thereafter, to order suit to be commenced against such delinquent and his sureties; (*Act* January 25, 1828.)

SUMMING UP. (*See* JUDGE-ADVOCATE.)

SUPERINTENDENT. The chief of the corps of engineers present at the Military Academy is the superintendent; (*Act* March 16, 1802.) The selection of the commander of the corps of engineers shall not be confined to said corps; (*Act* April 29, 1812.) Officers of engineers may be transferred at the discretion of the President from one corps to another, regard being paid to rank; (ART. 63.) The superintendent, "while serving as such by appointment from the President, shall have the local rank, pay, and allowances of colonel of engineers; and the commandant of cadets, while serving as such by appointment from the President, shall have the local rank, pay, and allowances of lieutenant-colonel of engineers;" (*Act* June 12, 1858.)

SUPERIOR. (*See* OBEDIENCE.)

SUPERNUMERARY. Graduates of Military Academy, where there are no vacancies among the commissioned officers of the army, may be attached as supernumeraries by brevet of the lowest grade of commissioned officer, not exceeding one to each company; (*Act* April 29, 1812.)

By *Act* March 3, 1847, the President was authorized to attach in the same manner as supernumerary officers of the lowest grade in any *corps* in the army, any non-commissioned officer who should distinguish himself in service, and be recommended by the commanding officer of his regiment.

SUPPLIES. The departments of supply to the army are 1. The Ordnance Department, which provides ordnance and ordnance stores; 2. The Quartermaster's Department, which furnishes quarters, forage, transportation, clothing, camp and garrison equipage; 3. The Subsistence Department, which furnishes subsistence; and 4. The Medical Department, which provides medicines and hospital stores. The Ordnance and Medical Departments, requiring special knowledge for their peculiar duties, could not be relieved of any part of the duties belonging to them respectively; but the want of connection between the Quartermaster's and Subsistence Departments may in war be attended with serious inconvenience, and no good reasons whatever, it is believed, exist for not uniting the two departments in one. Under the orders of one chief in the field, acting, of course, in subordination to the commander of the army, such a department might originate and direct such measures for the supply of the army as had not been provided for; control expenditures; insure a prompt and correct accountability for all disbursements and distributions, and do away with all antagonism of interest caused by the requirement that one department shall furnish subsistence stores, and the other transports. These with clothing and other supplies furnished by the Quartermaster, Ordnance, and Medical Departments, are the great wants of the soldier in active service. A well-armed and well-equipped soldier cannot dispense with food, transportation, and clothing, and the means of providing such necessities in war demand earnest thought, and are happily suggested in the following passages from the work of M. VAUCHELLE, *Cours d'Administration Militaire:*

"We have seen military administration in times of peace conducted upon complete principles and regulations; services regularly organized, and efficiently supported by the natural resources of a fertile and industrious country; sufficient funds always available; the immediate supervision and protection of the war ministry; independence assured to the control of military expenditure and consumption by well-defined laws; nothing wanting, in short, to satisfy all the wants of the army, and to provide them with regularity, order, and economy.

"It is not so, it cannot be so, in a state of war. In the field the frequency of movements, the rapidity of marches, the uncertainty of events, the ever-varying chances, the imperfection of means, the insufficiency of resources—the time ever too short for all that has to be provided and done—embarrass, retard, and paralyze administrative action. Every emergency exacts its immediately appropriate measure, and the least foreseen accident may in a moment frustrate the most wise arrangement, and upset the surest calculations. The duties of administration now assume an entirely new character; they become immense in their extent, limited only, indeed, by the intelligence of the administrator himself, who is charged with their execution.

"The first of all rules, that which the greatest captains, and the most enlightened administrators have never failed to enforce in their writings, and of which experience has everywhere proclaimed the value, is the formation of depots beforehand, and to such an extent that the army may not only be subsisted during the opening of the campaign, but as long after as the interests of military operations may require, or as distance may permit. A certain mistrust of the country about to become the seat of war is indeed prudent, for it is generally a country unknown to administration, or perhaps little or ill known, and which cannot fail to be opposed to its operations, since they are so apt to wound it in its interests or in its feelings. The subjects of which a knowledge appears the most important are: 1. The divisions of the territory into governments, provinces, counties, or departments, into districts, cantons, &c. 2. The organization of its territorial, military, civil, and financial administration. 3. Its natural products. 4. The periods of seed time and harvest of every description of grain, and the proportion between (local) produce and consumption. 5. The localities of large markets and fairs, the periods of these commercial gatherings, and the more important objects of their traffic. 6. The subsistence which might most conveniently be substituted in lieu of those established by our regulations, and the relative proportion to be established in such substitution. 7. The different branches of commerce and industry. 8. The means of re-mount, both as regards cavalry and general transport. 9. The manufacture of cloth, leather, and other material, suitable for the preparation of clothing, equipments, harness, &c. 10. The articles of consumption drawn from other countries, the designation of those countries, and the objects of exchange in importations and exportations. 11. The weights, measures, and coinage, with relative value to our own. 12. The current prices of articles of consumption. 13. Barracks, quarters, hospitals, magazines, and other establishments of administration, and their capacities, throughout the various towns and fortresses. 14. The most convenient spots for forming temporary establishments. 15. The principal points of communication by land and sea, with the distances between them, distinguishing the different routes, and indicating, as regards the roads, the spots at which they cease to be passable for carriages; and as regards rivers and canals, the places where they cease to be navigable. 16. In the large towns or fortresses the nature and quantities of the provisions stored therein, the means of grinding corn and baking, the principal mercantile firms, and the heads of large manufactories or workshops with whom it would be safe to deal for military supplies.

"One may easily conceive how useful such admirable statistics would be. On the outbreak of war the minister would feel no uncertainty either as to the nature or the extent of the arrangements he should have to make for himself, or as to the instructions to be given to his commissary-general. How many false moves would thus be avoided; how many useless and heavy expenses saved; how many unknown and lost resources would thus be discovered and employed for the benefit of the army and the relief of the country which has to support it. A commissariat should regulate its arrangements on the double chances of presumed success or failure, according to the peculiar nature of the war to be undertaken. In the case of success, then in proportion to the advance into the enemy's country, it should form its depots in the rear of the army, and establish by stages, on the line of operations, bakeries, magazines, hospitals, convalescent stations, regular convoys, &c., always taking care to select localities with reference to the most favorable means of communication and of defence. In the case of a reverse, the army falling back upon itself will thus find its administrative services secured by means of the supplies which prudence shall thus have collected. The rights of war, which are but the rights of the most powerful, tempered only by the interests of him who wields them, render an army, whatever it may be, absolute master of the provisions and other useful resources which exist, whether they have been provided as depots by the enemy, or destined for other purposes. Administration requires a numerous *personnel*, active, intelligent, and faithful, always ready to avail themselves of supplies for future use, for transmission elsewhere, or for immediate distribution to the troops, wherever they may be stationed. A commissariat requires an extensive and perfectly organized transport; this is the *sine qua non* to enable an army to subsist in the field. Transport is indispensable, and must be obtained at any price; it must, moreover, be *well adapted to the locality*, in order to be able to follow or rejoin bodies of troops in all directions. Thus it is to be understood that the country occupied must be expected to furnish a large proportion of the requisite transport. Although acting in the midst of a state of things essentially inimical to fixed regulations and established forms, the commissariat should prescribe for itself a strict and scrupulous system. In the face of so many pressing and urgent wants, which, if not supplied with regularity, may disturb the discipline and compromise even the honor of the army, it is not enough for the administrator to prove himself intelligent and economical in the dispensation of resources obtained with difficulty and labor; he should further, courageously attacking all abuses and repressing with severity all wastefulness and fraud, secure to himself the means of justifying his expenditure and distribution by authentic accounts, a duty but too rarely accomplished, but which should never be permitted to be neglected.

"War, it is said, should feed war; the axiom may be true, if not just, but in no case should it be pushed to extremes; circumstances may occur, indeed, to render its application impolitic and dangerous. Under no circumstances, however, can the enemy's country under occupation be altogether relieved from the charges of war; it must inevitably bear a large share, even though its contributions may occasionally be considered as advances only. But whatever their nature, these exactions from an enemy's country should be imposed with discernment and moderation, with reference to the population and the nature of the produce, the geographical position and the wealth of the country and, when possible, with consideration for the feelings of the vanquished. Pillage a country and you reduce the inhabitants to misery, to despair, to flight, and thus not only deprive yourself of assistance, but in the day of reverse find implacable and cruel adversaries."

All that can be done when a country yields nothing is to form depots wherever bodies of troops are likely to be stationed; to have the largest possible reserves at head-quarters; and to be prepared with a sufficient land transport establishment to carry all requisite supplies in the event of an advance or a change of position. But this is an exceptional state of things; in general the country can be placed under contribution, either voluntary or coercive, for the supply of provisions and forage, and the commissariat officer then enters upon his legitimate functions. Several measures are open to his adoption; he may avail himself of the enterprise of local contractors; he may make his purchases directly from the owners at the market price; he may fix an arbitrary rate for the different articles of supply; and lastly, he may levy contributions on the people and compel them to furnish according to their means the provisions required for the army. His own judgment must guide him in the choice of these measures. The employment of contractors, in time of peace undoubtedly advantageous, is attended with certain objections during a period of war. Sir Randolph Routh says truly, "the best and surest contractor is the country occupied by the troops and its natural resources carefully and duly economized;" and he proceeds to cite instances within his experience of the inconvenience arising from too great a confidence in contractors "who swarm about an army when it is prosperous to prey upon its wants, but are the first to fly in the event of a reverse."

The commissariat has to consult at once the wants of the army, the economy of the state, and the resources and feelings of the country in which he is acting. To seize supplies, unless from an enemy in arms, is to be deprecated; to pay for them more than their value, is equally objectionable; unnecessary force creates an ill feeling which may defeat the objects of administration; to submit to imposition enhances the difficulty of the service; but conciliation and fair dealing, backed by decision, will never fail to prove a good policy and enable the army to procure supplies without unnecessary expense to the public or uselessly exasperating the population. If the territory be that of a friendly or a neutral power, every effort should be made by the commissariat to arrive at a just estimate of its resources in grain, cattle, fuel, and other articles of supply, to ascertain their current market value, and having obtained all possible information on these points, the people should be invited, either through the local authorities, or the agency of private individuals, to furnish whatever is required, with the understanding that the usual price will be paid for the supplies brought in, and that the head-quarters of the army will prove a profitable market to them.

When confidence in the good faith of the purchaser has been once established, the population of a country occupied by a military force will be willing enough to sell, and should a disposition to hold back supplies in the hope of enhancing their value be shown, the interposition of the local authorities should be sought in preference to the adoption of arbitrary measures. Conciliation and firmness, temper and justice combined, will seldom fail to induce the inhabitants, even when their sympathies tend in another direction, to contribute to the extent of their means to the maintenance of the army quartered upon them.

Amid a hostile population a conquering army should exercise its power with every possible regard to justice. Fair treatment may reconcile a people to the presence of a conqueror, and induce it to submit to superior strength. No effort should be left untried to produce such a result, since a resort to force, although it may provide for immediate want, inevitably destroys the sources of supply. The best course to be adopted in levying supplies in an enemy's country is, having first ascertained the resources of the district, to demand, through the local authorities, the head men of villages, or other channels, that certain quantities of provisions should be brought at a given time to the head-quarters of the army, care being taken that the demand be not beyond the means of the district, and a fair price should be paid whenever a disposition is shown to comply promptly with these requisitions. Such a measure will rarely fail of effect, and when the inhabitants feel certain that there is no alternative between selling their produce and having it seized, they will submit to the necessities of war in its least aggravated form, and yield to a compulsion which, though it do violence to their national feelings, consults their individual interests. Nor is it only in the supply of provisions that the theatre of war should be laid under contribution; labor and transport may likewise be attained by means of judicious administrative arrangement. The stern rules of war justify the exaction of all the resources within its influence; it is for administration to render these exactions as little oppressive as possible when dealing with a class of people which, as a rule, is the most innocent of the causes of war, the most exposed to its ravages, and the least benefited by its results. In proportion as tact and moderation are displayed by the agents employed in levying supplies upon the population, so will the resources of the country become available and productive. Violence and wrong will convert the peaceable peasant into a desperate and implacable foe; conciliation and fair dealing may make him, if not an ally, at least a profitable neutral. Interests far beyond the hour may be involved in the action of military administration under such circumstances, and the seeds of rancor or good-will, sown to-day on the scene of contending armies, may bring forth fruit to influence the destinies of nations long after the combatants themselves have ceased to struggle.

If it be necessary at established stations that a prompt settlement should be effected for all services rendered to the army, and that every engagement entered into by the commissariat should be most scrupulously complied with, how much more so is this the case in the field. The love of gain—that mainspring of human action under all circumstances, and in all places—is seldom appealed to in vain; but the feeling must be supported by confidence; for one man who will run a risk for a remote prospect of reward, a hundred will toil for a certain remuneration, and it should be one of the first aims of administration to inspire all classes among which it is called upon to act, with a full and entire confidence in its good faith. A breach of faith involves more than immediate consequences: it permanently destroys *credit;* (FONBLANQUE.)

SURGEON. A staff-officer of the medical department. He has the rank of major, but "shall not in virtue of such rank be entitled to command in the line or other staff departments of the army;" (*Act* Feb. 11, 1847. *See* ARMY for the organization of the Medical Department.)

SURGEON-GENERAL. The chief of the Medical Department, with the rank of colonel, but subject to the same restriction of command as other officers of the Medical Department.

SURGERY, (*Military.*) Restricted to its rigorous signification, military surgery is the surgical practice in armies; but in its broad and ordinary acceptation embraces many other branches of art, comprehending the practice of medicine, sanitary precautions, hospital administration, ambulances, &c. The military surgeon must not only be a skilful physician and surgeon, but he must have a constitution sufficiently strong to resist the fatigues of war, and all inclemencies of weather; a solid judgment and a generous activity in giving prompt assistance to the wounded without distinction of rank or grade, and without even excluding enemies. He must have the courage to face dangers without the power, in all cases, of combating them; he must have great coolness in order to act and operate in the most difficult positions, whether amidst the movement of troops, the shock of arms, the cries of the wounded when crowded together, in a charge, in a retreat, in intrenchments, under the ramparts of a besieged place, or at a breach. He must have inventive ingenuity which will supply the wants of the wounded in extreme cases, and a compassionate heart, with strength of will which will inspire confidence in those with whom he is brought so closely in contact. The military surgeon, with his flying ambulance, throws himself into the field of battle, through the mêlée, under the fire of the enemy, runs the risk of being taken prisoner, being wounded, or being killed, and is worthy of all the honors that should be bestowed on bravery and skill in the performance of his high functions. Additional grades, as hospital-surgeons, surgeons of divisions, surgeons-in-chief, and inspector-generals of hospitals, &c., are required for every army in the field.

SURVEYS, (*Military or Expeditious.*) In military surveys the first thing done is to determine by triangulation, the principal points, to which surveys of details are subsequently referred.

Triangulation.—A base is chosen from the extremities of which a large extent of the ground to be surveyed may be seen. The length of the base ought to be in proportion to the extent of the triangle to be constructed; the equilateral form being preferable. This base is sometimes furnished by the regular charts from the topographical bureau. If not, it is measured by chain or by paces. From the base, pass by the fewest possible stations to two points occupying central positions in the survey, and suitable for stations; from these two points let lines radiate, cutting those that are known, and their intersections serve as points of data in details; multiply in this way the number of triangles, always diminishing the length of their sides; intermediate details may be determined by paces. Such is the process.

Plane Table.—For these operations a plane table and the ALIDADE are sufficient. The plane table used is of small dimensions. Commander Salneuve advises a plane table made of several rulers equal in length and breadth, united by parchment and strong cloth, upon which they are folded. When the plane table is used, the rulers are kept in the same plane by means of two other rulers which take a rectangular position towards the others by being revolved on one of their extremities and are then kept in place by means of a little hook at the other extremity of each. (*See* Fig. 217.)

When the work is ended or suspended, these two rulers are unhooked and turned so as to cover the last two of the parallel rulers, and the whole may then be folded and put in a pistol-holster. This plane table has besides an upright stick with an iron ferrule which supports the table horizontally on being stuck in the ground. If such a plane table has not been provided, however, it is easy to procure a small flat board, which will answer the same purpose.

Alidade.—For an alidade, a triangular rule of a double decimetre may be used, upon which are fastened two nails or pins which serve as sights. The problems resolved by means of the plane table and alidade are the following: 1. To determine the projection of an accessible point by means of two other points R and P given and also accessible. 2. To determine the same thing by means of two other points, only one of which, R, is accessible. 3. To determine the same thing by means of two others, R and P, which are inaccessible, but upon whose direction we can find stations. 4. To determine the same thing by means of two others not only inaccessible, but upon whose direction it is impossible to find a station. 5. The same determination by means of three points, R, P, Z, inaccessible.

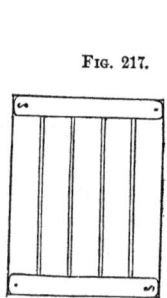

Fig. 217.

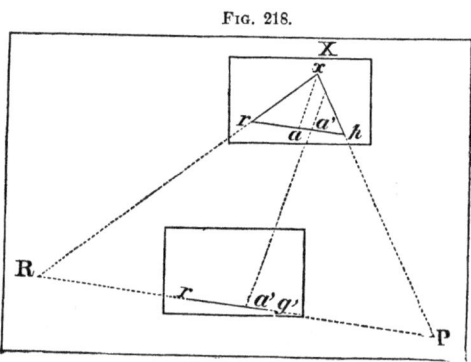

Fig. 218.

Adjusting.—The first thing to be done at each station is to adjust the plane table. Let R P be the base, (Fig. 218.) Assume $r\ h$ arbitrarily as its projection on a sheet of paper corresponding to the scale of say $\frac{1}{20000}$. At the station R place the instrument horizontally by means of a plumb-line, and let one of the sides of the alidade (using the same side for all operations) rest upon the projection, turning it until the side covers $r\,h$. Turn then the plane table until P is seen in the prolongation of the sights. The plane table is then adjusted, and it must be maintained in that position while at the station.

To adjust with reference to the meridian—trace this line upon the leaf and afterwards turn the side of the frame until it has the same direction; the problem is resolved by means of corresponding heights of the sun. Let R P be a side upon the ground, and $r\,p$ its provisional projection; (Fig. 219.) Erect upon the horizontal plane a vertical stile terminated by a plate of blackened iron, pierced with a little hole at its centre m, and disposed in such manner as to receive nearly perpendicularly the rays of the sun at noon. Project the centre m in m' upon the plan by a plumb-line, and from m' as centre describe several circumferences $n\,o'\,n'\,p\,o''$ p'.

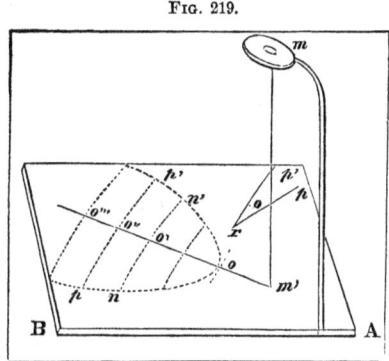

Fig. 219.

Observe the march of the sun a little before and a little after noon. Divide in two equal parts the circumferences intercepted by the solar spectrum: the middle points belong to the meridian, whose projection we thus have. Measure then the angle o made with the side of the frame A B; lay off by the line $r\,p'$ an equal angle; then turning the plane table until this line $r\,p'$ corresponds in direction with the line that it represents: the side of the frame indicates the meridian of the place. The questions may then be thus resolved:

First Question, (Fig. 218.)—Take a station at R; adjust upon P; look at x, the point sought, by turning the alidade around the point r, the projection of R, and trace $r\,x$ the projection of R x. Go to P; operate in the same manner, and the intersection of the two right lines $r\,x$ and $h\,x$ gives the projection of the point sought. This is the method of intersection.

Second Question.—Take a station at R; adjust upon P: radiate on X; go to X; adjust upon R, following the indefinite projection already traced; turn the alidade around p until P is seen: the intersection of the right line thus traced with the first gives X. This is a method of *offset*. (Fig. 220.)

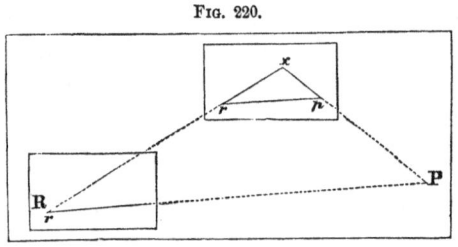

Fig. 220.

Third Question.—Assume upon $r\,h$ a point a', (Fig. 218;) supposed projection of A any point whatever; look at X; draw the assumed line through a': this line of direction will serve for adjusting when at X; for, although not the true projection of A X, it is necessarily parallel to it; go to X, adjust upon A; it is only necessary to draw two lines passing through R and P, turning the alidade on h and r: the point of intersection of these lines is the projection sought.

Fourth Question.—Take a fourth point Y, (Fig. 221,) at which the

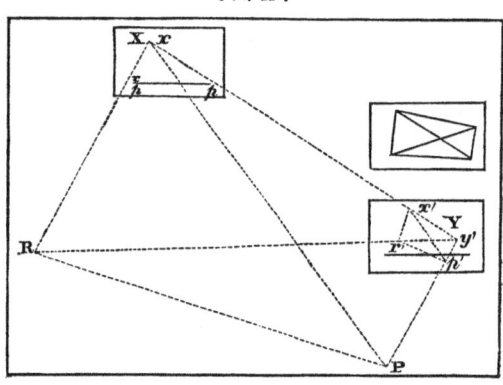

Fig. 221.

observer may place himself, and from which the other three may be seen; construct afterwards upon the leaf a quadrilateral similar to that made upon the ground, and then construct an equal quadrilateral upon $r\,p$: for this purpose, go to X, of which we have the arbitrary projection x'; look at R, P, and Y, tracing these directions upon paper; go to Y, of which we have also the projection y' upon the line leaving X and drawn through x'; adjust upon $x'\,y'$ and look at R and P; the points of intersection determine two angles r' and p' of a quadrilateral, of which the two others are x' and y', similar to that made upon the ground by R, P, X, and Y, and similar also to the projection sought; nothing more is necessary than to establish the relation on $r\,p$.

Fifth Question.—Measure at x (Fig. 222) the angles $r\,x\,p,\ p\,x\,z$, and make at r and z two angles $a\,r\,p,\ b\,z\,p$, which shall be respectively equal; through the middle of the lines $r\,p$ and $z\,p$ erect perpendiculars; at z and r raise also perpendiculars to $a\,r$ and $b\,z$; the points of

Fig. 222.

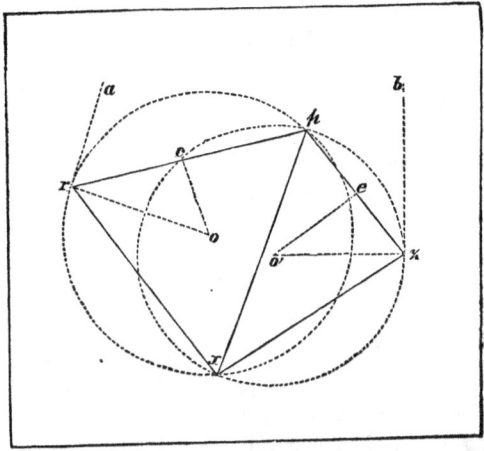

meeting o and o' are the centres of two circumferences which will intersect in x the projection sought: this is the method by capable segments. Or (Fig. 223) let r, p, z be the projections on the plan of three points

Fig. 223.

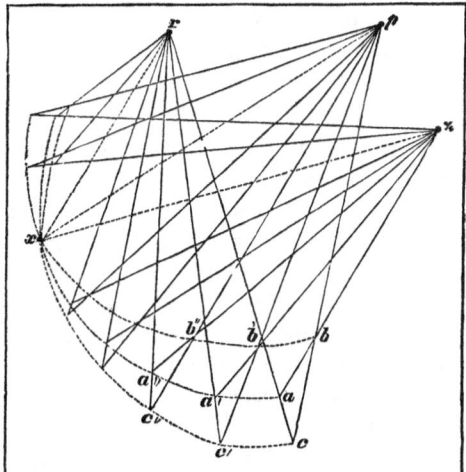

of ground R, P, Z. To determine X without change of station, adjust partially the plane table with reference to X. Afterwards, through R P Z or their projections, draw three right lines which will cut each other and form a triangle $a\,b\,c$. If the plan had been perfectly adjusted, the three lines would have cut each other at the same point x, the projection of X. It is necessary, then, to turn the plane table so that the three lines by their intersections will form another triangle $a'\ b'\ c'$ smaller than the first. Continue thus until the triangle is reduced to a point. The intersection of the three curves $a\ a'\ a''$, $b\ b'\ b''$, and $c\ c'\ c''$, gives the projection sought. These are the different means employed to determine the points of stations. When each is obtained, all details to the right and left of a direction may be filled up by sight and by paces; one of the sides of the triangle being taken for the direction, an angle made with it may be traced by means of the alidade, or else observed in number and degree with an instrument, and subsequently drawn with a protractor.

Observations of Angles.—A pocket sextant may be employed; or, in the survey of details, the operation may be accelerated by a compass. This instrument may be even used in the first triangulation, if the sides of the triangle are not too great. Time is gained, and the results are sufficiently satisfactory for an expeditious survey. The compass is nothing more than a magnetic needle in a rectangular box, at the bottom of which a limb turns in such a manner that the north and south line is exactly parallel to the larger side of the box. This instrument, when adapted to the plane table, greatly abridges the operations by the facility it gives for adjusting the survey. Thus the magnetic needle or magnetic meridian makes with the astronomical meridian an angle called the declination. If O be the declination, put the compass on the plane table in such a manner that the needle coincides with the north and south line. Turn it afterwards until the needle passes over the number of degrees equal to the declination O. Then the long side of the box is parallel to the meridian, and if it is wished that one of the sides of the survey should have this meridian direction, the needle is made to describe, by turning the plane table, an angle equal to that made by the side of the triangle with the side of the compass.

When without instruments, the adjustment of a survey may be determined by setting up vertically upon the plane table a pin or needle, and tracing by means of a watch the shadow of this pin at different hours of the day. The solar spectrum thus formed serves on subsequent days to adjust the plane table in the same manner. It is sufficient to do so to look at your watch and turn the table until the shade of the pin corresponds to the same indication of the hour. Or, the sides of hills in expeditious surveys are obtained as in regular surveys, by calculating them by means of the base and the angle of fall.

To level and measure angles without any instrument.—We may obtain the principal angles graphically as follows: plant a staff vertically in the ground by means of a plumb-line; trace upon a leaf of the sketch book a vertical line representing the height of the staff; rest upon the paper a rule, directing it upon the object whose relative height is to be determined; trace this line with a pencil, and the angle of depression or ascension is concluded from that made by the two lines, and the right angle formed by the vertical and horizontal. The difference of level between two points may also be determined by means of two staves of different heights: let the shorter be placed at the station and the other upon the direction of the point whose elevation is sought in such a position that the point of sight and the tops of the two staves may be on the same line. The difference of level is determined by the similitude of the two triangles. The angle of depression is found by placing the longer staff at the station. With the angle of elevation or depression, the height is always found approximately by means of a table of tangents: let h be the height, b the base, c the angle; then $h = b$ tan. c in right-angled triangles.

Or, when the angle of ascension or depression is known, differences of level may be determined graphically as follows: let A B (Fig. 224) represent a length of 1,000m. by the scale—lay off the line A C, making an angle of 25° with A B; draw the perpendicular B C: B C, multiplied by the denominator of the scale, will give a height corresponding to a base of 1,000m. at an angle of ascension of 25°. If lines then radiate from A, making angles of 5° with their adjacent radii, and the base is divided into parts of 50m., and perpendiculars are erected at the points of division, a figure is obtained by which all differences of level will be approximately determined.

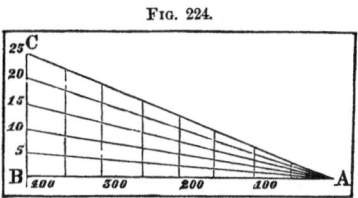

Fig. 224.

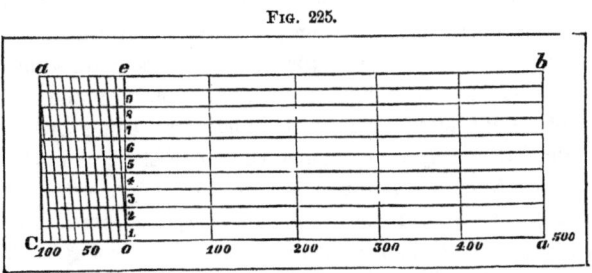

Fig. 225.

Scale.—All plans are accompanied by a graphic scale which makes known the length of lines on the ground by means of their representations upon the plan and reciprocally; (Fig. 225.) This figure represents a scale of $\frac{1}{20000}$, that is to say, a scale by which 1 metre on paper is equivalent to 20,000 metres on the ground.

Reduction of Plans.—It may be necessary to copy a plan and reduce the scale. This is done by tracing an outline in which the desired relation is preserved. The different parts are then reduced by means of an angle of reduction. This angle is constructed by tracing a line $a\,b$; (Fig. 226.) From b as a centre describe an arc of a circle with the radius $b\,c$ so chosen that $\dfrac{b\,c}{a\,b} = \dfrac{m}{n}$, being the relation between the two scales; draw then the tangent $a\,c$. It results from this that if $a\,d$, for example, is a line to be reduced, in describing from the point d an arc of a circle tangent to $a\,c$, e being the point of tangency, $d\,e$ will be the desired reduction.

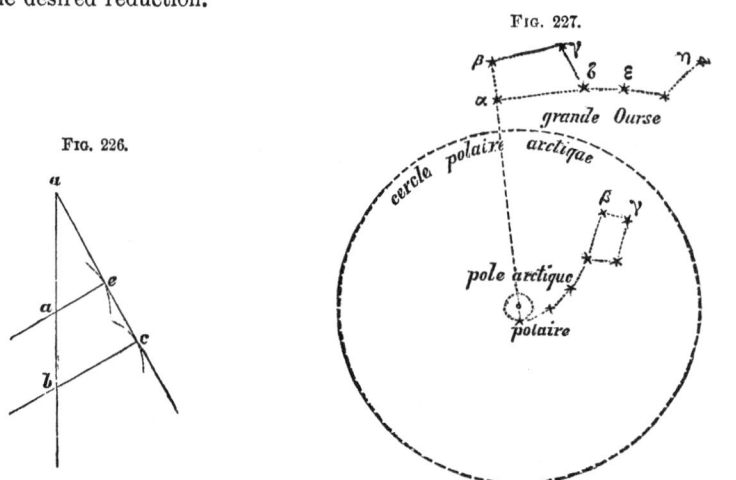

Fig. 226.

Fig. 227.

To trace a meridian at night.—The means of establishing the meridian by the solar spectrum have been indicated. The meridian may be determined at night by passing a plane through a plumb-line and the north star. The trace of this plane on a horizontal plane will be the projection of the meridian sought; the north star being only $1\frac{1}{2}°$ from the true pole. It is easy to recognize the north star; it is the seventh star of the little bear, and is found precisely in the prolongation of the two first stars of the Great Bear, (*Grande Ourse*,) a constellation disposed in symmetrical order as in Fig. 227; (*Aide Memoire d'Etat Major.*)

SUSPENSION. In cases where a court-martial may think proper to sentence a commissioned officer to be suspended from command, they shall have power to suspend his pay and emoluments for the same time, according to the nature and heinousness of the offence; (Art. 84.) Rank and command are distinct.

SUTLERS. All sutlers and retainers to the camp, and all persons whatsoever, serving with the armies in the field though not enlisted soldiers, are to be subjected to orders, according to the rules and discipline of war; (Art. 60.) All officers commanding in the field, forts, barracks, or garrisons of the United States, are hereby required to see that the persons permitted to sutle shall supply the soldiers with good and wholesome provisions or other articles at a reasonable price; (Art. 30.) Sutlers not to sell or keep their shops open after nine at night, or on Sundays during divine service or sermon; (Art. 29.) Exorbitant prices not to be exacted from sutlers by commanding officers for the hire of stalls or houses let out.

SWORD. The foot artillery sword resembles the Roman sword. The BLADE is 19 in. long, straight, two-edged; *Body* (or blade proper)—shoulder rounding, ridges, point bevels, edges; *Tang*, its riveting and rounding, three holes for the gripe rivets; HILT, (brass, in one piece,) cross, knob, and pommel of the cross; SCABBARD (harness leather jacked) blackened and varnished with mountings and ferrule.

The Infantry Sword.—BLADE, (straight, cut, and thrust,) back, edge, groove, bevel point; HILT (surmounting brass)—covering of gripe brass with grooves and ridges; GUARD in one piece; SCABBARD, (leather.) This sword is for the non-commissioned officers of foot troops; a similar one, without the guard plate, and with a blade 26 inches long, for musicians. The sword for officers not mounted is also of the same pattern, with ornamented gilt mountings, and a silver gripe; the inner half of the guard plate is made with a *hinge*. (See SABRE.)

SWORD-BAYONET. Short arms, as *carbines*, are sometimes furnished with a bayonet made in the form of a sword. The back of the handle has a groove, which fits upon a stud upon the barrel, and the cross-piece has a hole which fits the barrel. The bayonet is prevented from slipping off by a spring-catch; the sword-bayonet is ordinarily carried as a side arm, for which purpose it is well adapted, having a curved cutting edge as well as sharp point.

T

TABLES. (*See* Articles ARTILLERY; FIRING; RIFLED ORDNANCE; RATION; WEIGHTS.) (Consult *A Collection of Tables and Formulæ useful in Surveying, Geodesy, and Practical Astronomy, including Elements for the Projection of Maps,* by Capt. T. J. LEE, Top. Engineer; also *Ordnance Manual* for numerous useful tables.)

TABLE OF NATURAL SINES AND TANGENTS.

Deg.	Min.	Sine.	Tangent.	Deg.	Min.	Sine.	Tangent.
0	10	0029089	0029089	14	00	2419219	2493280
	15	0043633	0043634		15	2461533	2539676
	30	0087265	0087269		30	2503800	2586176
	45	0130896	0130907		45	2546019	2632780
1	00	0174524	0174551	15	00	2588190	2679492
	15	0218149	0218201		15	2630312	2726313
	30	0261769	0261859		30	2672384	2773245
	45	0305385	0305528		45	2714404	2820292
2	00	0348995	0349208	16	00	2756374	2867454
	15	0392598	0392901		15	2798290	2914734
	30	0436194	0436609		30	2840153	2962135
	45	0479781	0480334		45	2881963	3009658
3	00	0523360	0524078	17	00	2923717	3057307
	15	0566928	0567841		15	2965416	3105083
	30	0610485	0611626		30	3007058	3152988
	45	0654031	0655435		45	3048643	3201025
4	00	0697565	0699268	18	00	3090170	3249197
	15	0741085	0743128		15	3131638	3297505
	30	0784591	0787017		30	3173047	3345953
	45	0828082	0830936		45	3214395	3394543
5	00	0871557	0874887	19	00	3255682	3443276
	15	0915016	0918871		15	3296906	3492156
	30	0958458	0962890		30	3338069	3541186
	45	1001881	1006947		45	3379167	3590367
6	00	1045285	1051042	20	00	3420201	3639702
	15	1088669	1095178		15	3461171	3689195
	30	1132032	1139356		30	3502074	3738847
	45	1175374	1183578		45	3542910	3788661
7	00	1218693	1227846	21	00	3583679	3838640
	15	1261990	1272161		15	3624380	3888787
	30	1305262	1316525		30	3665012	3939105
	45	1348509	1360940		45	3705574	3989595
8	00	1391731	1405408	22	00	3746066	4040262
	15	1434926	1449931		15	3786486	4091108
	30	1478094	1494510		30	3826834	4142136
	45	1521234	1539147		45	3867110	4193348
9	00	1564345	1583844	23	00	3907311	4244748
	15	1607426	1628603		15	3947439	4296339
	30	1650476	1673426		30	3987491	4348124
	45	1693495	1718314		45	4027467	4400105
10	00	1736482	1763270	24	00	4067366	4452287
	15	1779435	1808295		15	4107189	4504672
	30	1822355	1853390		30	4146932	4557263
	45	1865240	1898559		45	4186597	4610063
11	00	1908090	1943803	25	00	4226183	4663077
	15	1950903	1989124		30	4305111	4769755
	30	1993679	2034523	26	00	4383711	4877326
	45	2036418	2080003		30	4461978	4985816
12	00	2079117	2125566	27	00	4539905	5095254
	15	2121777	2171213		30	4617486	5205671
	30	2164396	2216947	28	00	4694716	5317094
	45	2206974	2262769		30	4771588	5429557
13	00	2249511	2308682	29	00	4848096	5543091
	15	2292004	2354687		30	4924236	5657728
	30	2334454	2400788	30	00	5000000	5773503
	45	2376859	2446984		30	5075384	5890450

TABLE OF NATURAL SINES AND TANGENTS—(Continued.)

Deg.	Min.	Sine.	Tangent.	Deg.	Min.	Sine.	Tangent.
31	00	5150381	6008606	53	00	7986355	13270448
	30	5224986	6128008		30	8038569	13514224
32	00	5299193	6248694	54	00	8090170	13763819
	30	5372996	6370703		30	8141155	14019483
33	00	5446390	6494076	55	00	8191520	14281480
	30	5519370	6618856		30	8241262	14550090
34	00	5591929	6745085	56	00	8290376	14825610
	30	5664062	6872810		30	8338858	15108352
35	00	5735764	7002075	57	00	8386706	15398650
	30	5807030	7132931		30	8433914	15696856
36	00	5877853	7265425	58	00	8480481	16003345
	30	5948228	7399611		30	8526402	16318517
37	00	6018150	7535541	59	00	8571673	16642795
	30	6087614	7673270		30	8616292	16976631
38	00	6156615	7812856	60	00	8660254	17320508
	30	6225146	7954359	61	00	8746197	18040478
39	00	6293204	8097840	62	00	8829476	18807265
	30	6360782	8243364	63	00	8910065	19626105
40	00	6427876	8390996	64	00	8987940	20503038
	30	6494480	8540807	65	00	9063078	21445069
41	00	6560590	8692867	66	00	9135455	22460368
	30	6626200	8847253	67	00	9205049	23558524
42	00	6691306	9004040	68	00	9271839	24750869
	30	6755902	9163312	69	00	9335804	26050891
43	00	6819984	9325151	70	00	9396926	27474774
	30	6883546	9489646	71	00	9455186	29042109
44	00	6946584	9656888	72	00	9510565	30776835
	30	7009093	9826973	73	00	9563048	32708526
45	00	7071068	10000000	74	00	9612617	34874144
	30	7132504	10176074	75	00	9659258	37320508
46	00	7193398	10355303	76	00	9702957	40107809
	30	7253744	10537801	77	00	9743701	43314759
47	00	7313537	10723687	78	00	9781476	47046301
	30	7372773	10913085	79	00	9816272	51445540
48	00	7431448	11106125	80	00	9848078	56712818
	30	7489557	11302944	81	00	9876883	63137515
49	00	7547096	11503684	82	00	9902681	71153697
	30	7604060	11708496	83	00	9925462	81443464
50	00	7660444	11917536	84	00	9945219	95143645
	30	7716246	12130970	85	00	9961947	114300520
51	00	7771460	12348972	86	00	9975641	143006660
	30	7826082	12571723	87	00	9986295	190811370
52	00	7880108	12799416	88	00	9993908	286362530
	30	7933533	13032254	89	00	9998477	572899620
				90	00	10000000	Infinite.

Frigorific Mixtures.

Nitrate of ammonia 1, water 1; thermometer falls from 50° to 4°
Sulph. soda 8, muriatic acid 5 . . . . . 50 to 0
Phosphate of soda 9, nitrate of ammonia 6, diluted nitric acid, 4 . . . . 50 to —21
Common salt 1, snow or ice 2 . . . . . 32 to — 4
Cryst. chloride of lime 3, snow 2 . . . . 32 to —50

Elastic Force of Steam at Different Temperatures.
[From experiments of Committee of Franklin Institute.]
The unit is the atmospheric pressure, or 1 atmosphere = 30 in. of mercury.

Temp.	Press.	Temp.	Press.	Temp.	Press.	Temp.	Press.	Temp.	Press.
212	1	275	3	304½	5	326	7	345	9
235	1½	284	3½	310	5½	331	7½	349	9½
250	2	291½	4	315½	6	336	8	352½	10
264	2½	298½	4½	321	6½	340½	8½		

Freezing Points of Liquids.

Olive oil	36° Fahr.	Strong wines	20° Fahr.
Water	32	Sulphuric acid	1
Milk	30	Brandy	−7
Vinegar	28	Mercury	−39
Spirits of turpentine	16	Nitric acid	−55

Boiling Points of Liquids. (*Bar. 30 in.*)

Sulphuric ether	98°	Phosphorus	554°
Ammonia	140	Spirits of turpentine	560
Alcohol	174	Sulphur	570
Water, and essential oils	212	Sulphuric acid	590
Water, saturated with salt	224	Linseed oil	600
Nitric acid	248	Mercury	660

Liquids boil at a much lower temperature in vacuo, or under diminished pressure of the atmosphere. At the altitude of about 17,500 feet above the sea, where the barometer stands at 15.35 in., water boils at 180°.

RELATIVE STRENGTH OF THE ENGLISH, FRENCH, AND RUSSIAN NAVIES.

ENGLISH NAVY.

Class of Ship.	Steam.			Sailing.	Total of Steam and Sailing.
	Afloat.	Building or Converting.	Total.	Afloat.	
Liners	48	12	60	16	76
Frigates	34	16	50	13	63
Block Ships	9	...	9	...	9
Iron-cased Ships	...	4	4	...	4
Corvettes	16	5	21	3	24
Sloops	80	15	95	...	95
Small Vessels	27	...	27	...	27
Gun Vessels and Gun Boats	171	21	192	...	192
Floating Batteries	8	...	8	...	8
Transports	15	...	15	...	15
Mortar Vessels	4	...	4	...	4
Total	412	73	485	32	517

FRENCH NAVY.

Class of Ship.	Steam.			Sailing.	Total of Steam and Sailing.
	Afloat.	Building.	Total.	Afloat.	
Liners	33	4	37	9	46
Frigates	34	13	47	28	75
Iron-cased Ships	2	3	5	...	5
Corvettes	17	2	19	13	32
Avisos, &c.	86	3	89	46	135
Gun Boats	39	29	68	...	68
Floating Batteries	5	4	9	...	9
Transports	31	...	31	...	31
Total	247	58	305	96	401

RUSSIAN NAVY.

Class of Ship.	Steam.			Sailing.	Total of Steam and Sailing.
	Afloat.	Building.	Total.	Afloat.	
Liners	13	9	22	16	38
Frigates	18	3	21	...	21
Corvettes	11	11	22	...	22
Small Vessels	30	...	30	...	30
Gun Boats	112	25	137	...	137
Transports	8	...	8	...	8
Total	192	48	240	16	256

TACTICS—as distinguished from strategy, is the art of handling troops. Sect. 7 *Act* May 8, 1792, prescribes the tactics established by Congress in 1779, as the rules for the exercise and training of the militia.

Act of March 3, 1813, requests the President to cause to be prepared and laid before Congress a military system of discipline for the infantry of the army and militia of the United States.

Act of May 12, 1820, prescribes that the system of discipline and field-exercise, that is or may be ordered for the infantry, artillery, and riflemen of the regular army shall be the same for the respective corps of the militia.

Act of May 18, 1826, authorizes the Secretary of War to have prepared a complete system of cavalry tactics, and also a system of exercise and instruction of field-artillery, including manœuvres for light or horse artillery, for the use of the militia of the United States, to be reported for consideration or adoption by Congress at its next session.

Act of March 2, 1829, provides for the distribution of 60,000 copies of the abstract of infantry and light infantry and rifle tactics, and also 5,000 copies of the system of instruction for field-artillery prepared pursuant to *Act* of 1826.

Tactics of Gustavus Adolphus and his contemporaries.—Gustavus Adolphus, the greatest captain of his time, originated new principles in the art of war, which in their essence still subsist. His advent marks a fixed and certain epoch in the history of tactics. There are four ideas originated by him, which overthrew the tactics of his predecessors. 1. He gave in combats a greater, but not an absolute influence to the musket; and united in order of battle heavy and small arms. 2. He increased the mobility of his troops by breaking up heavy masses, and thus also diminished the destructive effects of an enemy's fire. 3. He ranged the different arms according to their intention, and thus established facility in manœuvring as well as their mutual capacity to aid each other. 4. He restored individual activity, which had all but ceased to exist, particularly in cavalry, since the invention of powder.

Gustavus Adolphus conceived and executed all his projects himself. He was at the same time an infantry, cavalry, and artillery soldier. He was a lover of mathematics and natural philosophy, and did not disdain to hold a pencil and compass. The order of battle of the Swedes consisted, according to circumstances, in a formation of two or three lines ranged parallel to each other or in echelons upon the wings, the cavalry behind the infantry or upon its wings. The cavalry was proportionably very numerous. It fought in four ranks. The infantry was ranged in six ranks. The batteries of artillery were massed and masked. In assaulting Germany, Gustavus had two hundred pieces.

Tactics before and during the war of the Spanish Succession.—At this epoch there were great men, but no one like Gustavus took a giant step in tactics. The art was at a stand during more than a hundred years notwithstanding the rapid succession of wars, and the reiterated occasions such wars offered to genius. In this world it is not events which produce changes, but superior minds which control events. Gradually, however, the musket became the only arm of infantry, and the pike was entirely discontinued. Thus the possibility of infantry, defending themselves against cavalry vanished, and in order to restore the equilibrium, the epicus or half-pike was introduced. Each infantry man carried one at the beginning of the 17th century. This order was general. It succeeded against the Turks, but cruelly impeded the mobility of infantry.

The bayonet appeared for the first time in the Netherlands in 1647, and essentially contributed to the discontinuance of the pike. At first this arm was very unhandy, as it was necessary to take it from the musket before firing. Under Charles XII. this was remedied, and in the Prussian army in 1732, the front rank was armed with a bayonet during the fire. In 1740 at the battle of Molwitz the three ranks were thus provided.

To appreciate the spirit of the tactics of this time, it is necessary to study the campaigns of Turenne and Luxembourg, and those of Prince Eugene and Marlborough. The principal characteristic of the tactics of this epoch consisted in the attack of the whole line at the same time, and consequently of the general opening of a battle upon all points at once. A part of a line was rarely maintained in position during the attack of other portions. The importance of echelons was not appreciated, or it was not known how to use them in the oblique order. Manœuvres, however, improved, but very slowly. Hence open fields of battle were generally preferred. If accidents of ground were sought, it was for the purpose of establishing lines of defence. Marches were executed, ordinarily, by many columns, each consisting of a single arm. There was therefore little reciprocity of action, and even in camps the same marked separation was preserved.

Tactics of Frederick the Great and his contemporaries.—Frederick found the art of war in a singular state. A great man—a born captain was indispensable to raise this art from the dust under which it had been trampled and all but stifled by a miserable formalism. The active genius, the living courage, the free will which had signalized the combats of ancient times had disappeared; the musket had become a powerful arm, but pedantry had seized upon the order of battle; all merit consisted in forms, and cavalry rendered useless in action had become only the furniture of parades.

The great merit of Frederick consisted in recognizing the spirit of his age, and giving it a new bent. When Frederick appeared in camp, he found the musket in general use. He occupied himself in perfecting it. He fixed the depth of infantry at three ranks, and thus were seen deployed those long and thin lines which later took with the art of moving them the denomination of tactics of lines.

Frederick required of his cavalry but two things: 1, Promptitude in surprising an enemy; and 2, United and violent attacks to overthrow and annihilate him. For these reasons he exacted the exclusive use of the sabre in cavalry, which soon disdained the gun as useless and unworthy of a true cavalier. All movements were executed regularly but rapidly. Frederick also occupied himself with perfecting artillery. He diminished the weight of field-pieces, and drew a marked line of separation between field and siege pieces.

The American Revolutionary War fixed attention specially upon the manner of fighting in dispersed order. This order of battle, in consequence of the difficulties of a wooded country, played here the principal part, and it may be affirmed that skill as marksmen—an important part of the true system of light infantry or rifle tactics—dates from that period.

Tactics during the French Revolution, and its immediate effects.—This epoch of tactics is distinguished by perfecting individual action, and renewing the force of infantry in the shock of battle, by dispensing with long thin lines which were in part replaced by the order in mass. From the French Revolution was born the principle that all citizens are equal, and all owe service to their country. As the first consequence of this principle arose the general and legal obligation of devoting one's self to the military service. This obligation put in movement an aggregate of moral forces which could not otherwise have been collected in armies. But in spite of the enthusiasm of the people, (at least at first,) the absence of military instruction and discipline was everywhere seen. It was necessary that generals should endeavor to create a new tactics.

Tactics then, for the first time, adapted itself to the national character of the soldier, and bent its forms to that character. It was impossible to harmonize the heavy tactics of lines with republican ardor. Instead, therefore, of losing their time in making soldiers machines, the *wise generals* preferred the machines already made. It was indispensable to create a more easy mechanism of sub-divisions, and they naturally determined upon formations in small masses, whilst the order in lines was gradually abandoned. Each republican, feeling himself called to defend his country, considered national interests as his own proper interest. It was not sufficient for him to occupy simply a place in the ranks, he wished to fight individually and with his own proper hands. The stamp of the *tirailleur* was thus impressed on every Frenchman by that ardent will, which was carefully maintained in giving full liberty to the highly pitched energy and courage of the soldier. But where it was necessary to break strength by strength, all were reunited in masses, and disputed the honor of dying in the foremost rank for the republic. These two systems (although they later took the name of systems) brought about the simple mechanism of the new French tactics, the essence of which is concentrated in the system of skirmishers and the system of masses.

A general tactics for all arms is a chimera. An army is composed of infantry, cavalry, artillery, and engineer soldiers. The three first are separate *arms*. Each of these arms must have its particular tactics. But the tactics of those arms, when united, is simply the proper use of each arm by the general-in-chief according to ever-varying circumstances. Each arm ought to think itself invincible. This moral element, or, what is the same thing, a courage developed by discipline, is the most essential quality of a soldier. No one will deny that this moral element is increased in offensive movements. The more infantry attacks with the bayonet, the more cavalry is employed in the charge, the more artillery is brought within range of grape, the greater will be the valor of the soldiers of all arms. Infantry is the great body or nucleus of all armies. An army which possesses good infantry may repair all its losses in war. Light infantry requires a more developed instruction, more corporal dexterity, more circumspection and intelligence than infantry of the line. To march in masses is the duty of the latter. To act in isolated positions under all circumstances of personal danger, is required of the former. All good infantry, whether light or heavy, is at home in close or distant combats. The distinctive characteristic of infantry of the line is a regular, bold, and decided march upon an enemy, in closed ranks, *en muraille*, with a heavy fire when commanded, and *sang-froid* under all circumstances. The distinctive characteristic of light infantry should be skilfulness as marksmen, circumspection, capacity to act independently, indefatigability in occupying an enemy for hours, and even days, incommoding him at long distances, destroying him at short, shunning pressure and attacking anew when pressure ceases, knowing no difficulties of ground, advancing boldly, but when too adventurous uniting smartly for safety, again to resume the independent movements of skirmishers as soon as the danger has disappeared.

In attack as in defence, infantry has three ways of fighting: 1, as skirmishers; 2, by the fire in masses; 3, by the bayonet. All three modes in their reciprocal action experience a great number of modifications, which must depend upon the skill of the tactician. He must thoroughly understand the advantages and disadvantages of the open and close order. He must be able to apply either the one or the other, according to circumstances, and always keep in view the practicability of passing from one to the other. Soldiers ranged in line elbow to elbow are, as it were, tied together, and the will of the whole is controlled by the commander. This is the order in line of battle. If the line be broken into companies or divisions, and ranged one behind the other, we have the order in column, and this order is important in manœuvring. (*See* MANŒUVRES IN COMBAT.)

The combat as skirmishers is in open or dispersed order. Almost all combats of infantry are begun by skirmishers. It is important, therefore, that infantry of the line as well as light infantry should be instructed as skirmishers. Nothing is so useful in concealing from an enemy our force and intentions than throwing forward skirmishers. If the skirmishers are skilful they may for a long time occupy an enemy, and meanwhile the great body of the army concealed behind the curtain thus formed may present themselves unexpectedly at a decisive point. (Consult *prescribed Tactics for Manœuvres of Infantry of the Line and Light Infantry; Cavalry Tactics; Artillery Tactics;* and *De la Tactique des Trois Armes, Infanterie, Cavalerie, Artillerie,* par C. DECKER.)

TAKE. In a military sense, to take is to make prisoner, or to capture. It has also a meaning in field movements, viz., to adopt any particular formation, as to "take open order."

To take ground to the right or left, is to extend a line, or to move troops in either of those directions.

To take down, is to commit to paper that which is spoken by another.

To take the field, is to encamp, to commence the operations of a campaign.

To take up the gauntlet, is to accept a challenge.

TAMBOUR—is a stockade or timber wall, loopholed, made with two faces, forming a salient angle at the gorge of a work, to serve as a retrenchment or to cover the staircase, with a ditch in front, and sometimes with a half roof sloping to the rear, to protect the defenders from hand-grenades and splinters of shells. (*See* BUILDINGS, *Defence of.*)

TAMP. To pack the excavation of a mine, after the charge has been deposited.

TAMPION OR TOMPION. Plug, stopper—iron and copper; lead plate for covering shot holes; muzzle cover of a mortar; small circular bit of hard wood, sheet iron, or stiff paper for covering the claying of a rocket; (BURNS.)

TANGENT—in trigonometry, is the straight line which touches a circular arc at one of its extremities, and is terminated by the production of the radius passing through the other extremity. The arc and its tangent have always a certain relation to each other, and when one is given in parts of the radius the other can always be computed by means of an infinite series. Let ϕ denote an arc, and tan. ϕ the tangent of the arc ϕ; we have the following series:

$$\phi = \tan. \phi - \tfrac{1}{3}\tan.^3\phi + \tfrac{1}{5}\tan.^5\phi - \tfrac{1}{7}\tan.^7\phi +, \&c.,$$

$$\tan. \phi = \phi + \frac{\phi^3}{3} + \frac{2\phi^5}{3.5} + \frac{17\phi^7}{32.5.7} + \frac{62\phi^9}{32.5.7.9} +, \&c.$$

For the manner of using sines, cosines, and tangents, *see* LOGARITHMS; SURVEYING; TABLE; TRIGONOMETRY.

TANG. The tang of the breech of a musket is the projecting part by which the barrel is secured to the stock.

TANGENT-SCALE—(sheet brass,) *flanch* 0.5 inch wide, cut to fit the base-ring of the piece; upper edge cut into notches for each $\tfrac{1}{8}$ degree of elevation.

TABLE OF TANGENT-SCALES FOR FIELD-GUNS AND HOWITZERS.

ELEVATION.	GUNS.		HOWITZERS.		
	6-pdr.	12-pdr.	12-pdr.	24-pdr.	32-pdr.
	in.	in.	in.	in.	in.
1° 15′	0.256	0.333	0.252	0.28	0.331
2°	1.025	1.334	0.945	1.138	1.310
3°	2.051	2.670	1.870	2.271	2.618
4°	3.077	4.006	2.791	3.400	3.920

TAR, &c. Charcoal is made in the simplest way by digging a hole in the earth, or choosing some old well or gigantic burrow, and filling it with piles of wood, arranging them so as to leave a kind of chimney down the centre. The top of the well is now covered over, excepting the chimney, down which a brand is dropped to set fire to the wood. The burning should proceed very gradually, and be governed by opening or shutting the chimney-top with a flat stone; for the wood should smoulder, and never attain to a bright red: it will take from two days to a week to make charcoal. The tarry products of the wood drain to the bottom of the well.

Tar is made by burning larch, fir, or pine, as though charcoal had to be made; dead or withered trees, and especially their roots, yield tar most copiously. A vast deal is easily obtained. It collects at the bottom of the pit, and a hole should be cleanly dug there into which it may drain. *Pitch* is tar boiled down. Turpentine is the juice that the living pine, fir, or larch tree secretes, in blisters under the bark; they are tapped to obtain it. Resin is turpentine boiled down. *Tar* is absolutely essential in a hot country to mix with the grease that is used for the wagon-wheels. Grease, alone, melts and runs away like water: the office of the tar is to give consistence. A very small proportion of tar suffices, but, without any at all, a wagon is soon brought to a standstill. It is, therefore, most essential to explorers to have a sufficient quantity in reserve. Tar is also of very great use in hot dry countries for daubing over the wheels, and the woodwork generally, of wagons. During the extreme heat, when the wood is ready to crack, all the paint should be scraped off it, and the tar applied plentifully. It will soak in deeply, and preserve the wood in excellent condition, both during the drought and the ensuing wet season. It is not necessary to take the wheels off, in order to grease the axles. It is sufficient to bore an auger-hole right through the substance of the nave, between the feet of two of the spokes, and to keep a plug in the hole. Then, in order to tar a wheel, turn it till the hole is uppermost; take the plug out, and pour the tar in; (GALTON's *Art of Travel*.)

TARGET. Practice at target-firing is essential to make a soldier. To obtain from the new small-arms the great results which they promise it is necessary: 1. That the soldier should know the different parts of the arm, or its nomenclature, how to take it apart and put it together, and the best method of keeping it in good order. This instruction should be given by sergeants and corporals under direction of the officers of the company; (*see* ARMS.) 2. The soldier must be taught the prescribed method of loading his arm. 3. The rules for firing must be known to him, that is to say, he must be taught the use of the hausse, or to regulate his arm according to the distance of the enemy; (*see* HAUSSE.) 4. He must be taught to estimate distances in order to apply the rules for firing; (*see* the method practised at Vincennes given p. 609.) 5. He must know how to aim. 6. He must hold the musket in the position his instructor prescribes, and aim with ease; preserve the body steady, but not constrained; resist the recoil; and not incline the rear sight to the right or to the left. If the rear sight, when raised and held upright, give the proper elevation for say 900 yards, and it then be inclined to the right although the aim is in such position taken with the 900 yards' sight, yet the elevation is actually lowered, and the bullet would, therefore, not only fly to the right of the object, but fall short from want of sufficient elevation. The more the sight is inclined, the greater will be the loss of elevation. Another cause of inaccuracy in aiming arises from aiming with a coarse front sight. Such an aim causes the line of sight to pass to the right or left of the front sight, and the ball consequently to go to the opposite side of the object from the side of the coarse sight by which we aimed. The elevations for different ranges being marked for a fine sight, therefore when it is necessary to use the coarse sight for a greater distance than the elevation used, the proper allowance must be made in aiming. 7. In pulling the trigger, in no manner to derange the musket. The soldier must acquire the habit of pulling the trigger when, in *raising* the piece, the sights cover the bull's-eye. Most of these details, it is obvious, will be better taught without wasting cartridges. When the soldier has been, however, sufficiently instructed in the simulated fire, to accustom him to the noise of the actual fire, it is necessary to begin with the explosion of caps, observing that he preserves his arm immovable as previously taught. To accustom him to the effect of the recoil, it is necessary to fire some blank cartridges.

Such are the gradual steps to be followed in practical firing, and by taking them better marksmen will be made than by passing men without previous preparation from the school of the soldier to target practice. After the soldier has been practised at firing at the target within the efficacious range of his arm, and has acquired the habit of estimating distances, without great errors; when he has been taught to fire at a mark changed at every fire, the distance of which he must estimate, he may be sent as a skirmisher against an enemy. He will know the range and use of his arm. He will appreciate its great power. The instruction of the soldier would not, however, be complete if he had been exercised only in firing singly. He must be accustomed to the *gêne* that he experiences in the ranks, to movements of his comrades, to the smoke which covers the front of the troops, to obeying the commands of the officer who directs the fire. The execution of the fire by platoon, by rank and by two ranks, upon squares, which indicate the effect of the fire, is a necessary instruction above all to officers, who learn in these exercises to direct and command firing, to estimate the relative value of different fires, and to judge of the importance of a simultaneous fire at proper moments. The whole instruction in firing may be given to the sergeants, corporals, and soldiers of a battalion without injury to other necessary instruction, and without hindrance of any duties in the course of a year.

The means of instruction adopted at Vincennes claim attention, in consequence of the manifest advantages of practising at ranges judged by the soldier himself. After attaining some proficiency as a marksman at specified distances, the soldier is taught to estimate distances as if before an enemy. From a squad of 16 men under a non-commissioned officer, four out of the 16 men are taken and posted at distances of 50, 100, 150, and 200 metres, facing the remainder of the men, who observe such details of each man's dress as can be distinguished at the several distances respectively. Having carefully noticed the differences which exist, the instructor practises the men at distances that are unknown to them, in order that they may apply the knowledge that they have gained by observation of dress at known distances. After the soldiers have been sufficiently practised in this way, their correctness in judging distances is subjected to another test. A man runs forward, and places a target at some distance unknown to the men; each man is then called upon in turn to name the distance, and the answers are recorded in a book. This kind of practice takes place at all distances, particularly between 500 and 1,000 paces, and is continued till all are moderately skilful. Firing then begins at distances unknown to the men, and those who are most successful are rewarded with promotion, and become the instructors of others. In order that the knowledge imparted at Vincennes may be extended to the whole army, at least one *sous-officier* is brought there from each regiment.

The new rifle musket and new rifle have an equal range, and greater precision than field-artillery, and a company of marksmen can produce an equal effect in the field at less cost than a battery of artillery. At 650 yards, for instance, almost every shot will take effect on horses and men attached to a battery. It will follow that the artillery must be more carefully covered in battle. (Consult *Instruction provisoire sur le tir à l'usage des bataillons de Chasseurs à pied.* See ARMS; FIRING; HAUSSE; STADIA.)

TATTOO OR TAPTOO. Drum-beat and roll-call at night.

TEAMSTERS. That to each regiment of dragoons, artillery, and mounted riflemen in the regular army there shall be added one principal teamster with the rank and compensation of quartermaster-sergeant, and to each company of the same, two teamsters, with the compensation of artificers; (*Act* March 3, 1847.)

TELEGRAPH, (*Universal.*) It consists of an upright post of moderate height, of two movable arms fixed on the same pivot near the top of it, and of a mark called an indicator on one side of it, merely to distinguish the low numbers 1, 2, 3, from the high numbers, 7, 6, 5. Fig. 228, A represents the telegraph exhibiting the sign 17, the other positions of which the arms are capable being dotted. Fig. 228, B represents the telegraph fitted up to make nocturnal signals. One lantern,

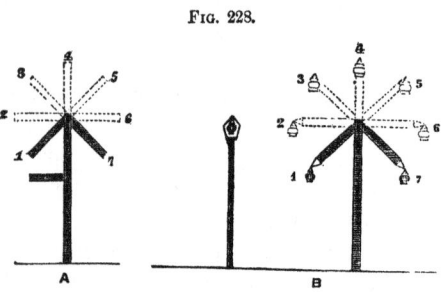

FIG. 228.

TEL.] MILITARY DICTIONARY. 611

called tne central light, is fixed to the same pivot upon which the arms move. Two other lanterns are attached to the extremities of the arms. A fourth lantern, used as an indicator, is fixed on the same horizontal

FIG. 229.
TABLE OF THE SIGNS OR COMBINATIONS.

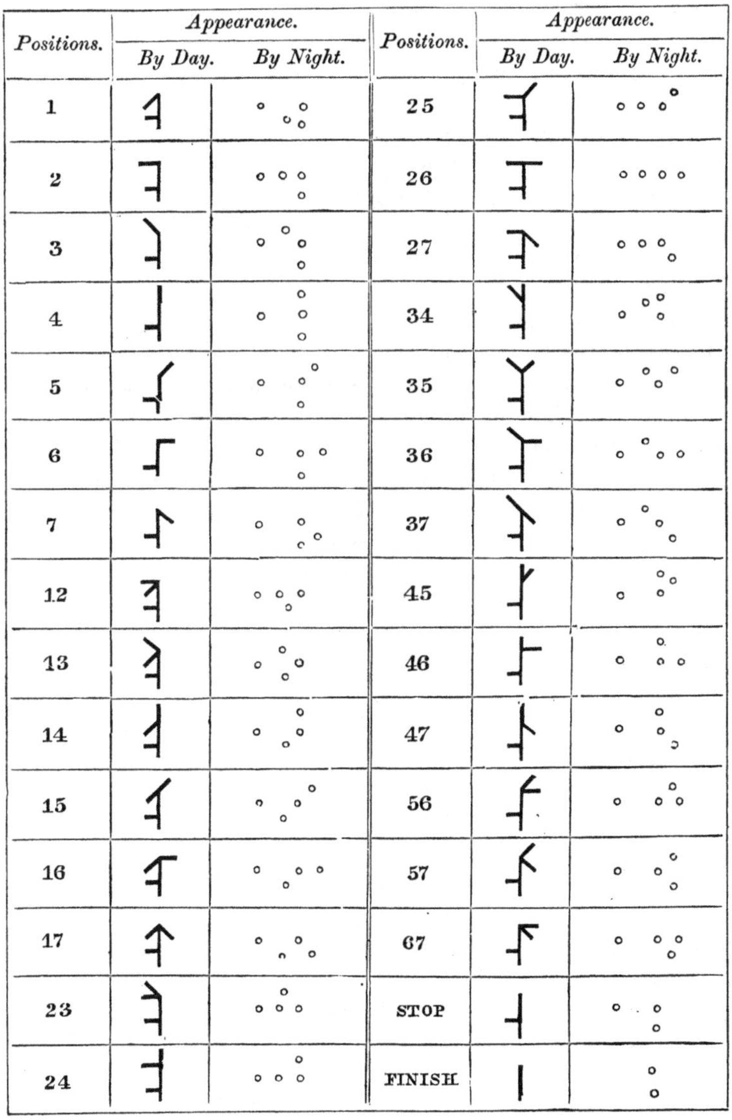

Positions.	Appearance.		Positions.	Appearance.	
	By Day.	By Night.		By Day.	By Night.
1			25		
2			26		
3			27		
4			34		
5			35		
6			36		
7			37		
12			45		
13			46		
14			47		
15			56		
16			57		
17			67		
23			STOP		
24			FINISH		

level with the central light at a distance from it equal to twice the length of the arm, and in the same plane nearly in which the arms revolve. Hence the whole apparatus consists of two fixed and of two movable lights—four in all. The number of telegraphic signs, combinations, or changes which this telegraph is capable of exhibiting is shown in Fig. 229, and one of those, No. 4, in the day telegraph is liable to be confounded with the post and should not, therefore, be used. The number is, however, amply sufficient for telegraphic communication whether by alphabet or by reference to a telegraphic dictionary of words and sentences. The indicator, both by day and night, is merely a mark and nothing more, and the central light by night and the post by day are also merely guides to the eye. The signs of the telegraph are in reality, therefore, only composed of combinations of two movable bodies by day and two lights by night. It has been ascertained by experiment that the arms for day signals should be about 1 foot in length per mile in order to be distinguished by a common portable telescope. By the above rule, a telegraphic arm of six feet in length may suffice for stations six miles apart, but it is better to add a little to these dimensions. The width of the arm need not exceed $\frac{2}{13}$ of its length. The indicator should be of the same width, but only $\frac{1}{2}$ of the arm in length. The height of the post should be such that movable objects near it should not obscure the indicator or arms when the telegraph is erected in the field. The telegraphs hitherto constructed on this principle are of two sizes: one having arms of $5\frac{1}{2}$ feet in length, with the lantern pivots placed $6\frac{1}{2}$ feet from the centre of motion; the other having arms $2\frac{1}{2}$ feet in length only, with the lantern pivots 3 feet 2 inches from the centre of motion. The latter are perfectly portable, as the whole apparatus does not weigh more than 34 lbs. In clear weather these small telegraphs make signals distinctly visible at a distance of three miles.

In cases of emergency, where the portable telegraph is not with an army, it has been ascertained by experiment that the most expeditious and satisfactory arrangement will always be to copy the regular construction as closely as circumstances will permit. A post, with two planks for the arms fixed externally on each side of the post, each worked merely by a couple of strings without pulleys, will constitute a day telegraph, and the addition of lanterns will convert the same simple apparatus into a night telegraph. In both cases the arms must be counterpoised by wood or iron, and also by weights in some rude manner, which must not impair the clearness of the telegraphic signs. (Consult *Aide Memoire to the Military Sciences by British Officers.* See SIGNALS.)

TENAILLE—is a low work, constructed in the main ditch, upon the lines of defence, between the bastions, before the curtain, composed of two faces, and sometimes of two flanks and a small curtain.

TENAILLONS—are works sometimes found constructed in an old fortress, on each side of the ravelin—the short faces being traced, on the prolongations of the faces of the ravelin, from the counterscarp of its ditch; the long faces being directed for flanking defence, to about the middle of the faces of the bastions.

TENAILLONS (*Demi*)—are very similar to tenaillons, excepting that their short faces are directed, perpendicular to the faces of the ravelin, about one-third or one-half down from the flanked angle.

TENT. (*See* CAMP.)

TERRE-PLEIN—is a name given to any space which is level, or nearly so; thus, the area on the rampart, between the banquette and the interior slope of the rampart, is called the terre-plein of the rampart.

TÊTE-DU-PONT. A field-intrenchment covering a bridge. (*See* REDAN.)

THEODOLITE. A surveying instrument for measuring the angular distances between objects projected on the plane of the horizon. In accurate surveying, when the instrument used for observing angles is a sextant or reflecting circle, or such that its plane must be brought into the plane of the three objects which form the angular points of the triangle to be measured, the altitudes of the two distant objects above the horizon of the observer must be determined, and a calculation is then necessary to reduce the observed angles to the plane of the horizon. With the theodolite this work is unnecessary. (Consult SIMMS' *Treatise on Mathematical Instruments;* DAVIES' *Surveying.*)

TIER SHOT. Grape shot sometimes so called.

TIGE ARMS. Sometimes called pillar breech arms. Arms with a stem of steel, screwed into the middle of the breech pin, around which the charge of powder is placed. The ball enters free and rests upon the top of the pin which is tempered, and a few blows with a heavy ramrod forces the ball to fill the grooves of the rifled arm. This invention was an improvement by Capt. Thouvenin on Delvignes' plan of having a chamber for the powder smaller than the bore. Capt. Minié's invention superseded the tige arms, by means of a bullet which is forced to fill the grooves by the action of the charge itself at the instant of explosion. (*See* ARMS; RIFLED ORDNANCE.)

TIMBER. Sawed or hewn timber is measured by the cubic foot, or more commonly by board measure, the unit of which is a superficial foot 1 inch thick. Usual rule for measuring round timber: multiply the length by the square of one-fourth the mean girth, for the solid contents, or $\frac{L\,C^2}{16}$; L being the length of the log, and C half the sum of the circumferences of the two ends. (Consult *Ordnance Manual*.)

TOISE—is 2.132 yards. Reduction of old French toises to metres; 1 metre = 39.37079 English inches.

Toises.		Metres.		Eng. Yards.
1	=	1.949	=	2.132
5	=	9.745	=	10.660
8	=	15.592	=	17.056
10	=	19.490	=	21.320
100	=	194.900	=	213.200
500	=	974.500	=	1,066.000
1,000	=	1,949.000	=	2,132.000

TOOLS. The French ordinance of 1831 prescribes the following camp tools: reaping-hook, scythe, axe, shovel, mattock, and bill-hook. Each tool has a leather case and a shoulder belt, in order that it may be carried by the men. (*See* UTENSILS.)

TOPOGRAPHICAL ENGINEERS. (*See* ENGINEERS, *Topographical*.)

TOPOGRAPHY—is the art of representing and describing in all its details the physical constitution, natural or artificial, of any determined portion of country; in making maps and giving a descriptive memoir. Military topography differs from geography in seeking to imitate sinuosities of ground; it represents graphically and describes technically commanding heights, water-courses, preferable sites for camps, different kinds of roads, the position of fords, extent of woods. It enumerates the resources that a country offers to troops and the difficulties which are interposed. By means of colored maps and other conventional signs, military topography presents before the eyes of a general much that is necessary to guide his operations. (Consult BARDIN. *See* RECONNOISSANCE; SURVEYS, *Military*.)

TOWER BASTION—is one which is constructed of masonry, at the angles of the interior polygon of some works; and has usually vaults or casemates under its terre-plein, to contain artillery, stores, &c.

TRACING. (*See* OUTLINE.)

TRADE. Licenses to trade with Indians shall not be granted to any but citizens of the United States, unless by express direction of the President; (*Act* April 29, 1816.) The superintendent of Indian affairs in the Territories, and Indian agents under the direction of the President of the United States, may grant licenses, not exceeding seven years, to trade with Indians; which licenses shall be granted to citizens of the United States and none others, taking from them bonds with securities, in the penal sum not exceeding five thousand dollars according to capital employed, and conditioned upon the due observance of the laws regulating trade and intercourse with Indian tribes. The superintendents and agents shall return to the Secretary of War, within each year, an abstract of the licenses granted, to be laid before Congress at the next session thereof; (*Act* May 6, 1822.)

Unlicensed trade punishable by forfeiture of merchandise, a fine not exceeding one hundred dollars, and imprisonment not exceeding thirty days; (*Act* March 30, 1802.) Receiving, or purchasing from any Indian, in the way of trade or barter a gun, any instrument of husbandry, or article of clothing, except skins or furs, punishable by forfeiture not exceeding fifty dollars and thirty days' imprisonment; (*Act* March 30, 1802.) The purchase of horses from Indians without license from the superintendent or other person authorized by the President to grant licenses, punishable with forfeiture not exceeding one hundred dollars for every horse purchased; (*Act* March 30, 1802.) No agent, superintendent, or other person authorized to grant licenses to trade or purchase horses shall have any interest or concern with any trade with Indians, excepting for and on account of the United States, under penalty of forfeiture not exceeding one thousand dollars and imprisonment not exceeding twelve months; (*Act* March 30, 1802. *See* WAR.)

TRAIL-HANDSPIKE—for field-carriages, 53 inches in length. (Hickory, or young oak.)

TRAIN. At the beginning of the French Revolution, artillery, engineer, and other supplies, and hospital trains were conducted by hired drivers. These men had neither military pride nor honor. They were cowardly and insubordinate, deserted in combats, cut the traces of their horses, and sought personal safety by abandoning equipages. On march and in camp or cantonments they were not unfrequently drunk and neglected their horses. These evils were corrected by enrolling them under the name of soldiers of the artillery train and equipages. They were given officers, a uniform and arms, and have since rivalled other corps of the army in zeal, courage, and devotedness. The artillery train now forms a part of the artillery, and is commanded by artillery officers. The train of provisions and ambulances is composed of squadrons and companies. The squadrons are commanded by a captain, and the companies by a lieutenant. Each soldier conducts two harnessed horses. He is armed with a pistol and a small sword.

In 1850 the corps of military equipages in France consisted of a central bureau for wagon parks at Vernon; of two arsenals of construction at Vernon and at Chateauroux; of three arsenals for repair in Algiers; and three companies of workmen. The soldiers properly belonging to the train made four squadrons. (Consult BARDIN and LE COUTRIER.) The quartermaster's department in our army is charged with wagon trains, but neither enlisted soldiers as workmen or drivers have yet been added to the department. (*See* CONVOY; QUARTERMASTER'S DEPARTMENT; WAGON.)

TRANSFERS. Officers of engineers are liable to be transferred, at the discretion of the President, from one corps to another, regard being paid to rank; (ART. 63.) During the recess of Congress, the President may, on the application of the Secretary of the proper department and not otherwise, direct, if in his opinion necessary for the public service, that a portion of the moneys appropriated for any one of the following branches of expenditure in the military department, viz.: For the subsistence of the army; for forage; for the medical and hospital department; for the quartermaster's department—be applied to any other of the above-mentioned branches of expenditure in the same [military] department; (*Act* March 3, 1809.) No appropriation for the service of one year shall be transferred to another branch of expenditure of a different year; (*Act* May 1, 1820.)

Nothing in the act of March 3, 1809, shall authorize the President to direct any sum appropriated for fortification, arsenals, armories, custom-houses, docks, navy-yards or buildings of any sort, or to munitions of war, or to the pay of the army or navy, to be applied to any other object of public expenditure; (*Act* March 3, 1817.) But the President, under the restrictions of the act of May 1, 1820, may transfer from one head of appropriations for fortifications to that of another for like objects; (*Act* July 2, 1836.)

TRANSPORTATION. (*See* QUARTERMASTER'S DEPARTMENT; SUPPLIES; TRAIN; WAGON.)

By Sea.—For transportation by sea, make an inventory of the number of articles, the weight of each, and the total weight of each kind, leaving room for remarks. In estimating the weight, increase the total by one half the weight of the small articles, such as accoutrements, tools, &c., which occupy considerable space in proportion to their weight, and apply for vessels sufficient for the transportation of the whole weight. Inventories of articles on each vessel should be made in duplicate, one copy being kept by the master of the vessel, the other by the person having the stores in charge. (*See* EMBARKATION.)

Horses.—The following arrangements on the English horse-transport steamer Himalaya, Capt. McClellan, gives as a model: Two rows of stalls, with the rear ends 2' at least from the vessel's side, are arranged on each deck. These stalls (Fig. 230) are each furnished with movable side-boards, a movable breast-board, and a fixed tail-board, all padded; the side-boards on both sides, the tail-board next to the horse and nearly to the bottom of the stall, and the breast-board on top and on the side next the horse. The padding used consists of felt, or raw hide, (the

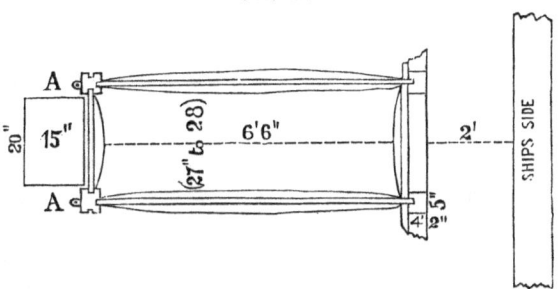

FIG. 230.

latter objectionable on account of the odor,) stuffed with cow's hair wherever the animal can gnaw it, with straw in other parts. It is from 2" to 3" thick. The feed-troughs are of wood, bound on the edges with sheet-iron or zinc, and attached to the breast-boards with two hooks. The breast and side-boards ship in grooves. Fig. 230 represents the horizontal projection of one stall. In front of each head-post a haltering-ring A is placed, and over this near the top of the post is a hook, to which the sea-halter is hung when not in use. The feed-troughs, head-boards, and stalls are whitewashed and numbered.

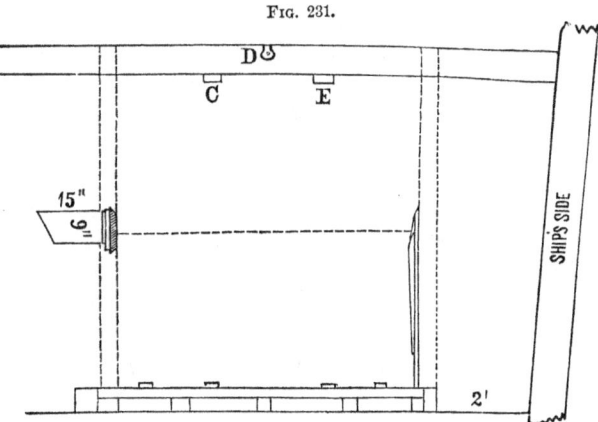

FIG. 231.

618　　　　　　　MILITARY DICTIONARY.　　　　　　[TRA.

Fig. 231 represents a section of one of these stalls through the axis. The flooring is raised above the deck on battens, and is divided into separate platforms for every two stalls, so that it can easily be raised to clean the deck beneath; 4 strong battens are nailed across to give the animals a foot-hold.

Fig. 232 is a section through the side-boards of a stall, and shows the dimensions of the timbers and height of side-boards, as well as the manner of inserting them in their grooves. *B* is the hook for hanging

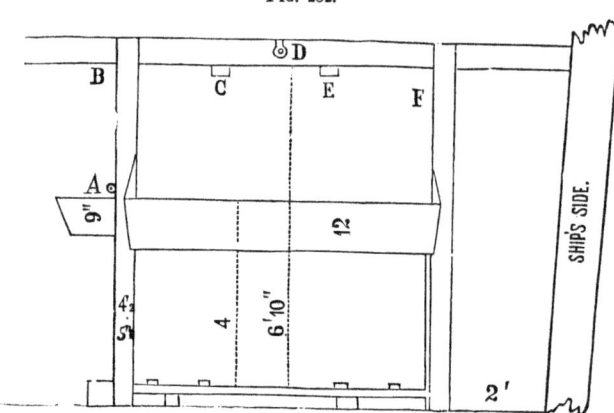

FIG. 232.

up the sea-halter. This halter is made of double canvas, 2′ wide, and has two ropes, which, being fastened one to each post, keep the animal's head still, and prevent him from interfering with his neighbor. *C* and *E* are battens for securing the ropes of the slings, shown in Fig. 233. *D*, bolts, for the same purpose, when the sling is of the form represented in Fig. 234. On the spar deck, the stalls are under sheds, every 8 stalls forming a separate set, so that they can readily be moved about when the decks are to be cleaned. Water-proof curtains are provided for the front and rear; a passage way of at least 2′ is left between the sheds and the bulwarks. When practicable, a staging is erected alongside, that the horses may be walked on and off the vessel; when this cannot be done, they are hoisted on board in the sling, a small donkey engine being used for the purpose. In this way, horses may be shipped or unloaded at the rate of one per minute. The slings are of canvas, of the shape and dimensions represented in Figs. 233 and 234. For hoisting in and out the horses, the sling is provided with a breast strap and breeching. On the main and orlop decks the sling ropes are attached to the bolts; on the spar deck to battens. It was intended to adopt the sling represented in Fig. 234, as diminishing vibration. At sea, the sling is used only when the animals show signs of weakness in bad weather, in which case about 1″ play is given to the sling, as it is only intended to prevent the horses from falling. To place the horses in the stalls, all the side-boards are removed except the one at the end of the row; a horse is then walked along to the last stall, and the other side-board put in, and so on with all the rest. They should be placed in the same order that they are accustomed to stand in the stable or at the picket rope. If it becomes necessary to remove a horse from his stall during the voyage, the breast-board is taken away, and he is walked out. All wooden parts are washed with some disinfecting compound, or simply whitewashed. Chloride of zinc is freely used. The decks are washed every day, and the stalls cleaned after every feed, especially at 7 P.M. From the spar and main decks, the stale passes off through the scuppers; from the orlop deck it passes to the hold, and is pumped out by the engine. On the Himalaya not the slightest disagreeable odor could be detected. The feed-troughs and horses' nostrils are washed every morning and evening with vinegar. A scraper, brush, and shovel are allowed to every eight stalls. A guard always remains over the horses, and in case of necessity a farrier or non-commissioned officer is sent for. Great attention is paid to ventilation. The orlop deck, although hotter than the others, appears to be the most favorable one for the horses.

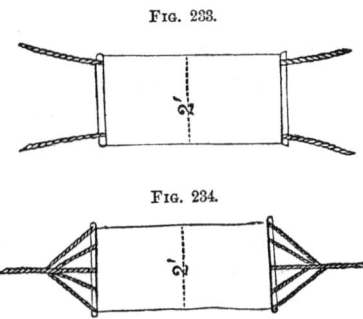

FIG. 233.

FIG. 234.

So long as cleanliness is preserved, the commander of the vessel does not interfere as to the hours for feeding, which are usually at 6 and 11 A.M. and 5½ P.M. If any horse refuses his food, the fact is at once reported. A supply of forage is always carried on board the ship. The horses drink condensed steam. The ration at sea was established at 10 lbs. of hay, 6 lbs. of oats, half peck of bran, and 6 galls. of water, as a maximum; but it is generally considered this is too great, and that $\frac{2}{3}$ the allowance except the water, would be ample, as it is found there is great danger from over-feeding at sea. No grain is given the day the horses come on board, but simply a mash of bran, which is considered the best habitual food at sea. For the men, bunks and hammocks are generally used. Standing bunks are found to be very objectionable, on account of the difficulty of keeping them clean. Hammocks are regarded as preferable for men in good health, while many officers consider it best to provide neither hammocks nor bunks, but to allow the men to lie down on the fore-decks, with their blankets and overcoats. When the transports are numerous, each one should have on the starboard and larboard, and on a broad pendant at the top of the mainmast, an easily distinguished number. By means of these numbers, which are marked on the bills of lading, the disposable resources of the expedition are known at any time. Vessels carrying some particular flag should be specially appropriated for the transportation of powder, fire-works, and ammunition, which may be separated from the pieces.

Disembarkation.—If it becomes necessary to transship, or leave any articles upon the vessels, the fact should be carefully noted on the manifests. The ships' crews load and unload, using for these purposes the yard-arms and tackle. It is ordinarily sufficient to furnish them with rollers and skids, in order to place the articles convenient to the tackle. Under some circumstances, it becomes necessary to establish bridge abutments, sheers, gins, &c. For the want of the ordinary means, a temporary crane may be established. To do this a long mortise is cut in a beam about $\frac{1}{3}$ of the distance from its end, and upon the ground is fixed a framework, furnished with a strong vertical pin. The beam is laid on this frame with the pin in the mortise, like an ordinary pintle, but in such a way that the ends of the beam can be raised and lowered. The shortest part of the beam is then turned towards the load, and the different weights being slung to it, are raised by lowering the opposite end, previously raised to make the lashing shorter. The beam is then turned around on its pintle until the weight is in the proper position, when it is lowered gently and unlashed. If a tree or beam fit for the purpose cannot be obtained, several small pieces may be lashed and pinned together.

Railroad Transportation.—In railroad transportation, when several trains are required, they should be in proportion to the power of the engine employed, and full loads should be placed on them. The men are provided, before starting, with provisions to last during the trip, which should be cooked and carried in the haversack. The canteens are filled with water; the French, in warm weather, mix brandy with it. As the horses can eat in the wagons, even whilst the train is in motion, hay (pressed if possible) should be distributed at the rate of about 8, 14, or 24 lbs. per horse, according as the trip is to last less than 12, between 12 and 24, or more than 24 hours. A feed of oats (half a ration, 6 lbs.) is carried in bags, and placed in the baggage wagons. It should not be given to the horses on the road, but after they have arrived at the terminus. The horses are carried in cattle-cars, or, if possible, in box-cars, which are covered. They are provided with bars at the doors to prevent the horses from backing out when the doors are opened. By taking care to keep the horses quiet, however, these bars may be dispensed with. The saddles, &c., the valises of the driver, and the bags of oats, are placed in the baggage cars, which should be provided with brakes. The "*materiel*" is carried on trucks or common platform cars. The troops should be at the station at least two hours before starting. The horses should have finished feeding about two hours previous to their arrival at the station, as they are then more docile. The baggage should arrive half an hour before the troops, under charge of an officer, and be loaded under the direction of the employés of the road.

The cars for artillery should be arranged as near as possible in the following order: 1st, a baggage wagon; 2d, a truck carrying the beams, platforms, &c.; 3d, the horse-cars; 4th, the cars for the men, one at least of which should be provided with a brake; 5th, trucks loaded with materiel; 6th, baggage cars (with brakes) loaded with saddles, &c. Cars with brakes should always be placed at the head and tail of the train. Guards should be detailed and so stationed on the train as to preserve order both when in motion and during stoppages. The commanding officer should pay especial regard to the wishes of those having the train in charge, and enforce an observance of the road regulations in his command. On arriving at the station, the commander at once divides his command and materiel into the portions to occupy the different cars.

Horses.—An officer is detailed to superintend the embarkation of the horses. He furnishes each car with two bundles of litter, and places forage along the long side of the car opposite to the door. A non-commissioned officer is charged with loading the saddles, &c. The men are, under an officer, formed into detachments proportional to the importance of the *materiel* to be embarked.

As soon as a truck has received its load, the wheels of the different trains are locked together with cord from .5 to .6 inch in diameter, chocks are placed under the wheels and nailed to the floor, and the stability of the whole secured by tying the carriages to the rings of the truck. Straw ropes, or other means, are made use of to prevent friction between the parts.

The men, with their knapsacks and arms, are divided, under the superintendence of an officer, into portions corresponding to the capacity of the cars. Each division is conducted promptly to the car it is to occupy, the men entering first going to the end farthest from the door, and so on. They seat themselves, holding their arms between their legs, the stock or scabbard resting on the floor. Fire-arms should never be laid on the seats or stood in corners, except when leaving the cars at the principal stopping places and stations.

Inspecting.—Immediately before starting, the commanding officer and conductor of the train inspect the cars to ascertain that every thing is in order. They should see that the couplings of the car containing the "*materiel*" are short enough to insure the contact of the buffers. The officers then enter the car assigned to them.

Regulations.—The men are strictly prohibited putting their heads or arms out of the car while it is in motion; passing from one car to another; uttering loud cries of any kind; and from leaving the cars at the station before the signal for doing so is given. The men with the horses, keep them from putting their heads outside the car. They feed them with hay from the hand, until they get used to the motion, hold them by the bridle or halter, and quiet their fears whilst the locomotive is whistling. In case of any accident, they make a signal outside the car, by waving a handkerchief. If at any station the commander deems it necessary for the men to leave the cars, after the time indicated by the conductor, he informs the officers of the length of the halt. The officers remain in the vicinity of the cars containing their men, in order to direct and govern their movements. The guard posts sentinels wherever it is necessary, especially at the doors, to prevent the men from gathering near or opening them. At a given signal on the bugle, the men leave the cars in order, and without side-arms. The men in the horse-cars get out over the side. If it becomes necessary to open the doors of these cars, the door-bars are first placed in position. About the middle of the trip, as near as possible, the police-guard and men with the horses, are relieved. At each halt of more than ten minutes, the commander, or some other officer, and the conductor inspect the cars and especially those which carry the ammunition wagons. Five minutes before starting a bugle-call gives the signal for entering the cars. At the station immediately preceding the terminus, the horses are bridled, and the forage is collected and formed into one bundle for each car. During feeding time there should be at least one man to every two horse-cars. In general, oats should be distributed only after the horses leave the cars. Hay is fed by hand by the drivers whilst the train is in motion. In ordinary weather, the horses are watered only when the trip exceeds twelve hours; and even in this case they need but little, and a single ordinary-size pailful suffices for two horses.

Unloading.—To prevent accidents, it is well to provide one or several movable bridges for discharging the horses, which are carried on the train. They are about sixteen feet long, a little wider than the car door, and are provided with hand-rails or ropes, movable at will. The bridge is supported at its upper extremity by a movable trestle of a height corresponding to the sill of the door, and the cars are unloaded by passing them in succession in front of this bridge; or, by fixing to the forepart of the bridge two strong flanges of iron which rest upon the floor of the car, the bridge may be applied in succession to each of the cars to be unloaded.

The non-commissioned officers in charge of the freight cars, immediately on arriving at the station, unload it as originally divided in the cars by the inverse means used to load it. As soon as the horse-cars reach the proper position, the men fix the movable bridges, open the doors, and bring the horses out in the inverse order in which they entered. If the horses have to be taken out of the same door they entered, the first two are backed out, and the rest follow after making a half turn. As soon as a rear team is disengaged it is taken to the place where the harness is deposited, and harnessed to a carriage which is conducted to the park, where the harnessing is completed. (Consult GIBBON; MCCLELLAN.)

TRAVELLING ALLOWANCE. Where any commissioned officer is obliged to incur any extra expense in travelling, and sitting on general courts-martial, he shall be allowed a reasonable compensation for such extra expense actually incurred, not exceeding one dollar and twenty-five cents per day to officers who are not entitled to forage, and not exceeding one dollar per day to such as shall be entitled to forage; (*Act* March 16, 1802.) (*See* ORDNANCE; TRAVELLING FORGE.)

An officer, who travels not less than ten miles from his station, without troops, escort of military stores, and under special orders in the case from a superior, or summons to attend a military court, shall receive ten cents a mile; or if he prefer it, the actual cost of his transportation, and of his field-allowance of baggage for the whole journey, provided he has travelled in the customary reasonable manner; (*Regulations for the Quartermaster's Department.*)

Whenever any officer or soldier shall be discharged from the service, except by way of punishment for any offence, he shall be allowed his pay and rations, or an equivalent in money, for such term of time as shall be sufficient for him to travel from the place of his discharge to the place of his residence, computing at the rate of twenty miles to a day; (*Act* March 16, 1802.)

TRAVELLING-FORGE. (*See* ORDNANCE.)

TRAVELLING-KITCHEN. Marshal Saxe, it is believed, first suggested the idea of cooking while marching, so as to economize the strength of soldiers; have their food well cooked in all weather, and avoid the numerous diseases caused by bad cooking, and want of rest. Colonel Cavalli, of the Sardinian artillery, has with the same laudable motive embraced a kitchen-cart in the improvements suggested by him to replace the wagons now in use, (*see* WAGON;) and an attempt is here made to elaborate the same idea of a travelling-kitchen, designed for baking, making soup, and other cooking, while on a march.

Fig. 235 represents a cart, $12\frac{1}{2}$ feet long, mounted on two 6-feet

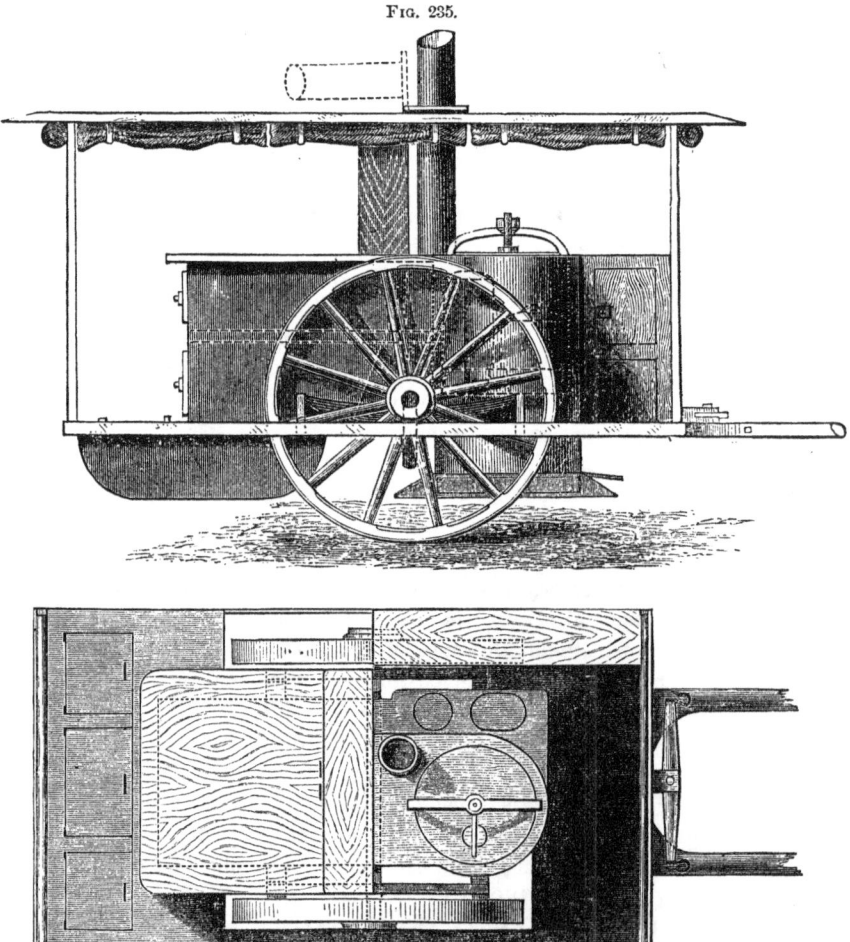

FIG. 235.

wheels, and covered with a very light canvas roof with leather-cloth curtains. A large range or stove forms the body of the vehicle; its grate is below the floor, its doors opening on a level with it. A *Papin's digester* is inclosed above the grate, in a flue whence the heat may pass around the double-oven in the rear, or straight up chimney, as regulated by dampers. At the side of the digester, over the grate, is a range, suited to various cooking vessels. The top of the oven forms a table nearly 5 feet square, at which three cooks may work, standing upon the rear platform. A foot-board passes from this platform to the front platform, where the driver and a cook may stand. Stores may be placed in the lockers at the side of the range, and under the rear foot board. The chimney may be turned down, above the roof, to pass under trees, &c., and may be of any height to secure a good draft. By bending the axle like that of an omnibus, the vehicle may be hung without danger of top-heaviness. Cooking vessels, more bulky than heavy, may be suspended from the roof, over the range, when not in use. The digester may have a capacity of 100 gallons, and an oven, of 60 to 75 cubic feet, would be quite adequate to the cooking for 250 men; or the dimensions of the cart may be smaller, and each company of 100 men might have its own travelling-kitchen, which would also furnish oven and cooking utensils for camp.

TRAVERSES—are portions of parapets, which cross the breadth of the covered-way, at the salient and re-entering places of arms. Other traverses are also placed between these, where necessary, to afford proper protection. Traverses are thrown up, to bar enfilade fire, along any line of work or passage which is liable to it.

TRAVERSE TABLE—is the tabulated form in which the northing, southing, easting, and westing are made on each individual course and distance in a traverse, for the purpose of finding readily, by inspection of the table, the difference of latitude and departure of any particular course and distance. Traverse tables afford a simple means of land-surveying, with compass and chain. If the sum of each adjacent pair of distances perpendicular to a meridian (*departures*) without survey, be multiplied by the northing or southing between them, in succession round the figure in the same order, the difference between the sum of the *north* products and the sum of the *south* products will be double the area of the tract. The *meridian distance* of a course is the distance of the middle part of that course from an assumed meridian. Hence, the double meridian distance of the first course is equal to its departure. And the double meridian distance of any course is equal to the double meridian distance of the preceding course, plus its departure, plus the departure of the course itself, having regard to the algebraic sign of each.

Then to find the area: 1. Multiply the double meridian distance of each course by its northing or southing. 2. Place all the *plus* products in one column, and all the *minus* products in another. 3. Add up each column separately and take their difference. This difference will be *double* the area of the land. In balancing the work, the error for each particular course is found by the proportion: as the sum of the courses is to the error of latitude, (or departure,) so is each particular course to its correction. When a bearing is due east or west, the error of latitude is nothing, and the course must be subtracted from the sum of the courses before balancing the columns of latitude. And so with the departures. Let it be required to find the contents of a piece of land, of which the following are the field-notes:

Sta.	Course.	Dis.	Sta.	Course.	Dis.
1	N. $46\frac{1}{2}$° W.	20 chains.	4	S. 56° E.	27.60 chains.
2	N. $51\frac{3}{4}$ E.	13.80 "	5	S. $33\frac{1}{2}$ W.	18.80 "
3	East	21.25 "	6	N. $74\frac{1}{2}$° W.	30.95 "

CALCULATION.

Stations.	Courses.	Dist. Chains.	Diff. Lat. N. +	Diff. Lat. S. −	Departures. E. +	Departures. W. −	Balanced. Lat.	Balanced. Dep.	D. M. D. +	Area. +	Area. −
1	N. $46\frac{1}{4}$ W.	20.00	13.77	—	—	14.51	+ 13.38	−14.56	14.56	202.0928	
2	N. $51\frac{3}{4}$ E.	13.80	8.54	—	10.84	—	+ 8.61	+10.81	10.81	93.0741	
3	East.	21.25	—	—	21.25	—	—	+21.20	42.82		
4	S. 56° E.	27.60	—	15.44	22.88	—	−15.29	+22.82	86.84	—	1327.7836
5	S. $33\frac{1}{4}$ W.	18.80	—	15.72	—	10.31	−15.63	−10.36	99.30	—	1552.0590
6	N. $74\frac{1}{4}$ W.	30.95	8.27	—	—	29.83	+ 8.43	−29.94	59.03	497.6229	
Sums........		132.40	30.58	31.16	54.97	54.65				792.7898	2579.8426
				30.58		54.65					792.7898
Error in northing,				.58		.32 Error in westing.					2)2087.0528
											1043.5264

Answer—104 Acres, 1 Rod, 16 Perches.

(Consult *Tables and Formulæ* by Capt. T. J. Lee, Top. Engineer.)

TREATY. No purchase, grant, license, or other conveyance of lands or of any title or claim thereto from any Indian nation, or tribe of Indians within the bounds of the United States, shall be of any validity in law or equity, unless the same be made by treaty or convention, entered into pursuant to the constitution. Penalty not exceeding forfeiture of $1,000 and 12 months' imprisonment for violation of this act. Provided, nevertheless, that any agent or agents of any State, who may be present at any treaty made by United States authority, in the presence and with the approbation of the United

States commissioners, may propose to, and adjust with, the Indians the compensation to be made to them for land claims within such States, extinguished by the treaty; (*Act* of Congress.)

TRENCH. The communications, boyaux or zigzags, as well as the parallels or places of arms opened by besiegers against a fortification are trenches. They are from 6 to 10 feet wide and about 3 feet deep. To *open* the trenches, is to break ground for the purpose of carrying on approaches towards a besieged place.

TRESTLE OR TRESSEL. The form of a trestle is the same as a *carpenter's horse*, that is, a horizontal beam supported by four legs. (*See* BRIDGE.) The horizontal beam, termed the cap or ridge beam in trestles used for field-bridges, is usually of eight-inch scantling, and from twelve to sixteen feet long. The legs are of four and a half inch scantling; they have a spread towards the bottom, the distance between them across being equal to half the height, and lengthwise of the cap, their inclination is one-twelfth of the height. They are fastened to the cap, about 18 inches from the ends, by nails; the side of the cap and the top of the leg being properly prepared for a strong, accurate fit. The legs are connected either in pairs, or else all four by horizontal pieces of three-inch scantling; sometimes diagonal pieces, going from the top of one leg to the bottom of the opposite one, are used.

Bridges or trestles are principally useful in crossing small streams not more than six feet deep. The trestles should not be placed farther apart than sixteen feet between the ridge beams; the balks should jut at least one foot beyond the ridge beams. The action of the current is counteracted by attaching each trestle to two cables stretched across the stream above and below the bridge. Another plan consists in making a network of tough twigs or cords around the legs near the bottom, and filling it in with broken stone. (Consult MAHAN.)

TRIALS—shall be carried on only between the hours of eight in the morning and three in the afternoon, except in cases which, in the opinion of the officer ordering the court, require immediate example; (ART. 75.) No officer, non-commissioned officer, or soldier shall be tried a second time for the same offence; (ART. 87.) And no person shall be liable to be tried and punished by a general court-martial for any offence which shall appear to have been committed more than two years before the issuing of the order for such trial, unless the person, by reason of having absented himself, or some other manifest impediment, shall not have been amenable to justice within that period; (ART. 88.)

All trials before courts-martial, like those in civil courts, are conducted publicly; and in order that this publicity may in no case be attended with tumult or indecorum of any kind, the court is authorized, by the Rules and Articles of War, to punish at its discretion, all riotous and disorderly proceedings or menacing words, signs, or gestures, used in its presence; (Art. 76.)

The day and place of meeting of a general court-martial having been published in orders, the officers appointed as members, the parties and witnesses, must attend accordingly. The judge-advocate, at the opening, calls over the names of the members, who arrange themselves on the right or left of the president, according to rank; (Art. 61.)

The members of the court having taken their seats and disposed of any preliminary matter, the prisoner, prosecutor, and witnesses are called into court. The prisoner is attended by a guard, or by an officer, as his rank or the nature of the charge may dictate; but during the trial, should be unfettered and free from any bonds or shackles, unless there be danger of escape or rescue. Accommodation is usually afforded, at detached tables, for the prosecutor and prisoner; also for any friend or legal adviser of the prisoner or prosecutor, whose assistance has been desired during the trial; but the prisoner only can address the court, it being an admitted maxim, that counsel are not to interfere in the proceedings or to offer the slightest remark, much less to plead or argue. The judge-advocate, by direction of the president, first reads, in an audible voice, the order for holding the court. He then calls over the names of the members, commencing with the president, who is always the highest in rank. He then demands of the prisoner, whether he has any exception or cause of challenge against any of the members present, and if he have, he is required to state his cause of challenge, confining his challenge to one member at a time; (Art. 71.) After hearing the prisoner's objections, the president must order the court to be cleared, when the members will deliberate on and determine the relevancy or validity of the objection; the member challenged retiring during the discussion.

Sufficient causes for challenge are:—the expression of an opinion relative to the subject to be investigated; having been a member of a court of inquiry which gave an opinion; or of another general court-martial, in which the circumstances were directly investigated; or of another general court-martial in which the circumstances were investigated incidentally and an opinion formed thereon; prejudice, malice, or the like. The privilege of challenge is not confined to the prisoner; for there may be sources of prejudice in favor of the prisoner as well as against him, and urgent motives that may sway to acquit, as well as condemn. When the prisoner and prosecutor decline to challenge any of the members, or where the causes of challenge have been disallowed, the judge-advocate proceeds to administer to the members of the court, the oath prescribed by the 69th Article of War, which is in the following words: "You, A. B., do swear, that you will well and truly try and determine, according to evidence, the matter now before you, between the United States of America and the prisoner to be tried; and that you will duly administer justice according to the provisions of 'an act establishing rules and articles for the government of the armies of the United States,' without partiality, favor or affection: and if any doubt shall arise, not explained by said articles, according to your understanding and the custom of war in like cases: and you do further swear, that you will not divulge the sentence of the court, until it shall be published by the proper authority: neither will you disclose or discover the vote or opinion of any particular member of the court-martial, unless required to give evidence thereof, as a witness, by a court of justice in due course of law. So help you God." The oath is taken by each member holding up his right hand and repeating the words after the judge-advocate. After the oath has been administered to all the members, the president administers to the judge-advocate, the particular oath of secrecy to be observed by him, and which, as prescribed by Article 69, is as follows: "You, A. B., do swear that you will not disclose or discover the vote or opinion of any particular member of the court-martial, unless required to give evidence thereof as a witness, by a court of justice in due course of law, nor divulge the sentence of the court to any but the proper authority, until it shall be duly disclosed by the same. So help you God."

The oath taken by the president and members contains a twofold obligation to secrecy: 1st, That they will not divulge the sentence of the court, until it shall be published by proper authority; and, 2d, That they shall not disclose or discover the vote or opinion of any particular member of the court-martial, unless required to give evidence thereof by a court of justice, in a due course of law. Both these obligations have their foundation in reason and good policy.

No sentence of a general court-martial is complete or final, until it has been duly approved. Until that period it is, strictly speaking, no more than an opinion, which is subject to alteration or revisal. In this interval, the communication of that opinion could answer no ends of justice, but might, in many cases, tend to frustrate them. The obligation to perpetual secrecy, with regard to the votes or opinions of the particular members of the court, is likewise founded on the wisest policy. The officers who compose a military tribunal are, in a great degree, dependent for their preferment on the President. They are even, in some measure, under the influence of their commander-in-chief —considerations which might impair justice. This danger is, therefore, best obviated by the confidence and security which every member possesses, that his particular opinion is never to be divulged. Another reason is, that the individual members of the court may not be exposed to the resentment of parties and their connections, which can hardly fail to be excited by those sentences, which courts-martial are obliged to award. It may be necessary for officers, in the course of their duty, daily, to associate and frequently to be sent on the same command or service, with a person against whom they have given an unfavorable vote or opinion on a court-martial. The publicity of these votes or opinions would create the most dangerous animosities, equally fatal to the peace and security of individuals, and prejudicial to the public service.

The oath which is taken by the judge-advocate, contains the same obligation to secrecy, except so far as it relates to the person who has the approving or disapproving of the sentence of the court. It is not inconsistent with his oath or duty, for the judge-advocate to communicate to the proper authority, his views of the proceedings of the court.

The judge-advocate is, however, bound by oath, as well as the members of the court, to maintain the strictest secrecy with regard to the votes or opinions of individuals for the reasons above stated. The oath taken by the members of the court commences with these words: "You, A. B., do swear that you will well and truly try and determine, according to evidence, the matter now before you, between the United States of America and the prisoner to be tried;" (Art. 69.) The expression, "prisoner," in the singular number, seems to imply that the swearing, and consequently the trial, should in each case be separate. That course should therefore be pursued.

Application to delay the assembling of the court, from the absence or indisposition of the witnesses, the illness of the parties, or other cause, should be made, when practicable, to the authority convening the court; but application to put off or suspend the trial may be urged with a court-martial, subsequent to the swearing of the members. It may be supported by affidavit, and the court, in allowing it to prevail, must be satisfied, if the cause be absence of a witness, that the testimony proposed to be offered is material, and that the applicant cannot have substantial justice without it. The points, therefore, which each witness is intended to prove, must be set forth in the application, and it must also be shown that the absence of the witness is not attributable to any neglect of the applicant.

A precise period of delay must be applied for, and it must be made to appear that there is reasonable expectation of procuring the attendance of the witness by the stated time; or, if the absence of a witness be attributed to his illness, a surgeon, by oral testimony, or by affidavit, must state the inability of the witness to the court, the nature of his disease, and the time which will probably elapse before the witness may be able to give his testimony. The court must obviously be adjourned at any period of its proceedings, prior to the final close of the prosecution and defence, on satisfactory proof, by a medical officer, that the prisoner is in such a state, that actual danger to his health would arise from his attendance in court; and where the prisoner is so ill as to render it probable that his inability to attend the court will be of such continuance as to operate to the inconvenience of the service, either by the detention of the members of the court from their regiments, or from other cause, the court may be dissolved by the authority which convened it. Though the prisoner may have been arraigned, and the trial proceeded with, the prisoner, on recovery, would be amenable to trial by another court. The illness of the prosecutor would, in few cases, justify the suspension of the trial, excepting, perhaps, for a very limited period; all prosecutions before courts-martial being considered at the suit of the United States, or an individual State, as the case may be. The court being regularly constituted, and every preliminary form gone through, the judge-advocate, as prosecutor for the United States, desires the prisoner to listen to the charge or charges brought against him, which he reads with an audible voice, and then the prisoner is asked, whether he is guilty or not guilty of the matter of accusation.

The charge being sufficient, or not objected to, the prisoner must plead either: 1st, Guilty; or 2d, Specially to the jurisdiction, or in bar; or 3d, The general plea of *not guilty*, which is the usual course where the prisoner makes a defence.

If from obstinacy and design the prisoner stand mute, or answer foreign to the purpose, the court may proceed to trial and judgment, as if the prisoner had regularly pleaded *not guilty*, (ART. 70;) but if the prisoner plead *guilty*, the court will proceed to determine what punishment shall be awarded, and to pronounce sentence thereon. Preparatory to this, in all cases where the punishment of the offence charged is discretionary, and especially where the discretion includes a wide range and great variety of punishment, and the specifications do not show all the circumstances attending the offence, the court should receive and report, in its proceedings, any evidence the judge-advocate may offer, for the purpose of illustrating the actual character of the offence, notwithstanding the party accused may have pleaded guilty; such evidence being necessary to an enlightened exercise of the discretion of the court, in measuring the punishment, as well as for the approving authority. If there be any exception to this rule, it is where the specification is so full and precise as to disclose all the circumstances of mitigation or aggravation which accompany the offence. When that is the case, or when the punishment is fixed, and no discretion is allowed, explanatory testimony cannot be needed.

Special pleas are either to the jurisdiction of the court, or in bar of the charge. If an officer or soldier be arraigned by a court not legally constituted, either as to the authority by which it is assembled, or as to the number and rank of its members, or other similar causes, a prisoner may except to the jurisdiction of the court-martial. Special pleas in bar go to the merits of the case, and set forth a reason why, even admitting the charge to be true, it should be dismissed, and the prisoner discharged. A former acquittal or conviction of the same offence would obviously be a valid bar, except in case of appeal from a regimental to a general court-martial. Though the facts in issue should be charged to have happened more than two years prior to the date of the order for the assembling of the court-martial, yet it is not the province of the court, unless objection be made, to inquire into the cause of the impediment in the outset. It would be to presume the illegality of the court, whereas the court should assume that manifest impediment to earlier trial did exist, and leave the facts to be developed by witnesses in the ordinary course. A pardon may be pleaded in bar. If full, it at once destroys the end and purpose of charge, by remitting that punishment which the prosecution seeks to inflict; if conditional, the performance of the condition must be known; thus, a soldier arraigned for desertion, must plead a general pardon, and prove that he surrendered himself within the stipulated period.

No officer or soldier, being acquitted or convicted of an offence, is liable to be tried a second time for the same. But this provision applies solely to trials for the same incidental act and crime, and to such persons as have, in the first instance, been legally tried. If any irregularity take place on the trial rendering it illegal and void, the prisoner must be discharged, and be regarded as standing in the same situation as before the commencement of these illegal proceedings. The same charge may, therefore, be again preferred against the prisoner who cannot plead the previous illegal trial in bar.

A prisoner cannot plead in bar that he has not been furnished with a copy of the charges, or that the copy furnished him differed from that on which he has been arraigned. It is customary and proper to furnish him with a correct copy; but the omission shall not make void, though it may postpone the trial. If the special plea in bar be such that, if true, the charge should be dismissed and the prisoner discharged, the judge-advocate should be called on to answer it. If he does not admit it to be true, the prisoner must produce evidence to the points alleged therein; and if, on deliberation, the plea be found true, the facts being recorded, the court will adjourn and the president submit the proceedings to the officer by whose order the court was convened, with a view to the immediate discharge of the prisoner. The ordinary plea is *not guilty*, in which case the trial proceeds. The judge advocate cautions all witnesses on the trial to withdraw, and to return to court, only on being called. He then proceeds to the examination of witnesses, and to the reading and proof of any written evidence he may have to bring forward. After a prisoner has been arraigned on specific charges, it is irregular for a court-martial to admit any additional charge against him, even though he may not have entered on his defence. The trial on the charges first preferred, must be regularly concluded, when, if necessary, the prisoner may be tried on any further accusation brought against him. On the trial of cases not capital, before courts-martial, the deposition of witnesses not in the line or staff of the army, may be taken before some justice of the peace, and read in evidence, provided, the prosecutor and person accused are present at the same, or are duly notified thereof. The examination of witnesses is invariably in the presence of the court; because, the countenance, looks, and gestures of a witness add to, or take away from, the weight of his testimony. It is usually by interrogation, sometimes by narration; in either case, the judge-advocate records the evidence, as nearly as possible, in the express words of the witness. All evidence, whatever, should be recorded on the proceedings, in the order in which it is received by the court. A question to a witness is registered before enunciation; when once entered, it cannot be expunged, except by the consent of the parties before the court; if not permitted to be put to the witness, it still appears on the proceedings accompanied by the decision of the court. The examination in chief of each particular witness being ended, the cross-examination usually follows, though it is optional with the prisoner to defer it to the final close of the examination in chief. The re-examination by the prosecutor, on such new points as the prisoner may have made, succeeds the cross-examination, and finally, the court puts such questions as in its judgment may tend to elicit the truth.

It is customary, when deemed necessary by the court, or desired by a witness, to read over to him, immediately before he leaves the court, the record of his evidence, which he is desired to correct if erroneous, and, with this view, any remark or explanation is entered upon the proceedings. No erasure or obliteration is, however, admitted, as it is essentially necessary that the authority which has to review the sentence, should have the most ample means of judging, not only of any discrepancy in the statements of a witness, but of any incident which may be made the subject of remark, by either party in addressing the court.

Although a list of witnesses, summoned by the judge-advocate, is furnished to the court on assembling, it is not held imperative on the prosecutor to examine such witness; if he should not do so, however, the prisoner has a right to call any of them. Should the prisoner, having closed his cross-examination, think proper subsequently to recall a prosecutor's witness in his defence, the examination is held to be in chief, and the witness is subject to cross-examination by the prosecutor. Although either party may have concluded his case, or the regular examination of a witness, yet should a material question have been omitted, it is usually submitted by the party to the president, for the consideration of the court, which generally permits it to be put. The prisoner being placed on his defence, may proceed at once to the examination of witnesses; firstly, to meet the charge, and secondly, to speak as to character, reserving his address to the court, until the conclusion of such examination. The prisoner, having finished the examination in chief of each witness, the prosecution cross-examines; the prisoner re-examines, to the extent allowed to the prosecutor, that is, on such new points as the cross-examination may have touched on, and the court puts any questions deemed necessary. The prisoner, having finally closed his examination of witnesses, and selecting this period to address the court, offers such statement or argument as he may deem conducive to weaken the force of the prosecution, by placing his conduct in the most favorable light, accounting for or palliating facts, confuting or removing any imputation as to motives; answering the arguments of the prosecutor, contrasting, comparing, and commenting on, any contradictory evidence; summing up the evidence on both sides, where the result promises to favor the defence, and, finally, presenting his deductions therefrom.

The utmost liberty consistent with the interest of parties not before the court and with the respect due to the court itself, should, at all times, be allowed a prisoner. As he has an undoubted right to impeach, by evidence, the character of the witnesses brought against him, so he is justified in contrasting and remarking on their testimony, and on the motives by which they, or the prosecutor, may have been influenced. All coarse and insulting language is, however, to be avoided, nor ought invective to be indulged in, as the most pointed defence may be couched in the most decorous language. The court will prevent the prisoner from adverting to parties not before the court, or only alluded to in evidence, further than may be actually necessary to his own exculpation. It may sometimes happen, that the party accused may find it absolutely necessary, in defence of himself, to throw blame and even criminality on others, who are no parties to the trial; nor can a prisoner be refused that liberty, which is essential to his own justification. It is sufficient for the party aggrieved, that the law can furnish ample redress against all calumnious or unjust accusations. The court is bound to hear whatever address, in his defence, the accused may think fit to offer, not being in itself contemptuous or disrespectful.

It is competent to a court, if it think proper, to caution the prisoner, as he proceeds, that, in its opinion, such a line of defence as he may be pursuing would probably not weigh with the court, nor operate in his favor; but, to decide against hearing him state arguments, which, notwithstanding such caution, he might persist in putting forward, as grounds of justification, or extenuation, (such arguments not being illegal in themselves,) is going beyond what any court would be warranted in doing. It occasionally happens, that, on presenting to the court a written address, the prisoner is unequal to the task of reading it, from indisposition or nervous excitement; on such occasions, the judge-advocate is sometimes requested by the president to read it; but, as the impression which might be anticipated to be made by it, may, in the judgment of the prisoner, be affected more or less by the manner of its delivery, courts-martial generally feel disposed to concede to the accused the indulgence of permitting it to be read by any friend named by him, particularly if that friend be a military man, or if the judge-advocate be the actual prosecutor. Courts-martial are particularly guarded in adhering to the custom of resisting every attempt on the part of counsel to address them. A lawyer is not recognized by a court-martial, though his presence is tolerated, as a friend of the prisoner, to assist him by advice in preparing questions for witnesses, in taking notes, and shaping his defence.

The prisoner having closed his defence, the prosecutor is entitled to reply, when witnesses have been examined on the defence, or where new facts are opened in the address. Thus, though no evidence may be brought forward by the prisoner, yet should he advert to any case, and, by drawing a parallel, attempt to draw his justification from it, the prosecutor will be permitted to observe on the case so cited. When the court allows the prosecutor to reply, it generally grants him a reasonable time to prepare it; and, upon his reading it, the trial ceases.

Should the prisoner have examined witnesses to points not touched on in the prosecution, or should he have entered on an examination impeaching the credibility of the prosecutor's evidence, the prosecutor is allowed to examine witnesses to the new matter; the court being careful to confine him within the limits of this rule, which extends to the re-establishing the character of his witnesses, to impeaching those of the defence, and to rebutting the new matter brought forward by the prisoner, supported by evidence. He cannot be allowed to examine on any points, which, in their nature, he might have foreseen previously to the defence of the prisoner. The prosecutor will not be permitted to bring forward evidence to rebut or counteract the effect of matter elicited by his own cross-examination; but is strictly confined to new matter introduced by the prisoner, and supported by his examination in chief. A defence, resting on motives, or qualifying the imputation attaching to facts, generally lets in evidence in reply; as, in such cases, the prisoner usually adverts, by evidence, to matter which it would have been impossible for the prosecutor to anticipate. The admissibility of evidence, in reply, may generally be determined by the answer to the questions: Could the prosecutor have foreseen this? Is it evidently new matter? Is the object of the further inquiry to re-establish the character of the witnesses impeached by evidence (not by declamation) in the course of the defence, or is it to impeach the character of the prisoner's witnesses? Cross-examination of such new witnesses, to an extent limited by the examination in chief, that is, confined to such points or matter as the prosecutor shall have examined on, is allowed on the part of the prisoner. (*See* CHALLENGE; COURT-MARTIAL; JURISDICTION. Consult MACOMB.)

TRIGGER. It has blade, tang or finger-piece, and hole for screw. (*See* ARMS.)

TRIGONOMETRY. Ordinary trigonometrical tables contain the logarithm of the sines, cosines, tangents, and cotangents for every ten seconds; but if the values of any one of the four be computed for the different angles between 0 and 90°, the values of all the others will be obtained at the same time. Thus, since cos. $A = \sin. (90° - A)$, a table of the values of the sine is also a table of the values of the cosine; and since tan. A = sin. A ÷ cos. A, the logarithm of the tangent of any angle is obtained by subtracting the logarithm of the cosine from the logarithm of the sine, and the logarithm of the cotangent by subtracting the logarithm of the sine from that of the cosine. It is usual to designate the semi-circumference of a circle whose radius is 1 by $\pi = 3.14159265$.

The solution of triangles is the proper object of trigonometry, and if tables contain the logarithms of the sines, cosines, tangents, and cotangents to every minute or smaller division of the quadrant, the means will be easy of applying such tables to each particular case; as, of the six parts of which a triangle consists, it is known from geometry that when any three except the three angles are known, all the rest are determined.

Plane Trigonometry.

A, B, C, the three angles; a, b, c, the three sides respectively opposite to them; R, the tabular radius; S, the area of the triangle; $p = \frac{1}{2}(a + b + c)$.

Right-angled Triangles: A being the right angle.

$$a = \sqrt{b^2 + c^2}\,;\quad b = c\,\frac{\tang. B}{R} = a\,\frac{\sin. B}{R}$$

Oblique-angled Triangles:

$$\frac{a}{\sin. A} = \frac{b}{\sin. B} = \frac{c}{\sin. C}$$

$$\tang. \tfrac{1}{2}(A - B) = \tang. \tfrac{1}{2}(A + B) \times \frac{a - b}{a + b}$$

$$c = \sqrt{\left((a - b)^2 + \frac{4ab\sin.^2 \tfrac{1}{2}C}{R^2}\right)} = \sqrt{\left(a^2 + b^2 - \frac{2ab\cos. C}{R}\right)}$$

$$\cos. \tfrac{1}{2}A = R\sqrt{\frac{p(p - a)}{bc}}\,;\quad \sin. \tfrac{1}{2}A = R\sqrt{\frac{(p - b)(p - c)}{bc}}$$

$$S = \tfrac{1}{2}ab\,\frac{\sin. C}{R} = \sqrt{p(p - a)(p - b)(p - c)}$$

General Formulæ:

$R. \sin.(a \pm b) = \sin. a \cos. b \pm \sin. b \cos. a.$
$R. \cos.(a \pm b) = \cos. a \cos. b \mp \sin. a \sin. b.$
$R. (\sin. a \pm \sin. b) = 2 \sin. \tfrac{1}{2}(a \pm b) \cos. \tfrac{1}{2}(a \mp b).$
$R. (\cos. a + \cos. b) = 2 \cos. \tfrac{1}{2}(a + b) \cos. \tfrac{1}{2}(a - b).$
$R. (\cos. a - \cos. b) = 2 \sin. \tfrac{1}{2}(a + b) \sin. \tfrac{1}{2}(a - b).$

$$\frac{\sin. a + \sin. b}{\sin. a - \sin. b} = \frac{\tang. \tfrac{1}{2}(a + b)\cot. \tfrac{1}{2}(a - b)}{R^2}$$

$$\sin. \tfrac{1}{2}a = \sqrt{\left(\frac{R^2 - R\cos. a}{2}\right)}\,;\quad \tang. \tfrac{1}{2}a = R\sqrt{\left(\frac{R - \cos. a}{R + \cos. a}\right)}$$

Chord of $A = 2 \sin. \tfrac{1}{2}A.$

TROOP. A company of cavalry. A particular beat of the drum.

TROPHY. Flags, colors, &c., captured from an enemy, and shown or treasured as a token of victory. Among the ancients, a trophy consisted of a pile or heap of arms taken from the vanquished troops, and raised by the conquerors on an eminence on the field of battle. As these were usually dedicated to some of the gods, it was considered sacrilege to demolish a trophy.

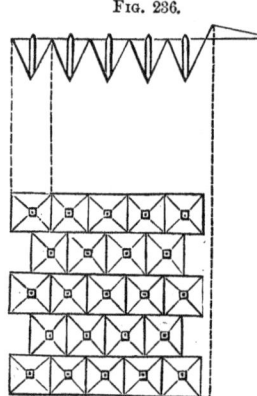

Fig. 236.

TROUS-DE-LOUP—or trapholes; are rows of pits in the form of inverted cones. They should be either $2\frac{1}{2}$ or 3 feet deep, so as not to be serviceable to the enemy's riflemen. They should be traced in a checkered form, and a strong pointed stake should be driven in the middle of each, (Fig. 236.) (*See* OBSTACLES.)

TRUCK. The casemate truck weighs 600 lbs., and is designed for transporting guns in casemate galleries. The *store truck* weighs 80 lbs; it is a common hand truck used for moving boxes.

TRUNNION. Short cylinder projecting from a piece of ordnance by which it rests upon its carriage. (*See* ORDNANCE.)

TRUSS. (*See* CARPENTRY.)

TUMBLER. (*See* ARMS; LOCK; MAYNARD'S *Primer*.)

U

UNDER. The correlative of over. (*See* COMMAND; OBEDIENCE; OVER; SUPERIOR.)

UNDRESS. Authorized habitual dress, not full uniform. The French designate the full dress as, *grande tenue ;* the undress, as *petite tenue*.

UNIFORM. Prescribed dress. The President shall have power to prescribe the uniform of the army; (ART. 100.)

UNMILITARY. Contrary to rules of discipline; unworthy of a soldier.

UNSPIKE. (*See* SPIKING.)

UPBRAID. Any officer or soldier, who shall upbraid another for refusing a challenge, shall himself be punished as a challenger; (ART. 28.)

UTENSILS—for camp and garrison are styled camp and garrison equipage, and are furnished by the quartermaster's department. The regulations allow: a general officer, three tents in the field, one axe and one hatchet; a field or staff officer above the rank of captain, two tents in the field, one axe and one hatchet; other staff officers or captains, one tent in the field, one axe and one hatchet; subalterns of a company, to every two, one tent in the field, one axe and one hatchet; to every 15 foot and 13 mounted men, one tent in the field, two spades, two axes, two pickaxes, two hatchets, two camp kettles, and five mess pans. Bed sacks are provided for troops in garrison, and iron pots may be furnished to them instead of camp kettles. Requisitions will be sent to the quartermaster-general for the authorized flags, colors, standards, guidons, drums, fifes, bugles and trumpets. The prescribed cooking utensils are evidently not adapted to field-service. The soldier is made too dependent on a baggage train. Some tools deemed necessary for service in the French army are also omitted in the enumeration of camp equipage furnished to the United States troops. (*See* TOOLS.)

V

VALUE. (*See* WEIGHTS.)

VANGUARD. Advanced guard.

VEDETTES OR **VIDETTES.** Sentries upon outposts, so placed that they can best observe the movements of an enemy, and communicate by signal to their respective posts and with each other.

VELOCITIES. (*See* INITIAL.) *Measurement of distances by sound.* The *velocity* of *sound*, in one second of time at 32° Fahrenheit in dry air, is about 1,090 English feet. For any higher temperature, add 1 foot for every degree of the thermometer above 32°. The measurement of distances by sound should always be made, if possible, in calm, *dry* weather. In cases of wind, the velocity per second must be corrected by the quantity, $f \cos. d$; f being the force of the wind in feet per second, and d the angle which its direction makes with that of the sound. Or, in general, in dry air,

$$v = 1{,}090 \text{ feet} + (t° - 32°) \pm f \cos. d.$$

VELOCITY AND FORCE OF WINDS.

Velocity in miles per hour.	A wind, when it does not exceed the velocity opposite to it, may be denominated	Velocity per second.	Force on a square foot.
		feet.	lbs.
6.8	a gentle, pleasant wind....................	10	0.129
13.6	a brisk gale...................................	20	0.915
19.5	a very brisk gale.............................	30	2.059
34.1	a high wind...................................	50	5.718
47.7	a very high wind.............................	70	11.207
54.5	a storm or tempest...........................	80	14.638
68.2	a great storm.................................	100	22.872
81.8	a hurricane....................................	120	32.926
102.3	a violent hurricane, that tears up trees, etc.	150	51.426

VENT. The opening or passage in fire-arms, by means of which the charge is ignited. The diameter of the vent is two-tenths of an inch in ordnance, except the eprouvette, which is one-tenth. The vents of brass guns are bored in vent pieces of wrought copper, which are screwed into the gun.

VERDICT. (*See* FINDING.)

VETERAN. An old soldier. Twenty years' service in the army entitles an enlisted soldier to the privileges of the army asylum. (*See* ASYLUM.)

VETERINARY. Veterinary surgeons are alone competent to treat grave cases of wounds and diseases in horses. Officers, however, may prevent accidents by watchfulness, recognize the existence of ailments, and by prompt care frequently relieve the horse entirely.

Limping.—The particular lameness is distinguished at a walk by observing that if a fore foot is lame, the horse raises the corresponding fore quarter before putting his foot to the ground. If a hind foot, he raises the hind quarter. At a trot, the contrary takes place. The horse should be watched in passing over ground of different degrees of hardness. For all lameness not connected with the shoe, prescribe rest, cold bathing, poultices. When there is pain in the joints, with swelling of the tissues, rub with spirits.

Lameness from shoeing may proceed: 1st, from *pricking*. If the nail be at once withdrawn, and the pricking is not deep, the lameness is not immediate. It is necessary, however, to enlarge the opening, introduce the essence of turpentine and dress with pledget, or lint coated with the same substance; act in the same way if the wound is old, after having taken out the nail, and cleared away to the bottom of the opening.

2. *Bleyme*, or inflammation in the foot of the horse between the sole and the bone. It is recognized by a red spot. Clear away the evil to the bottom, and dress as above. To prevent a return of the disease, it is perhaps necessary to clear away to the bottom of the offensive part for several successive shoeings.

3. *Solbature* is caused by the iron resting on the sole, or by a hard body introduced between the iron and the sole: clear the wounded part, apply a pledget coated with turpentine and retained in its place by a splint. Readjust the shoe.

4. *Burnt sole* is caused by an iron being applied when too hot and held too long. Act as in case of solbature.

These accidents from shoeing are all shown by limping. The precise seat of the accident is ascertained by pinching with the farrier's pincers. If the horse is to march, attach the shoe with but few nails, simply to hold it in its place.

Founder.—There is great heat in the foot without apparent cause. The horse walks with difficulty, resting on the heel; he shows discomfort, want of appetite, fever. It is necessary to unshoe him; cut the horn of the hoof to the quick towards the toe; even make it bleed; bathe with cold salt water; envelop the whole foot with linen soaked in vinegar to the crown; later, rub hard from the ham to the knees with essence of turpentine and camphorated spirits: diet, bran with water. The horse must not march.

Chaps, serosity of limbs.—These exact cleanliness, washing with warm water and a little spirit of wine, and towards the termination of the ailment, with sub-acetate of lead.

Injuries.—At the least appearance of tumor stop the development of inflammation by washing with fresh water, vinegared or salted. Strengthen the tissues by friction with brandy, united with soap or camphor. Take off the load. Put on the saddle in such a way as to leave a space between it and the tumor. If the ailment increases, notwithstanding these precautions, it is necessary to relieve the horse from all weight, continuing the washings and rubbings. If the tumor still increases, open it. When opened wash the wound once a day only; do not remove the pus entirely; prevent its contact with the air by means of oakum or lint. When the wound begins to heal, its cicatrization may be hastened by washing with sub-acetate of lead. When from their appearance tumors of the withers and loins seem to be soft and inclose red water, cut the hair smooth and apply a blister ointment, which it is rarely necessary to renew. When a horse is wounded under the tail, clean the wound and put in it the *unguentum populi.* For slight contusions from kicking, use twice a day the *unguentum populi,* and then rub the upper part with camphorated spirits. If the pain is severe, bleed and foment with warm mallows water.

Internal affections.—The ordinary symptoms are: dry and frequent cough, uneasiness and sadness, disgust of food, falling off; alteration of flank; hair not smooth; fever. Separate the horse from others; put him to diet on bran, attending to the prescriptions of the veterinary surgeon. Examine the eyes, gently reversing the eylids, pass the hand into the mouth of the horse; if the eye is red and the mouth very hot, bleed the horse, drawing from him 8 lbs. of blood; leave him two hours without eating; rub him down well, cover him and give him some injections; replace his allowance of oats with warm barbotage of barley-flour as much as possible. For want of appetite it is sufficient sometimes to sprinkle the forage with salt water. If the horse, in rising or lying down, looks at his flanks with an unquiet air he has colic. In this case it is often sufficient in order to cure him to rub hard with rumpled linen upon the belly, and apply injections of decoctions of mallows or lettuce. If an hour or two after the first trouble the colic is not over, call a veterinary surgeon; death may take place in a short time. If a horse tries often to urinate, and shows pain, it is retention of urine. Recourse must be had to emollient injections, and to nitrated drinks. In certain diseases of the breast prompt succor is necessary. In grave cases, in the absence of the veterinary, put blisters or setons upon the breast, and bleed.

The necessary tools, &c., are: syringes, bistouries, tape and needles to setons, dry oakum, camphorated spirits, soap, nitre, essence of turpentine, liquid, sub-acetate of lead, foot ointment, and *unguentum populi*. (*See* GLANDERS; HORSE. Consult *Memorial des Officiers d'Infanterie et de Cavalerie ;* SKINNER's *Youatt*.)

VICE-PRESIDENT OF THE UNITED STATES. Using contemptuous or disrespectful words against, punishable by cashiering or otherwise at the discretion of a court-martial; (ART. 5.)

VICTUALS. Whosoever shall relieve the enemy with money, victuals, or ammunition, or shall knowingly harbor or protect an enemy shall suffer death, or such other punishment as shall be ordered by the sentence of a court-martial; (ART. 56.)

VILLAGES. Cavalry, the better to preserve their horses, should occupy villages whenever the distance of the enemy, and the time necessary to repair to its post in battle, will permit. Their quarters should be preferably farmhouses or taverns having large stables. Posts are established by the colonel or commanding officer, and the squadrons conducted to their quarters by their respective captains. Where in an exceptional case regular distributions are not made, the resources which the household assigned as quarters presents are equally divided. About two hours after their arrival the squadrons in succession water their horses and then give forage. Cavalry and infantry also should, when thus cantoned near an enemy, occupy, wherever it can be done, houses which will hold an entire company or some constituent fraction of a company, and at break of day stand to their arms. When in the same cantonment, cavalry should watch over the safety of the cantonment by day and the infantry by night; and in the presence of an enemy they should be protected by an advance guard and natural or artificial obstacles.

VINEGAR. On board ship vinegar is essential to the comfort of horses, and should be freely used by sponging their mouths and noses repeatedly, and also their mangers. A small portion of vinegar drank with water supplies the waste of perspiration of men in the field. It is better than rum or whiskey; it allays thirst, and men who use it avoid the danger of drinking cold water when heated, and are not fevered as they are too apt to be by the use of spirituous liquors; (Dr. RUSH.)

VIOLENCE. Any officer or soldier who shall offer any violence against his superior officer, being in the execution of his office, on any pretence whatsoever, punished by death or otherwise, according to the nature of his offence; (ART. 9.) Violence to any person who brings provisions to the camp, garrisons, or quarters to the forces of the United States employed in any part out of the said States, punishable in like manner; (ART. 51.)

VOLUNTEERS. Whereas sundry corps of artillery, cavalry, and infantry now exist in several of the States, which by the laws, customs, or usages thereof, have not been incorporated with, or subject to, the general regulations of the militia; such corps shall retain their accustomed privileges, subject, nevertheless, to all other duties required by this act in like manner with the other militia; (*Act* May 8, 1792.) (*See* CALLING FORTH MILITIA; and MILITIA.)

This class of uniformed militia exists in every State of the Union. It is a regular, unpaid force, composed generally of men engaged in such private business operations, as must always prevent their being employed except in their immediate vicinage. But in cases of riot, or the defence of their own firesides, town or city, experience has shown it to be a most reliable organization. There is, however, another class of troops, also called volunteers, which have from time to time been raised by Congress for temporary purposes. Such troops are properly United States and not State troops. The manner in which their officers are to be appointed is therefore always designated by Congress. The act of May 28, 1798, authorized the President to appoint the company officers of such volunteers; the act of June 22, 1798, directed that the field-officers of such volunteers should be appointed by the President and Senate; the act of May 23, 1836, directed that the officers of volunteers then raised, should be appointed in the manner prescribed by law in the several States and Territories to which such companies, battalions, squadrons, regiments, brigades, or divisions shall respectively belong; the act of March 3, 1839, applies the same provision to the volunteers then authorized; the act of May 13, 1846, contains the same provision as to appointment of officers; and the act of June 26, 1846, authorizes the President, by and with the advice and consent of the Senate, to appoint such number of major-generals and brigadier-generals as the organization of such volunteer forces (raised by the act of May 13, 1846) into brigades and divisions, may render necessary; and in case the brigades or divisions of volunteers at any time in the service shall be reduced in number, the brigadier-generals and major-generals herein provided for shall be discharged in proportion to the reduction in the number of brigades and divisions.

There should, then, be no question that these volunteers are United States troops raised by Congress under its constitutional authority to raise and support armies; but, strangely enough, the officers have been usually commissioned by their respective States. It becomes, therefore, an important question to ascertain, if possible, by experience, whether the advantages which attend raising armies in this particular way are not greatly counterbalanced by its disadvantages; whether the efficiency of such an irregular force is in any degree commensurate with its cost; and whether deaths, diseases, discharges, and other casualties do not in such a force accumulate in such numbers as to deprive the Government of the moral right thus uselessly to sacrifice the citizens of the country.

The statistics of the Mexican war, published by Congress, (Doc. 24, House of Representatives, 31st. Congress, 1st Session,) furnish the following startling facts:—

REGULAR ARMY.	AGGREGATE FORCE.	LENGTH OF SERVICE.
Old establishment,	15,736	26 months.
Additional force,	11,186	15 "

The old establishment of the regular army, with an aggregate of 15,736 men during 26 months' service, lost by discharges for disability 1,782 men; by ordinary deaths, 2,623 men; and by deaths from wounds in battle, 792 men.

The additional regular force, with an aggregate of 11,186 men during 15 months' service, lost by discharges for disability 767 men; by ordinary deaths, 2,091 men; and by deaths from wounds in battle, 143 men.

The volunteer force, with an aggregate of 73,532 men during an average of 10 months' service, lost by discharges for disability 7,200 men; by ordinary deaths, 6,256 men; and by deaths from wounds in battle, 613 men.

The number of wounded in battle were: In the old establishment, 1,803 men; in the additional regular force, 272 men; and in the volunteers, 1,318 men. The number of deserters were, in the whole regular force, 2,849 men; and in the volunteer force, 3,876 men.

These statistics require no commentary to show the waste of life and money in employing volunteers. But without explanation they do not show the numbers of each description of force engaged in the different battles of Mexico, or how, with such a large aggregate of forces employed in Mexico, Taylor's battles were fought with never more than 6,000 men, and Scott had at his disposition only about 11,000 men for the march from Puebla and the capture of the city of Mexico. An analysis of the aggregates of forces engaged in those battles is therefore necessary, to ascertain by whom they were won, and this will lead to a subsequent inquiry, which will show why such ostentatious aggregates furnished so small a body of men for the great operations of the war.

Regular Army.—Ex-Doc. 24, House of Representatives, 31st Congress, 1st Session, shows that the old regular force on the frontier of Texas, May, 1846, at the commencement of the war was 3,554 men present and absent. This force alone under Taylor fought the battles on the Rio Grande, with an aggregate loss of killed in battle and died of wounds, of 72 men. There were wounded in the same affairs 147 men.

May 24, volunteers began to arrive on the Rio Grande. August 1, General Taylor reports that the volunteer forces ordered to report to him are much greater than he can employ, and regrets that one division of volunteers should not have been encamped at Pass Christian, where it could have been instructed; (Doc. 119, House of Representatives, 29th Congress, 2d Session.)

For the march from Camargo upon Monterey, General Taylor organized a force of volunteers of about 3,000 men, and about the same number of regulars. The volunteers lost 74 men killed and died of wounds in the battle of Monterey, Sept. 21, 22, and 23, 1846, and had 218 men wounded. The regulars lost in the same battle 68 men, and had 150 men wounded.

At the battle of Buena Vista, Feb. 22 and 23, there were engaged 517 regulars and about 4,400 volunteers. The loss of the regulars 8 killed and died of wounds, and 36 wounded; the loss of the volunteers 269 killed and died of wounds, and 372 wounded.

At the siege of Vera Cruz, March, 1847, there were 6,808 regulars and 6,662 volunteers. The loss of the regulars 10 killed and died of wounds, and 26 wounded; the volunteers lost 2 killed and 25 wounded.

At Cerro Gordo, April, 1847, there were 6,000 regulars and 2,500 volunteers. The loss of the regulars was 61 killed and died of wounds, and 201 wounded; the loss of the volunteers 38 killed and died of wounds, and 152 wounded.

At Contreras, Churubusco, San Antonio, and San Augustine, August 19 and 20, 1847, there was an aggregate of 9,681 of old and new regulars and marines, and 1,526 volunteers. The regulars lost in killed and died of wounds 137 men, and 653 wounded; the volunteers lost 52 men killed and died of wounds, and 212 wounded.

At Molino del Rey, September 8, 1847, there were 3,251 regulars engaged. Lost 195 men killed and died of wounds; 582 wounded.

At Chapultepec, and the capture of the city of Mexico, September 12, 13, and 14, 1847, the whole army for duty was 8,304 men. Regulars 7,035 men; volunteers 1,290 men. The regulars lost 144 killed and died of wounds, and 434 wounded; the volunteers 44 killed and died of wounds, and 239 wounded.

In all other incidental affairs and skirmishes, mostly with guerilla parties of the enemy during the whole war, the aggregate losses of the regulars were 65 killed and died of wounds, and 163 wounded; the loss of volunteers 62 killed and died of wounds, and 130 wounded.

Having thus analyzed the losses in battle of the regulars and volunteers, and given the numbers of each engaged in the important battles of the war, the inquiry recurs: why, with an aggregate of 73,000 volunteers and 26,922 regulars reported as being employed during the war, so small a body should have been at the disposition of commanders for marching against the enemy.

The first reason was undoubtedly the defective plan of campaign upon which the war was begun. Immediately after the victories of Palo Alto and Resaca de la Palma, the public mind was inflamed. The volunteer system caused great numbers to flock to the standard of the country. The pressure upon the Administration was great for their reception. General Taylor was flooded with volunteers for whom he could find no employment. A plan of campaign was therefore devised in Washington, for marching on New Mexico, marching on Chihuahua, marching on Monterey, and marching on California, with different detachments, thus hastily collected together without taking the necessary measures to organize and instruct the troops, and without first providing the *materiel* indispensable for such long marches. The plan was therefore defective in all those respects, but still more defective in its predominant idea of striking at remote frontiers of the enemy instead of marching on his capital. It was like pricking the fingers of man instead of pointing a dagger at some vital part.

The second and paramount reason why with such large aggregates of forces mustered into service so few were employed in battles, is the failure of the law to provide for a well-digested system of national defence prepared in peace, which would enable Congress and the Executive to meet any crisis in foreign affairs. This want caused the reception into service of 12,601 volunteers for 3 months at the beginning of the war with Mexico. These lost 16 men killed in battle and died of wounds; 129 by ordinary deaths, 922 by discharge, and 546 by desertion. Those killed in battle belonged to the Texas horse and foot, and they alone were engaged with an enemy.

Upon the declaration that war existed by the act of Mexico, Congress, however, authorized the President to accept volunteers for twelve months or for the war. He accordingly received 27,063 men of this class for twelve months. They lost during their service, killed in action or died of wounds, 439 men; by ordinary deaths 1,859 men; by discharges 4,636 men; and by desertion 600 men. Some of this class of volunteers rendered most effective service at Monterey, Buena Vista, Vera Cruz, and Cerro Gordo. The great mistake committed in regard to them was in receiving them for the short period of twelve months. Generally mustered into service in June, 1846, they were entitled to discharge in June, 1847, at a moment when their services were much needed, in order to strike a decisive blow at the capital of Mexico. Every effort was made to re-engage them, but without success; and General Scott, who had been employed to conduct military operations on the line from Vera Cruz to the capital, reluctantly put over 3,000 of these men in march from Jalapa to the United States in May, 1847, when he had ascertained that his column was not likely soon to be reinforced by more than 960 army recruits, and the services of those volunteers for the short remainder of their time could therefore no longer be usefully employed. Meanwhile the Administration, having late in 1846 awakened from its dream of conquering a peace, by directing blows against remote extremities of Mexico, had at last adopted the plan of striking at the vitals of their enemy. General Scott was put in command. Some volunteers were at once mustered into service for the war, but in insufficient numbers. Out of the whole force raised for the war, General Scott only received in time for his operations a regiment from New York, two from Pennsylvania, and one from South Carolina, and one company under Captain Wheat, who alone re-engaged themselves from the whole number of twelve-months volunteers; and these were the only regiments of volunteers, which took part in the battles in the valley of Mexico, and the capture of the city, September 14, 1847, which secured the conquest of peace. The whole volunteer force raised for service during the war with Mexico, (but with the exceptions stated, too late for important military operations,) were 33,596 men. They lost 152 men killed and died of wounds; ordinary deaths 4,420; discharges 3,890 men; and desertions 2,730 men. Of the 152 who were killed in battle or died of wounds, 134 belonged to the regiments mentioned as being with General Scott. It was not until Dec., 1847, months after the occupation of the capital, that other volunteers for the war reached Gen. Scott's head-quarters in the city of Mexico.

The same want of administrative ability was shown by the War Department in despatching regulars to the seat of war. Doc. 24, H. of R., 31st Congress, 1st Session, exhibits an aggregate of 15,736 men of the old regular regiments, and 11,186 men of the new regular regiments—in all 26,922 regulars—employed during the war, and yet the largest regular force employed at any one time against the enemy was less than 10,000 men. Let us endeavor to ascertain how this happened.

It has been seen that the whole regular force on the frontiers of Texas at the beginning of the war was 3,554 men, and that this force fought the battles of Palo Alto, Resaca de la Palma and Fort Brown in May, 1846. In September, this force had not been largely increased; for, at the battle of Monterey, Taylor had only about 3,000 regulars.

In February and March, 1847, the regular force employed both at Buena Vista and at Vera Cruz had been increased to 7,425 men. And in April, 1847, at Cerro Gordo, and on the line to Vera Cruz and at Tampico, the whole regular force did not exceed 8,000 men. These all belonged to the old regular regiments. Meantime, February 11, 1847, Congress passed an act for raising one regiment of dragoons, and nine regiments of infantry. But none of these troops reached Gen. Scott's head-quarters at Puebla, until July and August, 1847. The last detachment came up August 6, and Gen. Scott marched on the city of Mexico, August 7, 1847, with only 2,564 new regulars. The forces which took part in the battles in the valley of Mexico were then:

Old regular regiments	6,446 men.
New regular regiments	2,365 "
Marines	271 "
Volunteers	1,569 "
	10,651 men.

The greater part of the additional force of regulars raised for the war, as well as the very large numbers of volunteers raised for the same purpose, were not, it thus appears, put at the disposition of military commanders, until final success in battles had already been accomplished. The following tables, giving losses by regiments, &c., are from the report of the adjutant-general of Dec. 3, 1849:

MILITARY DICTIONARY.

GENERAL view of the subjects of inquiry, showing the aggregate of the regulars and volunteers employed during the Mexican war, with their average duration of service, and the casualties incident to each description of force.

	FORCES EMPLOYED & MUSTERED INTO SERVICE.		DISCHARGES.				DEATHS.									WOUNDED IN BATTLE.			Resignations.	Desertions.
							Killed in battle.		Died of wounds.		Total killed and died of wounds.		Ordinary.		Accidental.	Aggregate number of deaths—officers and men.					
	Aggregate number of officers and men.	Average length of service during the war.	By expiration of service.	For disability.	By order, and civil authority.	Aggregate number of discharges.	Officers.	Men.	Officers.	Men.	Officers.	Men.	Officers.	Men.			Officers.	Men.	Aggregate.		
		Mos																			
Old establishment....	15,736	26	1,561	1,782	373	3,716	41	422	22	307	63	729	49	2,574	139	3,554	118	1,685	1,803	37	2,247
Additional force......	11,186	15	12	767	114	893	5	62	5	71	10	133	36	2,055	80	2,264	36	236	272	92	602
Aggregate of regular army.	26,922		1,573	2,549	487	4,609	46	484	27	378	73	862	85	4,629	169	5,818	154	1,921	2,075	129	2,849
VOLUNTEER FORCE.																					
General staff.........	272				47	47	1				1		16			17				48	
Regiments and corps...	73,260	10	50,573	7,200	1,969	9,169	46	467		100	46	567	(*)	*6,256	192	7,061	129	1,189	1,318	279	3,876
Aggregate of regular and volunteer forces.......	100,454		52,146	9,749	2,503	18,825	93	951	27	478	120	1,429	101	10,885	861	12,896	283	3,110	3,393	456	6,725

* In the reports of the deaths of volunteers of ordinary disease, officers are not discriminated.

RECAPITULATION OF LOSS IN BATTLE OF THE REGULAR ARMY, BY REGIMENTS AND CORPS, IN THE CAMPAIGN OF 1846.

REGIMENTS AND CORPS.	KILLED IN BATTLE.			WOUNDED.			Aggregate in killed and wounded.	DIED OF WOUNDS.		
	Officers.	Men.	Total.	Officers.	Men.	Total.		Officers.	Men.	Total.
General staff............	......	......	......	1	......	1	1	......	......	......
Engineers..............	......	......	......	1	......	1	1	......	......	......
Topographical engineers...	......	......	......	2	......	2	2	1	......	1
1st regiment dragoons....	3	14	17	......	11	11	28	......	......	......
2d " "	2	17	19	......	19	19	38	......	......	......
1st regiment artillery.....	......	1	1	......	6	6	7	......	1	1
2d " "	......	2	2	1	8	9	11	......	1	1
3d " "	......	5	5	2	26	28	33	1	3	4
4th " "	......	5	5	1	16	17	22	......	4	4
1st regiment of infantry..	......	12	12	3	28	31	43	2	......	2
2d " "	......	......	......	......	......	......	......	......	......	......
3d " "	1	......	1	......	......	......	1	......	......	......
3d " "	5	15	20	3	36	39	59	1	1	2
4th " "	2	15	17	2	30	32	49	2	4	6
5th " "	......	9	9	4	36	40	49	......	3	3
7th " "	......	2	2	3	12	15	17	1	......	1
8th " "	2	10	12	8	51	59	71	......	10	10
Corps not specified......	3	......	3	......	......	......	3	......	......	......
Total in campaign of 1846.	18	107	125	31	279	310	435	8	27	35

RECAPITULATION OF LOSS OF REGULAR ARMY, IN THE CAMPAIGN OF 1847.

REGIMENTS AND CORPS.	KILLED IN BATTLE.			WOUNDED.			Aggregate in killed and wounded.	DIED OF WOUNDS.		
	Officers.	Men.	Total.	Officers.	Men.	Total.		Officers.	Men.	Total.
General staff..................	1	......	1	2	......	2	3	......	......	......
Surgeons.....................	......	......	......	2	......	2	2	1	......	1
Engineers....................	......	......	......	5	5	10	10	......	......	......
Topographical engineers.........	......	......	......	1	......	1	1	......	......	......
Ordnance....................	......	5	5	2	20	22	27	......	1	1
1st regiment dragoons..........	......	14	14	3	39	42	56	1	......	1
2d " "	1	8	9	6	31	37	46	......	5	5
3d " "	......	7	7	4	9	13	20	......	......	......
Regiment mounted riflemen.....	2	24	26	11	144	155	181	2	15	17
1st regiment artillery...........	4	25	29	5	108	113	142	1	1	2
2d " "	1	21	22	3	160	163	185	2	19	21
3d " "	3	13	16	2	82	84	100	......	13	13
4th " "	2	20	22	3	81	84	106	......	4	4
1st regiment infantry...........	......	......	......	......	......	......	......	......	......	......
2d " "	3	25	28	9	120	129	157	1	1	2
3d " "	......	20	20	4	101	105	125	......	4	4
4th " "	2	25	27	6	88	94	121	1	15	16
5th " "	5	36	41	8	170	178	219	2	11	13
6th " "	......	34	34	5	128	133	167	2	19	21
7th " "	2	18	20	3	59	62	82	......	13	13
8th " "	......	34	34	13	158	171	205	2	29	31
9th " "	1	10	11	5	64	69	80	......	......	......
10th " "	......	......	......	1	......	1	1	......	......	......
11th " "	2	10	12	4	42	46	58	......	......	......
12th " "	......	4	4	4	33	37	41	......	......	......
13th " "	......	......	......	......	......	......	......	......	......	......
14th " "	......	6	6	7	15	22	28	......	4	4
15th " "	2	11	13	6	97	103	116	1	......	1
16th " "	......	......	......	......	......	......	......	......	......	......
Voltigeurs....................	......	9	9	14	146	160	169	2	......	2
Marine corps.................	1	6	7	4	24	28	35	......	......	......
Artillery recruits..............	......	......	......	......	9	9	9	......	......	......
Corps not specified	1	12	13	......	11	11	24	......	......	......
United States navy............	......	1	1	1	9	10	11	......	......	......
Aggregate in the campaign of 1847.	33	398	431	143	1,953	2,096	2,527	18	154	172
Aggregate in the campaign of 1846.	18	107	125	31	279	310	435	8	27	35
Aggregate of regular army killed and wounded in 1846 and 1847.	51	505	556	174	2,232	2,406	2,962	26	181	207

RECAPITULATION OF LOSS IN BATTLE OF THE VOLUNTEER FORCES IN THE CAMPAIGN OF 1847, AND AGGREGATE OF THEIR LOSSES IN 1846.

REGIMENTS AND CORPS.	KILLED IN BATTLE.			WOUNDED.			Aggregate in killed and wounded.	DIED OF WOUNDS.		
	Officers.	Men.	Total.	Officers.	Men.	Total.		Officers.	Men.	Total.
Indiana brigade, staff...........	....	....	....	1	....	1	1	....	....	....
Mississippi rifles..............	2	38	40	5	51	56	96	....	3	3
1st regiment, Illinois..........	3	26	29	2	23	25	54	....	1	1
2d " "	10	37	47	6	68	74	121	....	2	2
3d " "	....	1	1	1	15	16	17	....	....	....
4th " "	1	3	4	5	39	44	48	....	8	8
Texas company.................	2	12	14	1	1	2	16	....	....	....
2d regiment, Indiana...........	3	29	32	8	28	36	68	....	....	....
3d " "	1	8	9	4	52	56	65	....	....	....
Regiment Kentucky cavalry....	1	26	27	4	29	33	60	....	....	....
1st regiment, Kentucky........	....	3	3	1	9	10	13	....	2	2
2d " "	3	41	44	3	54	57	101	....	2	2
Regiment Arkansas cavalry.....	2	15	17	1	31	32	49	....	2	2
1st regiment, Pennsylvania.....	....	17	17	2	47	49	66	....	....	....
2d " "	....	7	7	6	107	113	120	....	10	10
Regiment, South Carolina......	5	22	27	21	195	216	243	....	26	26
1st regiment, Tennessee.......	....	1	1	8	4	12	13	....	1	1
2d " "	2	13	15	1	38	39	54	....	3	3
Regiment, Georgia.............	....	2	2	....	8	8	10	....	....	....
2d Regiment, New York.......	2	25	27	19	137	156	183	....	19	19
California volunteers..........	....	....	....	1	1	2	2	....	....	....
Missouri volunteers............	1	6	7	5	38	43	50	....	4	4
Santa Fé volunteers............	....	....	....	....	4	4	4	....	....	....
Louisiana volunteers...........	....	5	5	1	5	6	11	....	....	....
Spy company...................	....	....	....	1	4	5	5	....	....	....
Maryland volunteers...........	1	6	7	1	3	4	11	....	....	....
Virginia volunteers............	....	1	1	....	4	4	5	....	....	....
Incidental loss.................	4	7	11	....	14	14	25	....	....	....
Corps not specified............	....	1	1	1	1	2	3	....	....	....
Chihuahua rangers.............	....	....	....	....	1	1	1	....	....	....
Texas rangers..................	....	....	....	1	4	5	5	....	....	....
1st regiment, New York.......	....	....	....	1	....	1	1	....	....	....
Sailors, marines, and California volunteers...................	1	2	3	....	4	4	7	....	....	....
New York and California volunteers........................	....	1	1	....	....	....	1	....	....	....
Aggregate in campaign of 1847.	44	355	399	111	1,019	1,130	1,529	....	83	83
Aggregate in campaign of 1846*.	5	70	75	20	207	227	302	1	9	10
Aggregate of volunteers killed and wounded in 1846 and 1847.	49	425	474	131	1,226	1,357	1,831	1	92	93

* The losses in 1846 were: April 28, 10 men of Capt. Walker's Texas rangers. September 21, 22, and 23, at Monterey, among the Maryland and District of Columbia battalion volunteers; 1st regiment, Tennessee; 1st regiment, Ohio; Louisiana volunteers; 1st regiment, Kentucky; Mississippi rifles and Texas volunteers. Dec. 6, California volunteers, under General Kearney. Dec. 26, Missouri volunteers, under Colonel Doniphan.

VOTES. (*See* FINDING.)

W

WAD—WADDING. Ring wads (or *grommets*, as they are called in the naval service) increase the accuracy of fire, and are preferred where the object is to keep the ball in its place. They consist of a ring of rope yarn, with two pieces of strong twine tied across it at right angles with each other. The ring is the full diameter of the bore. These wads may be attached with twine to the straps, or to the balls; or inserted like other wads after the ball. Wads, for firing hot shot, may be made of *hay*, wrapped with rope yarn, and are made in the same manner as junk wads for proving cannon. (Consult *Ordnance Manual.*)

WAGON-MASTER. The quartermaster-general is authorized to employ from time to time as many forage-masters and wagon-masters as he may deem necessary for the service, not exceeding twenty in the whole, who shall be entitled to receive forty dollars per month, and three rations a day, and forage for one horse; and neither of whom shall be interested or concerned directly or indirectly in any wagon or other means of transport, employed by the United States, nor in the purchase or sale of any property procured for or belonging to the United States, except as an agent of the United States; (*Act* July 5 1838, Sec. 10.)

WAGONS—are used by armies for the transportation of subsistence, other military stores, baggage, ammunition, sick and wounded. The different purposes for which they are used require differences in details which demand thought and study. In an able memoir, *Sur Divers Perfectionnements Militaires*, par J. CAVALLI, Col. d'Artillerie, (*Paris*, 1856,) it is proposed that all the different carriages for army transportation should be on *two large wheels*, and that there should be only two different models for the height of the wheels. The number of models for carriages is thus reduced to seven at most, which might be substituted for the wagons on four wheels now in use. The different vehicles used by the French in campaign, according to the recent work of M. LEON GUILLOT, *Sur Legislation et Administration Militaire*, are: the four-wheeled military wagon, made and lined with sheet-iron, specially intended for the transportation of bread and other important necessaries, but also adapted for ambulance purposes, as its interior admits the placing of four boards for the accommodation of the sick; the ammunition wagon and campaign forge for the artillery; and the ambulance wagon used in service by the French army in the East in 1854. The latter is suspended on six springs and has four wheels; it carries five persons, three upon the front seat, which is uncovered and rests on the wagon, while in the interior there are two places for reclining, each on a movable bed.

According to M. Vauchelle, the vehicles for administration purposes on four wheels should be the ordinary wagon and a light wagon, both covered with water-proof cloth; the first would serve for the transportation of bread, and also for medical and hospital stores, &c.; the second, suspended upon springs, should be specially devoted to hospital purposes. He would have, besides, ammunition wagons and field-forges; all on four wheels drawn by four horses, and conducted each by two soldiers. The maximum capacity of the wagon, according to Guillot, should be for 1,200 rations, weighing about 1,900 lbs. This burden is the mean between that for 1,000 rations weighing about 1,700 lbs. prescribed by Vauchelle, and that for 1,600 rations weighing about 2,650 lbs. adopted, notwithstanding the opposition of M. Vauchelle, by the French minister of war. If, for all these vehicles on four wheels drawn by four horses, there be substituted carts or two-wheeled vehicles, according to the models prepared by Cavalli, the four-wheeled vehicles carrying only 2,200 lbs., will give place to the carts carrying each 3,300 lbs.; that being only one-half of the burden of carts loaded in the same way now used in European commerce drawn by two horses. Under the proposed system, then, for an army of 100,000 men the number of vehicles, &c., would experience the following reductions:

  860 wagons would be reduced to . . 573 carts.
  3,268 conductors " " . 1,092
  3,820 horses   " " . . 1,277

If meat and forage are also to be transported for the army, and these articles are omitted in the foregoing calculations, then, supposing an army of 100,000 men has 30,000 horses, the proposed system would reduce

  2,567 wagons to . . 1,711 two large wheeled carts.
  9,804 conductors to 3,276 conductors.
  11,460 horses to . . 3,831 horses.

The carts proposed by Col. Cavalli are the following: 1. A dray for the transportation of heavy loads for the artillery and engineer trains, intended as a substitute for the platform or block carriage, and also for the siege truck. This dray weighs about 1,540 lbs.; it will carry a piece of ordnance weighing 7,500 lbs. suspended under it in place of its corbeille, and has been drawn by two horses at a trot from the glacis to the citadel, and by three horses over the ramps of the ditch of the citadel from which it had been lowered. The corbeille of this dray will carry 80 shells, and its flooring 60, weighing in all 8,000 lbs. The usual weight to each horse in the field, however, should not exceed 2,200 lbs. This dray is suitable for all heavy and embarrassing weights, and the division between the load below and that upon the superior bed is so arranged as to maintain the bars in a stable equilibrium without liability to be overturned, and without exerting any pressure upon the horse. 2. The ammunition cart, covered or uncovered, with two large wheels and having a movable water-proof cover, is designed to replace the present ammunition and battery wagons. It will hold 24 cases of powder, 120 lbs. each, of which about 18 would be filled up. The interior void of this cart is about 35 cubic feet. 3. The spring cart is of the same form as the ammunition cart, and differs from it only in being on springs. It is intended principally for the transportation of provisions and articles easily spoiled, as ammunition; and in cases of need as an ambulance. 4 and 5. Two models of carts for ambulances and other purposes drawn, one by two horses and the smaller by one horse; these two vehicles have also only two large wheels, and are not liable to upset. They are intended as substitutes for the ambulance wagon and other wagons. Two persons may be placed in front, and six behind, four of whom may recline on beds suitably arranged at the sides. The smaller cart will answer for two or three persons at most, only two of whom can recline. The smaller carts may also be issued to commanding officers and staff officers entitled to wagons. 6 and 7. The *kitchen-cart*—one to a battalion for 1,000 soups, or a smaller one for 250 soups. The two differ from each other only in length. They should be provided with boilers *a la Papin* with an interior fire-place. These constitute the body of the cart, the superior part of which is furnished with plank to be used as a table. At the extremity of the cart there are two foot boards upon which the cooks may rest while working during the march. Papin's digester is essential to cook well and rapidly. The interior arrangement of the fire-place which is suited to baking is very economical in fuel. The kitchen-cart is otherwise like the preceding. (*See* TRAVELLING-KITCHEN.)

Model No. 2, or even Nos. 4 and 5, will answer for the sutler. A field-forge may be readily placed in the rear of model No. 2, by means of a movable fire-place and bellows. It is proposed to harness to each vehicle intended as a transport two horses, in file; each cart has one conductor not mounted. The importance of the *travelling-kitchen* will be manifest to all soldiers. The cooking is done on the march. The soup is ready at the moment of halting. The strength of the soldier is economized; his food is well cooked in any weather; and numerous diseases, caused by bad food and want of rest, which too often decimate armies, will be avoided by its introduction into service. (*See* AMBULANCE; BAGGAGE; CONVOY.)

WAITERS. (*See* SERVANTS.)

WAR. The right of making war, as well as of authorizing reprisals, or other acts of vindictive retaliation, belongs in every civilized nation to the supreme power of the state. The exercise of this right is vested by the Constitution of the United States in Congress. A contest by force between independent sovereign states is called a *public* war. A *perfect* war is where one whole nation is at war with another nation, and all the members of both nations are authorized to commit hostilities against the other, within the restrictions prescribed by the general laws of war. An *imperfect* war is limited as to places, persons, and things—such were the limited hostilities authorized by the United States against France in 1798. Grotius calls a civil war, a *mixed* war; but the general usage of nations regards such a war as entitling both the contending parties to all the rights of war as against each other, and even as respects neutral nations. A formal declaration of war to the enemy was once considered necessary to legalize hostilities between nations. The usage now is to publish a manifesto, within the territory of the state declaring war, announcing the existence of hostilities, and the motives for commencing them.

During the second war between the United States and Great Britain, it was determined by the Supreme Court that enemy's property, found within the territory of the United States on the declaration of war, could not be seized and condemned as prize of war, without some legislative act expressly authorizing its confiscation. The court held that the law of Congress declaring war was not such an act. It is stated by Sir W. Scott to be the constant practice of Great Britain, on the breaking out of war, to condemn property seized before the war, if the enemy condemns, and to restore if the enemy restores.

One of the immediate consequences of the commencement of hostilities, is the interdiction of all commercial intercourse between the subjects of the states at war, without the express license of their respective governments. It follows, as a corollary from this principle, that every species of private contract made with an enemy's subjects during the war is unlawful, and this rule is applied to insurance on enemy's property and trade; to the drawing and negotiating of bills of exchange between the subjects of the powers at war; to the remission of funds in money or bills to the enemy's country; to commercial partnerships, which, if existing before the war, are dissolved by the mere force and act of the war itself, although as to other contracts it only suspends the remedy. But it is the modern usage not to confiscate in war the enemy's actions and credits, and the 10th article of the treaty between the United States and Great Britain, in 1794, stipulates, "that neither the debts due from individuals of the one nation to individuals of the other, nor shares, nor moneys which they may have in the public funds, or in the public or private banks, shall ever, in any event of war or national differences, be sequestered or confiscated; it being *unjust* and *impolitic* that debts and engagements contracted and made by individuals, having confidence in each other and in their respective governments, should ever be destroyed or impaired by national authority on account of national differences and discontents.

A person who removes to a foreign country, settles himself there, and engages in the trade of the country, furnishes by these acts such evidences of an intention permanently to reside there, as to stamp him with the national character of the state where he resides. In questions of *domicile* the chief point to be considered is the *animus manendi;* and if it sufficiently appears that the intention of removing was to make a permanent settlement, or for an indefinite time, the right of domicile is acquired by residence even of a few days. In general, the national character of a person, as neutral or enemy, is determined by that of his *domicile*; but the property of a person may acquire a hostile character, independently of his national character, derived from personal residence. Thus if a person enters into a house of trade in the enemy's country, or continues that connection during war, he cannot protect himself by mere residence in a neutral country; so also, the produce of an enemy's colony or other territory is to be considered as hostile property so long as it belongs to the owner of the soil whatever may be his residence.

In the modern law of nations, the right of *postliminy* is that by virtue of which persons and things taken by an enemy in war, are restored to their former state, when coming again under the power of the nation to which they belonged. The sovereign of a country is bound to protect the person and property of his subjects; and a subject, who has suffered the loss of his property by the violence of war, on being restored to his country can claim to be re-established in all his rights, and to recover his property. But this right does not extend in all cases to personal effects or movables, on account of difficulties of identification.

The rights of war in respect to an enemy are in general to be measured by the object of the war. No use of force is lawful except so far as it is necessary. Those who are actually in arms and continue to resist may be killed; but the inhabitants of the enemy's country who are not in arms, or who, being in arms, submit and surrender themselves may not be slain, because their destruction is not necessary for obtaining the just ends of the war. Those ends may be obtained by making prisoners of those taken in arms, or compelling them to give security that they will not bear arms against the victor for a limited period or during the war. The killing of prisoners can only be justified in those extreme cases where resistance on their part, or on the part of others, who come to their rescue, renders it impossible to keep them. *Cartels* for the mutual exchange of prisoners of war are regulated by special convention between the belligerent states, according to their respective interests and views of policy. Sometimes prisoners of war are permitted, by capitulation, to return to their own country upon condition not to serve again during the war, or until duly exchanged; and officers are frequently released upon their parole, subject to the same condition. By the modern usage of nations, commissaries are permitted to reside in the respective belligerent countries, to negotiate and carry into effect the arrangements necessary for the purpose.

All members of the enemy's state may lawfully be treated as enemies in a public war; but they are not all treated alike. The custom of civilized nations, founded on the general rule derived from natural law, that no use of force is lawful unless it is necessary to accomplish the purposes of war, has therefore exempted the persons of the sovereign and his family, the members of the civil government, women, children, cultivators of the earth, artisans, laborers, merchants, men of science and letters, and generally all public or private individuals engaged in the ordinary civil pursuits of life, from the direct effect of military operations, unless actually taken in arms, or guilty of some misconduct in violation of the usages of war. The application of the same principle has also limited and restrained the operations of war against the territory and other property of the enemy. By the modern usage of nations, which has now acquired the force of law, temples of religion, public edifices devoted to civil purposes only, monuments of art, and repositories of science are exempted from the general operations of war. Private property on land is also exempt from confiscation, excepting such as may become booty in special cases, as when taken from enemies in the field or in besieged towns, and military contributions levied upon the inhabitants of the hostile country. This exemption extends even to the case of an absolute and unqualified conquest of the enemy's country.

The exceptions to these general mitigations of the extreme rights of war, considered as a contest of force, all grow out of the same general principle of natural law, which authorizes us to use such a degree of violence and such only as may be necessary to secure the object of hostilities. Thus, if the progress of an enemy cannot be stopped, a frontier secured, or the approaches to a town cannot be made without laying waste the intermediate territory, the extreme case may justify a resort to measures not warranted by the ordinary purposes of war. But the whole international code is founded on *reciprocity*. Where, then, the established usages of war are violated by an enemy, and there are no other means of restraining his excesses, retaliation may be justly resorted to in order to compel the enemy to return to the observance of the law which he has violated. The effect of a state of war is to place all the subjects of each belligerent power in a state of mutual hostility. The law of nations has modified this maxim, by legalizing such acts of hostility only as are committed by those who are authorized by the express or implied command of the state. Such are the regularly commissioned naval and military forces of the state, and all others called out in its defence, or spontaneously defending themselves in case of urgent necessity, without any express authority for that purpose. The horrors of war would be greatly aggravated if every individual of the belligerent states were allowed to plunder and slay the enemy's subjects without being in any manner accountable for his conduct. Hence it is that in land wars, irregular bands of marauders are liable to be treated as lawless banditti, not entitled to the protection of the mitigated uses of war as practised by civilized nations.

The title to property lawfully taken in war may, upon general principles, be considered as immediately diverted from the original owner and transferred to the captor. As to personal property or movables on land, the title is lost to the former proprietor, as soon as the enemy has acquired a firm possession; which, as a general rule, is considered as taking place after the lapse of 24 hours, or after the booty has been carried into a place of safety, *infra præsidia* of the captor. In respect to ships and goods taken at sea, the sentence of a competent court is necessary; while, in respect to real property or immovables, the title acquired in war must be confirmed by a treaty of peace before it can be considered as completely valid. But it may be important to determine how far the possession of immovables, and the property arising out of such possession, extend. Grotius simply says that every kind of possession is not sufficient, but that it must be a *firm possession*, which he explains thus: "as if a country is so provided with permanent fortifications, that the advance party cannot enter it openly without first making himself master of them by force." Bynkershoek says: "Possession extends to every thing that is occupied, and what is occupied is placed within our power by the law of nature; but even that is considered as occupied, which is not touched on all sides with our hands or feet. * * * Hence it is not difficult to discern what may be considered as properly occupied in an occupied country. * * If, from the occupation of a strong place, dominion is exercised over the whole country, yet the victor is not considered in possession of those cities, walled towns, and fortresses, which the sovereign still retains."

There are various modes also in which the extreme rigor of the rights of war may be relaxed at the pleasure of the respective belligerents. 1. A general truce or armistice. This amounts to a temporary peace, and it requires either the previous special authority of the supreme power of the state, or a subsequent ratification by such power. 2. A partial truce or limited suspension of hostilities may be concluded between the military and naval officers of the respective belligerent states without any special authority for that purpose, where, from the nature and extent of their commands, such an authority is necessarily implied as essential to the fulfilment of their duties. The terms of the armistice should be free from all ambiguity. 3. Capitulations for the surrender of troops, fortresses, and particular districts of country fall naturally within the scope of the general powers intrusted to military commanders. 4. Passports, safe conducts, and licenses are documents granted in war to protect persons and property from the general operation of hostilities. A license is an act proceeding from the sovereign authority of the state, which alone is competent to decide on all the considerations of political and commercial expediency by which such an exception from the ordinary consequences of war must be controlled. 5. By rules laid down for the government of an army in an enemy's country in the new relation existing between the invading army and the citizens or subjects of the foreign country.

The martial law order of General Scott in Mexico, given in the article LAW, (*Martial,*) played so prominent a part in mitigating the horrors of war, as well as in aiding in the conquest of peace, that a concise history of that remarkable order will here find a fitting place. As early as May, 1846, General Scott presented for the consideration of the Secretary of War a *project* for a law, giving expressly to courts-martial in an enemy's country authority to punish offences, which in the United States are punishable by the ordinary criminal courts of the land. Congress did not, however, act upon the recommendation, and General Scott on the 8th of October, 1846, submitted to Mr. Secretary Marcy the draft of a letter which he recommends should be despatched to each commander of an army now operating against Mexico. "I am aware (he continues) that it presents grave topics for consideration, which is invited. It will be seen that I have endeavored to place all necessary restrictions on *martial law*. 1. By restricting it to a foreign hostile country; 2. To offences enumerated with some accuracy; 3. By assimilating councils of war to courts-martial; 4. By restricting punishments to the known laws of some one of the States of the Union;" (Doc. 59, *House of Representatives,* 30*th Congress,* 1*st Session.*) This *project* appears to have met with no favor from the Executive. In letters from General Taylor, dated October 6, and October 11, 1846, he reports the "most shameful atrocities" as having been committed without punishment, and he asks the Secretary of War "for instructions as to the proper disposition of the culprit" in a case of cold-blooded murder at Monterey. Mr. Marcy replied Nov. 25, 1846: "The competency of a military tribunal to take cognizance of such a case as you have presented in your communication of the 11th ult., viz., the murder of a Mexican soldier, and other offences not embraced in the express provisions of the Articles of War, was deemed so questionable, that application was made to Congress, at the last session, to bring them expressly within the jurisdiction of such a tribunal, but it was not acted upon. I am not prepared to say that, under the peculiar circumstances of the case, and particularly, by the non-existence of any civil authority to which the offender could be turned over, a military court could not rightfully act thereon; yet very serious doubts are entertained upon that point, and the Government does not advise that course. It seriously regrets that such flagrant offender cannot be dealt with in the manner he deserves. I see no other course for you to pursue than to release him from confinement and send him away from the army; and this is recommended."

The foregoing letter of the cautious War Secretary was written a few days after General Scott had been ordered to the theatre of war, to assume the direction of military operations; but in the opinion of the latter, "the good of the service, the honor of the United States, and the interests of humanity" demanded that the numerous grave offences not embraced in the Rules and Articles of War should not go unpunished; and accordingly, upon assuming command of the army in Mexico, he did not shrink from the responsibility which his station imposed. He issued his martial law order. Rigid justice was administered to American and Mexican under that order, and it, beyond all doubt, effected as important consequences as any act performed during his brilliant campaign ending with the conquest of peace. (Consult WHEATON's *Elements of International Law;* DUPONCEAU's *Bynkershoek;* GENERAL SCOTT's *Orders in Mexico.*)

WARRANT. A writ of authority. Warrant officers are such as are immediately below commissioned officers, exercising their authority by warrant only. Cadets are warrant officers. They may be tried by garrison courts-martial; but by the custom of war a court-martial cannot sentence a warrant officer to corporal punishment or reduction to the ranks.

WASHING. To each woman who may be allowed to a corps, not exceeding four to a company, one ration is given; (*Act* March 16, 1802.) They are washerwomen.

WASTE. Waste or spoil committed by troops, either in walks of trees, parks, warrens, fish-ponds, houses or gardens, corn-fields, inclosures of meadows, or maliciously destroying any property whatsoever belonging to the inhabitants of the United States, unless by the order of the then commander-in-chief of the armies of the United States, shall (besides such penalties as they are liable to by law) be punished, according to the nature and degree of the offence, by the judgment of a regimental or general court-martial; (ART. 54.) (*See* AMMUNITION.)

WATCH. The non-commissioned officers and men on board transports are usually divided into three watches, one of which must be constantly on deck.

WATCHWORD. (*See* PAROLE.)

WATER. Daily allowance for a man one gallon for all purposes. For a horse four gallons.

WEDGE—is one of the five simple mechanical powers. It is used sometimes for raising bodies, but more frequently for dividing or splitting them. The power is to the resistance acting perpendicularly on each side of the wedge, as the thickness of the back of the wedge is to the length of the side.

WEIGHTS AND MEASURES.

MEASURES OF LENGTH.

Inches.	Feet.	Yards.	Rods or Poles.	Furlongs.	Mile.
12	1				
36	3	1			
198	16½	5½	1		
7,920	660	220	40	1	
63,360	5,280	1,760	320	8	1

The inch was formerly divided into three parts, called *barley-corns*, and also into 12 parts called *lines*, neither of which denominations is now in common use. Scales and measuring rules are generally divided into *inches*, *quarters*, *eighths*, and *sixteenths*; or into *inches* and *decimal parts*; the latter of these divisions is used in the Ordnance Department.

For surveying land: 7.92 Inches = 1 link.   Gunter's chain.
                100 Links = 4 poles, or 22 yards, or 66 feet.

For map-making: Chains are often made of 50 links, each 1 foot in length.

For measuring ropes and soundings: 1 Fathom = 6 feet.
          1 Cable's length = 120 fathoms.

For measuring cloth: 1 Nail = 2¼ inches = 1-16th of a yard.
        1 Quarter = 4 nails.
        1 Yard = 4 quarters.
        1 Ell English = 5 quarters.

For measuring horses: 1 Hand = 4 inches.

Geographical measure: 1 Degree of a great circle of the earth = 69.77 miles.
        1 Geographical or nautical mile = 1-60th of a degree of the earth = 2,025 yards.
        1 Nautical league = 3 miles.

A standard measure has been adopted for the United States, copies of which are distributed to various parts of the country, for the purpose of establishing a uniform system. This standard is measured on a brass bar and copied from the British standard *yard*. For the proportion which it bears to the French *metre*, see below.

FOREIGN MEASURES OF LENGTH.

GREAT BRITAIN.—The Imperial standard yard of Great Britain, adopted in 1825, is referred to a natural standard, which is the distance between the axis of suspension and the centre of oscillation of a pendulum which shall vibrate seconds in vacuo, in London, at the level of the sea; that distance measured on a brass rod, at the temperature of 62° Fahr., is declared to be **39.1393** *imperial inches*.

FRANCE.—*Old system:*

  1 Point   = 0.0074 Eng. inches.
  1 Line = 12 points   = 0.08884 "
  1 Inch = 12 lines   = 1.06577 "
  1 Foot = 12 inches   = 12.7892 "
  1 Ell = 43 in. 10 lines = 46.716 " = 1.298 yd.
  1 Toise = 6 feet   = 76.735 " = 2.132 "
  1 Perch (Paris) = 18 feet.
  1 Perch (royal) = 22 "
  1 League, (common,) 25 to a degree = 2,280 toises = 4,861 yds., = 2.76 miles.
  1 League, (post,) = 2,000 toises = 4,264 yds., = 2.42 miles.
  1 Fathom (*Brasse*) = 5 feet French = 63.946 inches, or 5⅓ feet Eng., nearly.
  1 Cable length = 100 toises = 120 fathoms Fr., = 106⅔ fathoms English.
  1 Pace (pas) = ⅔ metre = 26.5 in. nearly.

TABLE FOR REDUCING OLD FRENCH MEASURES TO ENGLISH.

French feet.	English inches.	French feet or inches.	English feet or inches.	French lines.	English inches.	French points.	English inches.
1	12.7892	1	1.0658	1	0.0888	1	0.0074
2	25.5784	2	2.1315	2	0.1776	2	0.0148
3	38.3676	3	3.1973	3	0.2664	3	0.0222
4	51.1568	4	4.2631	4	0.3553	4	0.0296
5	63.9460	5	5.3288	5	0.4441	5	0.0370
6	76.7352	6	6.3946	6	0.5329	6	0.0444
7	89.5244	7	7.4604	7	0.6217	7	0.0518
8	102.3136	8	8.5261	8	0.7105	8	0.0592
9	115.1028	9	9.5919	9	0.7993	9	0.0666
10	127.8920	10	10.6577	10	0.8881	10	0.0740
11	140.6812	11	11.7234	11	0.9770	11	0.0814

NEW FRENCH SYSTEM.—The basis of the new French system of measures is the measure of a meridian of the earth, a quadrant of which is 10,000,000 *metres*, measured at the temperature of 32° Fahr. The multiples and divisions of it are decimal, viz.: 1 metre = 10 decimetres = 100 centimetres = 1,000 millimetres = 39.3707971 English inches, or 3.2809 feet.

Road Measure.—Myriametre = 10,000 metres. Kilometre = 1,000 metres. Decametre = 10 metres. Metre = 0.51317 toise.

TABLE FOR REDUCING METRES TO INCHES.

According to Capt. KATER's comparison, 1 metre = 39.37079 English inches.

Metres.	Inches.	Metres.	Inches.	Metres.	Inches.	Metres.	Inches.
0.001	0.039371	0.026	1.023641	0.051	2.007910	0.076	2.992180
2	0.078742	27	1.063011	52	2.047281	77	3.031551
3	0.118112	28	1.102382	53	2.086652	78	3.070922
4	0.157483	29	1.141753	54	2.126023	79	3.110292
5	0.196854	0.030	1.181124	55	2.165393	0.080	3.149663
6	0.236225	31	1.220494	56	2.204764	81	3.189034
7	0.275596	32	1.259865	57	2.244135	82	3.228405
8	0.314966	33	1.299236	58	2.283506	83	3.267776
9	0.354337	34	1.338607	59	2.322877	84	3.307146
0.010	0.393708	35	1.377978	0.060	2.362247	85	3.346517
11	0.433078	36	1.417348	61	2.401618	86	3.385888
12	0.472449	37	1.456719	62	2.440989	87	3.425259
13	0.511820	38	1.496090	63	2.480358	88	3.464630
14	0.551191	39	1.535461	64	2.519731	89	3.504000
15	0.590562	0.040	1.574832	65	2.559101	0.090	3.543371
16	0.629933	41	1.614202	66	2.598472	91	3.582742
17	0.669303	42	1.653573	67	2.637843	92	3.622113
18	0.708674	43	1.692944	68	2.677214	93	3.661483
19	0.748045	44	1.732315	69	2.716585	94	3.700854
0.020	0.787416	45	1.771686	0.070	2.755955	95	3.740225
21	0.826787	46	1.811056	71	2.795326	96	3.779596
22	0.866157	47	1.850427	72	2.834697	97	3.818967
23	0.905528	48	1.889798	73	2.874068	98	3.858337
24	0.944899	49	1.929169	74	2.913438	99	3.897708
25	0.984270	0.050	1.968540	75	2.952809	0.100	3.937079

		English.
Austria.—1 Foot	= 12.445 English inches	= 1.0371 feet.
1 Mile	= 4,000 toises	= 5 miles, nearly.
Prussia.—1 Rhineland foot	= 12.3557 English inches.	= 1.0296 feet.
1 Mile	= 8,552 yards, English	= 5 miles, nearly.
Russia.— 1 Foot	= 21.1874 English inches	= 1.7656 feet.

For the artillery, the English foot and inch are used.

1 Verst = 2,000 Russian feet = 1,177 yards.

Spain.— 1 Foot = 11.1284 English inches.
 1 Vara = 3 feet = 0.9274 English yard.
 1 League Royal = 25,000 Spanish feet = 4⅛ miles, nearly.
 1 Common league = 19,800 do. = 3½ "
 1 Judicial league = 15,000 do. = 2⅝ "
Mexico.— 1 Common league = 15,000 do. = 2⅝ "
Sweden.— 1 Foot = 11.6865 English inches.

MEASURES OF SURFACE.

Square measure.—144 Square inches = 1 square foot.
 9 Square feet = 1 square yard.
Land measure.—30¼ Square yards = 1 square perch or pole.
 40 Perches = 1 rood.
 160 Perches = 4 roods = 1 acre = 10 square chains (Gunter's) = 4,840 square yards = 70 yards square, nearly.
 640 Acres = 1 square mile.

French Superficial Measure.

Old system.—1 Square inch = 1.13587 English square inches.
 1 Arpent (Paris) = 100 square perches (Paris) or 900 square toises = 4,088 square yards, or 5-6ths of an acre, nearly.
 1 Arpent (woodland) = 100 square perches (royal) = 6,108 square yards, or 1 acre, 1 rood, 1 perch.
New, or *Decimal system.*—1 Are = 100 square metres = 119.603 square yards.
 1 Decare = 10 ares. 1 Hecatare = 100 ares.

MEASURES OF SOLIDITY.

Cubic or *Solid measure.*—1 Cubic foot = 1,728 cubic inches.
 1 Cubic yard = 46,656 " " = 27 cubic feet.
Measuring stone.—In different parts of the United States the *perch* of stone denotes a different quantity, but it is usually 24¾ cubic feet.
Measuring wood.—1 Cord is a prism 4 feet square and 8 feet long = 128 cubic feet.

French Solid Measures.

1 Cubic inch = 1.2106 cubic inches, English.
1 Cubic foot = 2091.85 cubic inches, English.
1 Cubic decimetre = 61.0271   "   "
1 Stere = 1 cubic metre = 61,027.1 cubic in. = 35.3166 cubic feet = 1.308 cubic yards.

MEASURES OF CAPACITY.

LIQUID MEASURE.

Gills.	Pints.	Quarts.	Gallons.
4	1		
8	2	1	
32	8	4	1

The standard gallon of the United States is the old wine gallon, which measures 231 cubic inches, and contains (as determined by Mr. Hassler) 58,373 Troy grains, or 8.3388822 avoirdupois pounds, of distilled water at the maximum density, (39°.83 Fahr.;) the barometer being at 30 inches.

A cubic foot contains 7.48 gallons.
A box 6 × 6 × 6.42 inches contains 1 gallon.
A box 4 × 4 × 3.61 inches contains 1 quart.

DRY MEASURE.

Pints.	Quarts.	Gallons.	Pecks.	Bushels.
2	1			
8	4	1		
16	8	2	1	
64	32	8	4	1

The standard bushel of the United States is the Winchester bushel, which measures 2,150.4 cubic inches, and contains 543,391.89 Troy grains, or 77.627413 lbs. avoirdupois, of distilled water, under the circumstances above stated.

A cubic yard contains 21.69 bushels.
A cylinder 14 in. diam. × 14 in. deep } contains 1 bushel.
Or a box 16 × 16.8 × 8 inches }
A box   12 × 11.2 × 8 inches   contains ½ bushel.
A box   8 × 8.4 × 8 inches   contains 1 peck.

N.B.—It will be observed that the pint, quart, and gallon of dry measure are not the same as for liquid measure.

FOREIGN MEASURES OF CAPACITY.

GREAT BRITAIN.—The British imperial gallon measures 277.274 cubic inches, containing ten pounds avoirdupois of distilled water weighed in air, at the temperature of 62°, the barometer being at 30 in. The same measure is used for liquids as for dry goods which are not measured by heaped measure; for the latter, the bushel is to be heaped in the form of a cone not less than 6 inches high, the base being $19\frac{1}{2}$ inches. The old distinctions of wine measure, ale and beer measure, and dry measure are discontinued.

For grain.—8 bushels = 1 quarter = 10.269 cubic feet.
  5 quarters = 1 load = 51.347 cubic feet.
For coal or *heaped measure.*—1 sack = 3 bushels = 4.89 cubic feet, nearly.
  1 chaldron = 12 sacks = 36 bushels = 58.68 cubic feet.
For timber.—1 load = 40 cubic feet.
Former wine gallon = 231 cubic inches.
Former ale gallon = 282 cubic inches.
Imperial gallon = 277.274 cubic inches, (as above.)

FRANCE.—1 Litre = 1 cubic decimetre = 61.0271 cubic inches = 1.057 U. S. quart = 1.761 imperial pint of Great Britain.
  1 Boisseau = 13 litres = 793.364 cubic in. = 3.4344 U. S. gals.
  1 Pinte = 0.931 litre = 56.816 cub. in. = 0.98383 U. S. quart.

SPAIN.—1 Wine arroba = 4.2455 U. S. gallons.
  1 Fanega (corn measure) = 1.593 U. S. bushels.

MEASURES OF WEIGHT.

AVOIRDUPOIS WEIGHT.

Drams.	Ounces.	Pounds.	Quarters.	Cwt.	Ton.
16	1				
256	16	1			
7,168	448	28	1		
28,672	1,792	112	4	1	
573,440	35,840	2,240	80	20	1

The standard avoirdupois pound of the United States, as determined by Mr. Hassler, is the weight of 27.7015 cubic inches of distilled water weighed in air, at the temperature of the maximum density, (39°.83;) the barometer being at 30 inches.

TROY WEIGHT.

Grains.	Dwt.	Ounces.	Pound.
24	1		
480	20	1	
5,760	240	12	1

The pound, ounce, and grain are the same in Apothecaries' and Troy weight; in the former, the ounce is divided into 8 drachms, the drachm into 3 scruples, and the scruple into 20 grains.

    7,000 Troy grains = 1 lb. avoirdupois.
    175 Troy pounds = 144 lbs. "
    175 Troy ounces = 192 oz. "
    437½ Troy grains = 1 oz. "

Foreign Weights.

GREAT BRITAIN.—The imperial avoirdupois pound is the weight of 27.7274 cubic inches of distilled water weighed in air, with brass weights, at the temperature of 62° Fahr.; barometer 30 inches. Therefore,

 1 cubic inch of distilled water at 62° weighs 252.458 grains.

 0.003961 cubic inch weighs 1 grain.

 22.815689 cubic inches weigh 1 Troy pound.

Horseman's weight: 1 stone = 14 pounds.

FRANCE.—*Old system:* 1 Livre = 16 onces = 1.0780 lb. avoirdupois.
       1 Once = 8 gros = 1.0780 oz. "
       1 Gros = 72 grains = 58.9548 grains Troy.
       1 Grain . . . . = 0.8188 " "

New system.—The basis of the system of weights is the weight, in vacuo, of a litre, or a cubic decimetre, of distilled water, at the temperature of 39°.2 Fahr.; $\frac{1}{1000}$th part of this weight is a *gramme*, the multiples of which are: 1 Decagramme = 10 grammes: 1 Hectogramme = 100 grammes: 1 Kilogramme = 1,000 grammes. The divisions are: 1 Decigramme = $\frac{1}{10}$th gramme: 1 Centigramme = $\frac{1}{100}$th gramme: 1 Milligramme = $\frac{1}{1000}$th gramme.

    1 Quintal = 100 kilogrammes.
    1 Millier = 1,000 kilogrammes = 1 ton sea weight, (French.)
    1 Kilogramme = 2.204737 pounds avoirdupois.
    1 Gramme = 15.433159 grains Troy = 0.03528 oz. avoirdupois.
    1 Pound avoirdupois = 0.4535685 kilogramme.
    1 Pound Troy = 0.3732223 kilogramme.

SPAIN.— 1 Pound = 1.0152 pounds avoirdupois.
SWEDEN.—1 Pound = 0.9376 " "
AUSTRIA.—1 Pound = 1.2351 " "
PRUSSIA.—1 Pound = 1.0333 " "

MEASURES OF VALUE.

All calculations of value in the military service of the United States are expressed in *Dollars* and *Cents*, although the denominations of *shillings* and *pence* are still in common use as a nominal currency in many of the States.

The standard of gold and silver is 900 parts of pure metal and 100 of alloy, in 1,000 parts of coin. The alloy of gold coin is 25 silver and 75 copper; the alloy of silver is copper.

Weight of Dollar .... = 412.5 grains Troy ⎫
"   Eagle .... = 258   "   ⎬ Other coins in proportion.
"   Cent .... = 168   "   ⎭

Relative Mint Value of United States and Foreign Coins.

GREAT BRITAIN..1 Guinea = 21 shillings............... =	5.059	dollars.
1 Sovereign, or 1 pound = 20 shillings . =	4.845	"
1 Crown = 5 shillings................ =	1.08	"
1 Shilling = 12 pence................ =	0.217	"
1 Penny .... =	0.018	"
FRANCE.........5 Francs .... =	0.932	"
1 Franc = 20 sous.................... =	0.185	"
1 Sous .... =	0.0093	"
SPAIN ....1 Doubloon, or 1 ounce............... =	15.57	"
AUSTRIA ....1 Ducat............................ =	2.275	"
1 Crown, or rix dollar .... =	0.97	"
20 Kreutzers........................... =	0.16	"
PRUSSIA ....1 Double Frederick.................. =	8.00	"
1 Thaler........................... =	0.693	"
RUSSIA ....1 Half-Imperial = 5 roubles........... =	3.967	"
1 Rouble........................... =	0.75	"
SWEDEN ....1 Ducat............................ =	2.267	"
1 Specie daler .... =	1.042	"
TURKEY........20 Piasters .... =	0.82	"

Dimensions of Drawing Paper.

Demy ....1 ft.	7½ in.	×	1 ft.	3½	inches.		
Medium ....1 "	10 "	×	1 "	6	"		
Royal ....2 "	0 "	×	1 "	7	"		
Super royal......2 "	3 "	×	1 "	7	"		
Imperial.........2 "	5 "	×	1 "	9¼	"		
Elephant ....2 "	3¾ "	×	1 "	10¼	"		
Columbier ....2 "	9¼ "	×	1 "	11	"		
Atlas ....2 "	9 "	×	2 "	2	"		
Double elephant...3 "	4 "	×	2 "	2	"		
Antiquarian......4 "	4 "	×	2 "	7	"		

(*Ordnance Manual.*)

WHEEL. In the simple wheel and axle, the power is to the weight as the radius r of the axle is to the radius R of the wheel. Or, $P = \dfrac{w\,r}{R}$.

In a system of wheels and pinions, the power is to the weight as the product of the radii (or number of teeth) $r\ r'\ r''$, &c., of the pinions is to the product of the radii (or number of teeth) $R\ R'\ R''$, &c., of the wheels:

$$P = w\,\frac{r\ r'\ r''}{R\ R'\ R''}.$$

WHIPPING. Abolished, except for desertion.

WHITING. To make whiting for accoutrements, it is necessary to boil many handfuls of bran enveloped in linen. Dissolve afterwards pipe-clay in this water. Whiten with it when cold. When the buff leather is greasy and does not receive the whiting, scrape it, and apply to it a solution of pipe-clay and Spanish whiting.

Another receipt, calculated for one hundred men, is the following: Pipe-clay, 3½ lbs.; Spanish whiting, 8 ounces; white lead, 4 ounces; glue, 1½ ounces; starch, 6 oz.; white soap, 5 oz. Put the pipe-clay and Spanish whiting in about five gallons of water; wash them and leave them to soak for six hours; 2d, throw out the first water, and replace it by 5½ gallons of pure water; add the white lead, glue, and white soap. Cook them together, taking care to stir constantly the composition. At the moment that the foam shows itself on the surface, withdraw the vessel from the fire without suffering the composition to boil; put then the starch in the whiting, and mix all well together.

WIDOWS AND ORPHANS. (*See* PENSION.)

WILLS, (NUNCUPATIVE.) A nuncupative will, so termed from naming an executor by word of mouth, is a verbal testamentary declaration or disposition. By the common law, it was as valid in respect to personal estate as a written testament. A will could not only be made by word of mouth, but the most solemn instrument in writing might be revoked orally. In a rude and uncultivated age, to have required a written will would have been a great hardship, but with the growth and progress of letters, the reason for permitting a verbal testament diminished in force, until finally an effort to establish such a will by means of gross fraud and perjury gave rise to the provisions of the statute of 29 Charles II., passed in 1676, termed the Statute of Frauds.

The only nuncupative wills now allowed are those made by soldiers and sailors. It appears from the preface to the Life of Sir Leoline Jenkins, that he claimed the merit, at the time of the preparation of the Statute of Frauds, of having obtained for the soldiers of the English army the full benefit of the testamentary privileges of the Roman army. The Roman soldier was indulged with very peculiar rights and immunities, in the way of exemption from the usual rules in respect to wills —*Inter arma silent leges.* In the camp and on the battle-field the testamentary law was silent. Amid the excitement and the perils of warfare the forms prescribed by law for the execution of a will were dispensed with, so that the soldier might declare his last wishes by word of mouth; or if wounded, he wrote with his blood on his shield, or with his sword in the dust; the disposition was held firm and sacred. Julius Cæsar authorized the making of the military testament in any mode, and without prescribed ceremonials. The example thus set was subsequently followed by Titus, Domitian, Nerva, and Trajan, until the usage became thoroughly established. It was extended also to the naval service, and officers, rowers, and sailors were in this respect esteemed as soldiers. This was the foundation of those privileges of soldiers in regard to nuncupative wills, which were allowed wherever the civil law prevailed, and have been very generally adopted among civilized nations. In France, the ordnance *De la Marine* of 1681, first gave special privileges to wills made at sea, and the ordnance of 1735 regulated the celebration of the military testament. The Code Civil has also adopted definite rules in regard to wills made at sea, in time of pestilence, or by soldiers in service. In Holland, when commerce began to be extended to distant voyages, the question arose whether wills made at sea were entitled to any peculiar immunity, and some jurists affirmed that they should be taken as military testaments. The matter was finally resolved in favor of their exemption in case of persons sailing to or returning from the Indies, by the ordinances of the West India Company in 1672 and 1675. In England, by the Statute of Frauds, passed about the same time, the full benefit of the privilege was given, without restriction, to all soldiers and sailors in actual service, and this liberal rule has continued to the present day.

Nuncupative wills, not being regulated by statute as to their mode of celebration or execution, the single question for the judgment of the court is, whether the nuncupation was made by a person entitled to that privilege. The restrictions of the Statute of Frauds were not applied to wills made by "any soldier being in actual military service, or any mariner or seaman being at sea." By the revised statutes of New York it was provided that nuncupative wills should not be valid, "unless made by a soldier while in actual military service, or by a mariner while at sea." The terms of the exception in the statute 1 Vict. c. 26, are, "any soldier being in actual military service, or any mariner or seaman being at sea." The phraseology is slightly different in these statutes; but the rule is substantially the same in all—that the nuncupation is only valid when made by a soldier in actual military service, or a mariner at sea, at the time of the testamentary act. It is not enough to be a soldier or a sailor, but there must be actual service. The military testament was first conceded by Julius Cæsar to all soldiers, but it was subsequently limited by Justinian to those engaged in an expedition—*solis qui in expeditionibus occupati sunt.* The exception was borrowed with the rule from the civil law, and the courts have invariably adhered to the principle that there cannot be actual warfare and the soldier not be engaged *in expeditione.* So also the nuncupation of a mariner to be valid must be made at sea. It is sometimes difficult to determine when the mariner is to be considered at sea. For example, Lord Hugh Seymour, the admiral of the station at Jamaica, made a codicil by nuncupation while staying at the house on shore appropriated to the admiral of the station. The codicil was rejected on the ground that he only visited his ship occasionally, while his family establishment and place of abode were on land at the official residence. But when a mariner belonging to a vessel lying in the harbor of Buenos Ayres, met with an accident when on shore by leave, made a nuncupative will, and died there, probate was granted for the reason that he was only casually absent from his ship. The will of a shipmaster made off Otaheite has also been allowed. The principle upon which the privilege of nuncupation is conceded applies to all persons engaged in the marine service, whatever may be their special duty or occupation on the vessel. As in the army the term " soldier " embraces every grade, from the private to the highest officer, and includes the gunner, the surgeon, or the general ; so in the marine, the term " mariner " applies to every person in the naval or mercantile service, from the common seaman to the captain or admiral. It is not limited or restricted to any special occupation on shipboard, but a purser, or any other person whose particular vocation does not relate to the sailing of the vessel, possesses the same right as the sailor. A cook is certainly as much a necessary part of the effective service of a vessel as the purser or the sailor ; and there would seem to be no reason why he should be excluded from the advantage of a rule designed for the benefit of men engaged in the marine, without reference to the particular branch of duty performed in the vessel. As well because the wills of soldiers and mariners were excepted from the operations of the provisions of the Statute of Frauds, as for the reason and ground of the exception, and the peculiar character of the military testament, it was never held requisite that their nuncupations should be made during the last sickness. Nor has any particular mode been prescribed in respect to the manner of making the testament. The very essence of the privilege consists in the absence of all ceremonies as legal requisites—or, as Merlin states the proposition, "their form was properly to have no form." It is true the Roman law prescribes two witnesses; but this, however, did not relate to the essence of the act, but only to the proof. In respect to evidence, we do not follow the civil or the canon law; no particular number of witnesses is required to verify an act judicially, and all the court demands is to be satisfied by sufficient evidence as to the substance of the last testamentary request or declaration of the deceased. This ascertained, the law holds it sacred, and carries it into effect with as much favor and regard as would be paid to the most formal instrument executed with every legal solemnity; (*Decision of the Surrogate of New York City*.) And so, according to numerous decisions, made in Great Britain, quoted by Prendergast, "whenever a military officer on full pay makes an informal will its validity can only be supported by showing the testator to have been on actual military service at the time the will was made. And the result of the decisions appears to be, that an officer serving with his regiment, or in command of troops in garrison or quarters, either in the United Kingdom or the colonies, is not deemed on *actual* military service. To satisfy the meaning of the act of parliament in that respect, he must be on an expedition, or on some duty associated with positive danger."

WINDAGE. The true windage is the difference between the true diameters of the bore and the ball. The loss of velocity caused by a given windage is directly as the windage and inversely as the diameter of the bore, very nearly.

WINDLASS. A machine used for many common purposes. It is a particular modification of the wheel and axle, the power being applied by means of a rectangular lever or *winch*.

WINGS. The right and left divisions of an army or battalion.

WITNESS. All persons who give evidence before a court-martial, are to be examined on oath or affirmation, as follows: You swear, or affirm (as the case may be) the evidence you shall give in the cause now in hearing, shall be the truth, the whole truth, and nothing but the truth. So help you God; (ART. 73.)

On the trial of cases not capital before courts-martial, the deposition of witnesses not in the line or staff of the army, may be read in evidence: Provided, the prosecutor and the person accused are present at the taking of the same, or are duly notified thereof; (ART. 74.) The list of witnesses for the prosecution is sometimes given to the prisoner, not as a right, but as a matter of convenience, when no evil result is apprehended from it; (HOUGH.)

The law has not given to courts-martial any power to compel the attendance of witnesses not of the line or staff of the army, even in capital cases. The want of such power might often defeat the ends of justice. A citizen witness is, however, paid his actual transportation or stage-fare and three dollars a day while attending the court and travelling to and from it, counting the travel at fifty miles a day. (*See* EVIDENCE; TRIAL.)

WOMEN. (*See* WASHING.)

WOOD. The most useful timbers in the United States are: the hickory, which is very tough and inflexible; white oak, tough and pliable; white ash, tough and elastic; black walnut, hard and fine-grained; white poplar, soft, light fine-grained wood; white pine and other pines, for building; cypress, soft, light, straight-grained, and grows to a large size; dogwood, hard and fine-grained. The timber growing in the centre of a forest is best.

WORK. (*See* FATIGUE DUTY.)

WORKING POWER. Working power of men: A foot soldier travels in one minute, in common time, 90 steps = 70 yards. In quick time, 100 steps = 86 yards. In double quick, 140 steps = 109 yards. He occupies in the ranks a front of 20 inches and a depth of 13 inches, without the knapsack; the interval between the ranks is 13 in.; 5 men can stand in a space of 1 square yard. Average weight of men, 150 lbs. each. A man travels, without a load, on level ground, during $8\frac{1}{2}$ hours a day, at the rate of 3.7 miles an hour, or $31\frac{1}{4}$ miles a day. He can carry 111 lbs. 11 miles a day. A porter, going short distances and returning unloaded, carries 135 lbs. 7 miles a day. He can carry in a wheel-barrow 150 lbs. 10 miles a day. The maximum power of a strong man, exerted for $2\frac{1}{2}$ minutes, may be stated at 18,000 lbs. raised 1 foot in a minute.

A man of ordinary strength exerts a force of 30 lbs. for 10 hours a day with a velocity of $2\frac{1}{2}$ feet in a second = 4,500 lbs. raised 1 foot in a minute = *one-fifth* the work of a horse. Daily allowance of water for a man 1 gallon, for all purposes. (*See* HORSE; MEASURE; &c.)

WORSHIP, (DIVINE.) It is earnestly recommended to all officers and soldiers diligently to attend divine service, and all officers who shall behave indecently or irreverently at any place of divine worship shall, if commissioned officers, be brought before a general court-martial there to be publicly and severely reprimanded by the president; if non-commissioned officers or soldiers, every person so offending shall, for his first offence, forfeit one-sixth of a dollar, to be deducted out of his next pay; for the second offence, he shall not only forfeit a like sum, but be confined 24 hours; and for every like offence, shall suffer and pay in like manner; which money, so forfeited, shall be applied by the captain or senior officer of the troop or company to the use of the sick soldiers of the company or troop to which the offender belongs; (ART. 2.)

WOUNDS. (*See* PENSIONS.)

WRONGS. (*See* INJURIES; REDRESSING WRONGS; REMEDY.)

Y

YARD. (*See* WEIGHTS AND MEASURES.)

Z

ZIGZAG OR BOYAUS—are defiladed trenches, run out from the parallels of attack, so as to form a covered road, by which the assailants can approach the fortress. (*See* SIEGE.)

www.ingramcontent.com/pod-product-compliance
Lightning Source LLC
Chambersburg PA
CBHW080527300426
44111CB00017B/2637